Landmark Experiments in Protein ⌐

Proteins are the workhorses of cells, performing most of the important functions that allow cells to use nutrients and grow, communicate among each other, and importantly, die if aberrant behavior is detected. How were proteins discovered? What is their role in cells? How do dysfunctional proteins give rise to cancers? *Landmark Experiments in Protein Science* explores the manner in which the inner workings of cells were elucidated, with a special emphasis on the role of proteins. Experiments are discussed in a manner as to understand what questions were being asked that prompted the experiments and what technical challenges were faced in the process; and results are presented and discussed using primary data and graphs.

Key Features

- Describes landmark experiments in cell biology and biochemistry.
- Discusses the "How" and "Why" of historically important experiments.
- Includes primary, original data and graphs.
- Emphasizes biological techniques that help understand how many of the experiments performed were possible.
- Documents, chronologically, how each result fed into the next experiments.

Landmark Experiments in Protein Science

Pascal Leclair

CRC Press
Taylor & Francis Group
Boca Raton London New York

CRC Press is an imprint of the
Taylor & Francis Group, an **informa** business

Credit: Shutterstock ID 2151963957. G-protein-coupled receptors (GPCRs) are the largest and most diverse group of membrane receptors in eukaryotes. 3D illustration.

First edition published 2024
by CRC Press
6000 Broken Sound Parkway NW, Suite 300, Boca Raton, FL 33487-2742

and by CRC Press
4 Park Square, Milton Park, Abingdon, Oxon, OX14 4RN

CRC Press is an imprint of Taylor & Francis Group, LLC

Library of Congress Cataloging-in-Publication Data

Names: Leclair, Pascal, author.
Title: Landmark experiments in protein science / Pascal Leclair.
Description: First edition. | Boca Raton : CRC Press, 2023. | Includes bibliographical references and index.
Identifiers: LCCN 2022059786 (print) | LCCN 2022059787 (ebook) | ISBN 9781032458687 (hbk) | ISBN 9781032458694 (pbk) | ISBN 9781003379058 (ebk)
Subjects: MESH: Proteins | Biochemistry--history | Cell Biology--history | Biomedical Research--history
Classification: LCC QP551 (print) | LCC QP551 (ebook) | NLM QU 11.1 | DDC 572/.6--dc23/eng/20230515
LC record available at https://lccn.loc.gov/2022059786
LC ebook record available at https://lccn.loc.gov/2022059787

ISBN: 978-1-032-45868-7 (hbk)
ISBN: 978-1-032-45869-4 (pbk)
ISBN: 978-1-003-37905-8 (ebk)

DOI: 10.1201/9781003379058

Typeset in Times
by KnowledgeWorks Global Ltd.

This book is dedicated to all the scientists who worked tirelessly for decades to elucidate the inner workings of the cell.

Contents

Preface

Throughout my life, I've often wondered how we've come to know what we know about pretty much everything. I remember, as a child, being in my parents' car on the way to my grandmother's place in the country and noticing the lampposts and trees go by the window, wondering what makes one different from the other. Later, when I started getting interested in biology, I thought learning about DNA was going to reveal everything I wanted to know about how our body works and particularly, why at times, it doesn't work and we get sick. It wasn't long before I realized that proteins—not DNA—were the workhorses of cells, performing all functions that allow cells to use nutrients and grow, communicate among themselves, and importantly, die if aberrant behavior is detected. And the more I learned about proteins, the more I marveled about how cells work and, particularly, at all the different ways proteins regulate the cellular machinery.

However, in my quest to learn more about the history of biology, I found that the one constant has been the feeling of not getting a clear understanding of HOW proteins were discovered: textbooks offer extensive descriptions of established processes while forgoing most historical aspects that enabled the discoveries they are describing. On the other hand, history books tend to focus on the landscape of the discoveries (i.e., the status of the industry, the mood of the scientists, various anecdotes, etc.); however, and understandably so, their descriptions of molecular processes are very limited. Specifically, I wanted to know WHAT experiments had been performed that enabled the discovery of proteins? HOW did we come to understand the various roles of proteins in cells? HOW did the scientists who tackled these questions come to the conclusions they did about their experiments? And as importantly, WHY did they decide to pursue these questions to begin with?

As such, *Landmark Experiments in Protein Science* is my attempt at answering these questions for myself. I tell the story in chronological order (as much as possible), which will hopefully allow the reader to appreciate how each discovery feeds into the next. The overall goal of my book is to show how each experiment and discovery came about: what question was being asked, how the problems that scientists were faced with were resolved—including technological innovations required to achieve their goals— the amazing insights that went into the analysis of the results, and finally, how these findings led to the next questions and discoveries. Understandably, given the extent of the literature on the various topics about which I write, the selection of papers I chose to discuss is but a drop in the bucket of science and many more could have been included given more space, time, and money. Similarly, the topics I chose to cover are those I felt could be applied to most—if not all—cells, and could not spare the space (or the time) to discuss topics that might be more specific in nature and only apply to a subset of cells.

Enjoy this journey through time and discover the passion, dedication, and amazing insights from scientists all over the world who, over the course of about two hundred years, patiently peeled away the complex layers of the cell to unveil its inner workings.

Pascal Leclair
University of British Columbia

Acknowledgments

I would like to thank everyone who read portions of the various drafts and offered comments and suggestions. I would also like to acknowledge all the individuals and publishers who have kindly granted permission to reprint material from their publications, as indicated in the text.

A special thanks goes out to my editor, Chuck Crumly, for taking an interest in this book, and to everyone at Taylor & Francis Group who helped make this book what it is.

Finally, I would like to thank my family who bore with me while I completed this book.

About the Author

Pascal Leclair earned a B.Sc. in biology and an M.Sc. in cell and developmental biology from the University of British Columbia, where he is currently a research technician. One of his projects involves elucidating the mechanism of action of caspase-independent cell death following ligation of the cell surface molecule, CD47. He is also part of the BRAvE initiative at BC Children's Hospital Research Institute in Vancouver, which is developing a precision medicine approach to inform better treatment for children who relapse from cancer. He lives near Vancouver, BC with his wife and twin boys.

1 Prelude to Biology
A History of Chemistry

Attempts at explaining the world around us date as far back as 3000 B.C. when Egyptians proposed the idea of the Ogdoad, a primordial force from which everything was formed. In ancient Greece, Empedocles suggested that all things were composed of four primal entities (earth, air, fire, and water; Plato would later call them elements), which combined by the action of opposing forces (love/hate, affinity/antipathy) to create an infinite number of forms. Around 350 B.C., Aristotle, who supported Empedocles' idea of the elements, suggested that all things were composed of different ratios of hotness, coldness, wetness, and dryness, such that a substance could be changed—or transmuted—into another simply by changing the ratios of these four elements. Although this idea seems strange now, it was not totally without support from observations of nature: water, a substance that is wet and cold, changed into a vapor (steam, which is wet and hot) by the addition of heat (or fire). This concept would guide all natural philosophers until the end of the 18th century, even though they were merely the musings of philosophers with no experimental evidence. However, science at the time was such that these musings were all that was at hand to explain the world we live in.

A first step toward evidence-based science came in the form of alchemy, which is generally regarded as the precursor to chemistry. This discipline was largely based on a widely accepted concept that all metals could be transmuted to gold under the right conditions as dictated by the concept of the four elements proposed by Aristotle; as such, for many centuries, alchemy was fueled by elusive attempts at finding the philosopher's stone, *a potent transmuting agent and universal medicine*.[1] Discovery of the philosopher's stone was perceived as such an attainable goal that most kingdoms employed alchemists tasked with finding it, and many others were drawn by the allures of wealth and fame—some of whom even made it their life's mission—and strived to discover the philosopher's stone for themselves. Less scrupulous individuals made healthy profits using other people's sense of greed and gullibility as a means to con them out of their money. Thus, the "profession" quickly attracted *charlatans and spurious adepts, who ... turned alchemy into one of the greatest popular frauds in history*.[2]

Despite these widespread abuses, alchemy was also seen by some as a genuine science tasked with deciphering the laws of nature, as expressed by Roger Bacon in the 15th century in his *Speculum Alchemiæ*:

> [...] *Alchimy* is a Science, teaching how to transforme any kind of mettall into another: and that by a proper medicine, as it appeareth by many Philosophers Bookes. *Alchimy*

therefore is a science teaching how to make and compound a certaine medicine, which is called *Elixir*, the which when it is cast upon mettalls or imperfect bodies, doth fully perfect them in the verie projection.

> [...] the naturall principles in the mynes, are *Argent-vive*, and *Sulphur*. All mettalls and minerals, whereof there be sundrie and divers kinds, are begotten of these two: but I must tel you, that nature alwaies intendeth and striveth to the perfection of Gold: but many accidents comming between, change the mettalls, as it is evidently to be seene in divers of the Philosophers bookes. For according to the puritie and impuritie of the two aforesaid principles, *Argent-vive*, and *Sulphur*, pure and impure mettals are ingedred: to wit, Gold, Silver, Steele, Leade, Copper, and Iron... [emphases in original][3]

John Read, who studied the history of alchemy, noted that

> [...] alchemy was a strange blend of logical thinking and mystical dreaming, of sound observation and wild superstition, of natural and moral ideas, and of objective facts and subjective conceptions. Nevertheless, the sympathetic student of the "divine Art" cannot but absorb something of the romance and wistfulness of alchemy's age-long struggle after the unattainable; nor can he remain unmindful of the immemorial traditions of alchemy's daughter, chemistry. At its best, alchemy was an earnest search after truth, in the light of that principle of the unity of matter which has been rediscovered by modern science.[1]

Egypt was known to be proficient in the arts of metallurgy, enameling, glass-tinting, and dyeing; *The Land of Khem* as it was known, is often considered the birthplace of alchemy and, in fact, alchemy takes its namesake from it: the country was referred to as *Al Khem* in the Islamic world and by *Alchemy* in the western world.[1] At the forefront of alchemy stood Abu Musa Jabir Ibn Hayyan, considered the father of chemistry, who developed experimental methods for the discipline, including calcination, sublimation, evaporation, and crystallization. He also isolated of a number of different acids that were of great use for future generations of chemists.[2]

In the 17th century, the Swiss physician and alchemist, Paracelsus, developed the concept of *iatrochemistry*, a subdiscipline of alchemy dedicated to the extension of life by the service of medicine, after years of studying a variety of disciplines as he traveled across Europe. As such, he proposed the theory of the *tria prima* that centered around the three "elements", salt, mercury, and sulfur which combined in different ratios to produce health or disease in an individual; thus, he believed that illnesses could be cured with

DOI: 10.1201/9781003379058-1

1

chemical treatments that restored an imbalance of these elements. Importantly, he proposed that the main task of the alchemist *should be to aid the apothecary and the physician, rather than to make gold* and insisted on the need for experimentation.[1] As such, followers of his doctrine put their ambitions of discovering riches by way of the philosopher's stone behind them and returned to their laboratories in search of drugs that could be purified and used to help ease human suffering. Paracelsus himself formulated a number of treatments for ailments, such as arsenic preparations for skin diseases and mercury salts for the treatment of syphilis, the latter of which remained the preferred treatment until the 19th century. Although his ideas and concepts were mostly wrong, Paracelsus is credited with unleashing alchemy from the clutches of charlatans and infusing it with humanistic purpose.

In 1661, the English natural philosopher, Robert Boyle, published *The Sceptical Chymist*, a treatise on the distinction between chemistry and alchemy, in which he argued that the theories of the latter—the four elements and the *tria prima*—were inadequate to explain the phenomena of chemistry. This book is seen as an important turning point for the transition of alchemy to chemistry, as it contained one of the first suggestions that matter was made up of atoms and molecules and provided the first definition of an element:

> I now mean by Elements, as those Chymists that speak plainest do by their Principles, certain Primitive and Simple, or perfectly unmingled bodies; which not being made of any other bodies, or of one another, are the Ingredients of all those call'd perfectly mixt Bodies are immediately compounded, and into which they are ultimately resolved.[4]

* * * * *

Although fire was discovered 300,000 to 400,000 years ago, the chemical mechanism of this phenomenon remained elusive until relatively recently, even though many tried to unravel its secrets over the years. Plato thought that all burnable matter contained a mystical inflammable principle, which alchemists later suggested to be the spirit of sulfur, or indeed, sulfur itself. Paracelsus explained this phenomenon with his *tria prima*: burnable substances were such because they contained sulfur, gave off a flame because of their mercury, and the ash they left behind was due to their salt content.[2] However, John Becher, having witnessed the burning of substances that did not contain sulfur, suggested an alternate hypothesis in the 17th century: all burnable substances contained a *terra pinguis*, an inflammable earth that allowed them to burn. He would later call this substance *phlogiston*.[2] Becher explained that when a substance burned, its phlogiston was given off in the form of a flame and subsequently transferred to the atmosphere. Thus, matter that did not burn simply did not contain any phlogiston. According to Becher, "burnable" also included the vital act of breathing, *for did not the lungs constantly exhale phlogiston as food was consumed during digestion in human and animal bodies?*[2]

The theory of phlogiston was subsequently "proven" by many chemists who weighed matter before and after it was burned and found that the ashes left behind weighed less than the original unburned matter. However, many others found that, in fact, the ashes increased in weight, to which Becher replied:

> Yes you fools, but you do not know that my phlogiston may sometimes possess the power of levity—it weighs less than nothing. Naturally, then, the ash of your metals weighs more than the metals you burned. Something, minus another thing which weighs less than nothing, weighs more than that original something.[2]

Despite the ever-changing properties of phlogiston, it nevertheless explained many facts that chemists struggled to explain: metals turned into powders when burned because they lost their phlogiston; the metallic properties of powdered metals were instantly restored by the addition of charcoal—a highly inflammable substance—because it transferred its phlogiston to the powder. As such, the concept of phlogiston persisted for more than two hundred years! Wrong though phlogiston might have been, it served an important purpose in the story of chemistry in that it urged chemists to investigate and determine the true mechanism of the phenomenon of fire, which, in turn, led to the demise of Becher's phlogiston.

In the mid-18th century—one hundred years after Becher proposed his phlogiston theory—an English natural philosopher, Joseph Priestley, was led to a career in science following a meeting with the American polymath, Benjamin Franklin, who sparked in him an interest in electricity.[2] In writing a history of this subject using reading material provided by Franklin, Priestley endeavored *to ascertain several facts which were disputed, and this led me by degrees into a larger field of original experiments in which I spared no expense that I could possibly furnish.*[5] For example, he found that the bubbles released in vats during beer-making contained a gas that had the power to extinguish burning chips of woods. He wondered if this could be the same gas that his contemporary, Joseph Black, had found to be released upon burning limestone 15 years earlier and which he called "fixed air". Priestley determined that the gas did not efficiently dissolve in water and his experiments produced the first *glass of exceedingly pleasant sparkling water*, which we now know as soda water (a weakly acidic solution of carbon dioxide in water).[2]

At this time, gases liberated during chemical reactions were clumsily and inefficiently collected in opaque, balloon-like bladders, so Priestley set out to improve this technique: he filled a glass bottle with mercury and inverted it into a vessel which also contained the liquid metal (Figure 1.1). A tube was then connected from the solution from which gas would presumably be evolved to the mouth of the bottle, such that any gas emitted would be bubbled through the

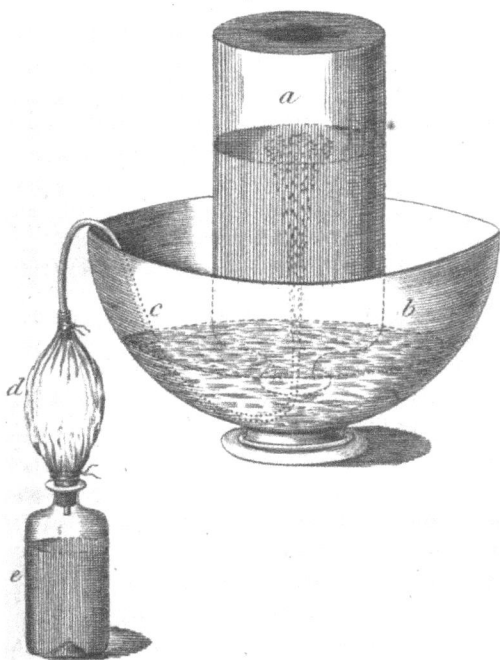

FIGURE 1.1 Priestley's apparatus for collecting gases. A sample was placed into a container, *e*, and the emitted gas was bubbled through and above a solution of mercury in an inverted bottle, a.[6]

mercury and rise to the top of a transparent bottle, thereby displacing the mercury and trapping the gas. Using his new apparatus, Priestley heated a wide variety of substances and collected the gases that were evolved.[6] During one particular experiment in 1774, he extracted air from *mercurius calcinatus per se*, a red powder obtained from heating mercury, as many had done before him, and, as expected, found that *air was expelled from it readily*.[6] However, on this fateful day, he glanced upon a nearby lighted candle and wondered what would happen if he placed it inside the bottle containing the expelled air, wholly expecting that it would be extinguished. To his surprise, he found that the *candle burned in this air with a remarkably vigorous flame … and it consumed very fast!*[6] Since the air could readily accept phlogiston from the burning matter, he hypothesized that this air—which he called *dephlogisticated air*—accounted for *the vital powers of the atmosphere*.[2]

Priestley set out to determine *how wholesome [...] this dephlogisticated air* really was, so he captured mice from around his home and placed them in sealed vessels containing either common air or dephlogisticated air.[2,6] Surprisingly, he found that mice breathing common air collapsed a mere 15 minutes later, whereas mice breathing dephlogisticated air survived at least twice as long! As such, he believed that the capacity for air to be breathed depended on the amount of phlogiston it contained. Thus, phlogisticated air could not sustain life since it was saturated with phlogiston—remember that combustion was thought to release phlogiston into the surrounding air. Therefore, combustion had two requirements: first, that

the body one wished to burn contain phlogiston, and second, that the atmosphere surrounding it had the capacity to accept it upon its release from the burning substance. As such, dephlogisticated air was pure and allowed for an increased capacity for combustion. Following a number of experiments on mice, he decided to try dephlogisticated air himself and reported that *my breath felt peculiarly light and easy for some time afterwards*, and prophetically suggested that *in time this pure air may become a fashionable article in luxury*.[6] However, he also warned that:

> as a candle burns out much faster in dephlogisticated than in common air, so we might, as may be said, *live out too fast*, and the animal powers be too soon exhausted in this pure kind of air. A moralist, at least, may say, that the air which nature has provided for us is as good as we deserve. [emphasis in original][6]

* * * * *

Around the time Priestley was experimenting with different airs, Henry Cavendish, a shy millionaire aristocrat for whom a life in politics was expected but who decided to dedicate his life to science, was also investigating the composition of air. In 1766, he published three papers describing the isolation of a new substance from common air. To accomplish this, he dissolved a number of metals in different acids and found that all combinations, save for dissolution in nitrous acid, resulted in the generation of a highly inflammable air which caused loud explosions upon being subjected to a flame (Figure 1.2). In addition, Cavendish found that the ratio in which this air was mixed with common air correlated to the loudness of the explosion and the size of the flame: the loudest bang occurred with a mixture of 6 parts of the inflammable air to 4 parts common air; with less inflammable air, the explosion became softer and

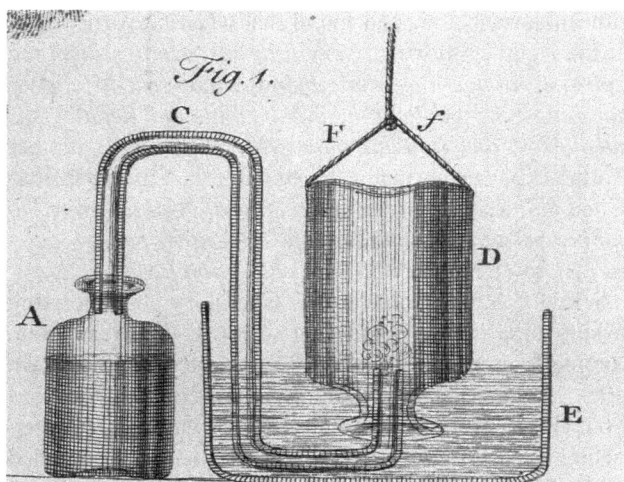

FIGURE 1.2 The apparatus used by Cavendish in his studies. Bottle A was filled with an acidic solution and a metal wire was added to it. The gas released was captured in bottle D via tube C. Water in E prevented the expelled gas from leaking out.[7]

the flame *spread gradually through the bottle* which contained the gases, whereas with more inflammable air, the bang was not very loud and *the mixture continued burning a short time in the bottle, after the sound was over.*[7] In addition, his results showed that the inflammable air alone did not burn, which indicated that a certain proportion of common air was required for the inflammable air to react with the flame. Using another apparatus of his own design, Cavendish accurately determined the density of this new air and found that it was more than ten times lighter than common air.[7] Since the inflammable air generated by a variety of metals had the same characteristics, that is, the explosion they caused and their density were identical, Cavendish concluded that:

> It seems likely from hence, that, when either of the above-mentioned metallic substances are dissolved in [*acid solutions*], their phlogiston flies off, without having its nature changed by the acid, and forms the inflammable air; but that, when they are dissolved in the nitrous acid [...] their phlogiston unites to part of the acid used for their solution, and flies off with it in fumes, the phlogiston losing its inflammable property by the union.[7]

Almost 20 years later, in 1784, Cavendish published another manuscript describing experiments on different airs that were subjected to an electrical spark. These experiments were inspired by those reported by Priestley, who had found that the walls of a glass vessel became covered with dew when inflammable air was subjected to an electrical spark. Cavendish confirmed these findings and determined that *almost all the inflammable air, and near one-fifth of the common air [...] are condensed into dew*, with an equal loss in weight of the air.[8] Furthermore, Cavendish noted that the dew appeared to be *plain water, and consequently that almost all the inflammable air, and about one-fifth of the common air, are turned into pure water.* He then repeated the experiment in which dephlogisticated air was mixed with inflammable air and found that *when they are mixed* [in the right] *proportion, very little air remains after the explosion, almost the whole being condensed, it follows, that almost the whole of the inflammable and dephlogisticated air is converted into pure water.* Following a number of other experiments to test this hypothesis, Cavendish concluded *that water consists of dephlogisticated air united to phlogiston; and that inflammable air is either pure phlogiston [...], or else water united to phlogiston [...].*[8]

While Cavendish was in his laboratory characterizing various airs, another prominent scientist, the Frenchman, Antoine-Laurent de Lavoisier, was performing his own series of experiments on airs and coming to conclusions about chemical reactions that would revolutionize chemistry and usher it into a new era; as such, he has been called the "father of modern chemistry". Lavoisier came from a privileged French family and intended on studying law. However, he was turned-on to science by one of his college professors, who showed great enthusiasm during his class demonstrations, and by a chance meeting with the great

Swedish scientist, Linnæus.[2] So adept was Lavoisier to his new chosen discipline that the French Academy of Sciences had already heard of him on a number of topics by the tender age of 25 and soon elected him as a member. In 1789, Lavoisier published the first textbook of modern chemistry, *Traité Élémentaire de Chimie*, in which he summarized the results of years of experiments which he had performed—though many were those initially reported by Priestley and Cavendish and which he attempted to reproduce; however, Lavoisier came to strikingly different conclusions about the results of these experiments.[2]

For example, upon repeating one of Priestley's experiments in which mercury was constantly heated for 12 days, at which point *the calcination of mercury did not at all increase*, he confirmed that the *air which remained after the calcination of the mercury [...], which was reduced to $5/6$ of its former bulk, was no longer fit either for respiration or for combustion.*[9] However, Lavoisier noticed that granules of red matter were formed during calcination, so he collected and placed them *into a small glass retort, having a proper apparatus for receiving such liquid or gaseous product, as might be extracted*, and reapplied fire until the retort became *almost red hot*, at which point *"the red matter began gradually to decrease in bulk, and in a few minutes after it disappeared altogether"* (Figure 1.3). Further analysis determined that the air collected during this experiment was *greatly more capable of supporting both respiration and combustion than atmospherical air* (this was the same air that Priestley called dephlogisticated air). Lavoisier summarized these experiments thusly:

> In reflecting upon the circumstances of this experiment, we readily perceive, that the mercury, during its calcination, absorbs the salubrious and respirable part of the air,

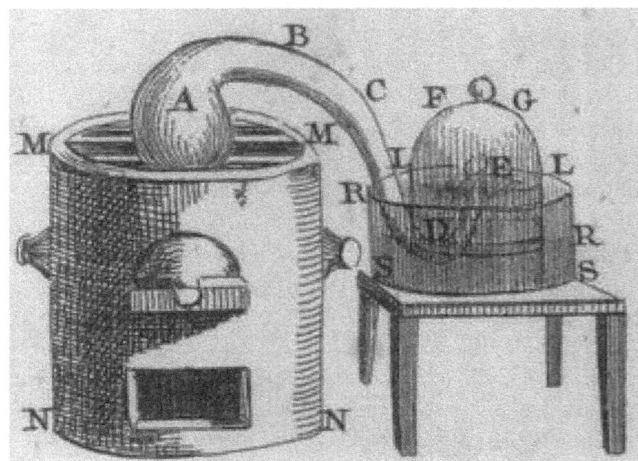

FIGURE 1.3 The apparatus used by Lavoisier to obtain dephlogisticated air from the combustion of mercury. Mercury was placed in the vessel RRSS and its height was marked. Mercury granules in A were heated by furnace MMNN and kept burning for 12 days. The expelled gas was collected in the bell glass FG.[9]

or, to speak more strictly, the base of this respirable part; that the remaining air is a species of mephitis, incapable of supporting combustion or respiration; and, consequently, that atmospheric air is composed of two elastic fluids of different and opposite qualities. As a proof of this important truth, if we recombine these two elastic fluids, which we have separately obtained in the above experiment [...] we reproduce an air precisely similar to that of the atmosphere, and possessing nearly the same power of supporting combustion and respiration, and of contributing to the calcination of metals.[9]

Lavoisier repeated this experiment using the combustion of phosphorus instead of mercury and showed that the airs which resulted from this reaction had the same characteristics as those produced by mercury: one increased the rate of respiration, while the other inhibited it. However, when the experiment was repeated using nitric acid, the air produced was not fit for respiration and extinguished a flame. In addition, when this air was combined in a ratio of 73:27 with respirable air, *an elastic fluid precisely similar to atmospheric air in all its properties* was formed.[9]

Lavoisier also noted that the terminology used to describe the different airs, that is, respirable, non-respirable, or noxious, were not adequate and required simpler words that better described these ideas. Thus, in 1787, Lavoisier presented to the Académie des Sciences a manuscript—prepared jointly with Louis-Bernard Guyton de Morveau, Claude Louis Bertholet, and Antoine François de Fourcroy, and with the consultation of many members of the Académie—in which a new system of nomenclature for chemical substances was proposed.[10] This proposal first addressed the two components of atmospheric air:

We have already seen, that the atmospheric fluid, or common air, is composed of two gases or aeriform fluids; one of which is capable, by respiration, of contributing to support animal life; and in it metals are calcinable, and combustible bodies may burn: The other, on the contrary, is endowed with directly opposite qualities; it cannot be breathed by animals, neither will it admit of the combustion of inflammable bodies, nor of the calcination of metals. We have given to the base of the former, which is the respirable portion of atmospheric air, the name of oxygen, from [...] acidum [*acid*], and [...] gignor [*to produce*], because one of the most general properties of this base is to form acids, by combining with many different substances. The union of this base with [...] what was formerly named pure, or vital, or highly respirable air, we now call oxygen gas [...]

The chemical properties of the noxious portion of atmospheric air being hitherto but little known, we have been satisfied to derive the name of its base from its known quality of killing such animals as are forced to breathe it, giving it the name of azot, from the Greek privative particle α and [...] vita [*life*]; hence the name of the noxious part of the atmospheric air is azotic gas [...] [*now known as nitrogen gas in English*]

The principal merit of the nomenclature we have adopted is, that, when once the simple elementary subject

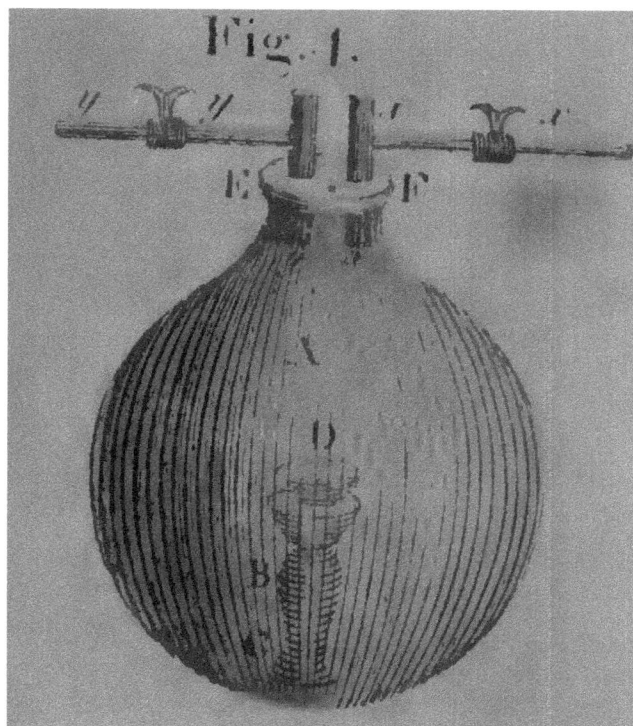

FIGURE 1.4 Vessel used by Lavoisier to show that metals are oxidized by oxygen. A glass vessel was equipped with a support, BC, upon which a cup, D, rested. Phosphorus or other metals were added to the cup and two tight-fitting tubes, xxx and yyy, were fitted through a stopper EF to supply a determined amount of oxygen. The metal was then set on fire and the oxygen consumed was determined by weighing the vessel before and after the reaction.[9]

is distinguished by an appropriate term, the names of all its compounds derive readily, and necessarily, from this first denomination.[9]

Thus, Lavoisier continued his experiments on airs by burning phosphorus in a vessel (Figure 1.4) and found that:

[i]ts combustion was extremely rapid, being attended by a very brilliant flame, and a considerable disengagement of light and heat [...]; at the same time, the whole inside of the glass became covered with light white flakes of concrete phosphoric acid.[9]

He then measured the weight of the flaky residue produced in the reaction and found that it

exactly equalled the sum of the weights of the phosphorus consumed and oxygen absorbed [... *Thus*] the only action of the phosphorus is to separate the oxygen from the caloric [*heat*], with which it was before united.

This experiment proves, in the most convincing manner, that, at a certain degree of temperature, oxygen possesses a stronger elective attraction, or affinity, for phosphorus than for caloric; and that, in consequence of this, the phosphorus attracts the base of oxygen gas from the caloric, which, being set free, spreads itself over the surrounding bodies.[9]

In addition, he found that the properties of this flaky phosphorus were significantly different from its previous form:

> from being insoluble in water, it becomes not only soluble, but so greedy of moisture, as to attract the humidity of the air with astonishing rapidity: By this means it is converted into a liquid, considerably more dense, and of more specific gravity, than water. In the state of phosphorus before combustion, it had scarcely any sensible taste; by its union with oxygen it acquires an extremely sharp and sour taste; in a word, from one of the class of combustible bodies, it is changed into an incombustible substance, and becomes one of those bodies called acids.

This property of a combustible substance to be converted into an acid, by the addition of oxygen, we shall presently find belongs to a great number of bodies: Wherefore, strict logic requires that we should adopt a common term for indicating all these operations which produce analogous results. This is the true way to simplify the study of science, as it would be quite impossible to bear all its specifical details in the memory, if they were not classically arranged. For this reason, we shall distinguish the conversion of phosphorus into an acid, by its union with oxygen, and in general every combination of oxygen with a combustible substance, by the term of oxygenation; From this I shall adopt the verb to oxygenate; and of consequence shall say, that in oxygenating phosphorus we convert it into an acid.[9]

He repeated this experiment using charcoal and sulfur and found similar results. However, in the case of charcoal, the product of the reaction did not produce a solid but remained in the gaseous form; this was the same substance that Joseph Black had discovered years before and named "fixed air", and that Priestley had used to make his soda water. However, Lavoisier felt that *since it is now ascertained to be an acid, formed like all others by the oxygenation of its peculiar base, it is obvious that the name of fixed air is quite ineligible.*[9] Lavoisier noted that these last three experiments

> may suffice for giving a clear and accurate conception of the manner in which acids are formed. By these, it may be clearly seen, that oxygen is an element common to them all, and which constitutes or produces their acidity; and that they differ from each other, according to the several natures of the oxygenated or acidified substances.
>
> [It] becomes extremely easy, from the principles laid down in the preceding chapter, to establish a systematic nomenclature for the acids: The word acid being used as a generic term, each acid falls to be distinguished in language, as in nature, by the name of its base or radical. Thus, we give the generic name of acids to the products of the combustion or oxygenation of phosphorus, of sulphur, and of carbon; and these products are reflectively named, the phosphoric acid, sulphuric acid, and the carbonic acid.[9]

He also noted that:

> There is [...] a remarkable circumstance in the oxygenation of combustible bodies, and of a part of such bodies as are convertible into acids, that they are susceptible of

different degrees of saturation with oxygen, and that the resulting acid, though formed by the union of the same elements, are possessed of different properties, depending upon that difference of proportion. Of this, the phosphoric acid, and more especially, the sulphuric, furnish us with examples. When sulphur is combined with a small portion of oxygen, it forms, in this first or lower degree of oxygenation, a volatile acid, having a penetrating odour, and possessed of very peculiar qualities. By a larger proportion of oxygen, it is changed into a fixed heavy acid, without any odour, and which, by combination with other bodies, gives products quite different from those furnished by the former.[9]

For such circumstances when substances are capable of different degrees of saturation by oxygen, he suggested using the suffix "ous" for the first, or lesser degree of oxygenation and "ic" for the second degree, which, in the case of sulfur, results in the terms sulfurous acid and sulfuric acid. Contrary to mercury and gases, metals which are solid at room temperature do not form acids during combustion but rather *are only changed into intermediate substances, which, though approaching to the nature of salts, have not acquired all the saline properties.* As such, he proposed to rename the expression *metallic calx*, originally used not only to identify *metals in this state, but to every body which has been long exposed to the action of fire without being melted*, to *oxyd* (later changed to oxide).

Another area explored by Lavoisier in this manuscript was the decomposition of water into carbonic acid. As such, he first used a device in which he boiled a known quantity of distilled water, the evaporation of which was passed through a furnace and out to a condenser (Figure 1.5). Lavoisier placed either charcoal or iron inside the tube (EF in Figure 1.5) where it was heated by the action of the furnace, to determine the influence of these substances on the vapors passing through it, or vice versa. In a control experiment designed to confirm that the apparatus itself did not change the composition of the water and in which the tube passing through the furnace remained empty, he found that *a quantity of water has passed over into the bottle [...], exactly equal to what was before contained in the retort [...], without any disengagement of gas whatsoever.*[9] In contrast, placing a piece of charcoal in the tube resulted in *a considerable quantity of gas [...]* [being] *disengaged*, which he determined was composed of carbonic acid and a highly inflammable gas, while the condensed water collected out from the furnace had lost weight compared to the starting quantity. His calculations showed that the amount of gas produced was equal to the amount of water consumed plus the charcoal burnt in the reaction, which led him to conclude that the water and charcoal combined to produce carbonic acid and the inflammable gas.[9]

He then repeated the experiment, placing iron in the tube instead of charcoal. He found that the only gas formed in this experiment was the inflammable gas and that, in contrast to the previous experiment, the mass of the iron had *increased* by an amount equivalent to that lost from the

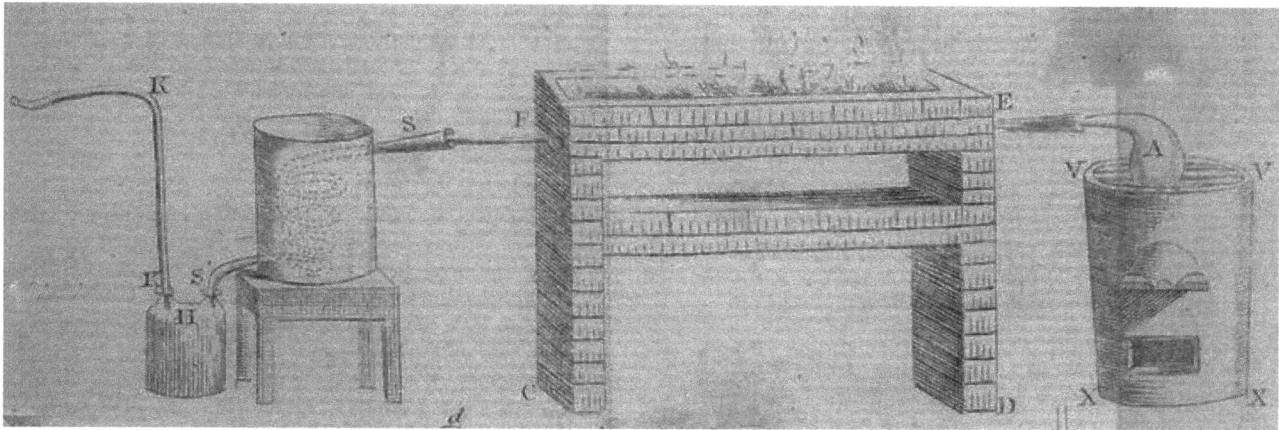

FIGURE 1.5 The apparatus used by Lavoisier to confirm the composition of water as hydrogen and oxygen. Water inside a glass retort, A, was made to boil by a furnace, VVXX, the vapors of which were directed through a second furnace, CDEF, via a tube. The vapor exiting from the furnace was condensed back to water in the "worm" and collected in the bottle, H, whereas the disengaged gases were collected via tube KK.[9]

water minus the weight of the inflammable gas. This suggested to him that the reaction was a *true oxydation of iron by means of water, exactly similar to that produced in air by the assistance of heat.*[9] Thus, from these experiments, Lavoisier concluded that:

[…] water is composed of oxygen combined with the base of an inflammable gas, in the respective proportion of 85 parts by weight of the former, to 15 parts of the latter.

Thus water, besides the oxygen, which is one of its elements, as it is of many other substances, contains another element as its constituent base or radical, and for this proper principle or element we must find an appropriate term. None that we could think of seemed better adapted than the word *hydrogen*, which signifies the *generative principle of water*, from […] aqua *[water]*, and […] gignor *[producer]*. We call the combination of this element with *[heat]* hydrogen gas; and the term hydrogen expresses the base of that gas, or the radical of water.

This experiment furnishes us with a new combustible body, or, in other words, a body which has so much affinity with oxygen as to draw it from its connection with caloric, and to decompose oxygen gas […] In this state of gas *[hydrogen]* is about $1/13$ of the weight of an equal bulk of atmospheric air; it is not absorbed by water […] and it is incapable of being used for respiration, without producing instant death.[9]

Explaining the results of Cavendish's experiments involving the use of varying ratios of common air and dephlogisticated air, Lavoisier noted that:

As the property of burning, which this gas possesses in common with all other combustible bodies, is merely the power of decomposing air, and carrying off its oxygen from the caloric with which it is combined, it is easily understood that it cannot burn, unless in contact with air or oxygen gas. Hence, when we set fire to a bottle full of this gas, it burns gently, first at the neck of the bottle, and then in the inside of it, in proportion as the external

air gets in: This combustion is slow and successive, and only takes place at the surface of contact between the two gases.[9]

To confirm that water was indeed formed from the combination of oxygen and hydrogen, Lavoisier fitted a glass bottle with tubes connected to oxygen and hydrogen sources, as well as *a metallic wire […] having a knob at its extremity […] intended for giving an electrical spark […] on purpose to set fire to the hydrogen gas* (Figure 1.6).[9] Thus, he filled the bottle with oxygen and then, *by means of pressure […]* [forced in] *a small stream of hydrogen gas […] to which we immediately set fire by an electrical spark* and found that

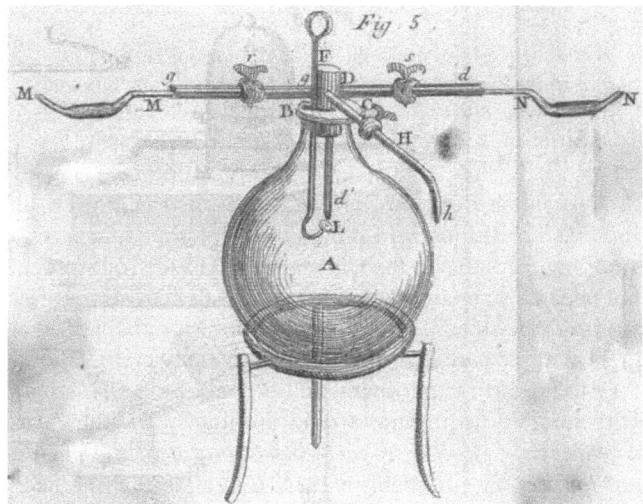

FIGURE 1.6 Lavoisier's apparatus to confirm the formation of water by the combination of oxygen and hydrogen. A large crystal balloon was fitted with four tubes: Hh was connected to a pump which allowed the evacuation of air; gg supplied oxygen, while dd' supplied hydrogen; GL contained a metallic wire with a knob which could produce a spark.[9]

water was indeed produced in an amount that was equal to the quantity of oxygen and hydrogen admitted into the bottle. Given these results, Lavoisier concluded that:

> we may now affirm, that we have ascertained with as much certainty as is possible in physical or chemical subjects, that water is not a simple elementary substance, but is composed of two elements, oxygen and hydrogen; which elements, when existing separately, have so strong an affinity for caloric, as only to subsist under the form of gas in the common temperature and pressure of our atmosphere.[9]

Therefore, the whole of Lavoisier's results effectively debunked the theory of phlogiston and welcomed chemistry into a new age! Soon after the publication of this manuscript, Lavoisier met his end at the hands of the guillotine during the French Revolution. Given the exceptional contribution to science at such a young age, a contemporary was quoted as saying that *it only took them a moment to cut off his head, and one hundred years may not be enough to produce another like it.*[11]

* * * * *

Although great strides were made by Lavoisier in terms of changing the nomenclature of chemical substances, the symbols used to represent them remained obscure and were legible by few. A noteworthy attempt at improving chemical symbols came in the early 1800s, when John Dalton published *A New System of Chemical Philosophy*, a two-volume manuscript in which, among other things, he suggested a new manner in which elements and molecules could be represented (Figure 1.7b).[12] Although it clearly took inspiration from the old system (Figure 1.7a), it was a vast improvement over the use of obscure symbols which were *of very little utility*, and which

> owed their origin, no doubt, to the mysterious relation supposed by the alchemists to exist between the metals and the planets, and to the desire which they had of expressing themselves in a manner incomprehensible to the public.[13]

Jöns Jakob Berzelius, a Swedish orphan who grew up in harsh conditions working on his stepfather's farm, was studying medicine at the University of Upsala, in Sweden, but became so enamored with analytical chemistry—after reading a book on Lavoisier's antiphlogiston theory—that he bribed the caregiver for access to the university laboratory to perform experiments in his spare time.[2] This man with humble beginnings would eventually be known as a respected *lawgiver, a veritable autocrat of the chemical laboratory.*[2] A such, in 1814, Berzelius saw the need for a new system of symbols to describe elements and molecules,

> solely to facilitate the expression of chemical proportions, and to enable us to indicate, without long periphrases, the relative number of volumes of the different constituents contained in each compound body. By determining the

weight of the elementary volumes, these figures will enable us to express the numeric result of an analysis as simply, and in a manner as easily remembered, as the algebraic formulas in mechanical philosophy.[13]

As such, he proposed that:

> The chemical signs ought to be letters, for the greater facility of writing, and not to disfigure a printed book [...] I *shall take, therefore, for the chemical sign, the initial letter of the Latin name of each elemental substance*: but as several have the same initial letter, I *shall distinguish them in the following manner:*—I. In the class which I call *metalloids.*, I shall employ the initial letter only, even when this letter is common to the metalloid and to some metal. 2. In the class of metals, I shall distinguish those that have the same initials with another metal, or a metalloid, by writing the first two letters of the word. 3. If the first two letters be common to two metals, I shall, in that case, add to the initial letter the first consonant which they have not in common.[13] [emphasis in original]

In addition, Berzelius proposed the use of superscripts to represent the ratio of each element in a molecule; however, this was deemed awkward and it was later changed to a subscript. As an example of the vast improvement in the legibility of written chemistry, whereas Lavoisier would have represented an equation consisting of iron, water, oxygen, and nitrous oxide as:

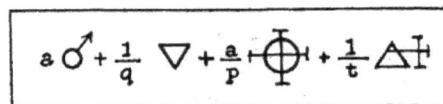

Berzelius would use the much simplified, $Fe + 2H_2O + 3O_2 + 4N_2O$![12] Naturally, some objected to the new symbols: Dalton himself had said that *A young student might as soon learn Hebrew as make himself acquainted with* these symbols. Nevertheless, it quickly made its way into textbooks and soon became standard practice.

* * * * *

Static electricity discharge has been known since ancient times; in fact, the word electricity has its root in the Greek word for amber, *elektron*, pieces of which were known to attract light objects after being rubbed together. However, the ability to store and release this electrical energy at will was not accomplished until the 18th century, when the Leyden jar was invented: this was essentially a jar into which a metal rod was inserted through a stopper at the mouth of the jar and immersed in a salt solution which filled the jar. The jar was charged, via the metal rod, using static electricity, and it would discharge when an object came in contact with the rod; as such, the major disadvantage of this device was that only a single charge could be stored, such that the apparatus would have to be recharged after each discharge. A solution to this problem came in 1800, when Alessandro Volta produced the first electric pile, or battery, which was capable of continuously

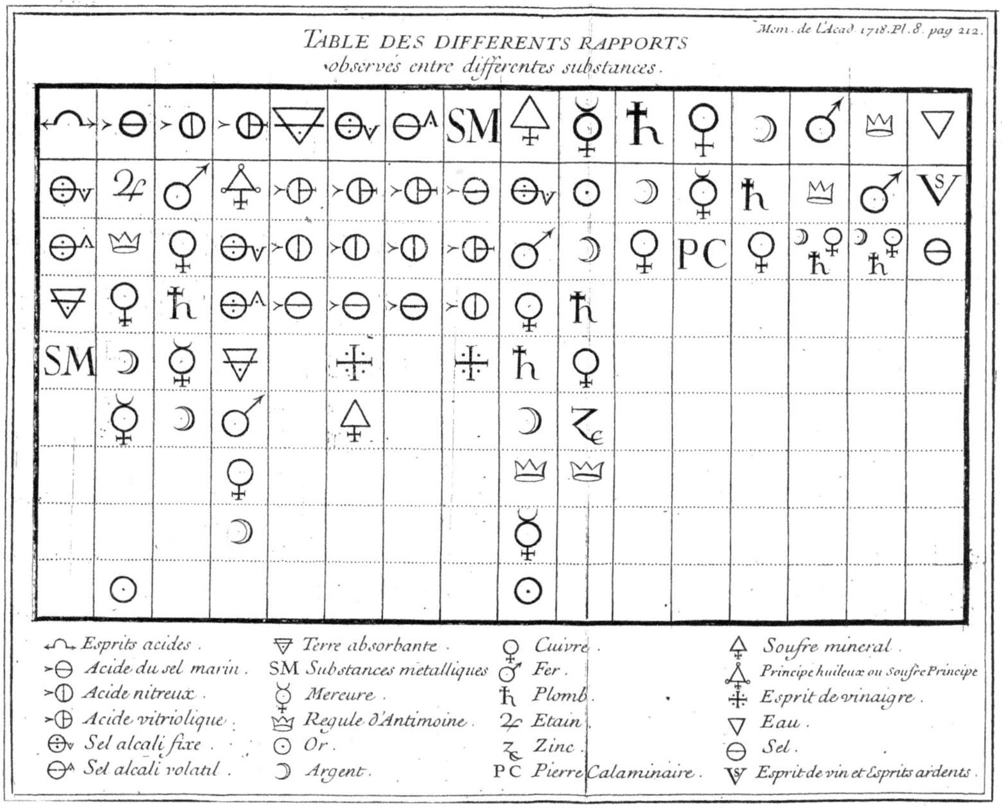

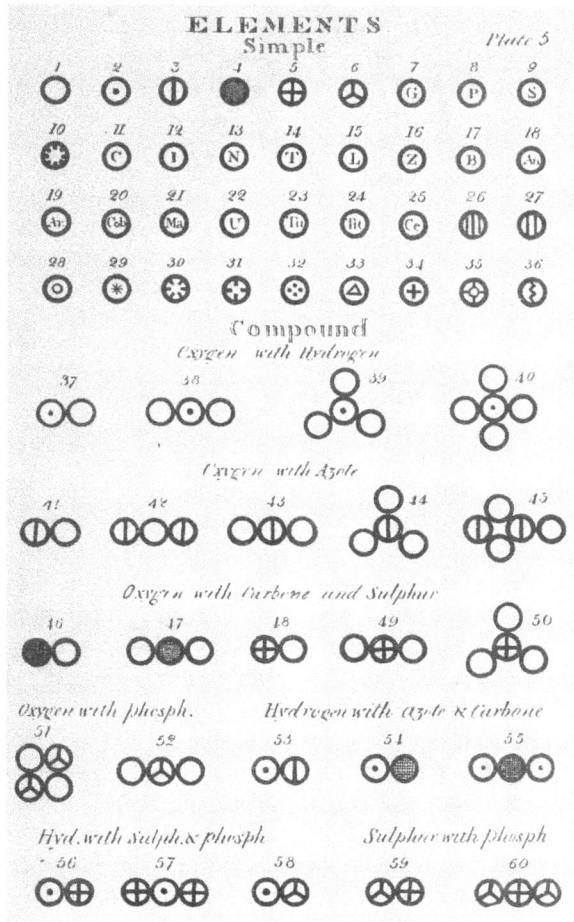

TABLE DES DIFFERENTS RAPPORTS
observes entre differentes substances.

Mem. de L'Acad. 1718.Pl.8. pag. 212.

(a)

ELEMENTS
Simple
Plate 5

Compound

Oxygen with Hydrogen

Oxygen with Azote

Oxygen with Carbone and Sulphur

Oxygen with phosph.   Hydrogen with Azote & Carbone

Hyd. with Sulph. & phosph   Sulpher with phosph

(b)

FIGURE 1.7 (a) Example of chemical symbols used before Dalton and Berzelius. Gold, ⊙; water, ∇; Salt, ⊖. (b) Dalton's proposed symbols for the description of elements and simple molecules. Oxygen, ○; hydrogen, ⊙; sulfur, ⊕.[12,14]

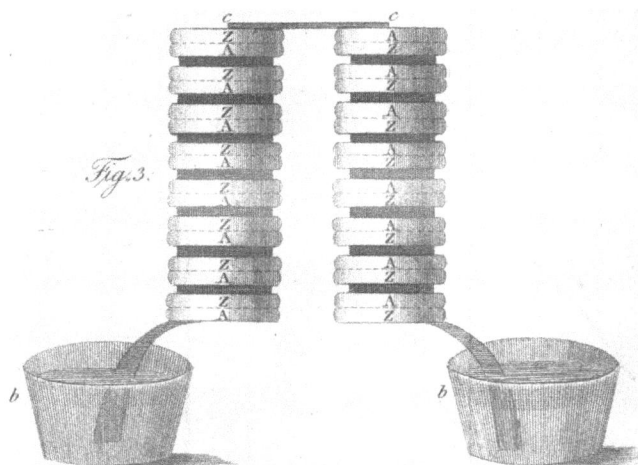

FIGURE 1.8   Alessandro Volta's electric pile. Zinc discs, Z, were placed on top of silver discs, A, and individual stacks were separated by saltwater-soaked cardboard. Each stack was connected to a salt solution, b, and to each other by a conductive bridge, c.[15]

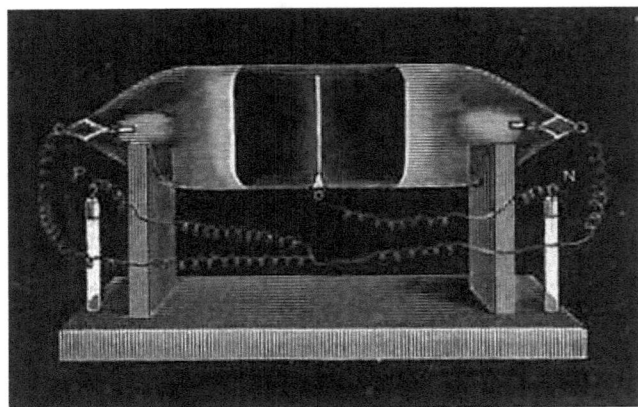

FIGURE 1.9   Crook's vacuum tube. The negative pole is in the center and the positive poles at each end. The dark space in the center was created as a result of the partial evacuation of the tube. A complete vacuum would have resulted in complete darkness except at the poles.[16]

supplying a charge for an extended period of time without the need to be recharged.[15]

This device was initially made by placing one piece of silver, one inch in diameter, on a table and adding a piece of zinc of the same size on top of it, followed by a piece of cardboard soaked in salt water; then, another set of these three items were added on top, continuing as such to form a stack of metal discs separated by saltwater-soaked cardboard *as tall as can be supported without falling over* (Figure 1.8).[15] Volta found that a stack containing 20 of these units was capable of generating *a slight commotion* when wetted fingers touched each extremity—similar to that induced by a Leyden jar, save that it be continuous. A more pronounced effect could be produced if the extremities were connected to a water basin and the fingers, the hand, or better yet, the whole arm, was immersed while touching the other extremity with a wet piece of metal. Volta found that the greater the number of units in the stack, the larger the electric shock was. In time, it was determined that the electricity generated in these stacks was dependent on the two metals that were used, each given combination producing a characteristic charge.

In 1834, the great Michael Faraday placed two metal plates, each attached to the electrodes of an electrical source, inside a closed tube and evacuated the air as best he could. In doing so, he found that the tube began to glow! And although he could not determine the reason for this phenomenon, he called the matter producing the glow *radiant matter*. More than 20 years later, advancements in technology allowed the German, Julius Plücker, to increase the vacuum inside the tube and as such, observe a *green phosphorescence on the glass in the neighbourhood of the negative electrode*;[16] however, it would be yet another 20 years before anyone could begin to make sense of this phenomenon. In 1879, William Crooks reported a series of experiments he performed in which he passed high levels

of electricity through a glass tube that had been nearly completely evacuated. He noticed that a dark space surrounded the negative pole which increased and decreased as the vacuum was varied (Figure 1.9). He suggested that the glow observed by Faraday was the result of electrical particles colliding with gas molecules inside the tube. Since the evacuation of the tube reduced the gas molecules present, a dark space would appear where no molecules were present:

> […] if we exhaust the air or gas contained in a closed vessel, the number of molecules becomes diminished, and the distance through which anyone of them can move without coming in contact with another is increased, the length of the mean free path being inversely proportional to the number of molecules present. The further this process is carried the longer becomes the average distance a molecule can travel before entering into collision […]
>
> We may naturally infer that the dark space is the mean free path of the molecules of the residual gas […][16]

As such, he found that if exhaustion was carried out far enough, *the dark space around the negative pole* [widens] *out till it entirely fills the tube* [...] *if no residual gas is left, the molecules will have their velocity arrested by the sides of the glass* [...] *and produce phosphorescence.*[16] Interestingly, he also found that a solid object placed between the glass and the electrode cast a shadow on the glass (Figure 1.10a). In addition, experiments in which he housed electrodes in a V-shaped tube revealed that radiant matter moved as a beam since it did not turn corners, the phosphorescence appearing at the bottom of the V-shape (Figure 1.10b). He also found that radiant matter could perform mechanical work since a paddlewheel would start turning when it was inserted in the path of the radiant matter.

Crooks then wondered if a magnet would have any effect on radiant matter. He found that the phosphorescence,

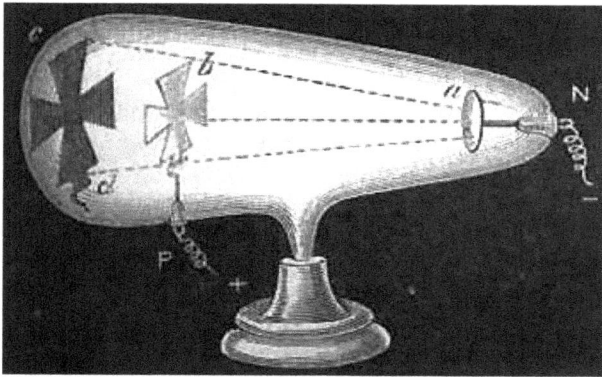

(a)

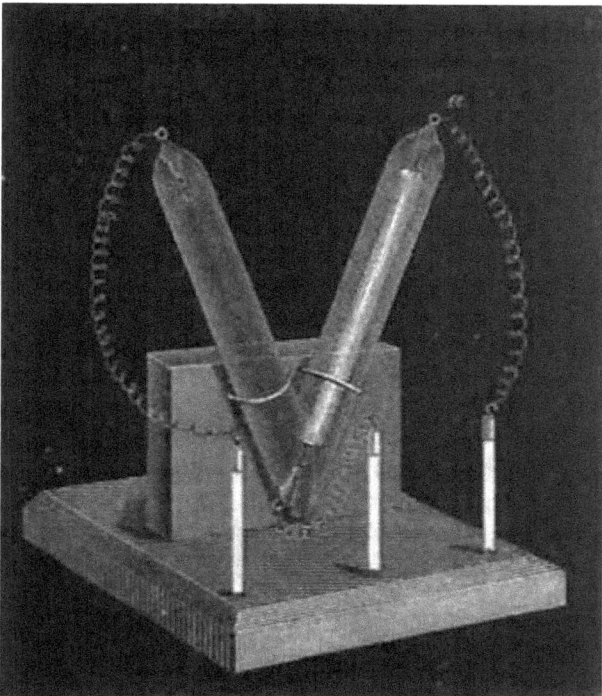

(b)

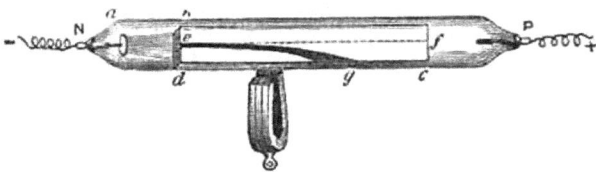

(c)

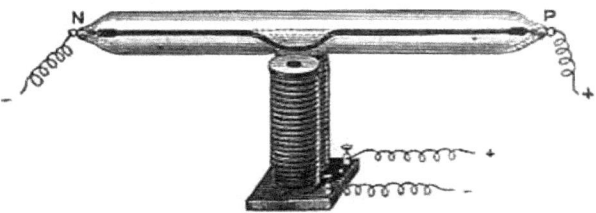

(d)

FIGURE 1.10 (a) A metal object placed in the path of radiant matter casts a shadow. (b) Radiant matter did not turn corners since the phosphorescence appeared at the bottom of this V-shaped tube. (c) Radiant light bent in direction of a magnet, whereas it merely dipped *in the center toward the magnet* in a partial vacuum (d).[16]

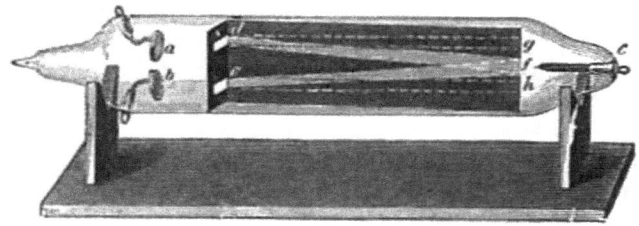

FIGURE 1.11 Radiant matter is composed of negatively electrified molecules. A tube consisting of two negative electrodes, a and b, and one positive electrode, c. The solid lines indicate the path of radiant light emitted from each electrode when fired one at a time. Results indicated that electrode a's line of fire, d-f, was deflected to d-g when the second electrode, b, was turned on.[16]

which was normally seen on the glass directly across from the electrode,

> becomes curved under the magnetic influence waving about like a flexible wand as *[he moved]* the magnet to and fro [...] The molecules shot from the negative pole may be likened to a discharge of iron bullet from a mitrailleuse, and the magnet beneath will represent the earth curving the trajectory of the shot by gravitation. *[Figure 1.10c].*[16]

Interestingly, the path was not completely altered if the vacuum was decreased (Figure 1.10d), but rather,

> dips in the center toward the magnet [...] Here the action is temporary. The dip takes place under the magnetic influence; the line of discharge then rises and pursues its path to the positive pole. In high exhaustion, however, after the stream of radiant matter had dipped to the magnet it did not recover itself, but continued its path in the altered direction.[16]

Crooks also fitted an evacuated tube with two negative electrodes and one positive electrode to determine if radiant matter carried a current, reasoning that *if the streams of radiant matter carry an electric current, they will act like two parallel conducting wires and attract one another; but if they are simply built up of negative electrified molecules, they will repel each other* (Figure 1.11).[16] He found the latter to be the case. Finally, Crooks determined that this phenomenon was not dependent on the gas inside the tube itself but rather, it was consistently dependent on the degree to which the gas had been evacuated from the tube, and concluded his report like so:

> In studying this fourth state of matter we seem, at length, to have within our grasp and obedient to our control the little indivisible particles which, with good warrant, are supposed to constitute the physical basis of the universe. We have seen that, in some of its properties, radiant matter is as material as this table, while in other properties it almost assumes the character of radiant energy. We have actually touched on the border-land where matter and force seem to merge into one another, the shadowy realm between known and unknown, which for me has

always had peculiar temptations. I venture to think that the greatest scientific problems of the future will find their solution in this border-land, and even beyond; here it seems to me, lie ultimate realities, subtitle, far-reaching, wonderful.[16]

By the end of the century, many scientists had investigated this strange matter, which, by this time had been renamed cathode rays (the negative electrode is called the cathode), and J.J. Thomson commented that:

according to the almost unanimous opinion of German physicists [cathode rays] are due to some process in the aether to which—inasmuch as in a uniform magnetic field their course is circular and not rectilinear, no phenomenon hitherto observed is analogous [...][17]

By this time, Jean Perrin had confirmed that these particles were composed of a negative charge since a *charge of negative electricity* was received when cathode rays were fired; however, many were still not convinced.[17] Therefore, Thomson repeated Perrin's experiment but modified it in a manner that would be suitable to address the objections: like Perrin, he used two coaxial cylinders with slits in them (to let the stream of cathode rays through) and attached them to a tube that was offset from the path of the rays coming from the cathode; he also mounted an electrometer at the end of the coaxial tubes to detect and measure charges (Figure 1.12a). By using a magnet to twist the rays so as to divert its path, Thomson found that a charge on the electrometer was only detected when the path of the rays was deflected such as it entered the tubes through the slits, thereby showing that *this negative electrification is indissolubly connected with the cathode rays.*[17] One of the objections to cathode rays being negatively charged particles was that they were shown to not be affected by electrostatic forces. Therefore, Thomson addressed this by designing a tube with two parallel metal plates in the center, between which the rays would pass before falling

on the glass opposite the cathode; a scale was also pasted on the glass to measure the angle of deflection of the ray (Figure 1.12b). Initially, he came to the same results previously shown, that is, electrostatic forces had no effect on the ray. However, he found that *the absence of deflection was due to the conductivity conferred on the rarefied gas by the cathode rays*, and determined that further exhaustion of the gas decreased this effect to such a degree that *the rays were deflected when the two aluminum plates were connected with the terminals of a battery of small storage-cells.*

Thomson thus concluded that

As the cathode rays carry a charge of negative electricity, are deflected by an electrostatic force as if they were negatively electrified, and are acted on by a magnetic force in just the way in which this force would act on a negatively electrified body moving along the path of these rays, I can see no escape from the conclusion that they are charges of negative electricity carried by particles of matter. The question next arises, What are these particles? are they atoms, or molecules, or matter in a still finer state of subdivision?[17]

To shed light on some of these questions, Thomson calculated, using a variety of experimental measurements obtained using 4 different methods, the mass-to-charge ratio, *m/e*, of the particles in question and determined that it was substantially smaller than even the smallest known value for this parameter: that of the hydrogen ion during electrolysis! In his conclusion, Thomson noted that

The two fundamental points about these carriers seem to me to be (1) that these carriers are the same whatever the gas through which the discharge passes, (2) that the mean free paths depend upon nothing but the density of the medium traversed by these rays [...]

The explanation which seems to me to account in the most simple and straightforward manner for the facts is founded on a view of the constitution of the chemical elements which has been favourably entertained by many

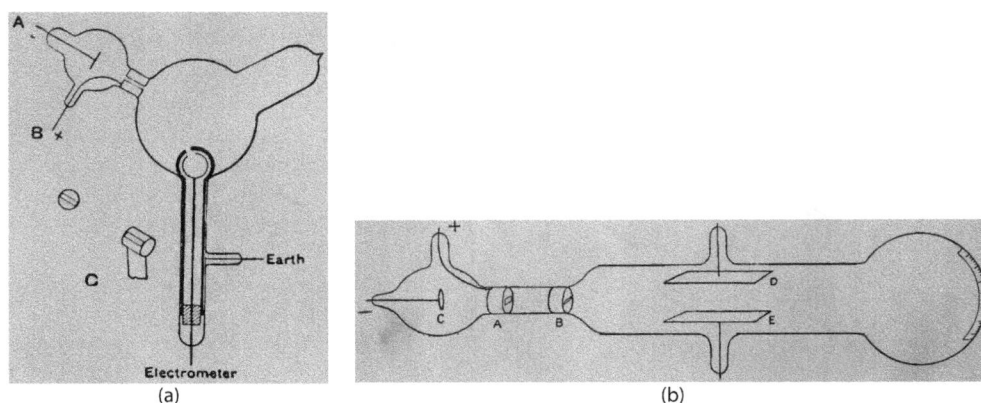

FIGURE 1.12   (a) Thomson's apparatus to deflect cathode rays. The cathode, A, was discharged and deflected by a magnet. A negative charge was detected only when it was appropriately deflected to enter the slit leading to the electrometer. (b) Thomson's cathode ray tube. The path of discharge from the cathode, C, was made to pass between two parallel electrodes, D and E. A scale was mounted to the end of the tube to measure the deflection imparted by the charge across the electrodes.[17]

chemists: this view is that the atoms of the different chemical elements are different aggregations of atoms of the same kind. In the form in which this hypothesis was enunciated by Prout, the atoms of the different elements were hydrogen atoms; in this precise form the hypothesis is not tenable, but if we substitute for hydrogen some unknown primordial substance X, there is nothing known which is inconsistent with this hypothesis [...]

If, in the very intense electric field in the neighbourhood of the cathode, the molecules of the gas are dissociated and are split up, not into the ordinary chemical atoms, but into these primordial atoms, which we shall for brevity call corpuscles; and if these corpuscles are charged with electricity and projected from the cathode by the electric field, they would behave exactly like the cathode rays. They would evidently give a value of m/e which is independent of the nature of the gas and its pressure, for the carriers are the same whatever the gas may be; again, the mean free paths of these corpuscles would depend solely upon the density of the medium through which they pass. For the molecules of the medium are composed of a number of such corpuscles separated by considerable spaces; now the collision between a single corpuscle and the molecule will not be between the corpuscles and the molecule as a whole, but between this corpuscle and the individual corpuscles which form the molecule; thus the number of collisions the particle makes as it moves through a crowd of these molecules will be proportional, not to the number of the molecules in the crowd, but to the number of the individual corpuscles. The mean free path is inversely proportional to the number of collisions in unit time, and so is inversely proportional to the number of corpuscles in unit volume; now as these corpuscles are all of the same mass, the number of corpuscles in unit volume will be proportional to the mass of unit volume, that is the mean free path will be inversely proportional to the density of the gas [...]

Thus on this view we have in the cathode rays matter in a new state, a state in which the subdivision of matter is carried very much further than in the ordinary gaseous state: a state in which all matter—that is, matter derived from different sources such as hydrogen, oxygen, &c.—is of one and the same kind; this matter being the substance from which all the chemical elements are built up.[17]

What Thomson had found and named "corpuscle" was in fact, the electron. And since atoms were known to be electrically neutral, the discovery of electrons indicated that positively charged particles, later called protons, must also exist, and Thomson proposed, in 1904, the *plum pudding model* of the atom, in which *the atoms of the elements consist of a number of negatively electrified corpuscles enclosed in a sphere of uniform positive electrification*, similar to the plums in plum pudding.[18] The same year, Hantaro Nagaoka, in Japan, proposed an alternate hypothesis, the *planetary model* of the atom, in which electrons revolved around a massive nucleus of protons and were held together by electrostatic forces.[19] Ernest Rutherford subsequently proved Nagaoka's hypothesis to be correct in 1911.[20]

* * * * *

In the 19th century, chemical bonds were represented by a line between elements, such as H–O–H for water, *which expressed in a concise way many chemical facts, but which had only qualitative significance with regard to molecular structure.*[21] Even by the early years of the 20th century, little was known about how the different atoms in molecules were held together. However, in 1916, Gilbert Lewis advanced a theory that provided an initial explanation for these bonds which has stood the test of time: atoms were proposed to donate, accept, or share electrons when they assemble into molecules. He started by pointing out that in addition to the conventional classification of chemicals in terms of being inorganic or organic, molecules should also be classified in terms of their polarity, that is, whether or not they have electrical charge. For example, he noted that polar and nonpolar molecules have *striking differences in properties*, such that polar molecules were mobile, reactive, and could be ionized, whereas nonpolar molecules were immobile, inert, and did not ionize.[22] As such, Lewis suggested that:

we may very safely assume that the essential difference between the polar and the nonpolar molecule is that, in the former, one or more electrons are held by sufficiently weak constraints so that they may become separated from their former positions in the atom, and in the extreme case pass altogether to another atom, thus producing in the molecule a bipole or multipole of high electrical moment. Thus in an extremely polar molecule, such as that of sodium chloride, it is probable that at least in the great majority of molecules the chlorine atom has acquired a unit negative charge and therefore the sodium atom a unit positive charge, and that the process of ionization consists only in a further separation of these charged parts.

If then we consider the nonpolar molecule as one in which the electrons belonging to the individual atom are held by such constraints that they do not move far from their normal positions, while in the polar molecule the electrons, being more mobile, so move as to separate the molecule into positive and negative parts, then all the distinguishing properties of the two types of compounds become necessary consequences of this assumption, as we may readily show.

Thus polar compounds with their mobile parts fall readily into those combinations which represent the very few stable states, while the nonpolar molecules, in which the parts are held by firmer constraints, are inert and unreactive, and can therefore be built up into the numerous complicated structures of organic chemistry. Many organic compounds, especially those containing elements like oxygen and nitrogen, and those which are said to be unsaturated, show at least in some part of the molecule a decidedly polar character [...]

When a molecule owing to the displacement of an electron, or electrons, becomes a bipole (or multipole) of high electrical moment, that is, when its charged parts are separated by an appreciable distance, its force of attraction for another molecular bipole will be felt over a considerable intervening distance, and two or more such bipoles will frequently be drawn together into a single aggregate in which the positive part of one molecule is brought as near as possible to the negative part of another. The molecules of a polar substance will therefore not only exhibit an unusually high intermolecular attraction at a distance, but will frequently combine with one another and show the phenomenon known as association [...]

Moreover a polar substance will combine with other substances to form those aggregates which are sometimes known as molecular compounds or complexes, and it may so combine with substances which are not of themselves markedly polar, for in the presence of a polar substance all other substances become more polar.[22]

Lewis presented a number of *important laws of chemical behavior* in the form of six postulates, which proposed that each atom is composed of a "kernel"—protons and electrons—which usually possesses a net positive charge, that is, it contains more protons than electrons, and cannot be altered.[22] They also possess an outer shell with a given number of electrons, depending on the element, and which, in the case of neutral atoms, like neon or argon, balances the excess positive charge of +8 in their kernel. Lewis also proposed that all atoms strive to have a complete outer shell, such that atoms that do not contain 8 electrons have a tendency to donate or accept electrons in an effort to total 8 outer-shell electrons. On the other hand, elements in molecules physically associate via "covalent bonds"; in this case, atoms form molecules by permanently sharing electrons to produce complete outer shells in each atom. Lewis also introduced a new form of chemical symbols that bears his name, the Lewis structure, to easily visualize the phenomenon of electron sharing, which uses colons *to represent the two electrons which act as the connecting links between the two atoms*,[22] such as H:Ö:H for water. The ideas conveyed in his paper were the basis of the modern electronic theory of valence.

In the following years, many scientists contributed to the advancement of the theory of valence, and in 1939, the American scientist, Linus Pauling, published his seminal *The Nature of the Chemical Bond*, in which he summarized all the findings, including his own, concerning chemical bonds.[21] He started by defining the different types of known chemical bonds: electrostatic bonds, such as the ionic bond of NaCl; covalent bonds such as those in water; and metallic bonds, those which contain a metallic element and whose bonds have special properties that will not be discussed here. As such, Pauling defined the chemical bond in the following way:

There is a chemical bond between two atoms or groups of atoms in case that the forces acting between them are such as to lead to the formation of an aggregate with sufficient stability to make it convenient for the chemist to consider it as an independent molecular species.[21]

As to the ionic bond, Pauling noted that

The atoms of metallic elements lose their electrons easily, whereas those of nonmetallic elements tend to add additional electrons; in this way stable cations and anions may be formed, which may essentially retain their electronic structures as they approach one another to form a stable molecule or crystal. In the sodium chloride crystal

[...] there exist no discreet NaCl molecules. The crystal is instead composed of sodium cations, Na^+, and chloride anions, Cl^-, each of which strongly attracted to and held by the six oppositely charged ions that surround it octahedrally. We describe the interactions in this crystal by saying that each ion forms ionic bonds with its six neighbors, these bonds combining all of the ions in the crystal into one giant molecule.[21]

Importantly, Pauling brought to the masses the concept of hydrogen bonds, which would eventually be determined to be one of the most important bonds in biology. He stated that a hydrogen bond could occur between no more than two atoms, since hydrogen only has one electron to share with one atom, and the more electropositive side of the hydrogen atom could only associate with one other atom[21]. In addition, *only the most electronegative atoms should form hydrogen bonds, and the strength of the bond should increase with the increase in the electronegativity of the two bonded atoms*; as such, the usual atoms involved in hydrogen bonds are oxygen and nitrogen, two of the most electronegative elements in the periodic table. As mentioned earlier, the hydrogen bond has been shown to be one of the most important bonds in biology, despite the fact that—and actually, because of the fact that—its bond strength is very weak: this makes the energy required to make and break hydrogen bonds very low, and as such, this *bond is especially suited to play a part in reactions occurring at normal temperatures.*[21] As we will see in later chapters, the hydrogen bond is instrumental for proteins to keep their three-dimensional shapes and the DNA molecule is kept in its double helix form as a result of hydrogen bonding.

* * * * *

REFERENCES

1. Read J. Alchemy and alchemists. *Folklore.* 1933;44(3): 251–278. doi: https://doi.org/10.1080/0015587X.1933.9718503
2. Jaffe B. *Crucibles: The Story of Chemistry. From Ancient Alchemy to Nuclear Fission.* Dover Publications, Inc.; 1976.
3. Bacon R. *Mirror of Alchemy.* Printed for Richard Oliue; 1597. Accessed October 10, 2022. https://www.gutenberg.org/ebooks/58393
4. Boyle R. *The Sceptical Chymist.* Published 1661. Accessed October 10, 2022. https://www.gutenberg.org/ebooks/22914
5. Priestley J. *An Appeal to the Serious and Candid Professors of Christianity.* Richard Taylor, Red Lion Court, Fleet Street; 1827. Accessed November 7, 2022. https://archive.org/details/anappealtoserio01priegoog
6. Priestley J. *Experiments and Observations on Different Kinds of Air. Vol. II.* J. Johnson; 1775. Accessed November 7, 2022. https://www.gutenberg.org/ebooks/29734
7. Cavendish H XIX. Three papers, containing experiments on factitious air. *Philosophical Transactions of the Royal Society of London.* 1766;56:141–184. doi: https://doi.org/10.1098/rstl.1766.0019
8. Cavendish H XIII. Experiments on air. *Philosophical Transactions of the Royal Society of London.* 1784;74:119–153. doi: https://doi.org/10.1098/rstl.1784.0014

9. Lavoisier AL. *Elements of Chemistry, in a New Systematic Order, Containing All the Modern Discoveries.* Vol 1. 5th ed. Evert Duyckninck, J and T Ronalds, eds; 1806. Accessed November 7, 2022. https://www.gutenberg.org/ebooks/30775

10. de Morveau LBG, Lavoisier AL, Berthollet CL, Fourcroy AF *Méthode De Nomenclature Chimique.* Cuchet, Libraire; 1787. Accessed November 7, 2022. https://gallica.bnf.fr/ark:/12148/bpt6k1050402r.image

11. Delambre J-BJ. *Notices Sur La Vie et Les Ouvrages de M. Le Compte J.-L. Lagrange.*; 1867. Accessed October 10, 2022. https://gallica.bnf.fr/ark:/12148/bpt6k2155691/f4

12. Dalton J. *A New System of Chemical Philosophy Part I.* R. Bickerstaff; 1808. Accessed November 7, 2022. https://archive.org/details/newsystemofchemi01daltuoft/page/n3/mode/2up

13. Berzelius J. Essay on the cause of chemical proportions, and on some circumstances relating to them: Together with a short and easy method of expressing them. In: Robert Baldwin ed. *Magazine of Chemistry, Mineralogy, Mechanics, Natural History, Agriculture, and the Arts.* Vol III. Paternoster-Row; 1814:51–62.

14. L'Aîné G. Des differents rapports observés en chimie entre differentes substances. In: *Histoire de l'Académie Royale des Sciences.* Imprimerie Royale; 1718:202–213. Accessed November 7, 2022. https://gallica.bnf.fr/ark:/12148/bpt6k3519v/f337.item.r=#

15. Volta A XVII. On the electricity excited by the mere contact of conducting substances of different kinds. In a letter from Mr. Alexander Volta, F.R.S. Professor of natural philosophy in the University of Pavia, to the Rt. Hon. Sir Joseph Banks, Bart. K.B.P.R. *Philosophical Transactions of the Royal Society of London.* 1800;90:403–431. doi: https://doi.org/10.1098/rstl.1800.0018

16. Crookes W. On radiant matter; A lecture delivered to the British Association for the Advancement of Science, at Sheffield, Friday, August 22, 1879. *American Journal of Sciences.* 1879;s3-18(106):241–262. doi: https://doi.org/10.2475/ajs.s3-18.106.241

17. Thomson JJ. Cathode rays. *The London, Edinburgh, and Dublin Philosophical Magazine and Journal of Science.* 1897;44:293–316.

18. Thomson J XXIV. On the structure of the atom: An investigation of the stability and periods of oscillation of a number of corpuscles arranged at equal intervals around the circumference of a circle; with application of the results to the theory of atomic structure. *The London, Edinburgh, and Dublin Philosophical Magazine and Journal of Science.* 1904;7(39):237–265. doi: https://doi.org/10.1080/14786440409463107

19. Nagaoka H LV. Kinetics of a system of particles illustrating the line and the band spectrum and the phenomena of radioactivity. *The London, Edinburgh, and Dublin Philosophical Magazine and Journal of Science.* 1904;7(41):445–455. doi: https://doi.org/10.1080/14786440409463141

20. Rutherford E LXXIX. The scattering of α and β particles by matter and the structure of the atom. *The London, Edinburgh, and Dublin Philosophical Magazine and Journal of Science.* 1911;21(125):669–688. doi: https://doi.org/10.1080/14786440508637080

21. Pauling L. *The Nature of the Chemical Bond.* 3rd ed. Cornell University Press; 1960.

22. Lewis GN. The atom and the molecule. *Journal of American Chemical Society.* 1916;38(4):762–785. doi: https://doi.org/10.1021/ja02261a002

2 The Cell and Heredity

In the 17th century, if one could not see it, it did not exist. Therefore, anything smaller than that which could be seen with the naked eye was inconceivable to the imagination! It was the stuff of science fiction, and even then, likely reserved for the most imaginative of minds. No one could have imagined that the human body was made up of billions of microscopic, living entities, all with specific roles and specialized functions. And within these, even smaller molecules perform all the functions that a cell requires to survive. The first significant step toward the elucidation of the inner workings of cells—the discovery of cells themselves—was the invention of the microscope in the early 1620s. By that time, Galileo had invented the telescope to look at outer space and many thought a similar technology could be used to examine very small things. However, it was Robert Hooke—assistant to Robert Boyle at Oxford University and the first curator of experiments at the Royal Society of London—who made significant improvements to it in the mid-1600s and as such increased its usefulness for biological observations. In the preface to his *Micrographia* in 1665, Hooke expressed a desire for the *science of nature* to step away from being *a work of the Brain and the Fancy* and *return to the plainness and soundness of Observations on material and obvious things* with the help of *instruments that improve the sense*.[1] In addition, he shared his excitement about the discoveries now possible with the help of new technologies, such as the microscope:

> The next care to be taken, in respect of the Senses, is a supplying of their infirmities with Instruments, and, as it were, the adding of artificial Organs to the natural; this in one of them has been of late years accomplist with prodigious benefit to all sorts of useful knowledge, by the invention of Optical Glasses. By the means of Telescopes, there is nothing so far distant but may be represented to our view; and by the help of Microscopes, there is nothing so small, as to escape our inquiry; hence there is a new visible World discovered to the understanding. By this means the Heavens are open'd, and a vast number of new Stars, and new Motions, and new Productions appear in them, to which all the ancient Astronomers were utterly Strangers. By this the Earth itself, which lyes so neer us, under our feet, shews quite a new thing to us, and in every little particle of its matter; we now behold almost as great a variety of Creatures, as we were able before to reckon up in the whole Universe itself.[1]

However, Hooke needed to improve two particular aspects to make the microscope useful from a biological standpoint: first, the manner in which lenses were made was so imperfect that ten lenses would usually need to be made before one adequate quality lens was manufactured; second, the apertures of the instrument were so small that very little light could be shone on the specimen, therefore, the samples usually appeared *dark and indistinct*.[1] Thus, Hooke improved the issue of brightness by using a convex glass, *one of whose sides is made rough by being rubb'd on a flat Tool with very find sand*, through which light could be efficiently concentrated in the very small opening of the aperture, *which very much augment a convenient light*. After trying a variety of different configurations, he found that the type of microscope which yielded the best results was six or seven inches long and contained two lenses encased in a brass tube filled with *very clear water* (Figure 2.1a).

In 1665, Hooke was the first to report—in his seminal *Micrographia*—use of a microscope to examine various biological samples, including those of plants and animals.[1] Importantly, one of his samples included a thin cross-section of clear cork which was cut using *a Pen-knife sharpen'd as keen as a Razor*, leaving the surface *exceedingly smooth*. When he placed these cross-sections on a dark plate, he observed an array of pores, *much like a Honeycomb* separated by *Interstitia, or walls* (Figure 2.1b). When he cut another section in the opposite direction (longitudinally), so as to reveal the length of each pore, he found that each pore was of limited length and resembled *little boxes [...] or cells*. In addition, he determined that there were about sixty of these cells aligned end-to-end in 1/18 inch of cork, and therefore, one cubic inch should contain over *twelve hundred Millions, [...] a thing almost incredible, did not our Microscope assure us of it by ocular demonstration.*

Hooke correctly reasoned that cork was light due to all the cavities being filled with air, which also seemed to explain why solutions could not traverse through it, since this air was *perfectly enclosed in little Boxes or Cells distinct from one another*. However, he knew that these were simply conjectures since the power of his microscope was not such that these hypotheses could be tested, and concluded that:

> there seems no probable reason to the contrary, but that we might as readily render the true reason of all their Phænomena; as namely, what were the cause of the springiness, and toughness of some, both as to their flexibility and restitution. What, of the friability or brittleness of some others, and the like; but till such time as our Microscope, or some other means, enable us to discover the true Schematism and Texture of all kinds of bodies, we must grope, as it were, in the dark, and onely ghess at the true reasons of things by similitudes and comparisons.[1]

The term "cell" was initially used by Hooke to describe the structures he saw in cork using his microscope and it was subsequently adopted by plant physiologists; however, it was not so widely used by anatomists. Plant cells have a thick wall of cellulose (a type of complex sugar) all around them which is evidently visible by microscopy, but lack of

DOI: 10.1201/9781003379058-2

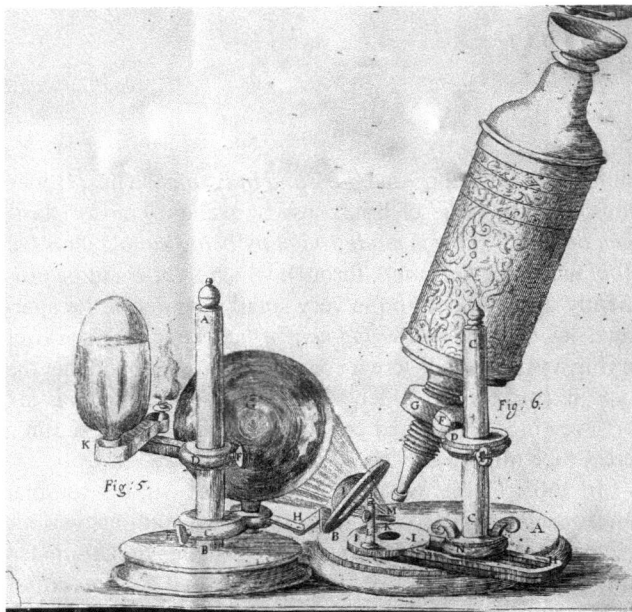

(a)

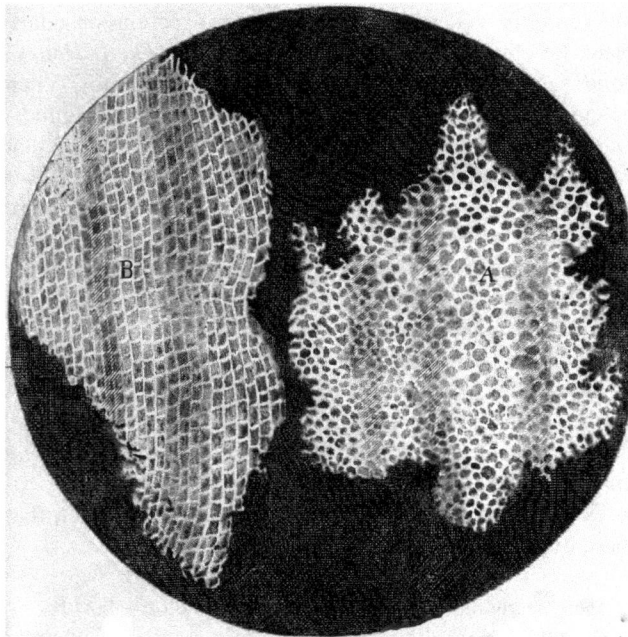

(b)

FIGURE 2.1 Hooke's microscope and his first observation of cells. (a) Light from a candle was concentrated on the specimen by passing it through a globe, G, filled with water. (b) A drawing of cork as seen through Hooke's microscope. On the left is shown a longitudinal section in which the length of the cells is visible and on the right, a cross-section which shows the pores.[1]

these structures in animal cells gave the illusion that there was nothing around the cell. Therefore, prior to the 1800s, when anatomists used the word "cell", they were alluding to what we now know as areolar tissue (such as lung tissue) which is mostly composed of air pockets enclosed by fibrillar tissue.[2]

Although generally attributed to Schwann and Schleiden, the cell theory actually began with Henri Dutrochet in an 1824 monograph in which he attempted to put an end to the disagreements regarding the *organization of plants*.[3] Some scientists had claimed that cells were separated by one continuous membrane that ran around each cell and throughout a specimen, while others reported seeing a separation between the cells, though these structures were sometimes cemented together, which gave the impression of continuity. Dutrochet attributed these differing observations to the fact that most studies were performed by observing thinly sliced plant samples through a microscope, an instrument which was subject to a number of *optical illusions*. Therefore, he proposed that samples should be examined from different perspectives and after having undergone a variety of modifications, such as chemicals or stains that color selective parts of a sample. As such, he boiled plant samples in nitric acid and found that the cells readily separated into distinct entities that maintained their form, which they had gained due to compressive forces from surrounding cells.[4] In addition, and contrary to popular belief, the cells did not share one continuous membrane throughout, but rather, each had its own separate wall that surrounded it. He also proposed that the growth of an organism results from both an increase in volume and the addition of smaller cells, and that these new cells, given time, increase in size and become similar in appearance and in development to those from which they came.

Several years later, in 1839, Theodore Schwann published a seminal monograph in which he compared the development of a variety of plants and animal cells into different tissues, with an aim *to prove the accordance of the elementary parts of animals with the cells of plants*.[5] Schwann was a German scientist who studied under the famed physiologist Johannes Müller, in whose laboratory he later became full-time researcher.[6] He then became professor of anatomy and physiology at the University of Liège, where he remained until his death in 1882.[7] In his monograph, Schwann acknowledged that his conclusions were largely based on the researches of his contemporary and friend, Matthias Schleiden, who had performed numerous plant studies, including observations on the cell nucleus.[8] Although the cell nucleus was first described in 1682 by the early microscopist, Antonie van Leeuwenhoek, it was not strictly defined as a separate entity, either in the form or function, from the rest of the cell.[9] Francis Bauer's early 19th-century drawings of plant cells, which clearly contained nuclei, were well known among Fellows of the Royal Society by the mid-century; but it was Robert Brown, the custodian of the botanical collections of the British Museum and later the president of the Linnean Society,[9] who named the structure in 1833: *This areola, or nucleus of the cell as perhaps it might be termed [...].*[10] His publication was also the first to formally define it as a separate entity:

In each cell of the epidermis [...], a single circular areola, generally somewhat more opake than the membrane of the cell, is observable. This areola, which is more or less distinctly granular, is slightly convex, and although it seems to be on the surface is in reality covered by the outer lamina of the cell. There is no regularity as to its place in the cell; it is not unfrequently however central or nearly so.[10]

Thus, in 1838, Schleiden suggested that the nucleus of a cell was a *universal elementary organ in vegetables* and hypothesized that it must hold an important place in the development of the cell;[2,8] however, since he could not always see the nucleus, he wrongly assumed that following the creation of a new cell, this structure was unnecessary and subsequently absorbed by the cell.[2] In addition, Schleiden noted that the nucleus contained *spots*, *rings*, or *points*, which he proposed, again erroneously, were the starting point for new cells (these would later be called nucleoli).[2,11]

These observations were shared with Theodore Schwann during dinner one evening and, in fact, it was following this conversation that Schwann realized he may be able to prove that animal and plant cells were similar in form and function.[6,11] Schwann began his monograph by reporting his examinations of notochords in frog and fish larvae, which he extracted using caustic potash (potassium hydroxide), and noted that the cells he observed were so similar to those of plants that they could scarcely be distinguished under the microscope (Figure 2.2). He continued by categorizing animal tissues based on the similarities between the cells that composed them, and in the third and final section of the monograph, Schwann summarized his findings and proposed that:

> The elementary parts of all tissues are formed of cells in an analogous, though very diversified manner, so that it may be asserted, that there is one universal principle of development for the elementary part of organisms, however different, and that this principle is the formation of cells.[5]

Rudolph Virchow, also known as the father of pathology, was a student of Müller alongside Schwann and was, in fact, influenced by Schwann's work on the cell theory.[2] Virchow gave a series of lectures on cellular pathology in 1858, in which he described his conclusions on the origin of a cell and stated that:

> [e]ven in pathology we can now go so far as to establish, as a general principle, that no development of any kind begins *de novo*, and consequently as to reject the theory of equivocal *[spontaneous]* generation just as much in the history of the development of individual parts as we do in that of entire organisms [...] Where a cell arises, there a cell must have previously existed (omnis cellula e cellula), just as an animal can spring only from an animal, a plant only from a plant.[12]

These findings remain a central principle of biology and is known as the Cell Theory:

• The cell is the most basic unit of life.
• All living things are composed of one or more cells.
• All cells arise from one individual cell.

＊ ＊ ＊ ＊ ＊

Between 1838 and 1840, Dr. Martin Barry, a graduate student of medicine at the University of Edinburg, published three groundbreaking memoirs in succession, in which he reported that each fertilized rabbit ovum gave rise to two daughter cells, then four, then eight, and continuing as such until the formation of the organism was complete. He also observed that the nucleus was presented to each new cell during cell division, and therefore, must be of vital importance to the cell, naming the nucleus as the center of origin *not only of the transitory contents of its own cell, but also of the two or three principal and last formed cells destined to succeed that cell.*[2] Wilhem Benedikt Hofmeister made similar observations about a decade later while studying plant fertilization, and his illustrations contained the first-known observations of chromosomes, though it was not until later in the century before they were adequately described and studied: German botanist, Edmond Russow, described them as "rods" in 1872, whereas the Belgian embryologist, Edouard van Beneden, used the term "stick" a few years later. Around the same time, the French embryologist, Edouard Balbini, observed that the nucleus disintegrated during cell division and gave rise to "bâtonnets étroits"—narrow sticks.

However, the first significant contribution about chromosomes to cell biology came from Walther Flemming in 1879 while studying salamander larvae. Here, the author tested a variety of reagents to fix fin and gill epithelium (skin) cells (which halts all biochemical reactions and preserves them in their structural state at the time of fixation), settling on picric acid. He subsequently stained them with a hematoxylin dye solution, which he found efficiently labelled the granular structure within the nucleus, and which he named *chromatin* from the Greek, χρῶμα (chróma, color).[13,14] As such, Flemming identified a number of distinct phases of the nucleus during the life cycle of a cell (Figure 2.3):

1. A *fine basketwork* of tightly wound, spired threads (prophase).
2. *Loose coils* in which the granules slowly become thicker and more isolated, and which were noted to be connected in pairs (prophase).
3. An *equatorial plate* in which the threads line up at the center of the cell parallel to the division of the cell, such that each daughter cell would receive half of the threads (metaphase).
4. An *astral form* in which bending of the threads is readily apparent. Subsequently, the threads were observed to split in half lengthwise almost exactly parallel (metaphase).
5. A *separation of the nuclear figure* in which the two nuclei move apart, the actual separation having already taken place in the previous stage (anaphase).
6. A *star form of the daughter nuclei* in which each half grouping of threads moves further and further apart toward its own pole. In addition, he noted the appearance of a constriction furrow (anaphase/telophase).

7. A *wreath and coil form of the daughter cell* in which the daughter cells form.
8. A *reticular form of the daughter nuclei* in which the daughter cells assume a configuration similar to that of the parent cell before division (cytokinesis).

This process was named karyomitosis (kary—Greek for nut—nucleus; mitos, Greek for threads) by Flemming, and the "threads" were subsequently given the name of *chromosomes* (colored bodies) due to their affinity for dyes.[2] The next year, Edouard van Beneden extended Flemming's

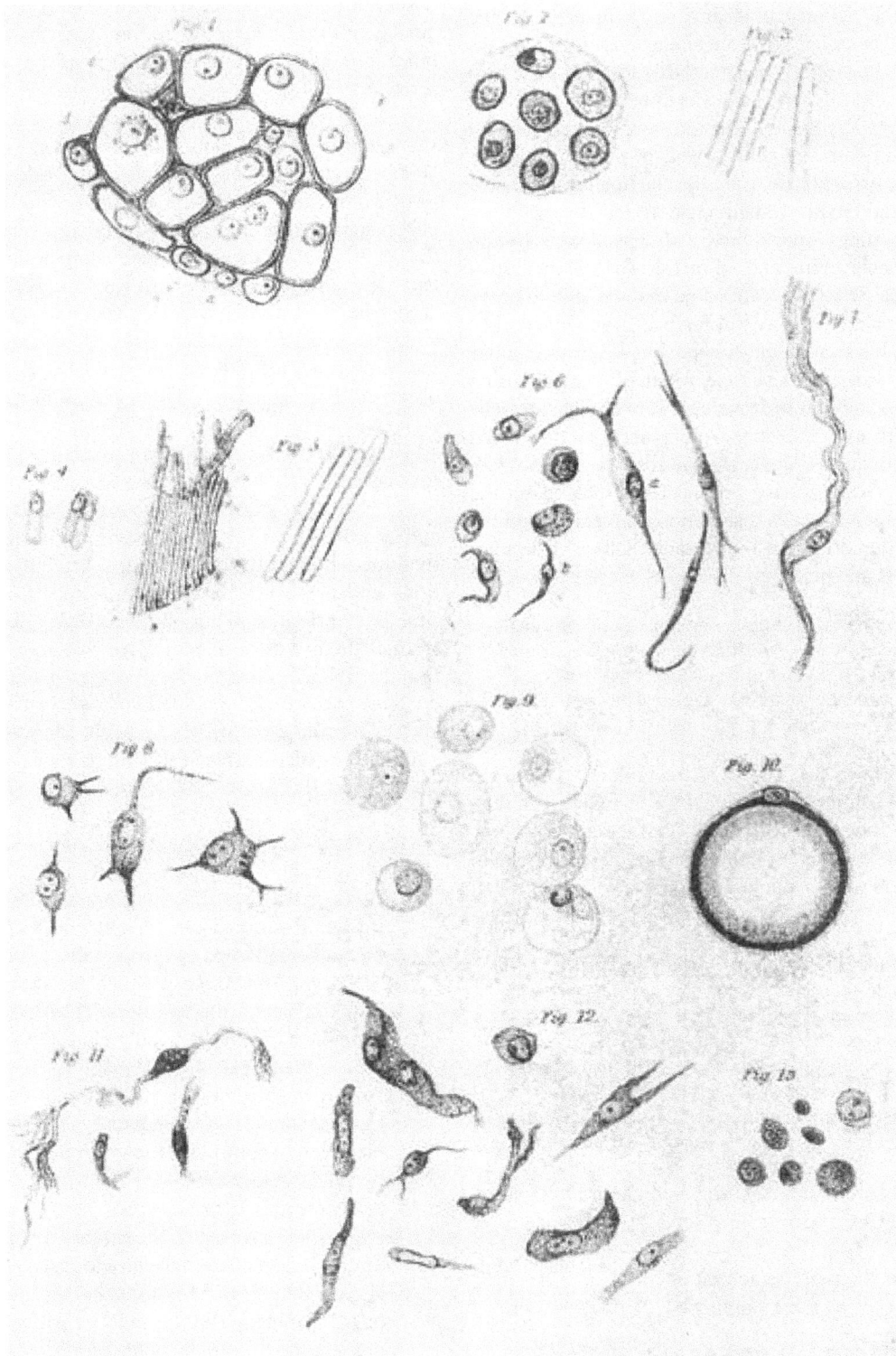

FIGURE 2.2 Drawings of various cells observed by Schwann.[5]

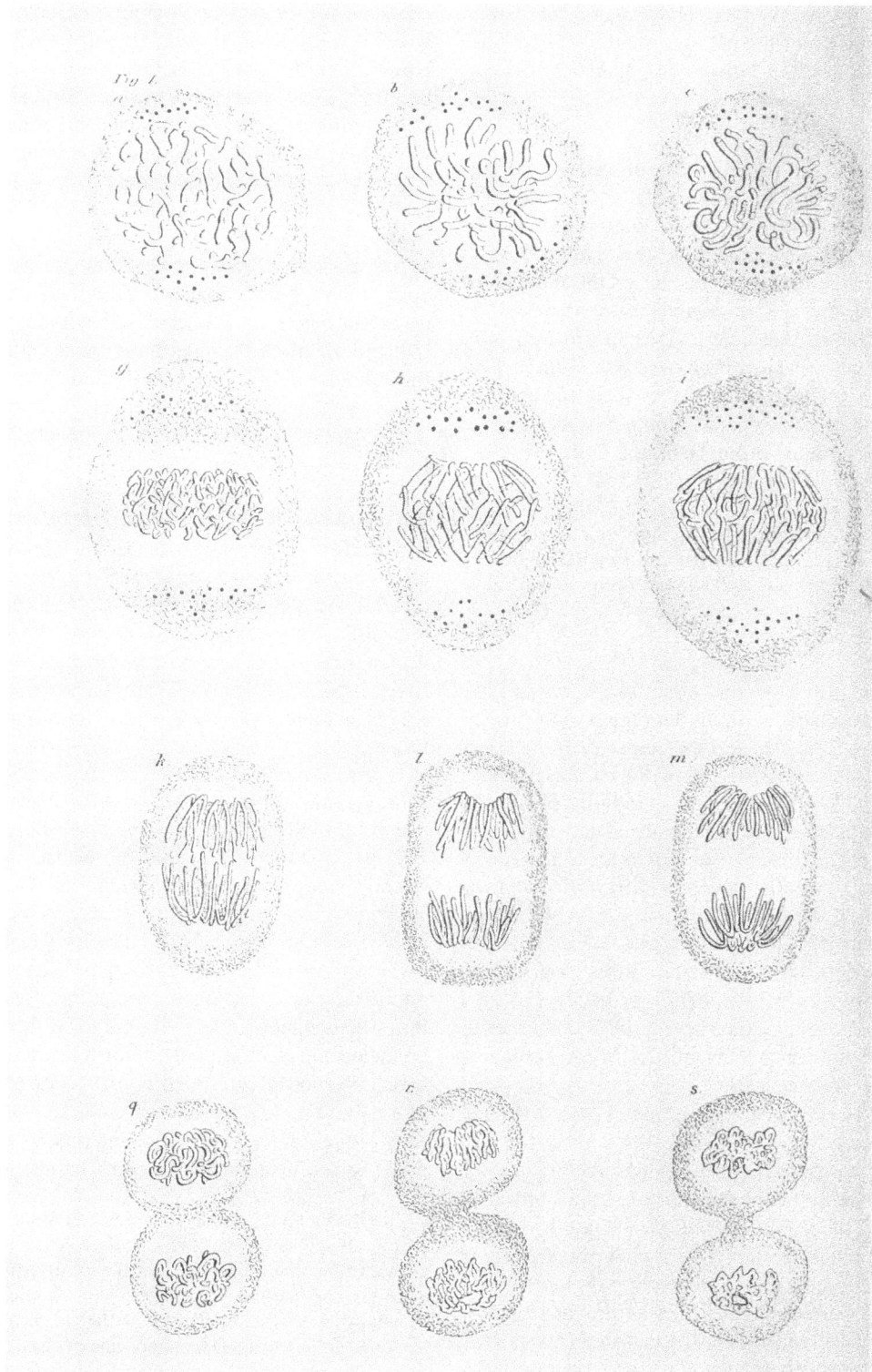

FIGURE 2.3 Stages of the cell cycle as illustrated by Walther Flemming. Seen here are prophase (a–c), metaphase (g–i), anaphase (k), telophase (l, m), and cytokinesis (q–s) (as the stages are now known).[15]

findings, studying number of chromosomes present at various times of an egg's cell cycle.[16,17] The author noted that in contrast to all other mature cells, both the egg and the spermatozoon contained exactly half as many chromosomes; the fertilization of the former by the latter results in the formation a cell with full complement of chromosome.[16] In addition, he found that the cell and its reconstituted nucleus began to divide in the same manner reported by Flemming, that is, first by duplicating their chromosomes and then undergoing mitosis, which led to the formation of two, four, eight, … cells, and finally a full-grown organism.

* * * * *

One of the major biological issues of the second half of the 1800s was the task of defining the nature and components of a cell, starting with the basic question of what should be called a cell. This issue engendered many debates due to the inherent differences between plant and animal cells as observed through microscopes; nonetheless, in 1856, Leydig proposed that a cell was a *substance primitively approaching a sphere in shape and containing a central body called a kernel* [nucleus].[18] As such, the cell was thought to consist of two separate components: a central nucleus surrounded by the protoplasm; the latter was characterized by a:

> translucent, viscid, or slimy material, dimly granular under the lower powers, minutely fibrillated under the highest powers of the microscope, which moves by contracting and expanding, and which possesses a highly complex chemical constitution.[2]

However, it was noted that the protoplasm was more of a concept than a substance, and that the precise constituents of this structure still needed to be elucidated. In addition, many questions arose concerning how the cell, as a whole, was held together. While researchers studying plant cells could clearly see a structure surrounding the cell which they called the cell membrane (we now refer to this structure as the cell wall, an entity which in itself is distinct from the cell membrane), these structures were not always seen in animal cells. This led Leydig to suggest that its existence may be cell-type dependent, while the protistologist, Max Schultz, thought that cells were held together due to their contents being immiscible and viewed cell membranes as an artifact which arose as a consequence of the hardening of the cell during its demise.[2,18]

Discovery of the presence of a cell membrane surrounding the protoplasm came via the study of osmosis, the passage of a solvent across a semipermeable membrane in an attempt to equalize the concentration of solute on either side of that membrane. The physiological role for osmosis had been known since the late 1770s when the volume of erythrocytes was found to vary depending on the solution in which the cells were immersed: they shrank in a solution of high salt and swelled in that of low salt.[18] It was later shown that the former phenomenon occurred due to water flowing out of the cell, whereas water entered the cell in the latter, allowing an equilibrium of intra- and extra-cellular salt concentrations to be maintained. However, the mechanism by which the protoplasm allowed the continuous flow of water in and out of the cell while acting as a barrier for other molecules remained elusive. The most widely accepted explanation was the suggestion that the protoplasm precipitated at the edge of the cell where it made contact with the surrounding environment, a phenomenon which was known to produce a semipermeable membrane.[18] However, in the late 1890s, Overton had immersed cells in hundreds of different types of solutions and found that they only shrank in polar solutions (such as water), whereas cell shape was maintained in apolar solutions (such as ether). In addition, he showed that apolar dyes entered cells much more readily than water-soluble dyes. Since apolar molecules would not be able to cross an apolar membrane due to the repulsion forces between the two like molecules, he correctly suggested that cell membranes might be made up of phospholipids and cholesterol (polar molecules), and that cell membranes were, therefore, a separate entity from the cellulose cell walls found in plant cells.[18]

Julius Bernstein then unified findings by Emil du Bois-Reymond and Walther Nernst and used them as the basis for his own line of study, which led to the suggestion that cells were surrounded by a semipermeable membrane. Bois-Reymond was another accomplished German scientist who had studied under Johannes Müller and who established electrophysiology as a discipline, performing experiments using instruments that he often had to build or improve on himself.[19] For example, he improved the galvanometer, a device used to measure electric currents, to enable the measurement of a minute flow of electrons through muscles and nerves and found that there existed a resting current in these tissues which would decrease or even reverse when the tissues were stimulated. This led him to suggest that there must be a potential difference between the interior and the exterior of the cell. The other finding that influenced Bernstein's line of experiments was that of Nernst, yet another great scientist who studied under Müller in Germany. He was studying electrochemical relationships between electricity and the movement of electrolytes when a colleague of his, the renowned Jacobus Henricus Von't Hoff, contacted him to develop a formula to predict the movement of solutes in a dilute solution. Nernst came up with an equation that predicted the electric potential difference in galvanic cells, which he showed depended on the concentration of solutes in a particular solution and on the temperature of the solution.[18] Importantly, Nernst realized that his new calculations could be applied to cellular electrochemistry, stating that:

> For the first time, based on in-depth analysis of the function of galvanic cells, we have calculated electromotive forces with fair approximation, starting from different physical quantities, and we have juxtaposed various principles of physics to one another, such as the electromotive force of liquid cells, transport numbers, and gas laws, among which no connection had been heretofore suspected.[20]

Thus, Bernstein conducted simple experiments in which he measured the current through frog muscles at different temperatures (Figure 2.4) and found that the resting current in these tissues increased linearly with absolute temperature, in agreement with Nernst's equation.[18,21,22] Since a greater potential difference was observed in biological cells compared to galvanic cells, Bernstein hypothesized,

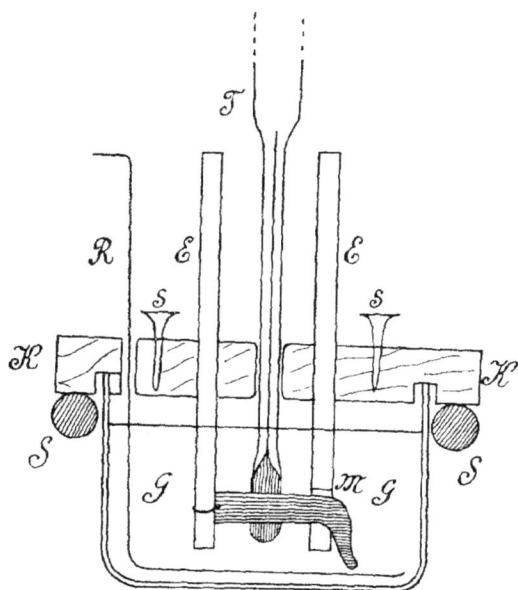

FIGURE 2.4 Bernstein's apparatus to measure electrical currents in frog muscles. The muscle (m) was placed in a glass jar filled with oil and electrodes (E) were attached to its cross-section (left electrode) and surface (right electrode) for electrical measurements.[22]

as others had previously implied, that an ion concentration gradient could be created by a semipermeable membrane surrounding the cell. Thus, Bernstein reexamined his experimental procedure in an attempt to explain this phenomenon and realized that the current generated was different if taken in the longitudinal versus the transverse sections of the muscle. To account for this, he wrote:

Let us imagine that these electrolytes diffuse unhindered from the axial cross section of the fibrils into the surrounding fluid, while they are prevented from diffusing through the longitudinal section by an intact plasmalemma which is impermeable to one kind of ion such as the anion (PO^{-4} etc.) to a greater or lesser degree. Then an electrical double layer would emerge at the surface of the fibril, with negative charges towards the inside and positive charges towards the outside. Indeed, this electrical double layer must also exist in the undamaged fiber, but would become apparent only in response to lesion or stimulation (negative variation). This assumption would imply a theory of pre-existence. As the semipermeable membrane plays an essential role in this theory, I will succinctly call it 'Membrane Theory'.[21]

Thus Bernstein's "Membrane Theory of Electrical Potentials" postulated that (1) cells consist of an electrolyte solution surrounded by a semipermeable membrane; (2) at rest, there exists a negative potential difference between the interior and exterior of the cell, resulting from a negative internal charge and a positive external charge; (3) during activity, the selective permeability of the membrane for potassium increases, creating a cellular influx of potassium and reducing the potential difference to low levels.[18,21]

Although the preceding experiments suggested the presence of a semipermeable membrane around the cell, they were not definitive proof. The first direct evidence that cell membranes were separate structures from protoplasm came in 1921, when Robert Chambers greatly improved microinjection techniques and injected cells with a variety of solutions.[23,24] The technique available to him at that time was to inject mercury into cells using a glass pipette. However, the pipettes left *nothing to be desired*, both in terms of their size and ease of making, and the method was very difficult since injection depended on the expansion of the mercury, which was difficult to control. Therefore, he devised an instrument that helped stabilize the needle while providing fine control over its position via the adjustment of screws (Figure 2.5).[24] As such, Chambers injected intact starfish eggs with neutral red—a dye that changes from red to yellow in a basic environment—and immersed them in a basic solution. He found that the cells remained red under these conditions, which indicated that the basic solution could not enter intact cells. However, when the solution was injected directly into the egg, a color change occurred at the area of injection and spread throughout the cell, stopping at a film surrounding the protoplasm. By this, Chambers concluded that there must be a structure surrounding the protoplasm that was made of different components from the protoplasm and which was not decomposed by these solutions.

Following Chambers' experiments and inspired by Overton's premise that cell membranes might be lipid-based, Gorter and Grendel set out to better characterize the cell membrane and elucidate its composition. To do so, they used a method developed by Irving Langmuir in 1917 and further improved by Neil K. Adam.[25,26] This method used the fact that oil molecules spread out and occupy a defined surface area when placed on top of a pool of water. Ingeniously, Langmuir used strips of paper to push oil molecules together and progressively pack them tighter and tighter until they formed a compact monolayer (Figure 2.6).[27] As such, when the lipids reached a state of maximal compactness—that is when they formed a tight monolayer—any additional pressure exerted a force on a stationary strip of paper, which straightened

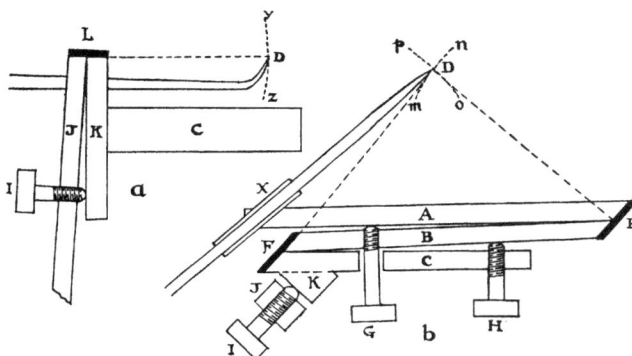

FIGURE 2.5 Robert Chambers' micromanipulator. The apparatus was furnished with multiple screws, I, G, and H, to adjust the position of the needle, D.[24]

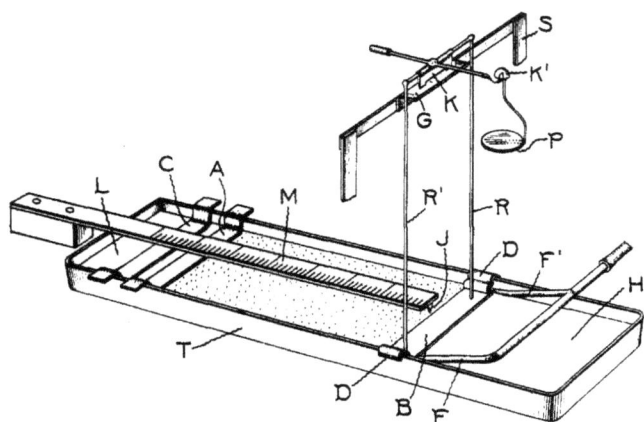

FIGURE 2.6 Langmuir's apparatus to determine surface tension. Water was placed in a tray and strips of paper (A, B, and C), slightly shorter than the width of the tray so as to move freely, were placed on its surface (the paper was coated with paraffin to avoid being soaked by the water). Glass rods (R, R′) were inserted through strip B and attached to a support (S) and a knife (K), upon which hung a pan (P) containing a selected weight. The weight caused the support and glass rods to tilt to the left, the degree of which was determined by the weight in the pan. Oil was then added between strips A and B, and strip A was slowly moved toward strip B until the lipids were compacted to the point of exerting a force on strip B, which tilted the glass rods to the right. The surface area between A and B was then measured with the ruler (M).[27]

glass rods inserted through the paper. The author found that beyond this point, the surface area occupied by the lipids was inversely proportional to the pressure exerted.

Using this technique, Gorter and Gretel isolated lipids from erythrocytes and added them to their apparatus, compacted them until they reached a tight monolayer, and then calculated the surface area they occupied. Comparing this value to the theoretical surface area of a cell, the authors determined that the lipids occupied twice as much area compared to cell surface, which implied that the cell membrane was composed of two layers of lipids—a lipid bilayer.[28] Given these results, they proposed that the molecules were arranged such that the hydrophobic tails of the lipids faced one another at the center of the bilayer, whereas the polar heads were directed toward the water. Although the lipid bilayer described by Gorter and Grendel would eventually be shown to be correct, it would not happen until the advent of superior techniques in the 1960s and 1970s.

* * * * *

The knowledge accumulated thus far was largely the result of the increase in resolving power of the microscope, thus giving scientists the unprecedented ability to peer inside cells that are too small for the naked eye to see. However, far removed from scientific hubs which had access to modern technologies—such as the microscope—to investigate nature, was an Augustinian monk that would become known as the "father of modern genetics". Gregor Mendel was recognized to have superior intelligence and

a voracious appetite for learning early in life.[29] His decision to not comply with his father's wishes to work on the family farm, opting instead to study at the monastery in the cultural center of the region, gave him access to a large library and researches of numerous scientists. There, he took classes on plant breeding and mathematics, both of which would later help him calculate genetic ratios from multiple breeding experiments and elucidate the underpinnings of heredity.

Farmers had long known the favorable outcomes resulting from generations of crossbreeding of sought-after traits. At the time, the favored theory for this phenomenon, called the "Blending Theory", proposed that parents' traits were diluted and blended together, and the offspring would inherit a mixture of these traits in a random manner.[29] Mendel was interested in testing this theory and studying the effects of crossbreeding on a number of simple pea traits, or phenotypes: color, size, shape, plant height, etc. Specifically, he wanted to ascertain if a precise law could be deduced to account for the manner in which traits were passed down to successive generations.[30] His experiments were painstakingly and meticulously performed over 8 years of crossbreeding thousands of peas and his analysis of the data shed light on the mechanism of genetic inheritance.[29]

To begin, Mendel had to determine which traits could be used for his experiments using only those that had clearly distinct phenotypes so as to be unambiguously identified.[30] He also needed to obtain true-breeding pea plants, that is, plants that always produced the same phenotype-of-interest generation after generation (e.g., only green peas or only tall plants), so that the propagation of each trait could be followed. In addition, he had to protect his plants from foreign pollination so as to not disturb his experiments. As such, he started crossbreeding two versions of a particular trait, a tall plant with a short plant, for example, to see which phenotype the progeny would take on from one generation of crossbreeding to the next. What he found was that the first generation of offspring, which he called the hybrid generation, only expressed one phenotype for each pair of phenotypes tested; for example, a true-breeding tall plant crossed with a true-breeding short plant yielded only tall plants (Figure 2.7). This first observation already discredited the blending theory, since one would have expected all plants to be of medium height according to this theory. Interestingly, he found that as for the height of plants, he invariably obtained progeny plants that were at least as tall as the tallest parent, such that each generation was taller than the previous.

Surprisingly, when Mendel self-pollinated his hybrid plants (the F1 generation), the phenotype of the next generation (F2) occurred in a given, recurring ratio: 3/4 of the progeny expressed the phenotype observed in the F1 generation, whereas 1/4 expressed the phenotype from the parent generation that was missing in the F1 generation. Therefore, he proposed the term "dominant" to describe *those characters which are transmitted entire, or almost unchanged in the hybridization, and therefore in themselves constitute the characters of the hybrid* and "recessive" for *those which*

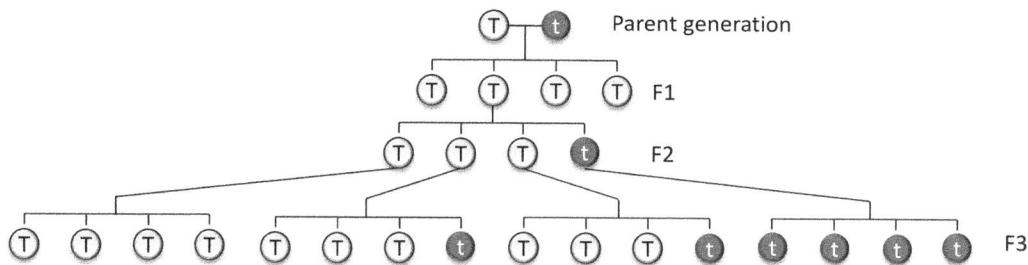

FIGURE 2.7 Example of Mendelian inheritance. Plant height is used here as an example; the tall trait is dominant, whereas the short is recessive.

become latent in the process.[29] In contrast, when the F2 plants were self-pollinated, their progeny had phenotypic ratios that depended on the parent: the F2 short plants (1/4 of all F2 progeny) only produced plants with the recessive trait after successive generations (therefore, produced true-breeding plants); 1/3 of tall plants (1/4 of all F2 progeny) only yielded plants with the dominant trait after multiple generations (also true-breeding plants); however, surprisingly, self-pollination of 2/3 of the F2 plants with the dominant trait (or half of F2) yielded an F3 generation in which 3/4 of plants expressed the dominant trait and 1/4 of plants expressed the recessive phenotype, a ratio identical to that found in the F2 generation! What these results led Mendel to conclude was that the recessive trait, although seemingly lost in the hybrid generation, was in fact not lost: it was simply not expressed.

Although the results of Mendel's experiments were groundbreaking in retrospect, they made little impact following their publication in 1866. It was only years after his death, in the early 1900s, that his findings were independently rediscovered by three scientists—Hugo de Vries, Carl Correns, and Erich Von Tschermak—who, unbeknownst to them at the start of their own research, were to replicate Mendel's findings and confirm the voracity of his conclusions.[31] The original work by Mendel was credited in these authors' publications and Mendel finally received his due credit. This rediscovery also led to a reexamination of Mendel's results and the formation of the laws of inheritance:[31]

- *Law of segregation*: during cell division, each copy of a gene separates so that each daughter cell receives one copy of each gene.
- *Law of independent assortment*: copies of different genes segregate independently from one another.
- *Law of dominance*: the genetic form that a gene can take can be either dominant or recessive.

Although it was clear by this time that the nucleus played a role in cell division, the extent of its importance, and in fact, its actual role, remained to be elucidated. In 1892, August Weismann proposed the "Theory of Heredity", which stated that certain cells of an organism, called germ cells or gametes (i.e., the egg and spermatozoon), were reserved strictly for reproduction purposes and that their chromatin held the heredity information transmitted from generation to generation.[32] However, it was Walter Sutton in 1903 who combined all that was known up that point—experiments and observations from van Beneden, Weismann, Bovari, Bateson, and others (the author states that he was unaware of Mendel's principle at the time his conclusions were reached, but that Mendel's results fully supported his conclusions)—to propose a role for chromosomes in heredity. As such, the author began by pointing out that:

It has long been admitted that we must look to the organization of the germ-cells for the ultimate determination of hereditary phenomena [...]
Nearly a year ago it became apparent to the author that the high degree of organization in the chromosome-group of the germ-cells as shown in Brachystola could scarcely be without definite significance in inheritance [...]
[...] many points were discovered which strongly indicate that the position of the bivalent chromosomes *[a matched pair of chromosomes from each parent]* in the equatorial plate of the reducing division is purely a matter of chance—that is, that any chromosome pair may lie with maternal or paternal chromatid indifferently toward either pole irrespective of the positions of other pairs—and hence that a large number of different combinations of maternal and paternal chromosomes are possible in the mature germ-products of an individual.[33]

As such, Sutton calculated that in an organism with a total chromosome content of 36—the largest number known to be present in an organism at the time—over 68 billion different combinations of genetic content could result from the fertilization of an egg. In contrast, chromosome content in humans—now known to be of 46—can generate on the order of 7,000 trillion possible genetic combinations! Sutton also addressed recent results by Boveri which indicated that chromosomes played a role in development, stating that:

[...] we should be able to find an exact correspondence between the behavior in inheritance of any chromosome and that of the characters associated with it in the organism.
Thus the phenomena of germ-cell division and of heredity are seen to have the same essential features, viz., purity of units (chromosomes, characters) and the independent transmission of the same; while as corollary, it follows in each case that each of the two antagonistic units

(chromosomes, characters) is contained by exactly half the gametes produced.

[…] it seems highly probable that homologous chromatin-entities are not usually of strictly uniform constitution, but present minor variations corresponding to the various expressions of the character they represent. In other words, it is probable that specific differences and individual variations are alike traceable to a common source, which is a difference in the constitution of homologous chromatin-entities. Slight differences in homologues would mean corresponding, slight variations in the character concerned—a correspondence which is actually seen in cases of inbreeding, where variation is well known to be minimized and where obviously in the case of many of the chromosome pairs both members must be derived from the same chromosome of a recent common ancestor and hence be practically identical.[33]

Finally, Sutton proposed that chromosomes were composed of smaller entities that defined the characteristics imparted by the chromosomes:

We have seen reason, in the foregoing considerations, to believe that there is a definite relation between chromosomes and allelomorphs or unit characters but we have not before inquired whether an entire chromosome or only a part of one is to be regarded as the basis of a single allelomorph. The answer must unquestionably be in favor of the latter possibility, for otherwise the number of distinct characters possessed by an individual could not exceed the number of chromosomes in the germ-products; which is undoubtedly contrary to fact. We must, therefore, assume that some chromosomes at least are related to a number of different allelomorphs. If then, the chromosomes permanently retain their individuality, it follows that all the allelomorphs represented by any one chromosome must be inherited together. On the other hand, it is not necessary to assume that all must be apparent in the organism, for here the question of dominance enters and it is not yet known that dominance is a function of an entire chromosome. It is conceivable that the chromosome may be divisible into smaller entities (somewhat as Weismann assumes), which represent the allelomorphs and may be dominant or recessive independently. In this way the same chromosome might at one time represent both dominant and recessive allelomorphs.

Such a conception infinitely increases the number of possible combinations of characters as actually seen in the individuals and unfortunately at the same time increases the difficulty of determining what characters are inherited together […].[33]

In the years to come, it would be shown that, indeed, each chromosome is composed of many distinct entities called genes, which are heritable and which determine the phenotypes of an organism. Within each gene are variations, or alleles, of the possible characteristics, called genotypes, which define a phenotype. Thus, the combination of alleles for each gene—one inherited from the father and one from the mother—determines the phenotype that is expressed for that particular gene (or a combination of genes, as many phenotypes are defined by more than one gene).

* * * * *

Fruit flies (*Drosophila melanogaster*) were reportedly used for the first the time in the laboratory setting by William E. Castle at the turn of the 20th century, the advantage of which was that they had a short life cycle, reproduced readily, and an initial culture could be obtained simply by leaving fermented fruit out in the open. Although initially working on mice and rats, Thomas Hunt Morgan switched to fruit flies around 1909 with hopes of inducing mutations in them.[34] Morgan was appointed professor of zoology at Columbia University in 1904 before moving to Caltech in 1928. At Columbia, he took on two undergraduate students, Calvin B. Bridges and Alfred H. Sturtevant, to help set-up what became known as the *fly room*; these students eventually became full-time research assistants with Morgan and remained there for the next 17 years. The atmosphere in the fly room was reportedly like few others, abound with *excitement and such a record of sustained enthusiasm*, which was influenced by *Morgan's own attitude, compounded of enthusiasm and combined with a strong critical sense, generosity, open-mindedness, and a remarkable sense of humor.*[34]

In an attempt to induce mutations, Morgan subjected his flies to numerous conditions and reagents: various temperatures, salts, sugars, acids, alkali, etc.[34] One year after the start of his fruit fly culture and having bred multiple generations, Morgan reported, in 1910, to have found a white-eyed, male mutant in a sea of red-eyed flies (called wild type, a term that Morgan coined to describe a fly that was not bred in his laboratory but obtained *in the wild*, and was therefore of typical genetic background).[35] Surprisingly, breeding this male mutant with a red-eyed female did not result in any white-eyed progeny. However, when the F1 was crossed with itself, the mutation reappeared in about 1/4 of the males (a Mendelian ratio for the phenotype in males). Crossing the original mutant fly with the F1 female flies resulted in the appearance of 1/4 white-eyed males and 1/4 white-eyed females, or 1/3 white-eyed flies in total! This last experiment indicated that females were not excluded from this phenotype, but were somehow less prone to it. Morgan called this phenomenon "sex-linked inheritance". A final experiment was performed to further shed light on this phenomenon: he crossed a white-eyed female with a red-eyed male, which led to most surprising results: all females had wild type and all males had white eyes!

Analyzing the results of these and other experiments, which included similar results for small wings and truncated wings (two sex-linked traits), he concluded that:

this difference exists because one of the factors for the sex-limited characters in question is absent from one of the female determining chromosomes, while the genes for the secondary sexual characters of the male are contained in other chromosomes, possibly in those that contain the male determinants.[36]

Morgan disagreed with the suggestion by Castle that there may be a *Y-element* to account for male characters in flies and proposed instead that female flies have two

X chromosomes and males have only one (in flies, as in mammals, females are XX and males are XY; but in contrast to mammals, a fly with only one X chromosome but without a second sex chromosome is male). Therefore, a female inheriting a recessive *factor* (these factors would eventually be shown to be genes) could have a dominant factor on its second X chromosome—resulting in the dominant phenotype—whereas a male, having only one X chromosome, would express the phenotype of whichever factor he received, whether it be wild type or mutant. Interestingly, Morgan also hypothesized that human females must have two X chromosomes and males only one, since the fruit fly's white-eye inheritance patterns closely resembled those of color-blindness in human males.[36]

In 1906, William Bateson and R.C. Punnett were studying inheritance in sweet peas and found that certain genes located on the same chromosome, which were expected to be inherited together since chromosomes were thought to segregate as a unit, were in fact inherited separately.[37] In other words, a pure line of dominant genes on a single chromosome, for example, purple color (P) and long pollen grain (L), was crossed to a recessive line, for example, red color (p) and round pollen grain (l). Mendelian laws for this cross predicted that the F2 progeny should have a 9:3:3:1 ratio (PL:Pl:pL:pl). However, the results were drastically different from this expected ratio and they could not determine the meaning of this or even suggest a mechanism by which this phenomenon might occur. It would be almost a decade before Morgan provided a hypothesis as to why this occurred.

Like Bateson and Punnett, Morgan also found traits in his flies that did not segregate according to the Mendelian ratio. To account for this discrepancy, he first made the assumption that genes were arranged in a linear fashion along each chromosome.[38] He was also aware of recent experiments that had been done since Bateson and Punnett's publication that showed that chromosomes sometimes exchanged genetic material.[39] Thus, Morgan proposed the "Theory of Linkage" to explain the degree to which genes on a chromosome may be inherited together or separated due to crossovers, proposing that it was dependent on the distance by which they were separated on a given chromosome (Figure 2.8).[38] As such, he showed that two genes located at either end of a chromosome had a 1:1 chance of segregating separately, whereas two genes that lie very close to each other may do so only once in 100, since *the chance of a break occurring between them is small in proportion to their nearness.*[38] Alfred Sturtevant, still an undergraduate student in Morgan's laboratory at the time, accurately mapped six genes on Drosophila's X chromosome by making numerous crosses and determining the odds that any given trait was expressed at the same time as another. These results verified and supported Morgan's hypothesis of chromosome crossovers and *strongly indicate that the factors investigated are arranged in a linear series, at least mathematically.*[40]

In 1915, Morgan, Sturtevant, Muller, and Bridges published a seminal monograph that consolidated everything that had been elucidated since the rediscovery of Mendel's experiment in the early 20th century,[41] starting with the mechanism by which Mendelian inheritance occurred: A specialized type of cell division, called meiosis, results in daughter cells, called gametes (ova and sperm), having only one copy of each chromosome.[41] Gametes of each species carry a distinct number of chromosomes (22 in humans, plus 1 sex chromosome), which combine upon fertilization and results in a cell having two copies of each gene: one from the mother and one from the father. The expression of a particular phenotype depends on the characteristics of the inherited alleles of the gene: in its simplest form, the phenotype of the dominant allele will be expressed unless the organism has two recessive alleles of that particular gene; alternatively (and much more commonly), a phenotype may be defined by the properties of two or more alleles. In addition, the fertilized egg carries either two X chromosomes, giving rise to a female, or one X and one Y chromosome, giving rise to a male (this is the case in most mammals but may be different for other organisms). Soon after fertilization, each chromosome is duplicated and aligned at the center of the cell, where one copy of each chromosome is pulled to either pole of the cell. The egg then undergoes cell division and each daughter cell receives a full complement of chromosomes.

In this first look at the beginnings of biology, we saw how Hooke's improvements to the microscope gave

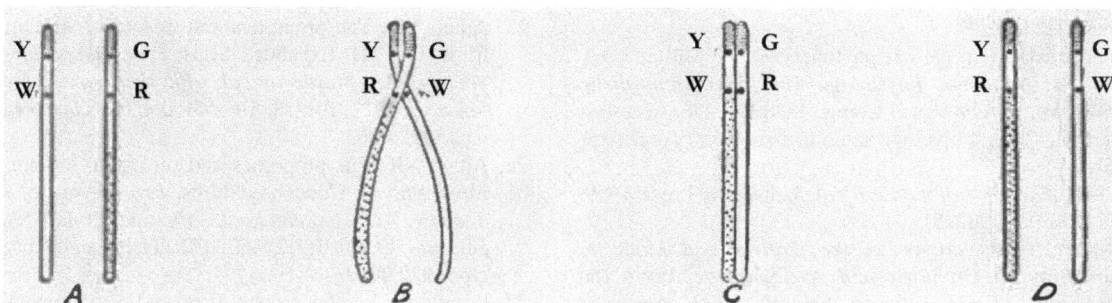

FIGURE 2.8   Morgan's theory of linkages. Gene pairs YW and GR are close enough on their respective chromosomes that they are rarely separated. However, if the two chromosomes cross-over between the two genes, genetic material can be exchanged, resulting in two chromosomes with the new pairs YR and GW. (Figure adapted from Morgan 1913).[38]

naturalists a powerful new tool to study nature. This led to the discovery of the simplest unit of life, the cell, which scientists at once began to investigate and found chromosomes residing in the nucleus. Around the same time, the Augustinian monk, Gregor Mendel, discovered, through years of laborious crossbreeding of plants, what would become known as the Mendelian Laws of Inheritance. Finally, Thomas Hunt Morgan and his students used the fruit fly to expand our knowledge of heredity and performed experiments that suggested that genes were the basic units involved in Mendelian inheritance and were aligned along the chromosome. Importantly, he and his students found and used sex-linked traits to confirm that exchanges of factors between chromosomes were possible by showing that genes on a given chromosome were linked to one another in a manner that depended on the distance between them.

* * * * *

REFERENCES

1. Hooke R. Micrographia: or some physiological descriptions of minute bodies made by magnifying glass with observations and inquiries thereupon. Published 1665. Accessed September 29, 2022. https://archive.org/details/micrographiaorso1670hook
2. Turner W. The cell theory, past and present. *Journal of Anatomy and Physiology.* 1890;24(Pt 2):253–287. Accessed March 31, 2019. http://www.ncbi.nlm.nih.gov/pubmed/17231856
3. Dutrochet Henri A du texte. *Recherches Anatomiques et Physiologiques Sur La Structure Intime Des Animaux et Des Végétaux et Sur Leur Motilité ([Reprod.])/Par M. H. Dutrochet,...* J.-B. Baillière; 1824. Accessed September 28, 2022. https://gallica.bnf.fr/ark:/12148/bpt6k97538k
4. Dutrochet R. The structural elements of plants. In: Gabriel ML, Fogel S, eds. *Great Experiments in Biology.* Prentice-Hall, Inc; 1955:6–9.
5. Schwann T. *Microscopial Researches into the Accordance in the Structure and Growth of Animals and Plants.* The Society Sydenham; 1847. Accessed September 29, 2022. https://www.biodiversitylibrary.org/bibliography/17276
6. Hunter GK. *Vital Forces: The Discovery of the Molecular Basis of Life.* Academic Press; 2000.
7. Hajdu SI. Introduction of the cell theory. *Annals of Clinical and Laboratory Science.* 2002;32(1):98–100. Accessed March 31, 2019. http://www.ncbi.nlm.nih.gov/pubmed/11848625
8. Schleiden MJ. Beiträge zur phytogenesis. In: Müller J, ed. *Archiv Für Anatomie, Physiologie Und Wissenschaftliche Medicin.* Verlag Von Veit et Comp.; 1838:137–176. Accessed March 31, 2019. https://www.biodiversitylibrary.org/item/49861
9. Harris H. *The Birth of the Cell.* Yale University Press; 1999. ISBN: 978-0300082951
10. Brown R. *Observations on the Organs and Mode of Fecundation in Orchidea and Asclepiadeae.* From the Transactions of the Linnean Society; 1833. Accessed September 29, 2022. https://www.google.ca/books/edition/Observations_on_the_Organs_and_Mode_of_F/iABLAAAAYAAJ?hl=en&gbpv=0

11. Locy WA. *Biology and Its Makers. With Portraits and Other Illustrations.* Henry Holt and Company; 1908. doi: https://doi.org/10.5962/bhl.title.5907
12. Virchow R. *Cellular Pathology as Based upon Physiological and Pathological Histology;* 1860. Accessed March 29, 2019. https://archive.org/details/dli.ministry.10993
13. Flemming W. Contributions to the knowledge of the cell and its life phenomena. In: Gabriel, Mordecai L.; Fogel S, eds. *Great Experiments in Biology.* Prentice-Hall, Inc; 1955:240–244.
14. Flemming W. Beiträge zur kenntniss der zelle und ihrer lebenserscheinungen. In: st. George V la Valette, Waldeyer W, eds. *Archiv Für Mikroskopische Anatomie.* Vol 16. Verlag von Max Cohen & Sohn; 1879:302–406. https://www.biodiversitylibrary.org/item/49519#page/7/mode/1up
15. Flemming W. *Zellsubstanz, Kern Und Zelltheilung.* F.C.W. Vogel; 1882. https://www.biodiversitylibrary.org/item/280108#page/7/mode/1up
16. van Beneden E. Researches on the maturation of the egg and fertilization. In: Gabriel ML, Fogel S, eds. *Great Experiments in Biology.* Prentice-Hall; 1955:245–247.
17. van Beneden É. *Recherches Sur La Maturation de l'oeuf, La Fecondation, et La Division Cellulaire.* Librairie Clemm; 1883. Accessed November 8, 2022. https://gallica.bnf.fr/ark:/12148/bpt6k97385454
18. Lombard J. Once upon a time the cell membranes: 175 years of cell boundary research. *Biology Direct.* 2014;9:32. doi: https://doi.org/10.1186/s13062-014-0032-7
19. Finkelstein G. Emil du Bois-Reymond vs Ludimar Hhermann. *Comptes Rendus Biologies.* 2006;329(5-6):340–347. doi: https://doi.org/10.1016/j.crvi.2006.03.005
20. de Palma A, Pareti G. Bernstein's long path to membrane theory: Radical change and conservation in nineteenth-century German electrophysiology. *Journal of History of Neurosciences.* 2011;20(4):306–337. doi: https://doi.org/10.1080/0964704X.2010.532024
21. Seyfarth EA. Julius Bernstein (1839–1917): Pioneer neurobiologist and biophysicist. *Biological Cybernetics.* 2006;94(1):2–8. doi: https://doi.org/10.1007/s00422-005-0031-y
22. Bernstein J. Untersuchungen zur Thermodynamik der bioelektrischen Ströme. *Archiv für die Gesamte Physiologie des Menschen und der Tiere.* 1902;92(10):521–562. doi: https://doi.org/10.1007/BF01790181
23. Chambers R. A micro injection study on the permeability of the starfish egg. *Journal of General Physiology.* 1922;5(2):189–193. doi: https://doi.org/10.1085/JGP.5.2.189
24. Chambers R. New micromanipulator and methods for the isolation of a single bacterium and the manipulation of living cells. *Journal of Infectious Diseases.* 1922;31(4):334–343. doi: https://doi.org/10.1093/infdis/31.4.334
25. Adam NK. The properties and molecular structure of thin films. Part III. Expanded films. *Proceedings of the Royal Society A: Mathematical, Physical and Engineering Sciences.* 1922;101(713):516–531. doi: https://doi.org/10.1098/rspa.1922.0063
26. Adam NK. The properties and molecular structure of thin films. Part II. Condensed films. *Proceedings of the Royal Society A: Mathematical, Physical and Engineering Sciences.* 1922;101(712):452–472. doi: https://doi.org/10.1098/rspa.1922.0057
27. Langmuir I. The constitution and fundamental properties of solids and liquids. II. Liquids. *Journal of American Chemical Society.* 1917;39(9):1848–1906. doi: https://doi.org/10.1021/ja02254a006

28. Gorter E, Grendel F On bimolecular layers of lipoids on the chromocytes of the blood. *J Exp Med*. 1925;41(4):439–443. doi: https://doi.org/10.1084/jem.41.4.439
29. de Castro M. Johann Gregor Mendel: Paragon of experimental science. *Molecular Genetics & Genomic Medicine*. 2016;4(1):3–8. doi: https://doi.org/10.1002/mgg3.199
30. Abbott S, Fairbanks DJ. Experiments on plant hybrids by Gregor Mendel. *Genetics*. 2016;204(2):407–422. doi: https://doi.org/10.1534/genetics.116.195198
31. Bateson W. *Mendel's Principle of Heredity*. University Press; 1909. doi:https://doi.org/10.5962/bhl.title.44575
32. Weismann A. *The Germ-Plasm. A Theory of Heredity*. Charles Scribner's Sons; 1893. doi: https://doi.org/10.5962/bhl.title.168967
33. Sutton WS. The chromosomes in heredity. *Biol Bull*. 1903;4(5):231–250. doi: https://doi.org/10.2307/1535741
34. Sturtevant A. Thomas Hunt Morgan 1866-1945. Published 1959. Accessed November 7, 2022. http://www.nasonline.org/member-directory/deceased-members/20001550.html
35. Morgan TH. Sex limited inheritance in drosophila. *Science (1979)*. 1910;32(812):120–122. doi: https://doi.org/10.1126/science.32.812.120
36. Morgan TH. The application of the conception of pure lines to sex-limited inheritance and to sexual dimorphism. *The American Naturalists*. 1911;45(530):65–78. doi: https://doi.org/10.1086/279195
37. Bateson W, Saunders ER, Punnett RC. *Experimental Studies in the Physiology of Heredity*. The Royal Society of London; 1905. Accessed April 2, 2019. http://archive.org/details/RoyalSociety.ReportsToTheEvolutionCommittee.ReportIi.Experimental
38. Morgan TH. *Heredity and Sex*. Columbia University Press; 1913. doi:https://doi.org/10.5962/bhl.title.6236
39. Janssens FA, Koszul R, Zickler D. The chiasmatype theory. A new interpretation of the maturation divisions. 1909. *Genetics*. 2012;191(2):319–346. doi: https://doi.org/10.1534/genetics.112.139725
40. Sturtevant A. The linear arrangement of six sex-linked factors in drosophila, as shown by their mode of association. *Journal of Experimental Zoology*. 1913;14(1):43–59. doi: https://doi.org/10.1002/jez.1400140104
41. Morgan TH, Sturtevant A, Muller HJ, Bridges CB. *The Mechanism of Mendelian Heredity*. Henry Holt and Company; 1915. doi: https://doi.org/10.5962/bhl.title.6001

3 Discovery of Proteins and Enzymes

At the end of the 1800s, little was known about the cell and its contents. Mendel had published his mostly ignored memoir on the ratios of trait inheritance, and stick-like structures were observed in the center of cells that readily took up stains. No one knew the function or the importance of this substance, but evidence of chromatin being the carrier of heritable information began to surface around this time and continued into the 20th century. However, it would not be until the 1940s before researchers provided a preponderance of the evidence that genes were located on chromosomes and were the units of inheritance.[1] Up until that time, proteins were thought to hold that honor since they were known to be found in the nucleus and were much more complex than DNA, a characteristic that lined up well with the incredible diversity of heritable traits.

Albumin, or egg white, is the oldest substance that is part of a family of molecules we would eventually identify as proteins, though it would be millennia before we referred to it as such. Proteins were formally discovered as far back as 1747 when an Italian chemist, Jacopo Bartolomeo Beccari, isolated a substance from wheat which resembled something that *did not seem possible to extract it except from animal matter.*[2] Since this substance was similar to hen's egg albumin, he referred to it as a plant albumin, though it was later renamed wheat gluten (from the Latin for glue) due to its sticky nature.[2] Thereafter, Kessel-Meyer described specific protocols designed to isolate this glutinous substance and determined the action of various solvents upon it, while Claude Berthollet reported that this matter released nitrogen and turned yellow when treated with nitric acid.[2]

In 1773, Antoine Augustin Parmentier published a monograph in which he repeated Beccari's experiment with wheat and in the process, provided a thorough characterization of wheat gluten. To isolate it, he added water to 2 pounds of flour to make a paste, over which he trickled a steady stream of water as long as the runoff was of a milky color. This left behind a few ounces of a substance which he described as being *similar to a tenacious membrane, elastic, insoluble in water, sticky and adhered well to dry material, without flavor, having an odor of glue, and of a yellowish color.*[3] He found that the substance was insoluble in mineral acids and spirits of wine but soluble in vinegar, upon whose addition, a "milky" solution was created. In addition, he determined that it burned with a strong odor reminiscent of animal matter and that it lost *its tenacity and coherence and [...] evidently suffered from a decided physical change*—in that it became a solid mass—when it was boiled in water.[2-4] In a subsequent manuscript, Parmentier noted that the amount of gluten extracted from various types of wheat was correlated to the wheat's color, such that

the darker species yielded more gluten. Finally, he reported that upon drying, gluten lost two-thirds of its weight, which indicated that water made up a large fraction of the substance's content.[2,4]

Around that time, chemists began to notice similarities between plant and animal substances, such as between vegetal oils and animal fats, between emulsions and milk, and between vegetal and animal sugars. This prompted the recognition that these organisms belonged to one *kingdom* that is, that of a *règne organique ou organisé* (organic or organized kingdom), as opposed to the inorganic and raw dead mass of minerals.[5] One such similarity was shown by Claude Berthollet, who found that both vegetal and animal substances released nitrogen or ammonium in large quantities when subjected to the action of fire.[4,5] Furthermore, the French chemist, Antoine-François de Fourcroy, saw physical similarities between the glutinous substance isolated by Beccari and Parmentier and the fibrous part of animal blood, stating that these were completely different from anything previously seen in plants.[4] In 1789, he published a monograph in which he extended these similarities, describing a substance isolated from plant sap which, like animal albumin, had a thick and stringy consistency, a bland taste, could be dissolved in ammonia, and coagulated upon boiling.[5] He first suspected the presence of this albuminous substance in plants when he realized that plant sap could not be desiccated with heat without also being apt to coagulation, a property only known to occur in albumins. To conclusively prove that this substance was indeed similar to animal albumin, he filtered the sap of watercress and let the eluate air dry, after which he added more water and filtered a second time to remove the starch that had precipitated. When he added the eluate to a bain-marie and applied heat, he found that the solution separated within a few minutes and created a great quantity of whitish lumps that presented the same properties as albumins. de Fourcroy described his methods for the isolation of albumins from a number of different plant tissues and conclusively showed that plants contained a substance akin to animal albumin.

In the following years, albuminous and glutinous substances were found to be present in a number of different plants and trees, and in various flours, respectively, but yielded little novel information on this new class of substances.[2] However, in the 1800s, Gerrit Mulder found that these new substances had one thing in common: they all consisted of the same chemical elements. Mulder was a young Dutch doctor who changed careers due to the toll the cholera epidemic had taken on him. Eventually accepting an appointment at Utrect University in the Netherlands, he started to concentrate on the composition of albumins and found that they all consisted of $C_{40}H_{62}N_{10}O_{12}$; the only

DOI: 10.1201/9781003379058-3

variation he found was in the content of sulfur and phosphorus, which were typically on the order of one or two with respect to the aforementioned ratio. In 1839, the term *protein* was suggested to Mulder by Jacob Berzelius—the elder statesman of chemistry who proposed the chemical symbols with which we are familiar today—to describe these new substances. In one of his many correspondences with Mulder, Berzelius writes:

> The word protein that I propose to you for the organix oxide of fibrin and albumin, I would wish to derive from *proteios* [Greek for "standing in front", or "in the lead"], because it appears to be the primitive or principle substance of animal nutrition that plants prepare for the herbivores, and which the latter furnish for carnivores.[6]

In one of his textbooks from the mid-1800s, Mulder described the importance of proteins as follows:

> In plants as well as in animals there is present a substance which is produced in the former, constitutes the part of the food of the latter, and plays an important role in both. It is one of the very complex compounds, which very easily alter their composition under various circumstances, and serves especially in the animal organism for the maintenance of chemical metabolism, which cannot be imagined without it; it is without doubt the most important of all the known substances of the organic kingdom, and without it life on our planet would probably not exist. It is found in all parts of plants [...] as well as [... in] the animal body. [...] Animals draw their most important proximate principals from the plant kingdom. [...] The herbivorous animals are, from this point of view, no different from the carnivores. Both are nourished by the same organic substances, proteins, which plays a major role in their economy.[6]

As noted above, the glutinous substance isolated from wheat by Beccari was first called albumin due to its similarity in texture to hen's albumin. However, as more of these substances were isolated in different parts of the plant and in animals and differences among them were uncovered, a trend began to emerge in which proteins could be classified in terms of their solubility in various solutions. As such, albumins were proteins that were soluble in water, coagulated on heating, and remained dissolved when added to a half-concentrated solution of ammonium sulfate.[2] In contrast, the globulins were classified as proteins which were insoluble in water but soluble in neutral saline solutions; these were further divided into the vitellins, which were soluble in saturated sodium chloride, and myosins, which were insoluble in this solution. The prolamins were the alcohol-soluble proteins, and finally, the glutelins, of which gluten was the only known member, were a category reserved for proteins which could not be dissolved in neutral or saline solutions, or in alcohol.

In the mid-1800s, most known chemical elements had only recently been discovered and only a few of these were used to make up the molecules known at the time, most of which were simple molecules, for example, H_2O,

NH_3, and NaOH. Therefore, there was no reason to think that proteins were any different: they were thought to be small molecules which formed colloids—a homogeneous substance consisting of particles dispersed in a second substance—to form the slimy protoplasmic substance that surrounded the cell nucleus. The colloid theory was proposed by Thomas Graham in 1861 to explain the movement of dissolved substances across semipermeable membranes, such as parchment paper or gelatinous starch, through which some substances moved readily while others moved at significantly reduced rates or not at all.[7] In doing so, Graham developed a new branch of chemistry, that of colloidal chemistry, which studied heterogeneous substances composed of small particles distributed within another medium; for example, foam (a gas within a liquid), fog (a liquid within a gas), and pigmented ink (a solid within a liquid) are all colloids.

Thomas Graham was a distinguished Scottish chemist and professor at University College London (then called London University) and a founding member and president of the Chemical Society.[8,9] The main area of study for most of his career was that of the diffusion of gases, where he established a law stating that the velocity of gases was proportional to the square root of their density.[9] Late in his life, Graham turned his attention to solutions and their rate of diffusion in water and across semipermeable membranes.[10] To that end, he used what he called the *jar-diffusion* method, in which he carefully pipetted a given salt solution to be diffused at the bottom of a large column of water, which he did slowly and meticulously so as to not disturb the water. After a fixed amount of time (usually days), he removed the water one stratum at a time from the top-down (a total of 16 strata) and evaporated the water of each fraction. He determined how far each salt could diffuse in water by quantifying how much salt was present in each stratum: the more salt present in the strata farthest away from the bottom of the jar, the more diffusible the salt.[11] In doing so, he found that, in contrast to salt solutions, most gelatinous substances—such as gelatin, albumin, gum, and caramel—diffused very little in water, remaining mostly at the bottom of the jar or in the lower strata.

Osmosis, the passage of water across a semipermeable membrane to equilibrate the solute concentration on either side of that membrane, was well established by the time Graham undertook these studies and he optimized this technique to investigate the degree to which solutes in solutions diffused across a membrane of wetted parchment paper. As such, he used a small glass bell jar commonly used as an osmometer and placed a piece of wetted parchment paper on its bottom. The solution to be dialyzed was added to the bell jar, which was then placed over a large volume of water and the contents were allowed to diffuse into the water over a fixed period of time (Figure 3.1); he called this novel type of diffusion "dialysis" (from the Greek, *dia*, "through", and *lysis*, "loosening or splitting"). He found that although a variety of salts could diffuse out of the jar through the membrane and into the water with ease, gelatinous substances, such as

FIGURE 3.1 Graham's glass bell jar dialyzer. A piece of wetted parchment paper was fixed to the bottom of a glass jar, upon which a sample was placed. The apparatus was then lowered atop a large volume of water.[11]

albumin and gelatin, were significantly inhibited from doing so; and some could not pass at all. For example, he showed that albumin diffused 1000 times less than sodium chloride, his standard of comparison since it readily diffused through the membrane.

From these experiments, Graham proposed the term *crystalloid* for the substances that could readily diffuse through the semipermeable membrane (since they formed crystals when dried), such as sodium chloride and other salts, and *colloid* (from the Greek word, glue) for those that could not, such as gelatin, albumin, and other proteins. He also observed that *[t]he colloidal character* [of proteins] *is not obliterated when liquified*, as compared to salts which readily dissolve in water.[11] However, we will see in the next sections that erroneous proposition that proteins were colloids would persist well into the 1900s before it was disposed of once and for all! Nonetheless, his method of dialysis was a great boon to the technique of protein purification, a fact he was well aware at the time of publication of his work, as he remarked in his paper that:

> [t]he purification of many colloid substances may be effected with great advantage by placing them on the dialyser. Accompanying crystalloids are eliminated, and the colloid is left behind in a state of purity. The purification of soluble colloids can rarely be effected by any other known means, and dialysis is evidently the appropriate mode of preparing such substances free from crystalloids.[11]

* * * * *

In 1752, a French scientist by the name of René-Antoine de Réaumur was interested in finding out how predatory birds could digest meat without a gizzard—an organ found in some animals, including birds—which contains small pebbles which aid in digestion by grinding foodstuff.[12] Ingeniously, he attached a piece of meat to the end of a small tube and inserted it into a bird's stomach through a fistula in its belly; the use of the tube allowed the piece of meat to be kept in suspension and therefore, prevented it from being affected by the churning action of the stomach. Surprisingly, he found that the meat was completely digested within 24 hours, which indicated that digestion occurred not only due to the physical action of the stomach upon its contents, but also as a result of a chemical process. Later in same century, the Italian Lazzaro Spallanzani used gastric juices from a variety of species (including samples of his own stomach!) to show that chemical digestion was not limited to birds. He also developed *in vitro* experiments to test the requirements for digestion, showing that gastric juices diluted in water were no longer capable of digestion, that the process was temperature dependent, that the time required for digestion was dependent on the quantity of gastric juices added to the meat, and that the active ingredient for digestion was not stable outside an organism's body.

In 1811, a Russian apothecary, Constatine Kirchhof, showed that sulfuric acid could convert starches into sugar or gum, depending on the extent of the reaction, without being used up in the reaction. This indicated that although the acid enabled—or catalyzed—the reaction, it did not take part in it. However, since this process was not efficient, Anselme Payen, the director of a sugar factory in Paris, and Jean Persoz, a scientist who had previously discovered the sugar dextrin, wanted to find an economical way of separating the inner starch of seeds (which could then be converted to dextrin) from the seed coat.[13] In 1833, they published their findings describing a process whereby germinating barley, oat, or corn were macerated in cold water and filtered under pressure, resulting in a clear liquid which contained *the active principle* [which converted the starch into sugar] *plus a small amount of nitrogenous matter, colored substance, and a quantity of sugar proportionate to the degree of germination.*[14] Addition of alcohol to this liquid effectively precipitated the active ingredient, forming an *insoluble flocculent deposit* which could be filtered and dried at low temperature. By repeating the process a number of times, a highly purified product could be obtained, which they named diastase from the Greek meaning separation (given that it could effectively separate inner starch from seed coats). Of particular importance, Payen and Persoz stressed that temperatures above 75°C and nearing boiling had to be avoided as activity would be lost under these conditions. Testing their extracted diastase was a simple matter of adding it to flour grains, *such that one part by weight is enough to render soluble in hot water the internal substance of one thousand parts of dry starch* in about 10 minutes.[14]

In the following years, a number of other substances were found to change one substance into another: emulsin—purified from almonds—could convert amygdalin to benzaldehyde, and pepsin curdled milk better than hydrochloric acid alone. Initially, *ferments* was the name given to these purified substances given the similitude of their action to that of alcoholic fermentation—the conversion of sugar to alcohol.[15] In addition, ferments seemed to be analogous to reactions previously identified by Kirchhoff using sulfuric acid, in which the substance that allowed a reaction to occur was not consumed in the reaction. In 1836, Jacob Berzelius, the Swedish chemist responsible for the coining the term *protein*, suggested the use of *catalytic force* since he did not believe it was *a force entirely independent of the electrochemical affinities of matter* but rather, only a *new manifestation*.[15]

In 1876, the term *enzyme*, meaning *in yeast*, was suggested by Wilhelm Kühne to describe ferments but it would soon be adopted to include all proteins that accelerate chemical reactions without being consumed in the reaction, including enzymes which took part in the process of digestion as studied by de Rémoir and Spallanzani in the 18th century and Payen and Persoz's diastase.[15]

* * * * *

Emil Fischer began studying carbohydrates in 1884 in an attempt to increase the purity of his crystals upon purification, when he found that yeast did not ferment all sugars equally. This led to the discovery of a most surprising phenomenon concerning the mechanism of interaction of proteins and gave rise to an analogy still in use today. Fischer was a German organic chemist who turned down a lucrative life as a timber merchant to pursue his passion for chemistry, studying under Kekulé (of benzene structure fame) and von Baeyer, a chemist dedicated to the synthesis of chemicals.[16] In 1892, he was appointed professor and director of the chemical institute at the Berlin University, where he served until his untimely death in 1919.[16] There, he succeeded in synthesizing several sugars (some of which, like mannose, were unknown at the time)—called isomers—that had the same chemical composition (e.g., $C_6O_6H_{12}$) but differed in the manner in which the atoms were connected (Figure 3.2).[17]

He also subjected his large library of sugars, both natural and synthetic, to a variety of yeast strains to determine if they could be fermented, a process whereby an organism converts sugars into alcohol and releases CO_2 as a byproduct. However, since some of these sugars were in very limited quantities, he devised a *small fermentation tube*

which contained barium hydroxide that became turbid when CO_2 was evolved:

In all cases, even when the sugar is not fermented, a small amount of carbon dioxide evolves, which covers the surface of the barium hydroxide with a thin layer of carbonate. Since this phenomenon occurs even when no sugar has been added to the solution, it is obviously caused by the small amount of carbohydrate present in the yeast itself or the extract. The situation is quite different, when the material is readily fermentable: the barium hydroxide is not only becoming strongly turbid, but is neutralized. Intermediate cases are these, where material has to be brought into a fermentable state first, as with the glucosides; fermentation proceeds slowly [...], yet here too, the amount of carbon dioxide developed is always large enough, that one cannot be in doubt about the real occurrence of fermentation.[17]

As such, he found that only certain types of sugars were fermentable by a given strain of yeast, even in cases where the geometry of the non-fermentable sugar only slightly differed from that of the fermentable sugar. For example, d-galactose and d-talose differ only in the orientation of one hydroxide group, yet a particular strain of yeast might ferment the former and not the latter. This phenomenon was also observed in other experiments in which a number of purified enzymes were used and this led to Fischer's seminal 1894 paper containing the first mention of his "lock and key" analogy, which was instantly adopted and to this day is used to describe the precise relationship between enzymes and their substrate:

The restricted action of the enzymes on glucosides may therefore be explained by the assumption that only in the case of similar *[chemical]* geometrical structure can the molecules so closely approach each other as to initiate a chemical action. To use a picture I would like to say that enzyme and glucoside have to fit together like lock and key in order to exert a chemical effect on each other. The finding that the activity of enzymes is limited by molecular geometry to so marked a degree, should be of some use in physiological research.[17]

* * * * *

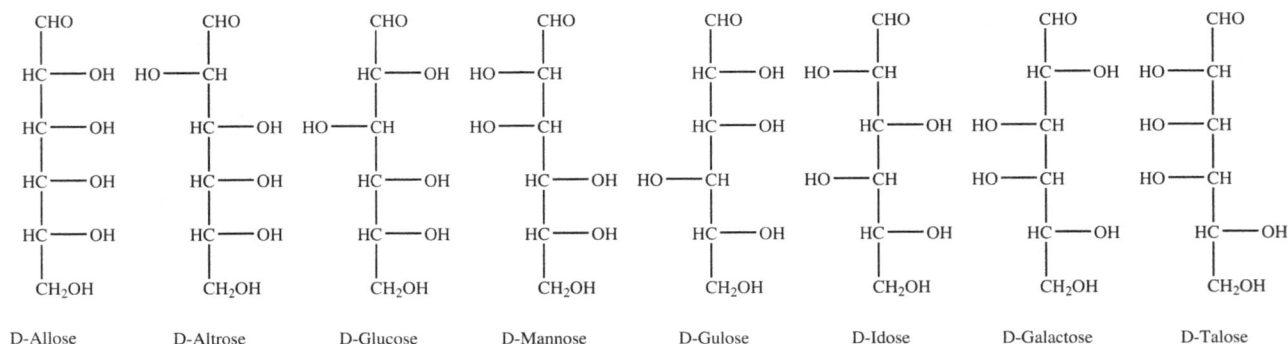

FIGURE 3.2 Hexose isomers synthesized by Emil Fischer.

By the 1890s, two types of ferments were recognized: unorganized ferments, such as diastase, which broke down molecules outside the cell (i.e., in a cell-free solution), and organized ferments, which broke down molecules in the presence of cells (such as yeast in alcoholic fermentation). Thus, there had been an ongoing, fervent debate about the requirements for fermentation: the *Enzyme theory of fermentation*—led by Justus Liebig and Felix Hoppe-Seyler—came about by analogy to enzymatic digestion such as diastase and invertase, which could convert starches into sugars and sugars into other sugars, respectively, and proposed that cells were not required in this process. In contrast, the *Vitalistic view*—headed by Baeyer, Theodor Schwann and supported by Louis Pasteur's experiments on anaerobic respiration (discussed in Chapter 5)—proposed that intact, living organisms, such as yeast, were required for fermentation to occur, since fermentation would stop if yeast cells were destroyed in a mixture.[18] An intermediate position was taken by Berthelot in 1860 when he suggested that fermentation occurred as a result of molecules produced by living yeast, but his experiments failed to produce such an extract and this largely went unheard. Similarly, Liebig had a prescient moment, a decade later, when he wrote that both views could be reconciled:

> It may be that the physiological process stands in no other relation to the fermentation process than the following: a substance is produced in the living cells which, through an operation similar to that of emulsin on salicilin and amygdalin, leads to the decomposition of sugar and other organic molecules; the physiological process would in this case be necessary to produce this substance but would stand in no further relation to the fermentation.[19]

As such, this chapter will conclude with the discovery of cell-free fermentation by two German brothers in 1897, which finally put an end to this debate. Hans, the elder sibling by 10 years, and Eduard Buchner grew up in Munich, where Hans studied medicine at Leipzig and was appointed chair and director of the Institute of Hygiene in 1894.[18] Eduard was to have a career in commerce as spurred on by his father, but his father's death, when Eduard was 12 years old, resulted in Hans becoming a surrogate father and mentor to his younger brother. Thus, encouraged by his older brother, Eduard pursued his passion for science and studied chemistry in Munich. He took a position in Adolf Baeyer's laboratory in 1884, working alongside Emil Fischer, where his research involved alcoholic fermentation. After receiving his Ph.D., he worked as a *privatdozent* in Baeyer's laboratory, where, with the aid of his mentor, Buchner received a special grant to set up his own laboratory with an aim to find out if there were *any special actions attributable to the contents of the yeast cell* during fermentation.[20] To investigate this, he had to obtain yeast cell extracts, a process which, at that time, was accomplished by the use of harsh chemical agents, such as sodium hydroxide, which damaged the delicate cell contents. Upon reading the available literature on yeast cell

extraction, Buchner outlined the conditions under which extraction had to occur: chemicals and high temperatures needed to be avoided and the procedure needed to be performed in as little time as possible to prevent degradation of the extracts.[20] According to Eduard, resolving these issues did not come easily:

> My first attempts to burst the cell membrane of yeast by freezing to −16°C and rapid thawing did not succeed. The membrane is much too elastic for it to be broken in this way. Direct grinding in a mortar had as little success, because the pestle could not get a purchase on such elastic stuff.[18]

However, after numerous discussions with his older brother and by suggestion from his mentor, he settled on the use of a recently developed and lesser-known method of crushing cells with sand, and enlisted the help of Hans' assistant, Martin Hahn, who devised an effective method to extract proteins from cells. His efforts resulted in a protocol where 1000 grams of yeast were ground for a few minutes with 1000 grams of quartz sand and 250 grams of diatomaceous earth (a natural, soft rock composed of fossilized diatoms) until it became *moist and plastic* (Figure 3.3a).[20,21] After adding water, the mixture was placed in a cloth and put under a hydraulic press under 400–500 atmospheres of pressure (Figure 3.3b), which could yield a total of about 500 ml of *juices*. A number of filtrations through filter paper provided them with a clear yellow liquid with a *pleasant yeasty smell*. However, these extracts tended to lose their enzymatic capabilities quickly so Buchner suggested that addition of a preservative might stabilize the solution.[18] Hahn tried a variety of different preservatives, including antiseptics, concentrated salt solutions, glycerin, and a 40% glucose solution. When Eduard visited the laboratory one day, he noticed bubbles emanating from the sugar solution and immediately ascribed this phenomenon as fermentation resulting from the presence of a minute amount of yeast in the solution (remember that Eduard was protégé of Baeyer's, who ascribed to the vitalistic view).

However, further investigation showed that this was not the case. First, others had shown that the amount of yeast that might be present in such extracts would not be enough to produce the amount of fermentation that Eduard had observed. In addition, the extracts could be filtered through a diatomite filter and biscuit-porcelain candles—which efficiently filtered out all yeast cells—without a decrease in its fermenting capabilities; third, the addition of antiseptics—which should have killed any and all yeast cells present in the extract—did not alter the activity of the extract; finally, yeast cells that were washed with alcohol or acetone, then washed with ether and air-dried—a process which inhibits their growth and creates *permanent yeast*—could nevertheless ferment sugars when it was added to them.[20] Thus, these experiments made it clear that the fermentation observed by Buchner proceeded in a cell-free yeast extract, and he concluded that *[t]he active agent in the expressed yeast juice appears rather to be a chemical substance, an enzyme, which I have called 'zymase'*.[20] These experiments

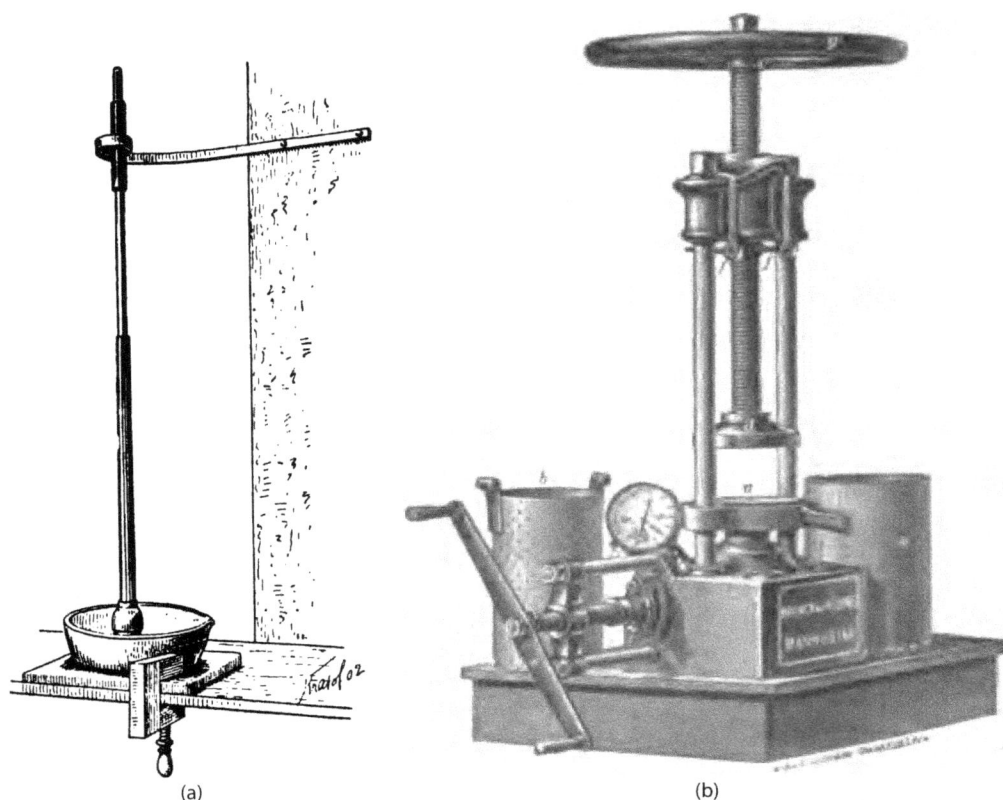

(a)   (b)

FIGURE 3.3 The apparatuses used by the Buchners to extract yeast juice. Yeast, sand, and diatomite were added to *a large mortar with a heavy long-shafted pestle* (a) to crush the yeast. The "doughy" mixture that resulted was wrapped in a canvas and the yeast juices were squeezed out by a hydraulic press (b).[22]

put to rest once and for all the vitalistic versus enzyme theory of fermentation debate; however, the general consensus was that no one had won nor lost: the process of fermentation was indeed a chemical process, but using molecules furnished by living organisms, as Berthelot and Liebig had ventured to propose many years before. As such, Buchner stated in his 1909 Nobel Prize lecture,

> The differences between the vitalistic view and the enzyme theory have been reconciled. Neither the physiologists nor the chemists can be considered the victors; nobody is ultimately the loser; for the views expressed in both directions of research have fully justified elements. The difference between enzymes and micro-organisms is clearly revealed when the latter are represented as the producers of the former, which we must conceive as complicated but inanimate chemical substances.[20]

* * * * *

REFERENCES

1. Dahm R. Friedrich Miescher and the discovery of DNA. *Developmental Biology.* 2005;278(2):274–288. doi: https://doi.org/10.1016/J.YDBIO.2004.11.028
2. Osborne TB. *The Vegetable Proteins.* Longmans, Green, and Co.; 1909. doi: https://doi.org/10.5962/bhl.title.28342
3. Parmentier AA. *Examen Chymique Des Pommes de Terre. Dans Lequel on Traite des Parties Constituantes Du Blé.* Chez Didot; 1773. Accessed November 7, 2022. https://gallica.bnf.fr/ark:/12148/bpt6k9647602f.texteImage
4. de Fourcroy A *Élemens d'histoire Naturelle et de Chimie.* Vol. 4. 3rd ed. Cuchet; 1789. Accessed November 7, 2022. https://books.google.ca/books?id=x–Cpj-7qnMC
5. de Fourcroy A. Sur l'existence de la matière albumineuse dans les végétaux. In: DeMorveau, Lavoisier, Monge, et al., eds. *Annales de Chimie, Ou Recueil de Mémoires Concernant La Chimie et Les Arts Qui En Dépendent.* Vol. 3ième. Joseph de Boffe; 1789:252–262. Accessed November 8, 2022. https://gallica.bnf.fr/ark:/12148/bpt6k6569445t
6. Fruton J. *Proteins, Enzymes, Genes. The Interplay of Chemistry and Biology.* Yale University Press; 1999.
7. Loeb J. *Proteins and the Theory of Colloidal Behavior.* McGraw-Hill Book Company, Inc; 1922. doi: https://doi.org/10.5962/bhl.title.17599
8. Tanford C, Reynolds J. *Nature's Robots: A History of Proteins.* Oxford University Press; 2001.
9. Wisniak J. Thomas graham. I. Contributions to thermodynamics, chemistry, and the occlusion of gases. *Educación Química.* 2013;24(3):316–325. doi: https://doi.org/10.1016/S0187-893X(13)72481-9
10. Kerker M. Classics and classicists of colloid and interface science. *Journal of Colloid and Interface Science.* 1987;116(1):296–299. doi: https://doi.org/10.1016/0021-9797(87)90123-8
11. Graham TX. Liquid diffusion applied to analysis. *Philosophical Transactions of the Royal Society of London.* 1861; 151:183–224. doi: https://doi.org/10.1098/rstl.1861.0011

12. Copeland RA. *Enzymes a Practical Introduction to Structure, Mechanism, and Data Analysis.* John Wiley & Sons, Inc.; 2000.

13. Payen A, Persoz JF. Mémoires sur la diastase, les principaux produits de ses réactions, et leurs applications aux arts industriels. In: Gay-Lussac MA, ed. *Annales de Chimie et de Physique.* Vol 53. Chez Crochard, Librairie; 1833:73–92. Accessed November 8, 2022. https://gallica.bnf.fr/ark:/12148/bpt6k6569011n

14. Payen A, Persoz JF. Memoir on diatase, the principle products of its reactions and their applications to industrial arts. In: Mordecai GL, Fogel S, eds. *Great Experiments in Biology.* Prentice-Hall, Inc.; 1955:25–27.

15. Hunter GK. *Vital Forces: The Discovery of the Molecular Basis of Life.* Academic Press; 2000.

16. Forster MO. Emil Fischer memorial lecture. *Journal of the Chemical Society, Transactions.* 1920;117(0):1157. doi: https://doi.org/10.1039/ct9201701157

17. Lichtenthaler FW. 100 years "schlüssel-schloss-prinzip": What made Emil Fischer use this analogy? *Angewandte Chemie International Edition in English.* 1995;33(23-24): 2364–2374. doi: https://doi.org/10.1002/anie.199423641

18. Kohler R. The background to Eduard Buchner's discovery of cell-free fermentation. *Journal of History of Biology.* 1971;4(1):35–61. doi: https://doi.org/10.1007/BF00356976

19. Finegold H. The Liebig-Pasteur controversy. *Journal of Chemical Education.* 1954;31(8):403. doi: https://doi.org/10.1021/ed031p403

20. Buchner E. Eduard Buchner—Nobel Lecture: cell-free fermentation. NobelPrize.org. Nobel Media AB © The Nobel Foundation 1907. https://www.nobelprize.org/prizes/chemistry/1907/buchner/lecture/ © Nobel Prize 1907

21. Buchner E. Alcoholic fermentation without yeast cells. In: Gabriel ML, Fogel S, eds. *Great Experiments in Biology.* Prentice-Hall, Inc.; 1955:27–30.

22. Buchner E, Buchner H, Hahn M. *Die Zymasegärung Untersuchungen Über Den Inhalt Der Hfezellen Und Die Biologische Seite Des Garungsproblems.* R. Oldenbourg; 1903. Accessed October 14, 2022. https://archive.org/details/diezymasegrungu00labogoog/page/n192/mode/thumb

4 Protein and DNA Subunits

In 1810, William Hyde Wollaston, an English chemist famous for discovering the chemical elements, palladium and rhodium, was studying the composition of human urine and found a type of calculus* that he did not recognize.[1] As such, he reported the following characteristics for this calculus: (1) all that remained of the calculus following destructive distillation was a *black spongy coal*; (2) burning with a blowpipe gave it the characteristic *smell of burned animal* typical of organic material; (3) ammonia was produced as a result of dry distillation; (4) water, alcohol, acetic acid, tartaric acid, citric acid, or saturated carbonate of ammonia could not dissolve it; (5) the acids, muriatic acid (hydrochloric acid), nitric acid, sulfuric acid, phosphoric acid, oxalic acid, and the bases, potash (a mixture of potassium carbonate and potassium hydroxide), soda (sodium carbonate), ammonia, and limewater (calcium hydroxide) could dissolve it. Wollaston also proposed that it must be an oxide given that it readily reacted with acids and bases, and since *the calculi that have yet to be observed have been taken from the bladder, it may be convenient to give it the name of cystic oxide†*.[1]

Several years later, the elder, Berzelius, felt that the term oxide was inappropriately used in this case: the presence of an oxygen atom in this organic molecule would not differentiate it from other organic molecules, as most contain oxygen.[1] Therefore, he proposed the name cystine for this new substance. Later, it was discovered that this molecule was rare among organic molecules, in that it also contained small amounts of sulfur. This led one scientist to suggest that cystine was, in fact, a double molecule connected by a disulfide bridge, and this would be shown to be correct: when treated with hydrochloric acid, cystine split into two molecules with much the same composition as cystine but with very different properties.[1] He called this new molecule cysteine. It would be almost a century, in 1899, before cystine was isolated as a breakdown product of protein hydrolysis (Figure 4.1a).

In 1819, Joseph-Louis Proust reported that wheat could spontaneously ferment and produce a substance reminiscent of cheeses made from the milk of cattle that grazed in certain fields, that is, *odorant, savory, sharp, and often delicious.*[4] This reminded him of an old adage: *Pabuli sapor apparet in lacte* (the taste of fodder appears in milk). Therefore, he attempted to find the molecule which gave cheese its unique flavor. To that end, he skimmed sheep's

milk (in his manuscript, he specified that this was a sheep found in Madrid around springtime …) and drained it before placing it in a jar for 1½ years. The milk curds, which he described as having a *detestable odor*, were well washed and desiccated until a syrupy solution was obtained, which had an *excessively intense cheesy taste*. The addition of alcohol and subsequent evaporation yielded abundant white crystals, which, after additional purification steps, produced a substance he called *oxide caséeux* that had similar characteristics to animal substances; that is, it contained ammonia, oil, and lots of carbon.

Two years later, the French chemist, Henri Braconnot, wanted to determine if proteins (or albumins, as he knew them) behaved like starches when treated with acids; that is, he wondered if they would be converted to sugars.[2] Thus, Braconnot boiled meat with dilute sulfuric acid and isolated a substance which he named leucine (from the Greek, white) due to its shinny white appearance.[2,5] However, he did not realize that this was the same substance that Proust had previously isolated; in fact, Braconnot himself isolated Proust's oxide caséeux in 1827 and did not realize it was the same as his leucine. Nonetheless, following analysis of his own purified oxide caséeux, Braconnot felt that the substance had been ill-named so he renamed it *aposépédine*. It was Mulder who, upon repeating this series of experiments in 1839, realized that *aposépédine*, leucine, and oxide caséeux, were one and the same, but the name leucine stuck.[2] The same year he isolated leucine, Braconnot also discovered another molecule when he boiled gelatin in dilute sulfuric acid, and named it *sugar of gelatin* since he found it was sweet to the taste; it was renamed glycine in 1858 due to its chemical similarity to leucine.

By the late 19th century, several more molecules of this type had been isolated, all containing both a basic amino group (NH_2) and an acidic carboxylate group (COOH); thus, the term "amino acid" was given to this new family of molecules (Figure 4.1b). The fact that amino acids contained both acidic and basic groups explained why they all readily dissolved in both acidic and basic solutions.[6] By 1908, all but two amino acids had been discovered, the last one being threonine in 1936, which was found to be chemically similar to a type of sugar, d(-)-threose, and hence the name.[5,7]

Amino acids were proposed to be the building blocks of proteins early on in the history of proteins since hydrolysis of the latter yielded the former. However, no one had conclusively proven this, in that proteins had yet to be synthesized from amino acids. It was recognized that if protein hydrolysis yielded amino acids, the opposite reaction, that is, dehydration of amino acids, should result in protein synthesis.[8] Furthermore, it was hypothesized that protein synthesis

---

* Gravel; stone. In English we understand by *gravel,* small sand-like concretions, or stones, which pass from the kidneys through the ureters in a few days; and by stone, a calculous concretion in the kidneys or bladder, of too large a size to pass, without great difficulty. From Lexicon Medicum, 1860 (Hooper & Akerly, 1860).

† Belonging to the urinary or gallbladder. From Lexicon Medicum, 1860 (Hooper & Akerly, 1860).

DOI: 10.1201/9781003379058-4

required urea in addition to amino acids since hydrolysis products of some proteins yielded molecules related to urea. Indeed, a product consistent with characteristics of proteins resulted from the incubation of amino acids with urea; however, these results contributed little additional knowledge about the chemistry of protein synthesis.[8] The first real clue in this area of research emerged in the late 1800s, when Theodor Curtius observed that glycine ethyl ester formed a bi-amino acid ring structure, called glycine anhydride (Figure 4.2, top row).[8] However, elucidation of the mechanism of protein synthesis would have to wait several years for Emil Fischer to begin his studies with this specific aim in mind. As we saw in the previous chapter, Emil Fischer was already regarded as a first-class chemist when he started his work on protein synthesis in the late 1800s, having performed much work on the synthesis of sugars and having described the relationship of enzymes with their substrates as being akin to that of a *lock and key*.[9]

Fischer noticed that Curtius' glycine anhydride was chemically only one step away from being a two-amino-acid molecule, if only the ring structure could be broken and stabilized (Figure 4.2).[8,9] Thus, he treated the molecule with ethanol and found that this produced the desired glycyl–glycine peptide; however, this chemical reaction produced an intermediate with a very reactive amino terminal that did not remain in this state very long and readily condensed back to glycine anhydride.[8,9] Ingeniously, Fischer realized that he could synthesize a three-amino-acid peptide if an amino acid radical was inserted at the NH_2 group of glycyl–glycine to stabilize the molecule.[8] This process allowed Fischer to synthesize long chains of amino acids, which he termed as *polypeptide*, and in 1907, succeeded in synthesizing a polypeptide consisting of 18 amino acids.[9] Following analysis of these synthesized molecules, Fischer stated that their characteristics resembled those of proteins so much that they would have been classified as proteins if first encountered in nature.

* * * * *

Friedrich Miescher was a Swiss scientist born into a scientific family: his father and especially his uncle, Wilhelm His, were accomplished physicians and professors in Basel. Miescher was trained as an otologist in the mid-1800s but began working in a laboratory as he felt an urge to study *the theoretical foundations of life*.[10] As he had an interest in determining *the chemical composition of cells*, his initial aim was to study proteins since they were believed to be important in understanding cellular functions. However, in the process of isolating proteins from cells, he found material he felt was most likely attributed to cell nuclei, and which precipitated in acidic solutions and redissolved when alkaline solutions were added, and which had stronger acidic properties than proteins.[10,11] At the suggestion of his mentor, Felix Hoppe-Seyler, Miescher used leukocytes from fresh bandages collected daily from a nearby surgical clinic to determine the nature of this substance.[10,12] After

Amino acids that have been demonstrated to be products of the hydrolysis of proteins

	AMINO ACID	EARLIEST OBSERVATION OF THE AMINO ACID		AMINO ACID	EARLIEST OBSERVATION OF THE AMINO ACID AS A PRODUCT OF HYDROLYSIS OF PROTEINS	
1	Cystine	Wollaston	1810	Glycine	Braconnot	1820
2	Leucine	Proust	1819	Leucine	Braconnot	1820
3	Glycine	Braconnot	1820	Tyrosine	Bopp	1849
4	Aspartic acid	Plisson	1827	Serine	Cramer	1865
5	Tyrosine	Liebig	1846	Glutamic acid	Ritthausen	1866
6	Alanine	Strecker (synthesis)	1850	Aspartic acid	Ritthausen	1868
7	Valine	von Gorup-Besanez	1856	Phenylalanine	Schulze and Barbieri	1881
8	Serine	Cramer	1865	Alanine	Weyl (Schützenberger	1888 1879?)
9	Glutamic acid	Ritthausen	1866	Lysine	Drechsel	1889
10	Phenylalanine	Schulze	1879	Arginine	Hedin	1895
11	Arginine	Schulze	1886	Iodogorgoic acid	Drechsel	1896
12	Lysine	Drechsel	1889	Histidine	Kossel Hedin	1896
13	Iodogorgoic acid	Drechsel	1896	Cystine	Mörner	1899
14	Histidine	Kossel Hedin	1896	Valine	Fischer	1901
15	Proline	Willstätter (synthesis)	1900	Proline	Fischer	1901
16	Tryptophane	Hopkins and Cole	1901	Tryptophane	Hopkins and Cole	1901
17	Oxyproline	Fischer	1902	Oxyproline	Fischer	1902
18	Isoleucine	Ehrlich	1903	Isoleucine	Ehrlich	1903
19	Thyroxine	Kendall	1915	Thyroxine	Kendall	1915
20	Oxyglutamic acid	Dakin	1918	Oxyglutamic acid	Dakin	1918
21	Methionine	Mueller	1922	Methionine	Mueller	1922

(a)

FIGURE 4.1 (a) Amino acids discovered by synthesis or in nature (left), or via protein hydrolysis (right), as it was known in 1931.[2] (b) List of amino acids found in proteins.[3] (Figure (a) reprinted with permission from Ref. [2]. Copyright © 1931 American Chemical Society.) (*Continued*)

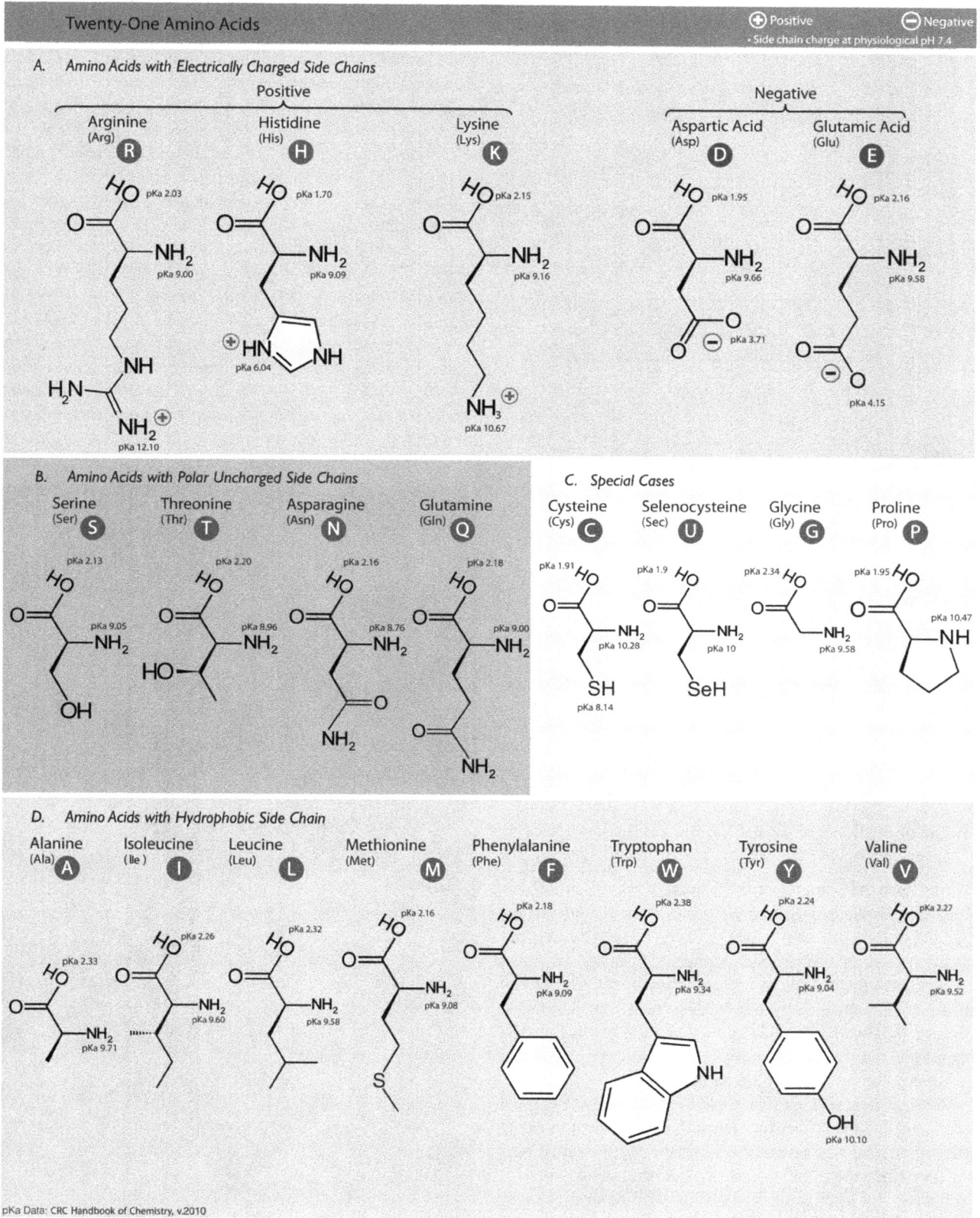

Twenty-One Amino Acids

Positive ⊕   Negative ⊖
• Side chain charge at physiological pH 7.4

A. Amino Acids with Electrically Charged Side Chains

Positive

Arginine (Arg) R
Histidine (His) H
Lysine (Lys) K

Negative

Aspartic Acid (Asp) D
Glutamic Acid (Glu) E

B. Amino Acids with Polar Uncharged Side Chains

Serine (Ser) S
Threonine (Thr) T
Asparagine (Asn) N
Glutamine (Gln) Q

C. Special Cases

Cysteine (Cys) C
Selenocysteine (Sec) U
Glycine (Gly) G
Proline (Pro) P

D. Amino Acids with Hydrophobic Side Chain

Alanine (Ala) A
Isoleucine (Ile) I
Leucine (Leu) L
Methionine (Met) M
Phenylalanine (Phe) F
Tryptophan (Trp) W
Tyrosine (Tyr) Y
Valine (Val) V

pKa Data: CRC Handbook of Chemistry, v.2010

Dan Cojocari, Department of Medical Biophysics, **University of Toronto**, 2010

(b)

FIGURE 4.1 (*Continued*)

$$NH_2.CH_2.COOC_2H_5 + NH_2.CH_2COOC_2H_5 =$$

Glycine ethyl ester

$$O=C \underset{NH.CH_2}{\overset{CH_2-NH}{<}} \! \! \! \! > C=O + 2\,C_2H_5OH$$

Glycine anhydrate
(diketopiperazine)

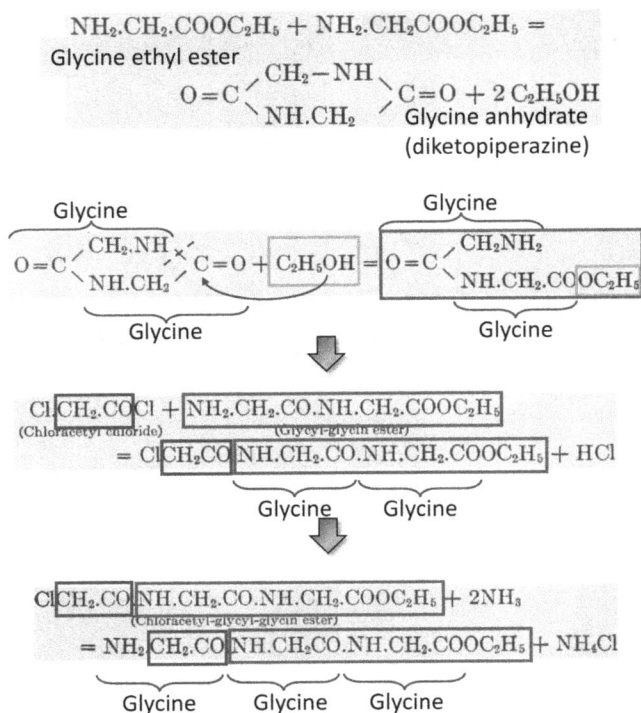

FIGURE 4.2 Scheme for the first synthesis of a peptide by Emil Fischer. Synthesis of Curtius' glycine anhydrate from two molecules of glycine ethyl ester is shown in the first line. Adding ethanol (green box) to glycine anhydrate breaks the NH–C bond, linearizing the molecule and yielding a dipeptide, glycyl–glycine ester (blue box), composed of two glycine molecules (red brackets). Adding a glycine derivative such as chloroacetyl chloride (glycine without the NH_2—indicated by a red box) to this intermediate allowed the incorporation of the carboxy terminal (CO) and side chain (H) of glycine, while addition of ammonia replaced the glycine derivative's last chloride with an amino group, resulting in the stable formation of a tripeptide, diglycyl–glycine ester, composed of three glycine amino acids. (Figure adapted from Ref [8].)

making sure the contents of the bandages had no indication of advanced decomposition, he washed them with different salt solutions in an attempt to extract the cells, settling on dilute sodium sulfate since, contrary to other solutions, it did not swell the cells to a slimy consistency. Following filtration to remove any cotton fibers from the bandages, Miescher allowed the cells to settle to the bottom of the tubes for a few hours, after which he could simply decant the supernatant to obtain pure cells.

Miescher next tackled the problem of isolating nuclei from these cells. He first tried dissolving the protoplasm in dilute hydrochloric acid for several days, but the results were unsatisfactory since the protein content of the cells often stuck to the nuclei. But after treating the cells with several changes of the acid solution for several weeks "at wintery temperatures" (remember, this was before refrigerators!) he was able to isolate fairly clean nuclei, as shown by the extract's lack of staining with iodine solution (a method used to stain proteins). Miescher then shook the solution vigorously in ether and water to separate the nuclei: those with protoplasm still

attached gathered at the interface between the two, whereas the relatively pure nuclei settled as a *fine powdery sediment* at the bottom of the tube in the aqueous phase.[12] Miescher then used dilute sodium carbonate to extract from the nuclei a yellow solution from which a substance could be precipitated when treated with excess acid; this precipitate was readily soluble in alkali but could not be dissolved acidic solutions.[12] This was the first purification of DNA!

Despite this success, a major concern for Miescher was the contamination of his samples with protoplasm; therefore, he turned to a substance known to dissolve proteins, pepsin (a component of gastric juices), which he extracted from pigs' stomach. He allowed the enzyme to digest his samples for an extended period of time—three or four times longer than usual—by which time a gray sediment had separated from the liquid which contained pure nuclei *without any trace of protoplasmic residue*. This powder yielded a substance that was *oily and light brown* upon washing with alcohol followed by evaporation (Figure 4.3).[12] After further

FIGURE 4.3 A vial of DNA isolated by Miescher—this one from salmon sperm. MUT | V. Marquardt. The object belongs to the Biochemical Specimen Collection of the University of Tübingen and is currently located in the Castle laboratory at Hohentübingen Castle.

analysis Miescher found that although it did not contain any sulfur, this substance contained an uncharacteristically large amount of phosphorus, an element not usually found in organic molecules at the time.[11] Therefore, Miescher concluded that *we are dealing with an entity sui generis not comparable to any hitherto known group*;[10] and since he found this novel substance mainly in the nucleus of cells, he called it nuclein (it was renamed *nucleic acid* by Altmann in 1889, who isolated the molecule in yeast).

Miescher's mentor, Felix Hoppe-Seyler, was so bewildered by his findings that upon submission of the manuscript to his journal in 1869, he insisted on personally reproducing Miescher's experiments before publishing the new research. Thus, Hoppe-Seyler and two of his students verified Miescher's results and extended his findings to include the isolation of nucleic acids from yeast and red blood cells, the results of which appeared in 1871 along with Miescher's paper.[11] Following these reports, Miescher became a department head in Basel, Switzerland, where his new laboratory had the advantage of being a short distance from the salmon-rich Rhine River.[13] It had been known from previous morphological studies that the head of spermatozoa was mostly composed of nuclei, so spawning season offered Miescher a seemingly endless supply of material to pursue his interests in nucleic acids.[13]

Picking up where Miescher left off, Albrecht Kossel, a German biochemist who started to work on nuclein in Hoppe-Seyler's laboratory in 1879, played an important role in the discovery of nucleotides, the basic subunits of nucleic acids. However, the breakdown products of nucleotides, nucleobases, would first be discovered, starting with guanine in 1844, which was isolated from bird excrement (known as guano, hence its name) (Figure 4.4). In 1882, Kossel confirmed it to be part of nucleic acids when he isolated it by acid hydrolysis of nuclein, a method that was also used in his subsequent isolation of adenine from beef pancreas nuclein in 1885.[11,13,14] These two molecules are part of a family of molecules known as purines, a term coined by Fischer in 1884 from the words *purum* (pure) and *uricum* (urine) since the first purine was isolated from uric acid, a component of urine now known to be a metabolic breakdown product of purine nucleotides.[15]

Thymine was discovered by Kossel and Neumann in 1893 after calf thymus and beef spleen were hydrolyzed by acids, and cytosine was discovered as a cleavage product of calf thymus one year later by the same scientists; uracil was first synthesized in 1884 by Behrend and subsequently isolated from wheat germ, calf thymus, herring sperm, and beef spleen nucleic acids early in the 1900s by a number of scientists, including, once again, Kossel.[11,13,16] Pinner, the first to propose the ring structure of these last three molecules, coined the term pyrimidine in 1885 to describe them by combining the words pyridine (the ring structure) and amidine (the segment containing two NH groups and the O between them), the two molecules that make up these bases.[14] Walter Jones commented in 1914 that the discovery of these molecules was considered to be of great significance

since it (1) provided a means to differentiate nucleic acids from amino acids; (2) enabled the study of distribution of these molecules without having to separate them from protoplasm; (3) refuted the claim that nucleic acids were simply a special kind of protein.[13]

In retrospect, elucidation of the nitrogenous bases of nucleic acids was a simple task to accomplish since they were the end product of the complete hydrolysis of nucleic acids using strong reagents. However, further characterization would not be so easily accomplished because although many different purines and pyrimidines exist in cells, only thymine, uracil, adenine, guanine, and cytosine make up the nitrogenous bases of nucleic acid. In addition, given the crude manner in which nucleic acids were purified, different sources yielded somewhat different components: inosinic acid (from the Greek—meat muscle) was initially isolated from meat extracts in 1847 by Justus Liebig and it was the first molecule to be characterized as a nucleotide (though, it was later discovered that it was an alteration of adenylic acid and not a component of nucleic acid). In 1893, Kossel started examining yeast nucleic acids and found, as was also later shown for the animal inosinic acid, that one of its components was a ribose sugar (a five-sided, cyclic carbohydrate).[13] In contrast, other animal nucleic acids were thought to be composed of a hexose sugar; still, other experimental results indicated that yeast nucleic acids contained both pentose and hexose sugars. In addition, when nucleic acids were purified from pancreas, only guanine was found as a hydrolysis product. Finally, studies using animal nucleic acids showed that only three of the five bases were in common with yeast, such that yeast uracil was replaced with thymine in animals. Jones summarized the contributing factors responsible for the confusion:

> Investigations have been conducted with impure material at a time when dependable methods of separating decomposition products were not available. Products of secondary decomposition were assumed to be primary, and substances formed from adherent impurities have been ascribed to the nucleic acids themselves.[13]

Such decompositions resulted in the destruction of the sugar component and/or alterations of the nitrogenous bases: purines were particularly susceptible to such alterations, such that guanine was modified to xanthine and adenine to hypoxanthine under the severe hydrolysis conditions often used.[13] In addition, it was found that cytidine could easily be converted into uridine, although they were otherwise fairly resistant to further decomposition.

Despite these difficulties, Pheobus Aaron Levene was able to elucidate a great deal about the structure of nucleic acids in the first decades of the 20th century. Coincidentally, he was born in Russia in 1869—the same year Miescher purified DNA for the first time—and studied medicine in St. Petersburg where he had as a mentor, among others, the famed Ivan Pavlov.[17] An interest for biochemistry was developed early in his studies when his chemistry teacher,

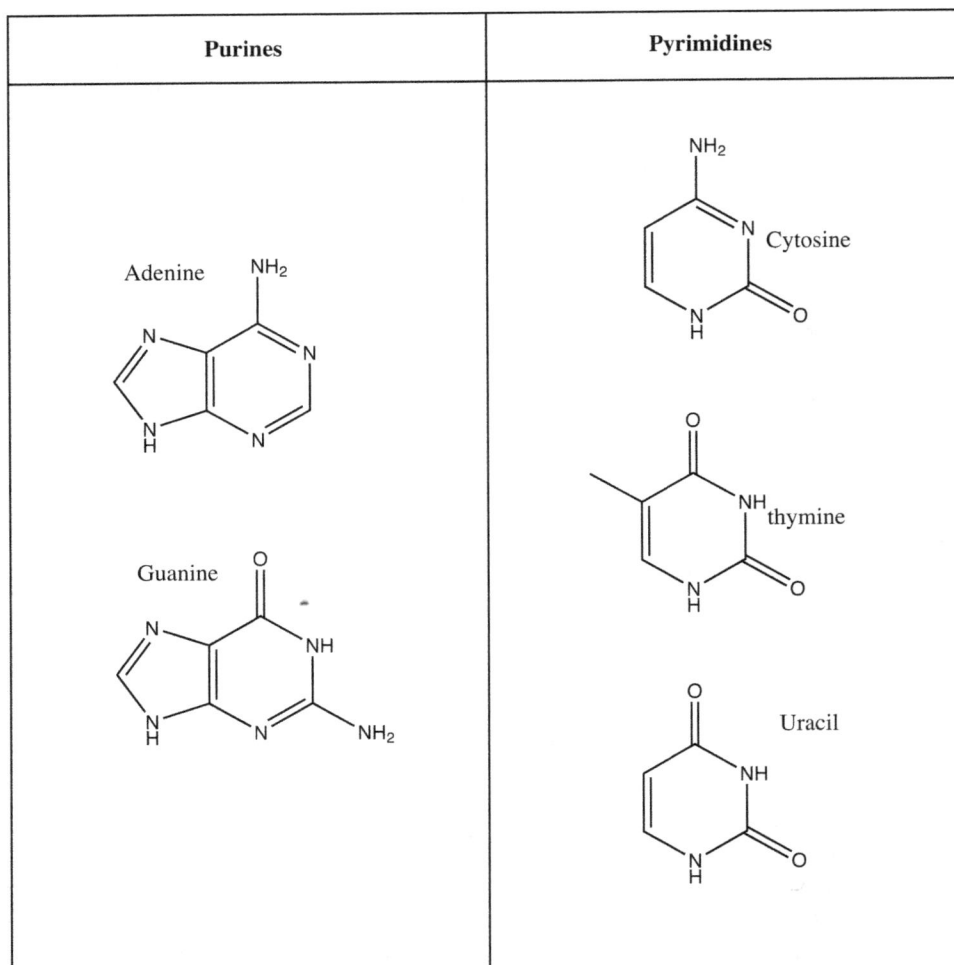

FIGURE 4.4   The chemical structure of the nitrogenous bases, or nucleobases, of nucleotides.

having noticed a particular talent in his pupil, allowed Levene the use of his laboratory. His family moved to New York in 1891 to escape anti-Semitic sentiments developing in Russia, where he practiced medicine for a few years until an opportunity arose where he could pursue his scientific interests. He studied nucleic acids, first in Dr. Ira van Gieson's laboratory in New York and then, for a short time, with Albrecht Kossel in Germany, as well as proteins and amino acids in Emil Fischer's laboratory in Berlin. Upon his return to New York in the early 1900s, he studied under Adolf Meyer, after which he was offered a post at the newly founded Rockefeller Institute for Medical Research, where he was put in charge of its Division of Chemistry in 1907.[17]

At that time, Haiser had already shown that acid hydrolysis broke down inosinic acid into a pentose sugar, phosphoric acid, and the base hypoxanthine (a breakdown product of adenine), whereas hydrolysis with milder agents produced a compound composed of phosphoric acid and pentose sugar.[11,13] In 1909, Levene and Jacobs took this a step further by showing that inosinic acid produced two different compounds depending on the type of hydrolysis used: acid hydrolysis yielded a compound composed of phosphoric acid

and a pentose sugar (as Haiser had shown), whereas neutral hydrolysis yielded a compound composed of a pentose sugar and hypoxanthine (Figure 4.5).[13] This was the first purification of these types of molecules and the authors called the purine-carbohydrate compound *nucleoside* (e.g., adenosine, guanosine, etc.), and the phosphate ester of a nucleoside, such as inosinic acid, a *nucleotide*.[18] These experiments, along with conclusions reached by Haiser years earlier, effectively proved the linkage order of the components of nucleotides to be phosphoric acid-pentose sugar-nitrogenous base (purine or pyrimidine). As expressed by Levene, this discovery allowed the isolation of *the sugars of the nucleic acids in crystalline form and thus* [... enabled the study of] *the structure of these peculiar sugars.*[11]

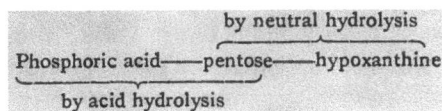

FIGURE 4.5   Nucleotides are composed of phosphoric acid, a pentose sugar, and a nitrogenous base.[13]

The first chemical studies using yeast nucleic acids were published by Kossel in 1893, in which he found that a carbohydrate, which he proposed was a pentose sugar, was formed upon hydrolytic cleavage of nucleic acids.[13] However, it would be another 15 years before Levene and Jacob confirmed this sugar to be a pentose, and specifically, that it was D-ribose, an unknown sugar at the time.[14] In the interim, four different sugars were proposed to be the sugar component of inosinic acid by four different groups and a "spirited debate" ensued about the validity of these claims. Since the chemical make-up of the sugar had already been determined ($C_5O_5H_{10}$), a total of eight different isomers could theoretically have been the correct sugar (Figure 4.6a).

In one of their 1909 publications, Levene and Jacob reported the isolation of pure crystals of the unknown pentose sugar of nucleotides, whose characteristics could now easily be assessed and compared to the candidate sugars. As such, they could immediately eliminate four of these sugars, since they had the wrong optical rotation, and two more because they were not aldo pentoses, a fact which was known about the new sugar.[13] Furthermore, confirmation of these deductions was provided by the fact that the unknown sugar yielded the osazone, D-arabinosazone, upon reaction with phenylhydrazine (a reaction developed by Emil Fischer

to characterize sugars), a property shared by both D-ribose and D-arabinose. Finally, the authors showed that oxidation of the unknown pentose yielded an inactive product (i.e., it did not rotate unpolarized light), as did D-ribose, whereas the D-arabinose yielded an active product, thus, indicating that the pentose sugar was in fact, D-ribose (Figure 4.6b). This was confirmed to be correct a few years later when D-ribose was synthesized by a different group and found to have the same characteristics as the new molecule identified by Levene and Jacobs.

Late in the 1890s, Haiser and Bang had found that pancreas guanylic acid could be hydrolyzed to produce not only phosphoric acid and a nitrogenous base (guanine), similar to what Levene and Jacobs found in yeast nucleotides, but also glycerophosphoric acid. However, others had isolated both guanylic and adenylic acids from pancreas nucleic acids, and yet another group found the same results as Bang had found but did not isolate any glycerophosphoric acid.[13] Bang had suggested that guanylic acid was composed of several nucleotides bound together, and therefore more akin to nucleic acids than to inosinic acid. The main argument for his conclusion came from analysis of his purifications and the fact that the acid tended to gelatinize when it was purified.[19] Although many disagreed with him, no one had conclusively shown the true nature of guanylic acid one way or another.

Thus Levene and Jacobs refined their protocol for the isolation of guanylic acid in pure crystalline form, which allowed them to accurately characterize the nucleotide.[19] Their method consisted of first purifying guanylic acid from pancreas as an impure lead salt, then transforming it into a sodium salt, and then into a mercury salt, which was insoluble in dilute acids. Since the purification of bases was usually performed in dilute acids, this last step provided a means to isolate guanylic acid from all other bases, thus removing potential contaminants. As such, Levene and Jacobs found that guanylic acid only began to gelatinize when minute amounts of contaminants were added to a purified solution. Combining their previous results on guanylic acid with these new findings enabled the precise examination of the acid's structure and components, which showed that guanylic acid was indeed a simple nucleotide, consisting of phosphoric acid, D-ribose, and the nitrogenous base, guanine.[19] However, since guanylic acid was more similar to inosinic acid and yeast nucleic acids than to thymus nucleic acids—in that ribose rather deoxyribose was a component of its nucleic acid—it was thought that the former's *constituent groups are those of plant nucleic acids, and they cannot therefore be directly derived from the nucleic acid of the animal tissues in which they occur.*[13]

This statement would eventually be proven wrong by Levene himself, who's efforts in elucidating the properties of nucleic acids resulted in many papers on thymus nucleic acids. These were first examined by Kossel and Neumann in 1893 and were thought to only contain thymic acid, but other groups had published results isolating guanosine from thymonucleic acid. However, since it had previously been shown that one of constituents of guanosine was a pentose

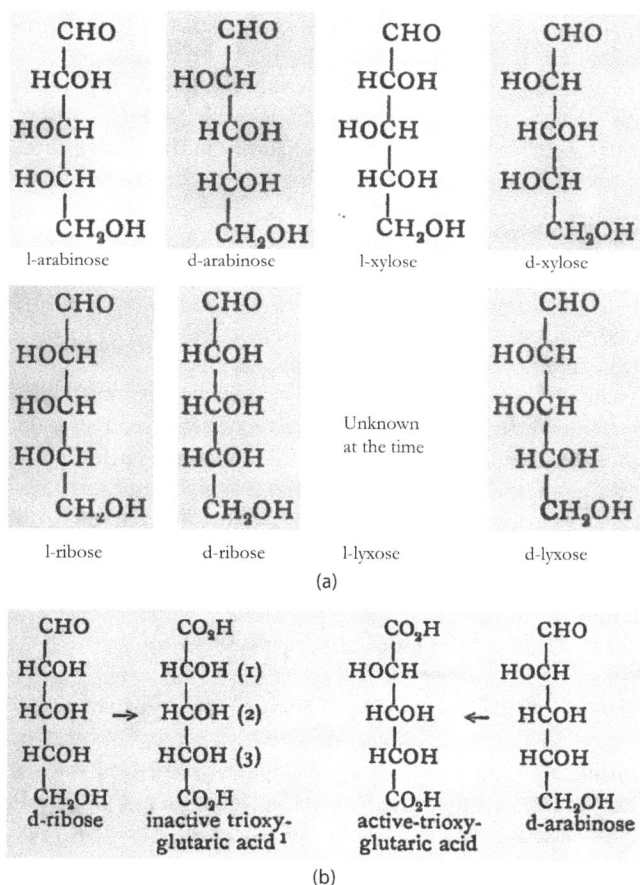

FIGURE 4.6 (a) The eight different pentose isomers analyzed by Levene and Jacob; (b) oxidation of D-ribose to yield an inactive product. (Figure adapted from Ref [13].)

sugar and thymonucleic acid was reported to only contain a hexose sugar, Levene and Jacobs suggested that the isolated guanosine was a result of contamination.[20] Thus, the authors set out to investigate these claims but found that *methods employed successfully on the other occasion led to no results in the experiments with thymus nucleic acid.*[20] In addition, they stated that from *the very start our progress was greatly retarded by the marked differences shown by this acid towards the chemical treatment to which the ribose nucleotides had so readily yielded*, and concluded that *the instability of the sugar is the cause of the failure of chemical methods.*[21]

Thus, Levene set out to determine if an enzyme might be able to accomplish a purer isolation of thymonucleic acid components. After considerable research, Levene and Medigreceanu found several enzymes from animal organs that could cleave nucleic acids into various constituents, some *capable of cleaving nucleic acids to nucleotides, others capable of dephosphorylating nucleotides (nucleotidases), and some capable of hydrolyzing nucleoses [sic] (nucleosidases).*[22–24] As such, they used an enzyme isolated from intestinal juices of dogs that had been shown to cleave nucleic acids into nucleosides, and indeed, isolated a guanine nucleoside from thymonucleic acid.[20] However, much to their surprise, their analysis revealed that this nucleoside was made up not of a pentose sugar as found in yeast and guanylic acid, but rather a hexose sugar similar to thymic acid. It would be almost another 20 years before the exact nature of this sugar would be determined.

In 1894, Kossel had found that hydrolysis of thymus nucleic acids with dilute acids yielded levulinic acid, a molecule known to be formed by the action of sulfuric acid on a hexose sugar, which suggested that thymonucleic acid was composed of a hexose precursor.[14] However, Levene and Mori later determined that this reaction was also characteristic of 2-deoxypentoses, a relatively new sugar at the time.[24,25] Therefore, reexamination of thymus nucleic acid was necessary. Levene's newly found enzymes were instrumental in the isolation of this sugar—which they provisionally named thyminose—and enabled investigation into the true nature of the molecule.[25,26] Several lines of evidence led them to conclude that it was in fact a deoxyribose sugar: (1) it changed color with Kiliani's reagent, which turns greenish-blue upon reaction with 2-desoxy sugars;[11,24,25] (2) a pine stick dropped in a solution of the sugar and exposed to hydrogen chloride vapors turned a blueish purple color;[25,26] (3) it consumed an amount of iodine consistent with an aldodesoxypentose upon oxidation;[25] (4) it changed color upon reaction with Schiff's reagent—a test used to identify aldehydes which can be obtained from deoxyribose sugars.[25] As such, Levene and London stated that *the nature of the sugar of the thymonucleic acid gives a ready explanation of those peculiarities of the acid which up to date seemed rather puzzling.*[24]

In 1912, Levene developed a method to enzymatically cleave nucleic acids, which involved passing a *solution of nucleic acid through a segment of* [a dog's] *gastro-intestinal tract* and collecting it *from an internal fistula.* However, following some preliminary experiments in collaboration with Ivan Pavlov, they determined that using a gastric fistula instead would avoid contamination from remnant food.[24] Therefore, they used this improved protocol to perform new experiments, though Levene and London admitted that the *procedure was repeatedly varied in the course of the work, and we are inclined to believe that the optimal conditions for the digestion of the nucleic acid still remain to be established.* Nonetheless, their protocol involved flowing 50 g of nucleic acids through the gastric fistula for 1–2 hours and collecting 350–700 ml of fluid. However, they soon found that *hydrolysis of the nucleic acid could be accomplished by adding portions of the* [gastric] *secretions daily to the solution of nucleic acid without passing through the gastrointestinal tract.*[24] In doing so, they observed that the *great resistance of thymonucleic acid as compared with yeast nucleic acid was very striking*, yielding at most 1.5 g of thymus guanoside versus about 5 g of yeast guanoside from 200 g of nucleic acid.

The first step in isolation of the nucleosides was the separation of the purines from the pyrimidines following extraction of the digest. This could be accomplished by simple filtration since the purines gelatinized due to the presence of guanosines, and the pyrimidines were therefore contained in the filtrate (filtrate 1). The purine fraction was then dissolved in hot water and barium hydroxide, which released the phosphate impurities and yielded guanoside in the form of a semigelatinous precipitate upon cooling. The precipitate was dissolved in water and an excess of lead acetate was added, forming a precipitate which consisted of nucleotides and which was filtered out. Ammonium hydroxide was then added to the filtrate to precipitate it and then boiled to dissolve the precipitate anew. Upon cooling, a *flocculent precipitate* [was] *formed* which was allowed to cool for 4–5 hours in the cold, separated by filtration, and dissolved again in water through which hydrogen sulfide was passed. In this reaction, the sulfide reacted with the lead in the nucleosides and formed a lead sulfide precipitate, which could be filtered to yield pure guanine nucleoside crystals.[24] The filtrate from the semigelatinous guanosine extraction was further purified into adenine nucleoside in the same manner as guanine nucleoside.

The pyrimidine filtrate from the original filtration (filtration 1) was concentrated to a small volume and precipitated with alcohol, followed by addition of barium hydroxide to the filtrate to release the phosphoric acid, and purified with lead sulfate in the same manner described above. The soluble lead salts were used to purify thymine nucleosides by removing the lead using hydrogen sulfide gas, and the filtrate was cooled and made acidic. Silver carbonate was added to precipitate impurities, which were removed by filtration, and the solution was neutralized using barium carbonate, both of which were removed when the solution was passed through a steam of hydrogen sulfide. This crystallized solution yielded thymine nucleoside. To the lead-insoluble fraction of the pyrimidine solution was added alcoholic picric acid which formed an *amorphous precipitate* which was filtered off and washed with ether.[24] This yielded crystals of cytosine nucleoside after several washes. Thus, Levene and his team discovered the presence of four

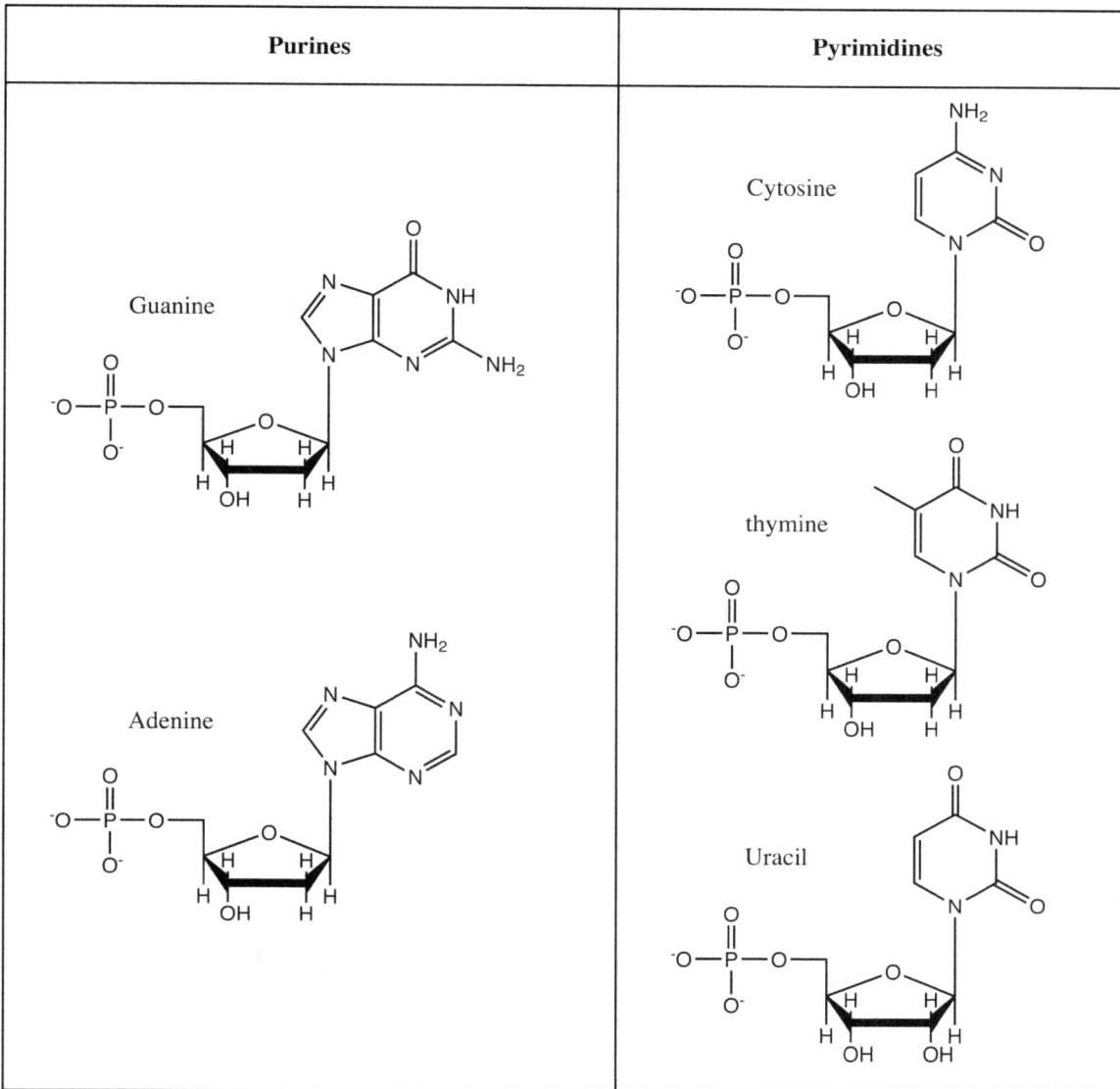

FIGURE 4.7 Chemical structure of nucleotides found in nucleic acids.

different nucleotides in thymus nucleic acids, which until that time, were thought to only contain thymic acid. They also showed that this nucleic acid was composed of a different sugar than yeast nucleic acid: it contained deoxyribose instead of ribose. And finally, their results showed that guanylic acid was not a special nucleic acid, but rather a simple nucleotide which was a component of both animal and yeast nucleic acids (Figure 4.7).

Following these results, it seemed clear that plant nucleic acid was composed of similar components to that of yeast, and both differed from that of animal nucleic acid. As such, it was assumed that ribonucleic acids (RNA) were specific for plants and deoxyribonucleic acids (DNA) were specific for animals although some objected to this classification (Table 4.1).[11]

TABLE 4.1

List of Nucleic Acids Found in Yeast/Plants vs. Animals, as Known in the Early 20th Century

Yeast/Plant Nucleic Acids	Animal Nucleic Acids
Adenine	Adenine
Guanine	Guanine
Cytosine	Cytosine
Uracil	Thymine
D-ribose	D-deoxyribose
Phosphoric acid	Phosphoric acid

* * * * *

REFERENCES

1. Wollaston WH. On cystic oxide, a new species of urinary calculus. *Proceedings of the Royal Society of London.* 1810;1:376–377. doi: https://doi.org/10.1098/rspl.1800.0212

2. Vickery HB, Schmidt CLA. The history of the discovery of the amino acids. *Chemical Reviews.* 1931;9(2):169–318. doi: https://doi.org/10.1021/cr60033a001

3. File: Amino acids.png - Wikimedia Commons. Accessed November 1, 2022. https://commons.wikimedia.org/wiki/File:Amino_acids.png

4. Proust JL. Recherches sur le principe qui assaisonne les fromages. In: Gay-Lussac J, Arago F, eds. *Annales se Chimie et de Physique.* Vol. 10. Crochard; 1819:29–49. Accessed September 30, 2022. https://gallica.bnf.fr/ark:/12148/bpt6k6570892b?rk=42918;4

5. Plimmer RA. *The Chemical Constitution of the Proteins.* Longmans, Green, and Co.; 1908. doi: https://doi.org/10.5962/bhl.title.21068

6. Vickery HB, Osborne TB. A review of hypotheses of the structure of proteins. *Physiological Reviews.* 1928;8(4): 393–446. doi: https://doi.org/10.1152/physrev.1928.8.4.393

7. Meyer CE, Rose WC. The spatial configuration of α-amino-β-hydroxy-n-butyric acid. *Journal of Biological Chemistry.* 1936;115(3):721–729. doi: https://doi.org/10.1016/S0021-9258(18)74711-X

8. Robertson BT. *The Physical Chemistry of the Protein.* Longmans, Green, and Co.; 1918. Accessed November 7, 2022. https://archive.org/details/in.ernet.dli.2015.214545

9. Forster MO. Emil Fischer memorial lecture. *Journal of the Chemical Society, Transactions.* 1920;117(0):1157. doi: https://doi.org/10.1039/ct9201701157

10. Dahm R. Friedrich Miescher and the discovery of DNA. *Developmental Biology.* 2005;278(2):274–288. doi: https://doi.org/10.1016/J.YDBIO.2004.11.028

11. Levene PA, Bass LW. *Nucleic Acids.* The Chemical Catalog Company, Inc.; 1931. Accessed November 7, 2022. https://babel.hathitrust.org/cgi/pt?id=uc1.b4165245&view=1up&seq=5

12. Miescher F. On the chemical composition of pus cells. In: Gabriel ML, Fogel S, eds. *Great Experiments in Biology.* 10th ed. Prentice-Hall, Inc; 1955:233–239.

13. Jones W. *Nucleic Acids: Their Chemical Properties and Physiological Conduct.* Longmans, Green, and Co.; 1914. Accessed November 7, 2022. https://archive.org/details/nucleicacidsthei00jonerich/page/n7/mode/2up

14. Bendich A. *The Nucleic Acids: Chemistry and Biology.* Vol. I, Chargaff Erwin, Davidson JN, eds. Academic Press; 1955. doi: https://doi.org/10.5962/bhl.title.6974

15. Fischer E. Ueber das Purin und seine Methylderivate. *Berichte der Deutschen Chemischen Gesellschaft.* 1898;31(3): 2550–2574. doi: https://doi.org/10.1002/CBER.18980310304

16. Chargaff E, Davidson JN. Introduction. In: *The Nucleic Acids: Chemistry and Biology*, Vol. 1. Academic Press; 1955:1–8.

17. van Slyke DD, Jacobs WA. Phoebus Aaron Theodor Levene 1869-1940. *National Academy of Sciences of the United States of America Biographical Memoirs.* 1944;23(4): 75–126. Accessed March 31, 2019. http://www.nasonline.org/publications/biographical-memoirs/memoir-pdfs/levene-phoebus-a.pdf

18. Cohen JS, Portugal FH. The search for the chemical structure of DNA. *Connecticut Medicine.* 1974;38(10):551–557. Accessed March 30, 2019. https://profiles.nlm.nih.gov/ps/access/CCAAHW.pdf

19. Levene PA, Jacobs WA. On guanylic acid: Second paper. *Journal of Biological Chemistry.* 1912;12(3):421–426. doi: https://doi.org/10.1016/S0021-9258(18)88678-1

20. Levene PA, Jacobs WA. Guaninehexoside obtained on hydrolysis of thymus nucleic acid. *Journal of Biological Chemistry.* 1912;12(3):377–379. doi: https://doi.org/10.1016/S0021-9258(18)88673-2

21. Levene PA, Jacobs WA. On the structure of thymus nucleic acid. *Journal of Biological Chemistry.* 1912;12(3):411–420. doi: https://doi.org/10.1016/S0021-9258(18)88677-X

22. Levene PA, Medigreceanu F. On nucleases: Second paper. *Journal of Biological Chemistry.* 1911;9(5):389–402. doi: https://doi.org/10.1016/S0021-9258(18)91455-9

23. Levene PA, Medigreceanu F. On nucleases. *Journal of Biological Chemistry.* 1911;9(1):65–83. doi: https://doi.org/10.1016/S0021-9258(18)91493-6

24. Levene PA, London ES. The structure of thymonucleic acid. *Journal of Biological Chemistry.* 1929;83(3):793–802. doi: https://doi.org/10.1016/S0021-9258(18)77108-1

25. Levene PA, Mori T. Ribodesose and xylodesose and their bearing on the structure of thyminose. *Journal of Biological Chemistry.* 1929;83(3):803–816. doi: https://doi.org/10.1016/S0021-9258(18)77109-3

26. Levene PA, London ES. Guaninedesoxypentoside from thymus nucleic acid. *Journal of Biological Chemistry.* 1929;81(3):711–712. doi: https://doi.org/10.1016/S0021-9258(18)63722-6

5 The Energy of Cells Part I
Glycolysis and the Krebs Cycle

Glycolysis comes from the Greek meaning "splitting of sugar" and it is a process whereby sugars get broken down in a chain of events that leads to the production of energy which cells can use to power cellular processes. As we will see in this chapter, in the absence of oxygen, the metabolism of sugar yields lactic acid in multicellular organisms and ethanol in fermenting yeast, but only a small amount of energy can be produced from this pathway and energy stocks are rapidly depleted. However, in presence of oxygen, pyruvic acid is produced instead, and this molecule is further processed in a second pathway, the Krebs cycle (also called the tricyclic or citric acid cycle) where most of the energy used by cells is produced, with CO_2 and water as by-products. The complexity of these pathways—multi-step processes which include more than two dozen enzymes, most of which acting sequentially on the breakdown of carbohydrates—led to much confusion in the first quarter of the 20th century (and in the writing of this chapter …). As such, to fully understand the metabolism of sugars into usable energy, we must first go back to the process of fermentation.

We saw in Chapter 3 that the Buchner brothers had put to rest the vitalistic versus enzyme theory of fermentation in the late 1800s, which resulted in the realization that cellular processes were performed not by an intrinsic force but rather by a complement of enzymes that catalyze chemical reactions. Specifically, they had ascribed the process of fermentation to one enzyme, zymase, which was found to be essential for fermentation, but could not bring it about on its own. They subsequently found that zymase was dependent on a second enzyme, which was named, aptly enough, co-enzyme.[1] However, the mechanism by which yeast fermented sugars—and why this phenomenon occurred in the first place—was still very much an obscure process. In 1861, the famed French chemist, Louis Pasteur, discovered two aspects of fermentation that would shed light on this process and bring about a new direction for the study of cell biology.

Despite the fact that at the time, all organisms were assumed to require oxygen to survive and all known ferments were yeast, Pasteur found a species of bacteria which not only performed better in the absence of oxygen, but indeed, oxygen was detrimental to its existence.[2] Since most bacteria required oxygen and could not ferment, Pasteur reasoned that perhaps the ability to survive without oxygen was linked to the ability to ferment. A follow-up paper testing this hypothesis in yeast unexpectedly revealed not only that it could in fact survive without oxygen, but that this condition was actually required for fermentation.[3] He also found that yeast grown under aerobic conditions (i.e., in the presence of oxygen) enjoyed a rate of multiplication

so *remarkable, hitherto unknown from the life of this little plant*, adding that *it would not be exaggerated to say that they multiply one hundred times faster in one condition* [aerobic] *compared to the other* [anaerobic, without oxygen].[3] However, Pasteur determined that although yeast could survive in the presence of oxygen, their power to ferment was lost—it consumed twenty times less sugar under aerobic conditions. Thus, as with bacteria, yeast also fermented strictly in the absence of oxygen, a phenomenon that became known as the Pasteur effect.[4] As such, Pasteur concluded:

> It must be admitted that yeast, so greedy of oxygen that it removes it from atmospheric air with great activity, has no more need of it and does without when one refuses it this gas in the free state, presenting it instead in profusion in the combined form in fermentable material; if one refuses it this gas in the free state, immediately the organism appears as an agent of sugar decomposition. With each respiratory gesture of the cells, there will be molecules of sugar whose equilibrium will be destroyed by the subtraction of a part of their oxygen. A phenomenon of decomposition will follow, and hence the ferment character which on the other hand will be missing when the organism assimilates free oxygen gas.[3,4]

In the early 1900s, a series of papers were published by Arthur Harden and William Young which reported the importance of phosphates for fermentation.[1] Leading up to this, Harden was seeking to determine the reasons for the relatively short duration of fermentation of yeast-juice (prepared using a similar method to Buchners') and thought that this might be due to the actions of a proteolytic enzyme on zymase. As such, he performed an experiment in which yeast-juice was incubated with serum (which he thought might stimulate fermentation) and found that, indeed, fermentation increased by 60–80%.[1] According to Harden, these results were:

> [...] the starting-point of a series of attempts to obtain a similar effect by different means, in the course of which a boiled and filtered solution of autolysed yeast-juice was used, in the hope that the products formed by the action of the tryptic enzyme on the proteins of the juice would, in accordance with the general rule, prove to be an effective inhibitor of that enzyme.[1]

As such, Harden and Young filtered yeast-juice and found that neither the eluate nor the filtrate alone could ferment sugar, but could do so if combined.[5] They also boiled each fraction and determined that fermentation required a

DOI: 10.1201/9781003379058-5

heat-labile *ferment* and a thermostable *co-ferment*.[6] Indeed, the authors found that yeast-juice that had been boiled—in an attempt to inactivate the proteolytic enzyme—and filtered increased the degree to which fermentation proceeded when added to yeast; however, they also determined *that the increase in the alcoholic fermentation was not directly dependent on the decrease in the action of the proteoclastic enzyme but was due to some independent cause*.[1] The authors later found that this increase in fermentation was due to the presence of inorganic phosphates (P_i) in the yeast-juice and these results could be replicated by simply adding phosphate to fermenting yeast.[7] Furthermore, they determined that addition of phosphate not only increased the initial burst of fermentation upon its addition, but also increased the total fermentation, mainly as a result of a slow but constant rate of fermentation which followed the initial burst and lasted for hours (Figure 5.1a). They also found

that the amount of CO_2 evolved and alcohol produced were equivalent to the phosphate added. Importantly, the authors filtered the fermentation products and found that nearly all the phosphate was retained on the filter *in a form which is not precipitated by ammoniacal magnesium citrate mixture*, conditions under which organic phosphate would be precipitated.[8] This indicated that the phosphate in question was bound to another molecule rather than present as soluble P_i.

In a subsequent paper, Harden and Young more precisely determined the fate of phosphates during fermentation. As such, they first identified the non-precipitable phosphate as hexosephosphate by performing standard tests for sugars.[6] Importantly, they found that when glucose was absent from the fermenting mixture (a condition under which limited fermentation occurred), free phosphate accumulated relatively quickly, while this phenomenon was delayed when glucose was added (Figure 5.1b). This suggested to them

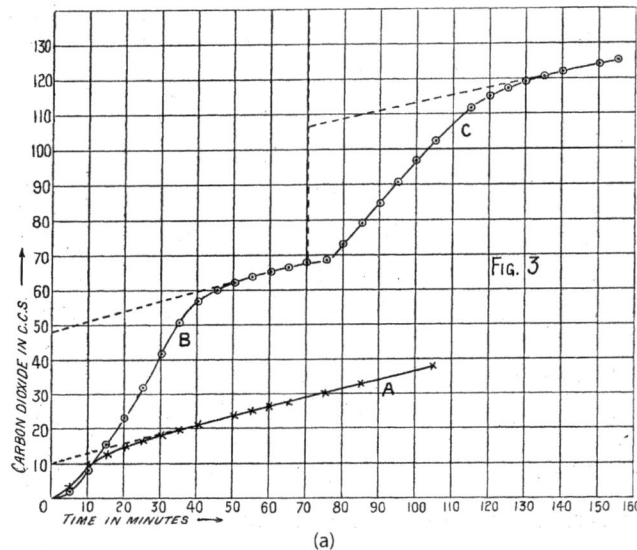

(a)

Experiment.	Material digested.	Time of digestion.	Free phosphate as $Mg_2P_2O_7$ per 25 c.c.
		days.	gramme.
12	(a) Yeast-juice alone	0	0·128
	(b) ,, ,,	2	0·284
	(c) ,, ,,	10	0·283
	(d) Yeast-juice + glucose	2	0·123
	(e) ,, ,,	10	0·255
13	(a) Yeast-juice alone	0	0·024
	(b) ,, ,,	1	0·053
	(c) ,, ,,	2	0·063
	(d) ,, ,,	4	0·069
	(e) Yeast-juice + glucose	1	0·011
	(f) ,, ,,	2	0·025
	(g) ,, ,,	4	0·051

(b)

FIGURE 5.1 (a) Addition of phosphate results not only in an initial burst of fermentation but also in a slow and constant increase. A, fermentation in the absence of added phosphate; B, fermentation in the presence of added phosphate; C, fermentation following a second addition of phosphate after B has come to a steady state rate of fermentation. (b) The presence of glucose delays the accumulation of soluble P_i. P_i accumulated quickly in the absence of glucose, whereas its accumulation was delayed in its presence. Note how free phosphate reached a maximum by day 1 or 2 in the absence of glucose, but by 10 days in its presence. (Figure adapted from ref. [6].)

$(1)\ \underset{\text{Glucose}}{2C_6H_{12}O_6} + \underset{\substack{\text{Source of} \\ \text{Phosphate}}}{2R_2HPO_4} = 2CO_2 + \underset{\text{Ethanol}}{2C_2H_6O} + \underset{\text{Hexosephosphate}}{C_6H_{10}O_4(PO_4R_2)_2} + 2H_2O$

$(2)\ \underset{\text{Hexosephosphate}}{C_6H_{10}O_4(PO_4R_2)_2} + 2H_2O = \underset{\text{Glucose}}{C_6H_{12}O_6} + \underset{\text{Phosphate}}{2R_2HPO_4}$

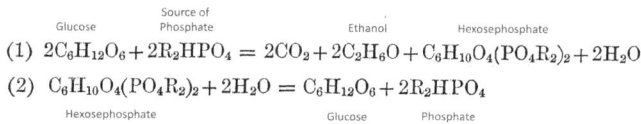

FIGURE 5.2 Equation proposed by Harden and Young for fermentation. Equation (1) does not proceed without the fermenting complex—the ferment and co-ferment. The rate of fermentation following the initial burst when glucose is added is determined by the fact that some of the hexosephosphate formed is subsequently hydrolyzed back to the initial reactants by an unknown reaction (equation 2). (Figure adapted from ref. [6].)

that phosphate might be recycled by recombining with glucose. In addition, the authors found no accumulation of free phosphate if a fermenting mixture was boiled and then returned to conditions favorable for fermentation, which indicated to them that an enzyme was required to cleave the phosphate from hexosediphosphate. Finally, the authors determined that although P_i could significantly increase the rate of fermentation, an excess of the substance was detrimental. Thus, P_i seemed to have a range of concentration at which its effect was beneficial. These experiments led Harden and Young to propose the equation in Figure 5.2 for the process of fermentation, noting that the:

first of these equations does not include the fermenting complex, without which, however, the change does not occur, and it is probable that both the glucose and the phosphate form an intermediate association with this complex, which then breaks down, giving rise to the substances on the right-hand side of the equation, and at the same time regenerating the fermenting complex.[6]

To confirm that an enzyme was responsible for the splitting of hexosephosphate into glucose and phosphate, Harden and Young added hexosephosphate to solutions capable or not of fermenting and assessed the amount of free phosphate and glucose produced after a period of time (Figure 5.3).[9] Results indicated that these products were significantly formed only in the presence of fermentation and when hexosephosphate was supplied, which indicated that they were produced during fermentation and that they originated from hexosephosphate; they proposed the name

hexosephosphatase for the enzyme responsible for the cleavage of hexosephosphate into P_i and glucose. In addition, they showed that these hydrolysis products were themselves capable of fermentation, thereby confirming that these products were recycled during fermentation. They also tested the fermentation products of different types of sugars and found that they all produced the same hexosephosphate sugar. Finally, Harden and Young isolated hexosephosphate and determined that it was a hexosediphosphoric acid (later known by its chemical name, fructose-1,6-bisphosphate; see Figure 5.11 for its chemical structure). From these data, Harden and Young concluded that:

[...] molecules of sugar which are involved in the reaction, of which the equation [in Figure 5.2] [...] may be decomposed into smaller groups, and that the hexosephosphate may be formed by a synthesis from these. As the formation of the hexosephosphate is invariably accompanied by that of an equivalent amount of carbon dioxide and alcohol, [this] explanation appears [...] probable, as it provides a source for the simultaneous production of these substances.[9]

* * * * *

The study of respiration—that is, the oxidation of various molecules resulting in the consumption of oxygen and the release of CO_2—actively began in the 18th century, when the famed chemist Antoine Lavoisier discovered that performing work increased oxygen consumption in both animals and man. However, it was almost another century before scientists attempted to quantify the amount of oxygen consumed and CO_2 produced.[10] Liebig, evaluating the available data, ascribed some of the oxygen consumption to the maintenance of tissues which undergo mechanical stress, such as muscles during contraction, and asserted that:

It signifies nothing what intermediate forms food may assume, what changes it may undergo in the body, the last change is uniformly the conversion of its carbon into carbonic acid, and of its hydrogen into water; the unassimilated nitrogen of the food, along with the unburned or unoxidised carbon, is expelled in the urine or in the solid excrements.[10]

Though a requirement for oxygen in muscle contraction was well accepted by the mid-1850s, it came to light that

Solutions.	Free phosphate as $Mg_2P_2O_7$.			Reducing substance as glucose.		
	Before.	After.	Produced.	Before.	After.	Produced.
1	0	0·0141	0·0141	0	0	0
2	0·0009	0·0030	0·0021	0·0244	0·0275	0·0031
3	0·0009	0·0968	0·0959	0·0244	0·0595	0·0351

FIGURE 5.3 Hexosephosphate is subject to hydrolysis by an enzyme to produce glucose. Solution 1: fermenting solution alone; solution 2: hexosephosphate alone; solution 3: fermenting solution and hexosephosphate. The data represent the amount of $Mg_2P_2O_7$ and glucose measured before and after incubation of each solution.

this phenomenon could also occur in the absence of oxygen: live frogs continued moving normally for a short time in a nitrogenous environment free of oxygen.[10] Berzelius studied the composition of muscle in hunted stags and identified an acid—lactic acid, which was similar to that found in milk—which he was convinced was present in proportion to the extent of exercise performed prior to analysis.[10] Glycogen, a branched polysaccharide composed of strings of glucose (which is used for storage of glucose in the liver and skeletal muscle), was also discovered in muscle around this time and was found, in contrast to lactic acid, to decrease as exercise progressed, one researcher writing that *the glycogen of muscle undergoes incessantly a lactic fermentation, and this in the living animal as in the corpse, is the only transformation of muscle glycogen.*[10]

Thus, while Harden and Young were working out the mechanism of yeast fermentation, Frédéric Battelli and his collaborator, Lina Stern, were studying respiration in tissues and found several molecules that could increase the rate of respiration. Up until the 1900s, tissues were studied directly as isolated from the body, which limited the conditions that could be varied in these investigations. As such, Battelli and Stern contemplated the possibility of developing a cell-free system—akin to what the Buchner brothers had developed with yeast—to better study respiration in tissues and started by optimizing a protocol for emulsions of muscular tissue. To this end, they added finely ground muscle to a number of different solutions—water, blood, or NaCl—to a flask which was *energetically agitated* in a water bath at constant temperature for the duration of the experiment (Figure 5.4).[11,12] Although a few other researchers had attempted to obtain a good cell-free system to study respiration in muscle, Battelli and Stern's method significantly increased the efficiency of the assay, which resulted in much greater rates of respiration and better control over the experiment's conditions. As such, the authors noted major differences in terms of oxygen consumption and CO_2 evolved in various tissues—muscles having not only the greatest respiratory activity, but also the largest variation between different types of muscles within a given organism and in similar muscles between species. In addition, they found that the respiratory activity of muscles decreased rapidly after the death of the animal, a phenomenon which seemed to coincide with the onset of *rigor mortis*. Thus, Battelli and Stern determined that a maximum rate of respiration occurred in red muscle emulsions obtained immediately after death and suspended in blood at a constant temperature of 38–40°C, and gradually decreased thereafter until it nearly stopped or stopped completely.[11,13] In addition, the authors determined that extracts that had been boiled absorbed very little oxygen and that replacing the atmosphere within the flask with pure oxygen significantly increased respiration.[13]

In subsequent papers, Battelli and Stern attempted to study the effects of various molecules on the rate of respiration and found that addition of phosphate to muscle emulsions considerably increased respiration, similar to what Harden and Young found in yeast.[13] They also determined that a number

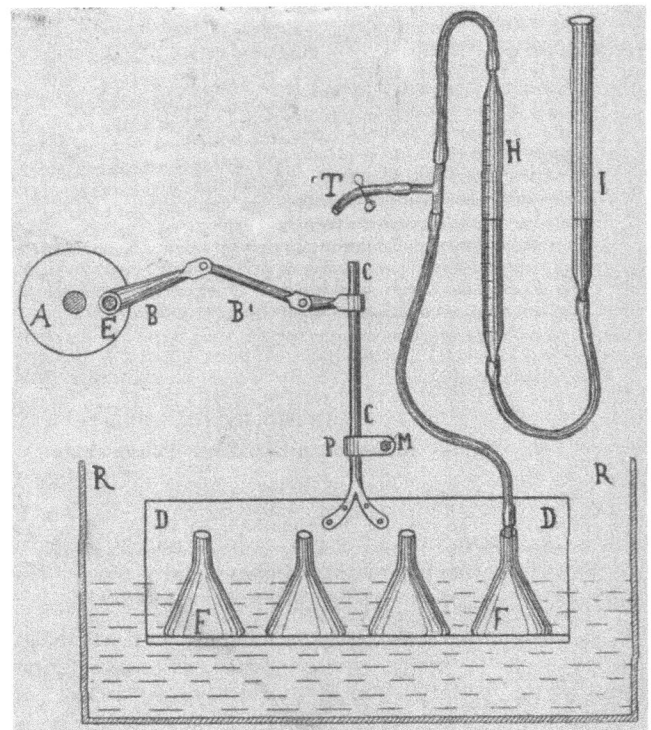

FIGURE 5.4 Battelli and Stern's apparatus for measuring CO_2 evolved by tissue emulsions. Flasks (F) containing the samples were immersed in a water bath (R), whose temperature was maintained constant in a manner not depicted. A wooden board (D)—mounted to another board which supported the flasks—was attached by way of levers to a wheel capable of rotating at 450 rpm (which translates to about 300 shakes of the flasks per minutes). CO_2 evolution was determined by a series manometers connected to a sample flask by a rubber tube.[12]

of reagents, such as arsenic, inhibited respiration of muscle extracts.[14] Finally, they filtered their muscle emulsions using a cloth, and found, much like Harden and Young, that neither fraction permitted respiration on their own, but could once more when recombined.[15] Furthermore, the authors found that the filtrate had a heat-resistant component, whereas the residue had a heat-labile phase. Thus, respiration in tissue emulsions seemed to behave in a similar manner as fermentation of yeast extracts in a number of ways.

In 1910, Battelli and Stern revisited experiments performed by Torsten Thunberg, who had previously reported an increase in oxygen absorption following addition of succinic, fumaric, malic, and citric acids in frog muscle (see Figure 5.5 for chemical formulas); however, he did not ascribe this phenomenon to the oxidation of the acids during respiration.[16] Similarly, Battelli and Stern found very little increase in respiration if the acids were added to muscle emulsions immediately after harvesting, as Thunberg had previously done. However, they determined that under the right conditions, these acids could indeed increase respiration by way of their oxidation, as reported by the amount of oxygen absorbed and CO_2 released.[16] In addition, in contrast to malic, fumaric, and citric acids, succinic acid could

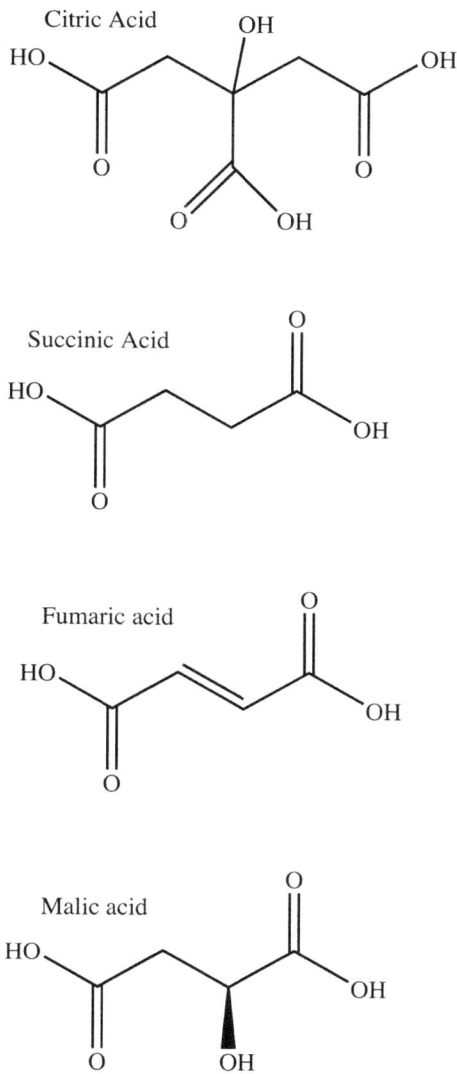

FIGURE 5.5 Chemical structures of acids involved in respiration. Note that fumaric acid is formed by the removal of hydrogens from succinic acid, and malic acid is formed by the addition of a hydroxyl group to fumaric acid.

$$CH_3CH_2OH + Mb + (Pt) \longrightarrow CH_3CHO^- + MbH_2$$

FIGURE 5.6 Redox reaction between alcohols and methylene blue. Alcohols are H^+ donors and Mb^+ is a H^+ acceptor in the absence of oxygen. Platinum (Pt) is required as a catalyst for the reaction to proceed. Mb is blue in color, while MbH_2 is colorless.

increased ability to react with other atoms or molecules—but Weiland instead suggested it occurred as a result of an activation of hydrogen, wherein the oxygen played the part of an acceptor for the activated hydrogen, thereby completing the reaction. Weiland demonstrated this by using methylene blue—a blue solution that changes to a colorless solution when bound by hydrogen (i.e., when reduced)—as a hydrogen acceptor in the absence of oxygen.[18] As such, methylene blue only becomes colorless when hydrogen is bubbled through it in the presence of platinum (which is required as a catalyst). Thus, Thunberg used the known oxidation of alcohols to aldehydes to study this phenomenon and determined that this reaction could indeed proceed in the absence of oxygen if methylene blue was used as a hydrogen acceptor (Figure 5.6). This indicated that what had previously been considered an oxidation reaction—that is, O_2 being activated by taking a H^+ from the alcohol—was in fact a reduction of alcohol—that is, the loss of an activated hydrogen proton—which left its sole electron behind and in which oxygen in the environment acted as an acceptor; Thunberg termed this phenomenon *dehydrogenation*.

As such, Thunberg set out to determine if the oxidation of succinic acid to fumaric acid involved the activation of its hydrogen by using Weiland's methylene blue technique. To that end, he filled two tubes with equal quantities of methylene blue and muscle extracts, adding succinic acid to only one of the tubes, evacuated them to remove the oxygen, and incubated the samples for a period of time. What he found was that the tube with succinic acid completely decolorized methylene blue, which indicated that it had lost one or more hydrogens in the process of being converted to fumaric acid.[18] Furthermore, using this new technique, Thunberg found that the mechanism of hydrogen activation was a widespread phenomenon in cells and this method became known as the "Thunberg technique" which used "Thunberg tubes".[18] From these experiments and others, he concluded that all foodstuffs were hydrogen donors and consequently, hydrogen was the main fuel of the body, stating that:

restore respiration in tissues that had completely lost their ability to respire, and to do so to the same extent as before respiration ceased.[17] Therefore, the authors suggested that the use of succinic acid was a better method for determining the full potential of tissues for oxidation and respiration, noting a *narrow parallelism* between a tissue's power to oxidize succinic acid and its respiratory activity.

* * * * *

In 1920, Thunberg developed a new, more sensitive, and more convenient method to test the oxidative potential of a compound, based on the emerging concept of reduction/ oxidation (or redox) reactions developed by Heinrich Weiland a few years earlier.[18] Prior to this, redox reactions were thought to occur as a result of the activation of oxygen— that is, oxygen was brought to a state whereby it had an

I consider the catabolism of the food stuffs to take place in a series of continuous de-hydrogenations, carried out by a series of dehydrogenases […] When oxygen is present the hydrogen split off by the dehydrogenases is transported to this oxygen with formation of water.

This oxidative catabolism of the food stuffs is characterised not only by this series of dehydrogenations but also by two other processes, viz. addition of water and the splitting off of carbon dioxide.[18]

In 1924, Alfred Fleisch was studying differences between normal muscle and cancer tissues when he found that cyanide, which was known to inhibit respiration—as reported by its inhibition of oxygen uptake—could nevertheless induce the reduction of methylene blue; therefore, he concluded that Thunberg's methylene blue technique was not equivalent to oxygen consumption.[19] As such, he investigated this phenomenon further as related to some of the known components of the respiratory pathway (see Figure 5.5). He found that addition of citric acid to muscle extracts produced a strong acceleration of the reduction of methylene blue and of oxygen uptake even in the presence of cyanide, whereas this poison completely suppressed the action of succinic acid on oxygen uptake. These results suggested that the oxidation of citric and succinic acids was catalyzed by different enzymes, as had previously been proposed by Battelli and Stern. In addition, Fleisch determined that addition of methylene blue to an incubation mixture of muscle extracts, succinic acid, and cyanide, could completely rescue the inhibition of oxygen uptake by cyanide, though with slower kinetics (Figure 5.7).[19] He also found that direct oxidation of succinic acid increased after the muscle was thoroughly washed, but

that the reduction of methylene blue in the presence of cyanide decreased, which indicated *that the facility of hydrogen transport to methylene blue is diminished by thorough washing but not transport to oxygen.* To explain this phenomenon, Fleisch suggested that *a special hydrogen transport factor is responsible for the transport of hydrogen* [from a molecule] *to methylene blue.*[19]

Fleisch also discussed the activation of hydrogen versus oxygen as related to the oxidation of carbohydrates, stating that *it seems to be extraordinary that this oxidation of succinic acid is inhibited by cyanide and restored by methylene blue.* He suggested that an enzyme, which he named succinodehydrogenase, catalyzed the oxidation of succinic acid to fumaric acid by removing two hydrogen atoms from former and transferring them to a hydrogen acceptor, whether oxygen during aerobic respiration or methylene blue in the absence of oxygen such as when using a Thunberg tube.[19] Fleisch proposed the equation in Figure 5.8 to explain these results, stating that reactions #1 and #2 were not sensitive to cyanide and therefore allowed the removal hydrogens from

Table V. $0 \cdot 2$ *g. of muscle* $+ 0 \cdot 3$ *cc.* $M/1000$ *methylene blue or* $0 \cdot 3$ *g. of sarcoma* $+ 0 \cdot 2$ *cc.* $M/1000$ *methylene blue, with in each case* $0 \cdot 2$ *cc. of* $M/2$ *phosphate,* p_H $7 \cdot 6$.

	Muscle.			Sarcoma			
Number of experiment	1	2	3	1	2	3	4
Succinic acid $M/5$	—	1	0·1	—	—	0·1	0·1
HCN $M/100$	—	—	0·2	—	0·3	—	0·3
H₂O	0·5	0·4	0·2	0·6	0·3	0·5	0·2
Beginning of decolorisation in minutes	*	4	4	33	33	16	16
End of decolorisation		18	18	83	200	37	83

* means more than 70 minutes.

(a)

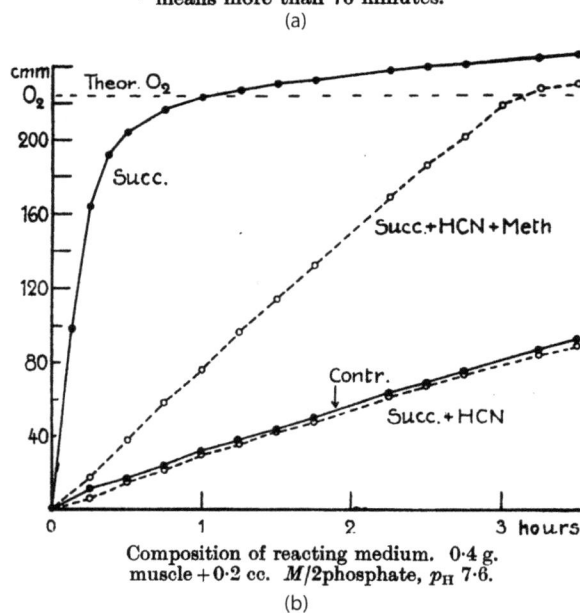

Composition of reacting medium. $0 \cdot 4$ g. muscle $+ 0 \cdot 2$ cc. $M/2$ phosphate, p_H $7 \cdot 6$.

(b)

FIGURE 5.7 Inhibition of succinic acid oxidation by cyanide can be rescued by addition of methylene blue. (a) Addition of succininc acid to muscle extracts decolors methylene blue (column 2), but this reaction is unaffected by cyanide (HCN, column 3). (b) In contrast, cyanide inhibits the uptake of oxygen following addition of succinic acid, and methylene blue partially rescues this phenotype. (Figure adapted from ref. [19].)

1) $$CH_2.COOH \atop CH_2.COOH + Enzyme = {CH.COOH \atop CH.COOH} + (Enzyme\ H_2)$$
(a)

(2) $(Enzyme\ H_2) + methylene\ blue = Enzyme + (methylene\ blue\ H_2)$
(b)

(3) $2\ (Enzyme\ H_2) + O_2 + X = 2\ Enzyme + 2H_2O + X$
(c)

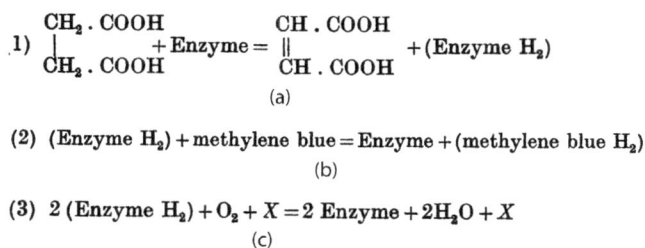

FIGURE 5.8 Fleisch's proposed reactions for the oxidation of succinic acid. Two hydrogens are removed from succinic acid by succino-dehydrogenase, yielding fumarate (equation 1). In the Thunberg tube (i.e., in the absence of oxygen), the enzyme transfers the hydrogens to methylene blue (equation 2), whereas oxygen is the acceptor when it is present (equation 3). However, oxygen also needs to be activated by an unknown catalyst X for this to occur.[19]

succinic acid in the presence of methylene blue, in which case a *hydrogen transport factor* played a crucial role. On the other hand, since reaction #3 was inhibited by cyanide, Fleisch suggested that a catalyst X must exist for the activation of oxygen, and concluded that both hydrogen and oxygen must be activated during the physiological oxidation of succinic acid. As he explained:

> Since cyanide is known to be a specific inhibitory substance for oxidising enzymes or oxygen activators we may believe that X is an oxygen activator. We come then to the view, that the hydrogen of (enzyme H_2) cannot be oxidised by molecular unactivated oxygen [...] If in an aerobic experiment the oxygen catalyser X, and therefore the oxidation of succinic acid, is inhibited by cyanide, there exists a possibility of restoring this oxidation by adding methylene blue. Then the reactions (1) and (2) proceed in spite of the presence of cyanide, and the methylene blue is reduced. Since molecular oxygen is present the autoxidisable methylene white formed is continuously re-oxidised. Thus methylene blue replaces the catalyser X; it forms a substitute for an oxidising enzyme.[19]

* * * * *

In 1912, Embden borrowed the technique developed by Buchner and used by Harden and Young to extract yeast-juice and applied it to dog muscle: he ground up minced muscle with sand and squeezed out the juice using a Buchner press. As such, he found an accumulation in lactic acid upon incubation of the press juice at 40°C for 2 hour and suggested the name "lactacidogen" (meaning lactic acid generator, later renamed fructose-6-phosphate) for its precursor, which he found contained phosphorus, nitrogen, a pentose sugar, and adenine.[20] Embden was also inspired to look for changes in inorganic phosphate concentrations in muscles since it had been known since the 1870s that phosphate accumulated in urine after exercise, and no doubt also by Harden and Young's findings about the importance of P_i in yeast fermentation. Indeed, Embden and his group found that P_i increased proportionally to lactic acid formation.[20] In 1913, Carl Neuberg found that pyruvic acid was

fermented to CO_2 and ethanol in a variety of yeast strains and suggested that it may be an intermediate in carbohydrate metabolism.[21] Following these experiments and much of the work performed by Harden and Young, Neuberg proposed a cyclical mechanism for the metabolism of carbohydrates whereby pyruvic acid was reformed in the process.[21] Though generally wrong, this scheme introduced for the first time the idea of a cyclical characteristic to respiration, though not the one suggested by Neuberg.

The development of cell-free muscle extracts by Battelli and Stern greatly increased the progress of discoveries in the glycolytic pathway and Otto Meyerhof started a seminal series of studies involving muscle metabolism during World War I. He performed experiments using frog or rabbit muscle extracts and determined that lactic acid was produced upon addition of a number of different sugars and that it was significantly increased with addition of phosphate or hexosephosphate.[20] He also found that much more lactic acid accumulated in muscle homogenates when oxygen was absent and that this accumulation decreased if oxygen was added after the build-up of lactic acid had begun, though not proportionately to oxygen consumed.[22] However, he did find that the decrease in glycogen was proportional to the build-up of lactic acid during anaerobic respiration, while it was proportional to oxygen uptake during aerobic respiration, the latter of which occurred without an accumulation of lactic acid. These results were wrongly interpreted as meaning that glycogen breakdown to lactic acid occurred during both aerobic and anaerobic respiration, but that lactic acid was resynthesized to glycogen during the former.[4]

Meyerhof also found that both free sugar (i.e., glucose) and polysaccharides, such as glycogen, were equally metabolized when he used fresh rabbit extracts, but that storage of extracts significantly decreased the potential to metabolize the former while it did not affect the latter. However, he found that the ability for glycolysis of free sugars was regained, and in fact exceeded that of glycogen, when a yeast component with the properties of an enzyme was added, which indicated that yeast extracts could bring about rapid phosphorylation of hexoses in muscle extracts (Figure 5.9).[23,24] Thus, Meyerhof suggested that *a factor essential for glucose esterification is more concentrated in yeast*, and proposed the name hexokinase since it was necessary for the phosphorylation of free hexoses.[23]

* * * * *

As far back as the late 1800s, Richard Altmann had observed cellular structures which he called "bioblasts", using microscopy and concluded that *they were 'elementary organisms' living inside the cells and carrying out vital functions.*[25] These structures were later renamed mitochondria from the Greek "mitos" (thread) and "chondros" (granule), given their appearance during spermatogenesis, and were soon shown to be ubiquitous in multicellular organisms. Plant researchers eventually came to the conclusion that mitochondria served as *a structural expression of the reducing substances concerned in*

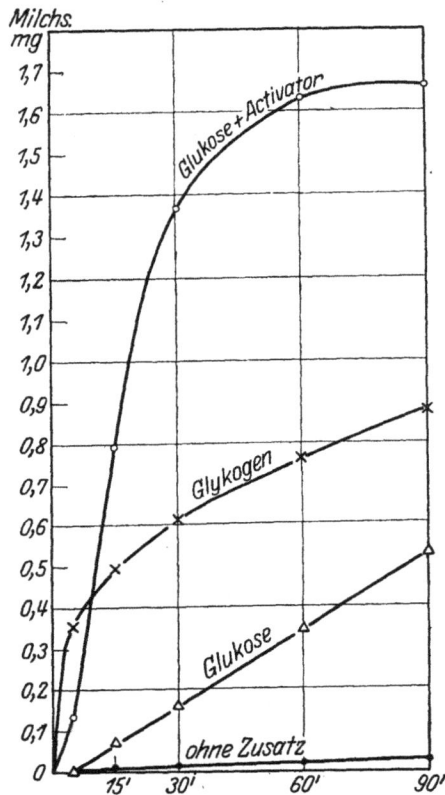

FIGURE 5.9 Increase in glycolytic activity of inactivated muscle extracts by addition of yeast extracts. Glucose, glycogen, or glucose + a yeast extract (activator) were added to a reaction mixture and the amount of lactic acid produced was assessed.[24] (Reprinted by permission from Springer Nature Customer Service Centre GmbH: Springer Nature, Die chemischen Vorgänge im Muskel und ihr Zusammenhang mit Arbeitsleistung und Wärmebildung by Otto Meyerhof Copyright © 1930.)

cellular respiration.[25] In the late 1800s, Charles MacMunn described a respiratory pigment in pigeon pectoral muscle which produced a specific spectrum of light when in a reduced stated, whereas the bands disappeared when oxidized.[26] MacMunn named the substance "myohaematin" or "histohaematin" and proposed that it was a derivative form of a family of molecules, but an entity distinct and separate from hemoglobin. Others, however, including the voice of chemistry at the time, Hoppe-Seyler, disagreed, and from this a consensus was formed that MacMunn's molecule was the same as ordinary hemoglobin. However, in 1925, David Keilin investigated a substance involved in respiration while studying respiration in parasitic insects and worms, and concluded that:

[...] the pigment myo- or histohaematin not only exists, but has much wider distribution and importance than was ever anticipated even by MacMunn [...] Considering that this pigment is not confined to muscles and tissues, but exists also in unicellular organisms, and further, that there is no evidence that it is a simple haematin in the proper sense of the word [...] I propose [...] to describe it under the name of *Cytochrome*, signifying merely "cellular pigment" [...]

The number and wide range of systematic distribution of the species which show this pigment clearly [...] is so great that it may safely be concluded that cytochrome is one of the most widely distributed respiratory pigments.[26]

In his studies, Keilin used thoracic bee muscles, which he obtained by freezing bee specimens and quickly thawing them out, after which he cut off the head and abdomen, *and by compressing the thorax laterally with the fingers the thoracic muscles are expelled in one mass through the anterior opening of the thorax.*[26] The muscles were then compressed between a slide and a coverslip, and mounted on a microspectroscope to determine the sample's light spectrum characteristics (Figure 5.10). As such, Keilin found that the muscle's spectrum disappeared when it was oxidized by bubbling oxygen through it. Furthermore, he found a number of substances—cyanide and pyrophosphate, and alcohols, formaldehyde, and acetone—that could inhibit either the oxidization or the reduction of cytochrome, respectively. Since the studies thus far could not determine whether cytochrome was normally found in the oxidized or reduced state *in vivo*, Keilin set out to address this question but was limited by the fact that excision of tissue quickly oxidized the sample. To resolve this, he used the common wax-moth that was bred in his laboratory. He removed the scales from its thorax, attached it to a slide by the ventral surface with gum Arabic—the moth still being alive at this point—and examined the surface using a microspectroscope.[26] As such, he observed that moths did not show any discernible spectrum when they were calm and quiet, but in those which thrashed and

[v]ibrated their wings in efforts to detach themselves from the slides, the bands of cytochrome gradually appeared, band a being very clear, and bands b and c appeared as almost fused into one band.[26]

These bands disappeared when the moths settled down. He also observed the appearance of a spectrum when a moth with oxidized cytochrome was exposed to cyanide (at which point the *insect became motionless*), but the bands disappeared again when fresh air was supplied and the

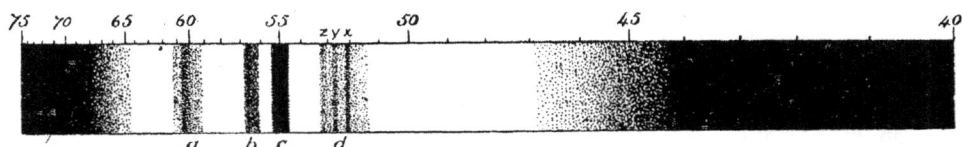

FIGURE 5.10 Light spectrum of bee thoracic muscles. Bands for cytochromes are indicated a–d.[26]

insect regained consciousness. The above results indicated that cytochrome was primarily present in the oxidized state, but some reduced cytochrome was also present in low concentrations.

Keilin also determined that cytochrome was present in the highest concentrations in muscles that were most used. For example, in the thoracic muscle of male winter moths, which are equipped with well-developed wings, cytochrome was easily identifiable, whereas it was almost absent in the female of the species which do not fly. Finally, Keilin suggested that the spectrum bands observed were derived from three separate components of cytochrome, which were named cytochrome *a*, cytochrome *b*, and cytochrome *c*.

* * * * *

In 1927, Embden and Zimmerman discovered an adenylic acid (adenosine monophosphate, AMP) in muscle that was different from that usually found in nucleic acids, in that it was readily deaminated by an enzyme. They also found that the ammonia released by deamination was the principal source of nitrogen in muscle. However, although the AMP they found was later shown to be an artifact, this nonetheless led to the discovery of one of the most important molecules in the cell: adenosine triphosphate (ATP).[27] Two groups were struck by Embden and Zimmerman's discovery and investigated it further. For their part, Fiske and Subbarow used muscle filtrate that was devoid of proteins and treated it with calcium chloride, unexpectedly finding that a purine nitrogen was precipitated.[28] They subsequently purified the precipitated purine derivative by repeated precipitation with mercuric acetate, calcium chloride, and silver nitrate, and found that it accounted for most of the nitrogen and phosphate in muscle that was not bound to any other known molecule. In addition, the authors determined that the isolated molecule was composed of adenine, a carbohydrate, and surprisingly, three phosphates. They also found that two of the phosphates were easily hydrolyzed and proposed that this *fact is doubtless sufficient to explain why Embden and Zimmerman obtained a nucleotide [...] which still retains the one resistant phosphoric acid group.*[28] The same year, Fiske and Subbarow proposed that creatine phosphate was the main source of energy during muscle contractions.[29] Consistent with this, Lundsgaard found that the breakdown of creatine phosphate was proportional to the extent of muscle contractions.[29] It was then that Lipmann first proposed the idea of a high-energy bond, stating that the *energy utilized in the mechanical set-up of muscles under all circumstances was derived from energy-rich phosphate bonds*, and specifically, that of creatine phosphate, which was *supplied constantly by glycolytic or oxidative foodstuff disintegration.*[29]

However, in 1934, Lohmann performed experiments that indicated that the primary source of energy may not come from creatine phosphate after all, but rather from ATP. He found that dialyzed muscle extracts could not efficiently hydrolyze creatine phosphate and suggested instead that

creatine phosphate was, in fact, hydrolyzed to make ATP; thus, he formulated the following equation for the splitting off of phosphate from phosphocreatine:

$$ATP = adenylic\ acid + 2H_3PO_4$$
$$Adenylic\ acid + 2\ phosphocreatine = 2\ creatine + ATP$$

Supporting the energy-yielding role of ATP in this reaction, Lohmann found that (1) creatine phosphate was only dephosphorylated when ATP was added to a mixture of dialyzed or inactivated extracts; (2) equation 2 proceeded only if ATPase (an enzyme which hydrolyzes ATP to ADP and P_i) was inactivated; and (3) equation 1 only proceeded under conditions where equation 2 was inhibited.[21] Since it was also known that phosphocreatine was produced at the expense of ATP, these findings also suggested that equation 2 was reversible and this was indeed shown soon after. Further studies by Lohmann revealed the chemical structure of ATP and demonstrated a step-wise dephosphorylation of the molecule, leading to the common designation of adenosine mono-, di-, and tri-phosphate used today.[21,29] To visually differentiate the energy-rich from the energy-poor bonds, Lipmann introduced the "squiggle" mark, such that the energy-poor bond in AMP was indicated as ad-ph, whereas the energy-rich phosphate bonds in ADP and ATP were indicated as ad-ph~ph and ad-ph~ph~ph, respectively.[29]

* * * * *

Much work was done between the time the Buchners discovered cell-free fermentation in the 1890s and the complete elucidation of the molecules and enzymes involved in the processes of yeast fermentation and glycolysis. And given the sheer volume of research that was required to achieve this, it is beyond the scope of this book to review the many experiments involved in elucidating the dozen or so reactions involved in these pathways. However, suffice it to say that it was eventually determined that the same enzymes were involved in fermentation and in glycolysis. As such, when free glucose is available, these processes begin with the phosphorylation of glucose—derived from the digestion of nutrients—to glucose-6-phosphate (G6P) by hexosekinase, which gives it a negative charge and traps it inside the cell (Figure 5.11). This step is required since glucose is membrane-permeable and without phosphorylation, it would freely exit cells before it can be processed. In contrast, glycogen—a molecule composed of chains of glucose molecules which is used as glucose storage in the liver and muscles—is used in the absence of free glucose (i.e., when blood sugar levels are low) by systematically phosphorylating terminal glucoses, which break off the glycogen chain, and then converting the free glucose phosphates to G6P. In either case, hexokinase converts G6P into fructose 6-phosphate, and then phosphofructokinase adds an additional phosphate to make fructose 1,6-bisphosphate.

This half of the pathway is known as the preparatory phase and it is energy dependent, with a cost of 2x ATP molecules per molecule of glucose, which are used as a

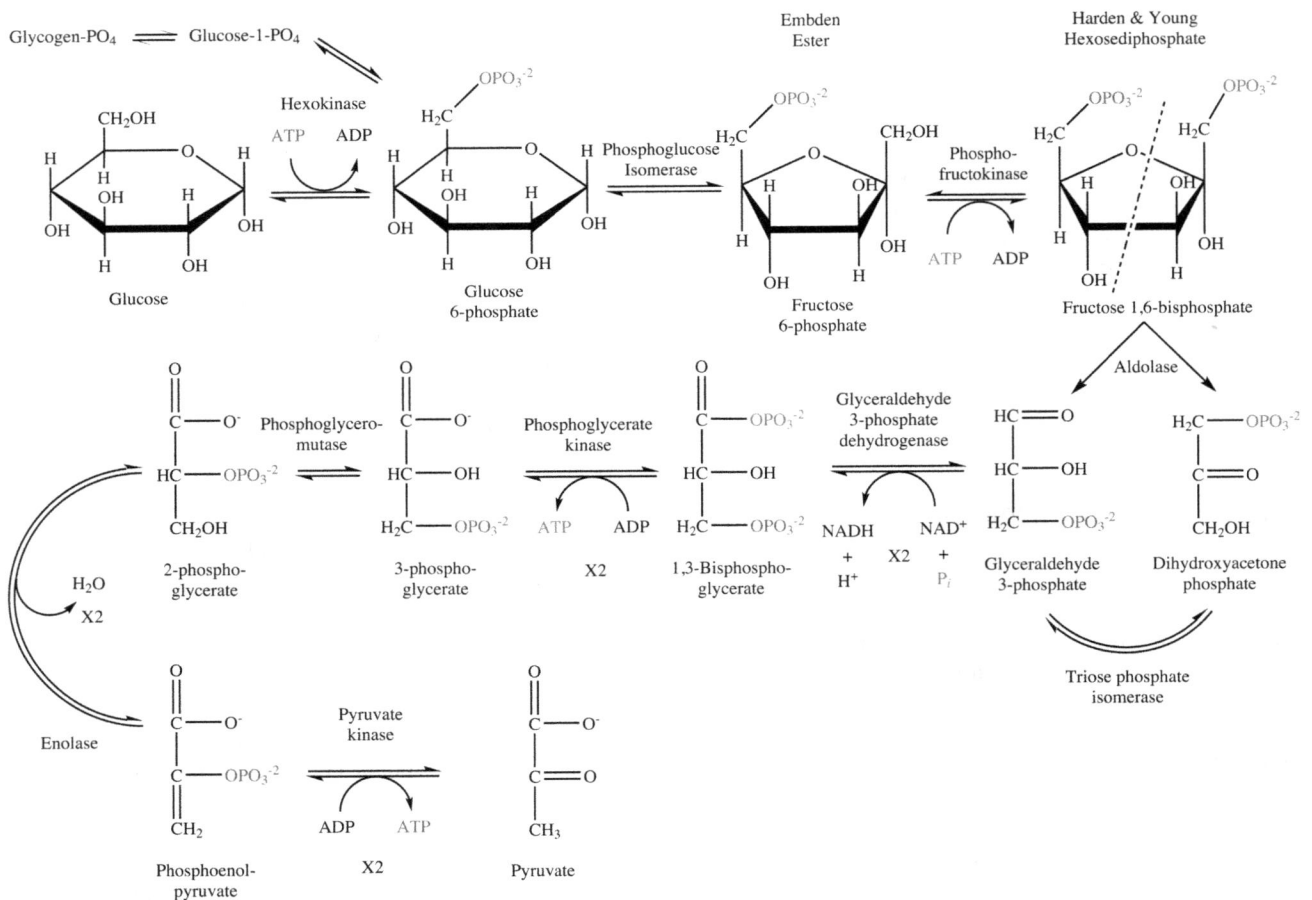

FIGURE 5.11 Reactions involved in glycolysis. If free sugar is low, such as during fasting, glucose molecules in glycogen break off after their phosphorylation by glycogen phosphorylase, yielding glucose-1-phosphate, which then undergoes another chemical reaction which moves the phosphate from the 1-carbon to the 6-carbon, yielding glucose-6-phosphate (G6P). On the other hand, in the presence of free glucose—such as after a meal—glucose is phosphorylated directly to G6P by hexokinase. Either way, G6P undergoes a number of reactions which converts it to pyruvic acid.

source of phosphate in these phosphorylation events. The next section is the pay-off phase and begins with the cleavage of fructose 1,6 bisphosphate into two three-carbon molecules called glyceraldehyde 3-phosphate. A series of enzymatic reactions then convert this molecule into pyruvate, with the production of 2x ATP molecules and 1x of another high-energy molecule, NADH (we will talk more about this molecule in Chapter 9). In the absence of oxygen, the end product of this pathway, pyruvic acid, is converted to waste products: lactic acid in animals and CO_2 and alcohol in yeast during fermentation. In the presence of oxygen (as is usually the case in animals), pyruvate is used in the next pathway, the Krebs cycle, to generate most of the energy required by the cells (see next section and Chapter 9).

* * * * *

Hans Krebs had studied medicine and was a practicing doctor, but research was where his heart was and he therefore sought and landed a position in Otto Warburg's laboratory in 1926, learning the techniques involved in studying sugar metabolism and learning how to do research.[30] Unfortunately, Warburg did not see Krebs' potential as a researcher and told him that he would not help him get a post after he was to leave the laboratory in 1930. Therefore, Krebs secured a position as a clinician and conducted research in his spare time, discovering the ornithine cycle—a pathway involved in making urea (a metabolic waste product found in urine)—from CO_2 and ammonia with the assistance of one of his medical students.[30] In 1933, after working as a clinician for a few years, Krebs lost his post in the wake of Hitler's election to power since Jews were forbidden to enter university grounds and he resettled in England the same year.

As a new investigator at the University of Sheffield in 1937, he and his graduate student, William Johnson, were studying how food was metabolized to CO_2 and water to produce energy. At this time, very little was known about what takes place during aerobic respiration beyond glycolysis, but several reactions had been shown to take place and a tentative pathway was put forth by Thunberg, though little experimental evidence fully supported it.[31] Albert

Szent-Györgyi had determined that succinate, fumarate, malate, and oxaloacetate (four-carbon molecules) accelerated respiration, while Carl Martius and Franz Knoop showed that citrate (a six-carbon molecule) was rearranged to isocitrate and then to α-ketogluterate, which, in turn, was known to form succinate. These results, published as Krebs was completing a series of experiments on respiration, helped him interpret his own results and allowed him to outline the main steps in oxidative respiration.[31]

Krebs and Johnson's initial finding was that the addition of citrate to minced pigeon muscle accelerated and prolonged the consumption of oxygen to an extent that could not be justified merely in terms of the complete oxidation of citrate.[32] This phenomenon was also found to be dependent on other substrates being present in the reaction and was further increased by addition of glycogen or fructose 1,6-bisphosphate. However, the authors determined that citrate was not consumed nor produced in these reactions, which suggested that it was recycled. Thus, the authors attempted to find conditions which resulted in the accumulation of citrate, which they found in the poison, arsenite, and in malonate, both known inhibitors of respiration.[32] In addition, oxygen consumption measurements indicated that the oxidation of citrate was not complete under these conditions, so they looked for intermediate products and found that *large quantities of α-ketoglutaric acid* [and succinic acid] *are present in those suspensions in which citric acid was oxidised in the presence of arsenite.*

The authors therefore proposed the following pathway to account for the oxidation of citric acid:

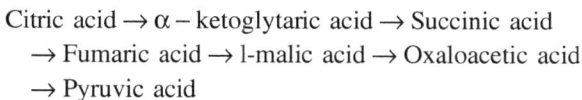

Citric acid → α – ketoglytaric acid → Succinic acid
    → Fumaric acid → l-malic acid → Oxaloacetic acid
    → Pyruvic acid

and concluded that:

If it is true that the oxidation of citric acid is a stage in the catalytic action of citric acid then it follows that citric acid must be regenerated eventually from one of the products of oxidation. We are thus led to examine whether citric acid can be resynthesised from any of the intermediates of the citric acid breakdown.[32]

To test their hypothesis, the authors performed a simple experiment in which they incubated pigeon muscle extracts with oxaloacetate (a four-carbon molecule) in the absence of oxygen and determined that citric acid (a six-carbon molecule) was indeed formed. However, this experiment also brought up the question of where the extra two carbons in citric acid were derived, and suggested that these might come from glycogen or a carbohydrate since addition of these substances were known to increase respiration; he named the unknown molecule which donates the two carbons, "triose", but left it open *whether triose reacts as*

such or as a derivative for example as a phosphate ester, or pyruvic acid or acetic acid.[32] Thus, citric acid was converted to oxaloacetic acid through a number of oxidation-reduction reactions, the latter of which was converted back to citric acid by addition of three carbons from a then-unknown three-carbon molecule (Figure 5.12a).

The remaining gap in this pathway, that is, the formation of citric acid from oxaloacetic acid and "triose", came primarily from the researches of two investigators: Fritz Lipmann and Severo Ochoa. These groups had shown a requirement for phosphate in the oxidation of pyruvate, the end product of glycolysis, but no phosphorylated intermediate could be detected. And although *phosphopyruvic acid had been excluded [...] evidence suggested that acetyl phosphate might be the intermediate.*[33] Lipmann found that acetyl phosphates were rapidly broken down under conditions normally used to isolate phosphorylated proteins—that is, in magnesia mixtures at alkaline pH; thus, he optimized the method for isolation of such molecules and determined they were stable at neutral pH in a calcium solution. This allowed him to efficiently isolate the unknown intermediate from contaminating inorganic phosphate and proteins, and he concluded that *the calcium precipitation method can be used to determine acetyl phosphate by difference.* As such, Lipmann incubated bacterial extracts with or without pyruvate for a period of time and found that (Figure 5.12b):

[...] large amounts of inorganic phosphate disappear [...] and are nearly equivalent to the extra oxygen consumed [...] The organic phosphate so formed behaves like acetyl phosphate; *i.e.*, it is split readily and completely by the acid required to determine phosphate directly.[33]

He later successfully isolated the intermediate in question and confirmed its identity as an acetyl phosphate derived from pyruvate.[34]

In the late 1930s, Klein and Harris had adapted a technique whereby sulfanilamides were used as an acetylation trap to identify acetylation reactions (Figure 5.13a) and results indicated that acetylation was somehow involved in the process of respiration.[35] In the 1940s, Lipmann extended these findings when he noticed that pigeon liver extracts lost their ability to acetylate sulfanilamide upon autolysis, but regained this ability after addition of boiled extracts from a number of different organs, indicating that a heat-stable substance catalyzed the transfer of an acetyl moiety.[36] Lipmann and his group subsequently isolated the substance responsible for this activity, and it was found to contain pantothenic acid linked to an adenylic acid derivative via a phosphate bridge, and an amino acid derivative; he named it coenzyme of acetylation (coenzyme A or CoA) (Figure 5.13b).[37,38] It was soon confirmed that CoA was a component involved in enzymatic acetylation and pyruvate metabolism in bacteria and in animals.[39–41] Then, Lipmann's and Ochoa's groups both showed that CoA was an essential component involved in the presentation of an "activated" acetyl radical into the citric acid cycle in the

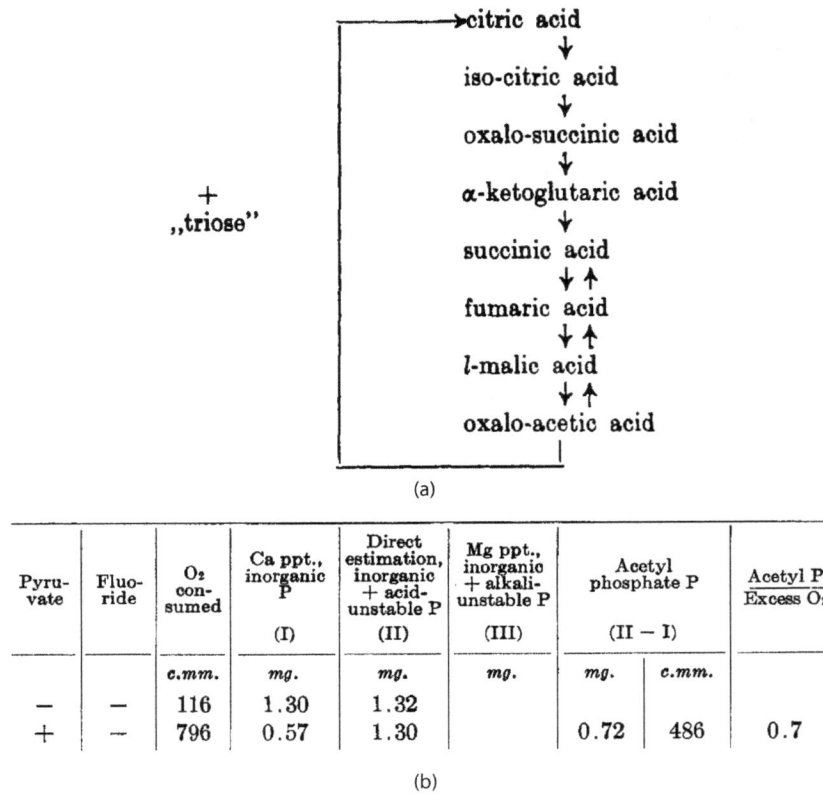

(a)

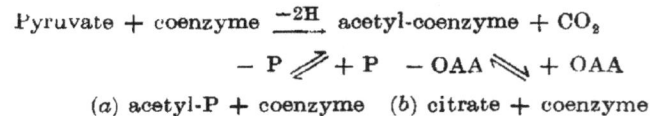

(b)

FIGURE 5.12 Pyruvate requires an intermediate form to enter the citric acid cycle. (a) The citric acid cycle as known by the late 1930s.[32] (b) Pyruvic acid was oxidized in a bacterial enzyme solution and the amount of acetyl phosphate was assessed by calcium precipitation of phosphate. Column (I) shows that a significant amount of P_i disappears with the addition of pyruvate, concomittant with an increase in O_2 consumed. The organic phosphate formed—acetyl phosphate—is shown by substraction of calcium-precipitated P_i from total phosphate (column (II)).[33] (Figure (a) adapted with permission from John Wiley and Sons, Krebs H, Johnson W. The role of citric acid in intermediate metabolism in animal tissues, FEBS Lett. Copyright © 1980 Published by Elsevier B.V.)

form of an acetyl-CoA–enzyme complex.[42–44] Specifically, Stern and Ochoa published a short communication in 1949 showing that CoA was required for the formation of citrate in pigeon liver extracts (Figure 5.14a).[45] The next year, Novelli and Lipmann extended these findings using yeast, *Escherichia coli*, and pigeon liver extracts and proposed a new reaction that differed from previous ones, *in that it represents a reaction in which the acetate is only activated in the methyl or tail group* (Figure 5.14b).[43]

Around the same time, Ochoa's and Lipmann's groups had both found evidence for reactions involving an acetyl phosphate intermediate in pyruvate metabolism that was dependent on the presence of CoA;[39,43,45,46] however, this point had not yet been directly proven. Thus, with an aim to directly test the requirement of acetyl phosphate in the metabolism of pyruvate, Strecker et al. used variations of C^{13}- and C^{14}-labeled compounds added to a fermentation reaction to test their incorporation into pyruvic acid, proposing that one of three reactions would occur (Figure 5.15a).[47] These experiments showed that acetate was not converted into an intermediate, acetyl phosphate or otherwise, which was subsequently incorporated into pyruvic acid (Figure 5.15b). However, the isotope in the final formate was diluted by almost 60% with a

concomitant incorporation into pyruvate (exchange between formate and pyruvate was already known to occur).[48] Thus, reaction #3 clearly proceeded and these results suggested the involvement of a new compound in the metabolism of pyruvic acid, shown here as CH_3COX.[47] Ochoa's group then proposed the following scheme for the metabolism of pyruvate:

$$\text{Pyruvate} + \text{coenzyme} \xrightarrow{-2H} \text{acetyl-coenzyme} + CO_2$$

$$-P \nearrow \quad + P \quad -OAA \searrow \quad + OAA$$

$$(a)\ \text{acetyl-P} + \text{coenzyme} \quad (b)\ \text{citrate} + \text{coenzyme}$$

and suggested that

[t]he primary dehydrogenation reaction may result in the formation of an acetyl-enzyme or acetyl-coenzyme intermediate which, in the presence of another enzyme, may be cleaved either (a) by phosphate with formation of acetyl phosphate, or (b) by oxalacetate with formation of citrate.[42]

The latter scheme was supported by their findings that bacterial extracts could *form citrate at a good rate **in the almost complete absence of orthophosphate**, provided the [...] system is supplemented with oxalacetate and condensing*

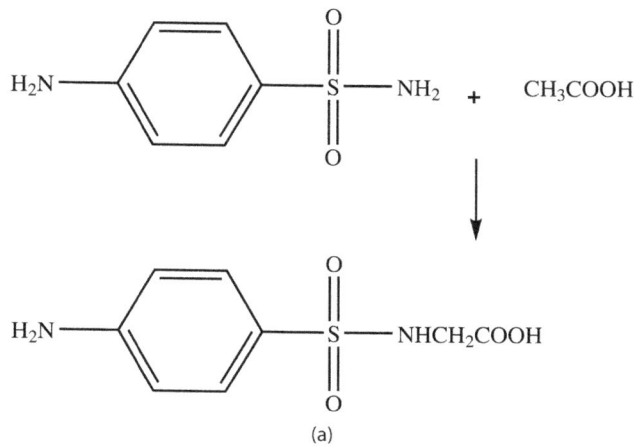

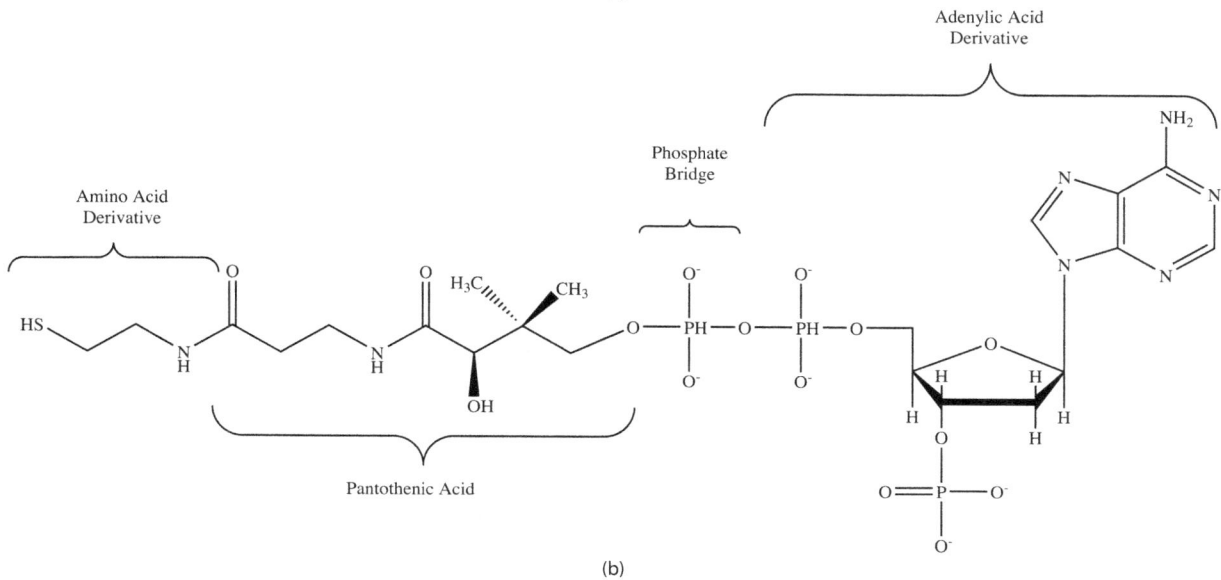

FIGURE 5.13 (a) Sulfanilamides react with acetyl groups to yield stable molecules. (b) Chemical structure of co-enzyme A.

	Citrate found					
	Complete	**No Co A**	**No ATP**	**No Mg^{++}**	**No OAA**	**No acetate**
With acetate............	0.86, 1.26	0.06	0.09	0.57	.0	0.06

(a)

(b)

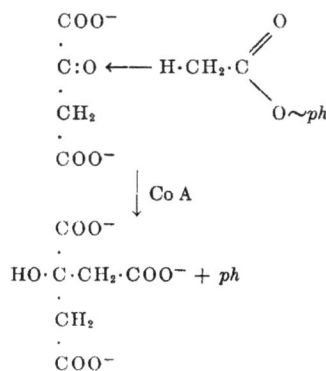

FIGURE 5.14 (a) CoA is required for citrate formation. Removing CoA from a reaction mixture prevents the formation of citrate by an unknown enzyme. (b) Novelli and Lipmann's new reaction for the synthesis of citrate. The "tail" group of acetate is activated by CoA which catalyzes a reaction with OAA.[43] (Figure (a) adapted from ref. [45].)

$$
\begin{array}{ll}
(1) & \text{HC}^{14}\text{OOH} + \begin{array}{c} \text{C}^{13}\text{H}_3\cdot\text{COOH} \\ \text{and} \\ \text{CH}_3\cdot\text{C}^{13}\text{O}\cdot\text{OP(OH)}_2 \end{array} \rightarrow \text{C}^{13}\text{H}_3\cdot\text{C}^{13}\text{O}\cdot\text{C}^{14}\text{OOH} \\
\\
(2) & \text{HC}^{14}\text{OOH} + \text{CH}_3\cdot\text{C}^{13}\text{O}\cdot\text{OP(OH)}_2 \rightarrow \text{CH}_3\cdot\text{C}^{13}\text{O}\cdot\text{C}^{14}\text{OOH} \\
\\
(3) & \text{HC}^{14}\text{OOH} + \text{CH}_3\cdot\text{COX} \rightarrow \text{CH}_3\cdot\text{CO}\cdot\text{C}^{14}\text{OOH}
\end{array}
$$

(a)

Formate Incorporation in Pyruvate As Compared to Acetyl Phosphate and Acetate

(CH₃ from acetate) (CO from acetyl phosphate) (COOH from formate) (Formate) (Acetate) (Acetyl phosphate)

Enzyme preparation No.*	Final pyruvate			HC^{14}OOH		$\text{C}^{13}\text{H}_3\cdot\text{COOH}$		$\text{CH}_3\cdot\text{C}^{13}\text{O}\cdot\text{PO}_3\text{H}_2$	
	CH₃—	—C=O	—COOH	Added	Final	Added	Final	Added	Final
	per cent C^{13}	*per cent* C^{13}	*c.p.m. per* mM	*c.p.m. per* mM	*c.p.m. per* mM	*per cent* C^{13}	*per cent* C^{13}	*per cent* C^{13}	*per cent* C^{13}
I	0.01	−0.01	3950	12,000	5240	2.07	1.79	2.03	1.66
II	−0.02	−0.01	2540	12,000	8050	2.07	1.80	2.03	

(b)

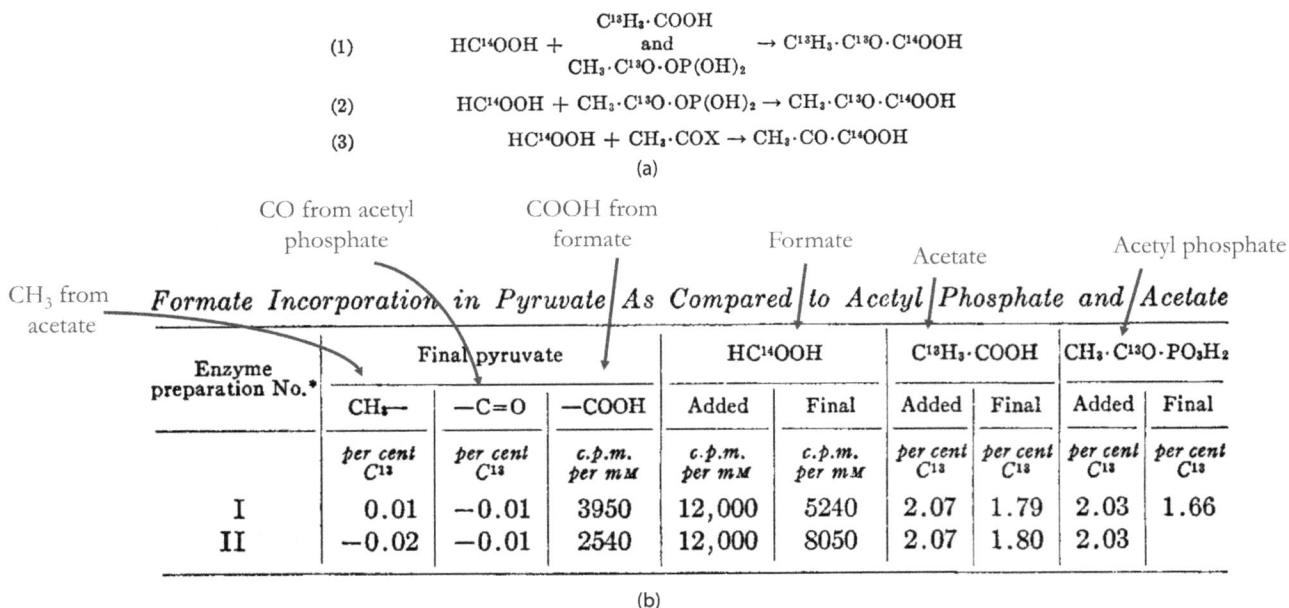

FIGURE 5.15 Only formate, and not acetate nor acetyl phosphate, is incorporated into pyruvic acid. (a) Reaction schemes to determine the mechanism of synthesis of pyruvate from formate and acetate: (1) incubation of C^{14}-labeled formate with C^{13}-labeled acetate (methyl carbon labeled) and acetyl phosphate (carboxyl carbon labeled) producing pyruvate with both the methyl and carboxyl carbons labeled with C^{13}; (2) incubation of C^{14}-labeled formate with C^{13}-labeled acetyl phosphate (carboxyl carbon labeled) producing pyruvate with its carboxyl carbon labeled with C^{13}; (3) C^{14}-labeled formate with and unknown compound producing pyruvate with unlabeled carbons. (b) Radiolabeled formate (HC^{14}OOH), acetate ($\text{C}^{13}\text{H}_3\cdot\text{COOH}$), and acetyl phosphate ($\text{CH}_3\text{C}^{13}\text{O}\cdot\text{PO}_3\text{H}_2$) were added to a reaction mixture and the species incorporated into pyruvate was assessed. Compared to acetyl phosphate and acetate, the isotope in formate decreased significantly and labeled –COOH was incorporated into pyruvate.[47]

enzyme [emphasis in the original] (Figure 5.16), and proposed that this new compound could be coenzyme A.[42]

Inspired by Utter et al.'s and Strecker et al.'s results showing an exchange between pyruvate and formate, Chantrenne

Bacterial enzyme	System	*A* Micromoles of citrate formed per ml.				*B* Micromoles formed per ml. in 40 min.	
		5 min.	10 min.	20 min.	40 min.	CO₂	Acetyl phosphate
E. coli (6 mgm. of protein)	Complete		0·61	1·42	3·02	2·9	1·62
	No phosphate*		0·64	1·51	3·38	0·6	~0·1
	No phosphate, no coenzyme A		0·16	0·34	0·58		
	No phosphate, no DPN				0·09		
	No DPN					0·2	~0·1
	No condensing enzyme				1·06		

FIGURE 5.16 Citrate formation can occur in the absence of phosphate. A complete bacterial enzyme solution (A) or one lacking oxaloacetate (B) was incubated for the indicated times in the presence or absence of the indicated substances, before assessing the amount citrate, CO₂, or acetyl phosphate produced. Note that citrate is formed in the absence of phosphate (second line). (Adapted by permission from Springer Nature Customer Service Centre GmbH: Springer Nature, Nature, Korkes S, Stern JR, Gunsalus IC, Ochoa S. Enzymatic synthesis of citrate from pyruvate and oxalacetate. Copyright © 1950.)

& Lipmann explored the possibility that the "activated" acetyl-enzyme was formed from a CoA–enzyme complex derived directly from pyruvate and forming a precursor to citric acid synthesis.[49] As such, the authors incubated radiolabeled formate with bacterial extracts and measured the amount of labelled pyruvate at the end of the reaction, results of which showed a nearly complete exchange after about 60 minutes (Figure 5.17).[49] In addition, they found that CoA was required for this reaction to proceed, as extracts devoid of CoA resulted in minimal exchange. To further confirm these results, Chantrenne and Lipmann next turned their attention to whether or not pyruvate could act as an acetyl donor in the formation of citric acid. As such, the authors used a method developed by their group, in which pigeon liver extracts were fractionated using a variety of acetone concentrations, which resulted in the separation and identification of a number of acetyl acceptors and donors. These were then used to test the ability of various substances to act as acetate donors or acceptors, using their newly formed acetyl donors as donors or sulfanilamide as an acetate trap.[46] As such, Chantrenne and Lipmann found that pyruvate was indeed an acetate donor, though not at an ideal rate since the conditions required for this assay were detrimental to the exchange of formate with pyruvate.[49] However, despite the less-than-optimal conditions used in their experiments, *the pyruvate-formate exchange system yielded enough acetyl-Co A-enzyme complex to bring about the acetylation of sulfanilamide.*[49] These results

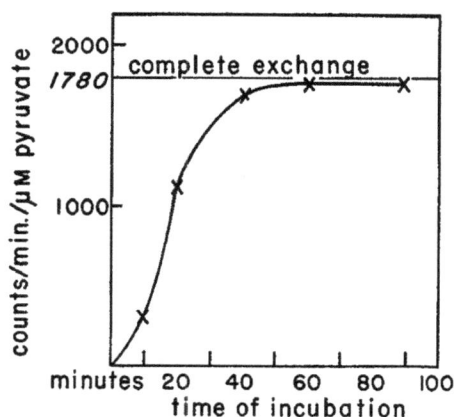

FIGURE 5.17 Exchange between formate and pyruvate. Radioactive formate was incubated with an enzyme solution, which included pyruvate, for a period of time and radioactive pyruvate was measured following its isolation. This reaction reached equilibrium within one hour.[49]

led the authors to propose the following reversible exchange reaction for formate and pyruvate:

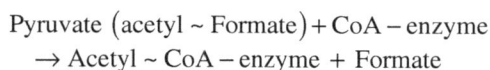

$$\text{Pyruvate} \left(\text{acetyl} \sim \text{Formate}\right) + \text{CoA} - \text{enzyme}$$
$$\rightarrow \text{Acetyl} \sim \text{CoA} - \text{enzyme} + \text{Formate}$$

noting that *the acetyl part of pyruvate becomes joined in energy-rich linkage to the Co A-enzyme and formate is liberated.*

Lipmann's group confirmed these results the next year and applied them to the acetyl phosphate system, which led them to propose that:

From data obtained during the last few years, the conclusion is ventured that the so called "active acetate," fed into a great variety of synthetic channels, is acetyl-CoA, which may be derived by a variety of enzyme reactions from a variety of acetyl-donating molecules. In turn, acetyl~CoA may donate its acetyl residue to a variety of acceptor systems.[50]

Ochoa's group subsequently purified the condensing enzyme catalyzing the reversible reaction of acetyl-CoA with oxaloacetate to yield citrate—the first enzyme involved in the tricarboxylic acid cycle to be purified in crystalline form—and confirmed its involvement in citric acid synthesis.[51,52] In addition, Lynen and coworkers determined that acetyl-CoA was an acetylated mercaptan ($CoA\text{-}S\text{-}CO\text{-}CH_3$), since sulfhydryl was released upon citrate synthesis.[52] Measurements of free energy during the reaction indicated that the energy-rich component of acetyl-CoA was contained within the acyl-mercaptide bond (i.e., $S\text{-}CO$), which led the authors to propose the following mechanism for the formation of citrate from oxaloacetate:

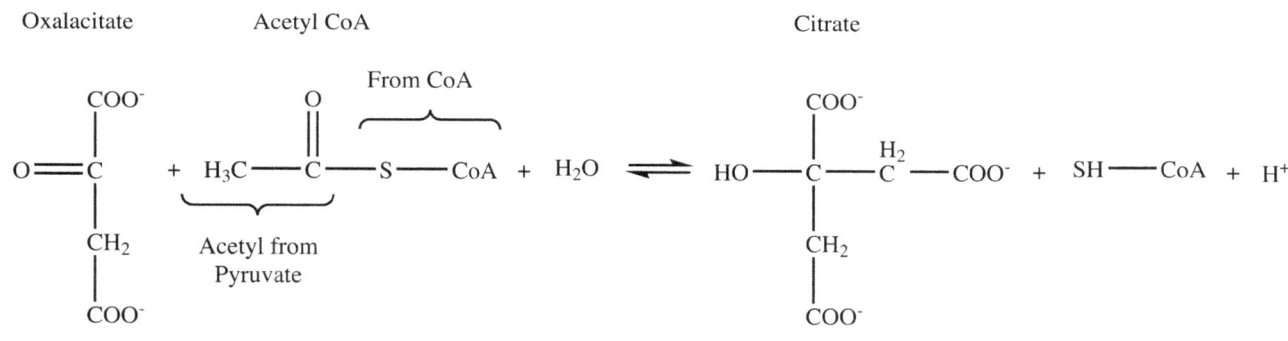

Although much of this work was performed in the context of acetate or acetyl phosphate as precursors to citric synthesis, it would later be shown that these reactions are not well distributed at all in animals, yeast, or in many types of bacteria. In addition, citrate formation from fatty acids and amino acids—also with the involvement of CoA—were shown to occur in many organisms, including humans. The predominant mechanism for citrate formation yielding the most net energy was determined to occur during aerobic respiration, where the end-product of glycolysis, pyruvate, is converted to the high-energy molecule, acetyl-CoA, which then condenses with oxaloacetate to form citric acid. This cycle, originally named the citric acid cycle, is also known as the tricarboxylic acid cycle and later renamed the Krebs cycle in honor of its discoverer.

In this chapter, we discussed the complicated discovery of the mechanism of glycolysis and the Krebs cycle, which together, supply the precursors for the energy needs of all cells. This pathway begins with glycolysis, which is a quick, but low energy-yielding process, and uses the multiple phosphorylation steps of glucose to generate 4x ATP and 2x NADH molecules at the expense of 2x ATPs (this manner of ATP synthesis is called substrate-level phosphorylation, Figure 5.18). However, in the presence of oxygen, the last molecule generated during glycolysis, pyruvate, is converted to acetyl-CoA (generating 2x NADH per glucose molecules), which enters the Krebs cycle by reacting with oxaloacetate to produce citrate. Citrate then undergoes a number of redox reactions which generates 6x $NADH^+$ and 2x $FADH_2$ per glucose molecule, which go on to donate protons to the electron transport chain in mitochondria, thereby generating a hydrogen gradient that is used to synthesize the vast majority of ATP (by a process called oxidative phosphorylation; more on this in Chapter 9). In total, respiration generates more than 30 ATP molecules per molecule of glucose that can be used as a source of energy or to phosphorylate proteins.

* * * * *

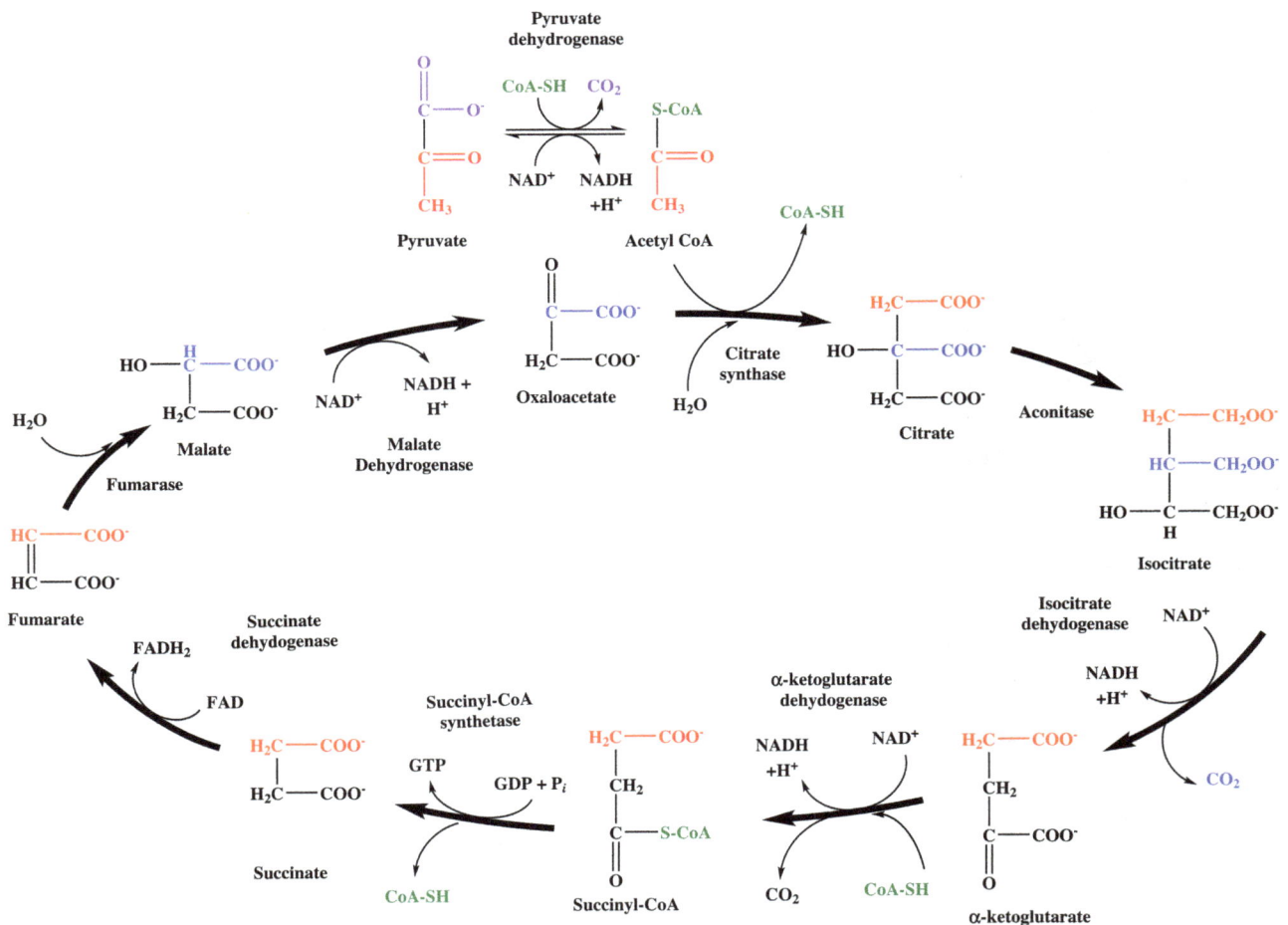

FIGURE 5.18 The Krebs cycle. The end product of glycolysis, pyruvate, combines with CoASH to form the high-energy molecule, acetyl-CoA, which then reacts with oxalacetate to form the six-carbon molecule, citrate. Citrate undergoes a number of redox reactions which produce NADH and $FADH_2$, both of which go on to produce ATP in mitochondria by oxidative phosphorylation. The three-carbon oxalacetate is then recycled in the Krebs cycle by reacting with acetyl-CoA as described above.

REFERENCES

1. Harden A. *Alcoholic Fermentation*. 2nd ed. Longmans, green and co.; 1914. Accessed March 25, 2019. https://archive.org/details/alcoholicferment00hardrich

2. Pasteur L. Animalcules infusoires vivant sans gaz oxygène libre et déterminant des fermentations. In: *Comptes Rendus Hebdomadaires des Séances de l'Académie des Sciences.* Vol 52. Mallet-Bachelier; 1861:344–347.

3. Pasteur L. Expériences et vues nouvelles sur la nature des fermentations. *Comptes Rendus Hebdomadairs Seances Académie des Sciences.* 1861;52:1260.

4. Needham DM. Respiration. In: *Machina Carnis: The Biochemistry of Muscular Contraction in Its Historical Development.* Cambridge University Press; 1971.

5. Harden A, Young WJ. Proceedings of the physiological society: November 12, 1904. *Journal of Physiology.* 1905; 32(suppl):i–v. doi: https://doi.org/10.1113/jphysiol.1905.sp001098

6. Harden A, Young WJ. The alcoholic ferment of yeast-juice. Part III. The function of phosphates in the fermentation of glucose by yeast-juice. *Proceedings of the Royal Society B: Biological Sciences.* 1908;80(540):299–311. doi: https://doi.org/10.1098/rspb.1908.0029

7. Harden A, Young WJ. The influence of phosphates on the fermentation of glucose by yeast juice. *Proceedings of the Chemical Society (London).* 1905;21(297):189–191. doi: https://doi.org/10.1039/pl9052100175

8. Harden A, Young WJ. The alcoholic ferment of yeast-juice. *Proceedings of the Royal Society B: Biological Sciences.* 1906;77(519):405–420. doi: https://doi.org/10.1098/rspb.1906.0029

9. Harden A, Young WJ. The alcoholic ferment of yeast-juice. Part V. The function of phosphates in alcoholic fermentation. *Proceedings of the Royal Society B: Biological Sciences.* 1910;82(556):321–330. doi: https://doi.org/10.1098/rspb.1910.0023

10. Needham DM. Muscle metabolism after the chemical revolution; lactic acid takes the stage. In: *Machina Carnis: The Biochemistry of Muscular Contraction in Its Historical Development.* Cambridge University Press; 1971.

11. Battelli F, Stern, L. Les échanges respiratoires dans les émulsions des tissus animaux. In: Masson, ed. *Comptes Rendus des Séances de La Société de Biologie et de Ses Filiales.* Vol 58. Libraires de l'Académie de Médecine; 1906:679–681.

12. Battelli F, Stern L. Recherches sur la respiration élémentaire des tissus. *Journal de Physiologie et de Pathologie Générale.* 1907;9:1–16.

13. Battelli F, Stern L. Recherches sur l'activité respiratoire des tissus. In: *Journal de Physiologie et de Pathologie Générale.* Vol 9. Masson et Cie, ed; 1907:34–49. Accessed March 30, 2019. https://gallica.bnf.fr/ark:/12148/bpt6k6446089w.r= %22Recherches sur l%27Activité Respiratoire des Tissus %22.?rk=21459;2

14. Battelli F, Stern L. Recherches sur le mécanisme des oxydations dans les tissus des animaux isolés. In: Cie M et, ed. *Comptes Rendus des Séances de La Société de Biologie et de Ses Filiales.* Vol 1. Libraires de l'Académie de Médecine; 1907:296–297.

15. Battelli F, Stern L. Recherches sur les processus des combustions élémentaires dans les muscles isolés. In: Masson et Cie, ed. *Comptes Rendus des Séances de La Société de Biologie et de Ses Filiales.* Vol 1. Libraires de l'Académie de Médecine; 1907:958–961.

16. Battelli F, Stern L. Oxydation des acides malique, fumarique et citrique par les tissus animaux. In: Masson et Cie, ed. *Comptes Rendus des Séances de La Société de Biologie et de Ses Filiales.* Vol 2. Libraires de l'Académie de Médecine; 1910:552–554.

17. Battelli F, Stern L. L'oxydation de l'acide succinique comme mesure du pouvoir oxydant dans la respiration principale des tissus animaux. In: Masson et Cie, ed. *Comptes Rendus des Séances de La Société de Biologie et de Ses Filiales.* Vol 2. Libraires de l'Académie de Médecine; 1910:554–556.

18. Thunberg T. The hydrogen-activating enzymes of the cells. *The Quarterly Review of Biology.* 1930;5(3):318–347. doi: https://doi.org/10.1086/394361

19. Fleisch A. Some oxidation processes of normal and cancer tissue. *The Biochemical Journal.* 1924;18(2):294–311. doi: https://doi.org/10.1042/bj0180294

20. Needham DM. The influence of brewing science on the study of muscle glycolysis; adenylic acid and the ammonia controversy. In: *Machina Carnis: The Biochemistry of Muscular Contraction in Its Historical Development.* University Press; 1971:782.

21. Needham DM. Adenosinetriphosphate as fuel and as phosphate-carrier. In: *Machina Carnis: The Biochemistry of Muscular Contraction in Its Historical Development.* Cambridge University Press; 1971:782.

22. Needham DM. The relationship between mechanical events, heat production and metabolism; studies between 1840 and 1930. In: *Machina Carnis: The Biochemistry of Muscular Contraction in Its Historical Development.* University Press; 1971:782.

23. Meyerhof O. Conversion of fermentable hexoses with a yeast catalyst (hexokinase). In: Kalckar H, ed. *Biological Phosphorylations: Development of Concepts.* Prentice-Hall; 1969:38. Accessed March 30, 2019. https://www.worldcat.org/title/biological-phosphorylations-development-of-concepts/oclc/462815362&referer=brief_results

24. Meyerhof O. Die chemischen Vorgänge im zellfreien Muskelextrakt. In: *Die Chemischen Vorgänge Im Muskel Und Ihr Zusammenhang Mit Arbeitsleistung Und Wärmebildung. Monographien Aus Dem Gesamtgebiet Der Physiologie Der Pflanzen Und Der Tiere.* Vol 22. Springer, Berlin, Heidelberg; 1930:140–175. doi: https://doi.org/10.1007/978-3-642-90668-8_5

25. Ernster L, Schatz G. Mitochondria: A historical review. *The Journal of Cell Biology.* 1981;91(3 Pt 2):227s–255s. doi: https://doi.org/10.1083/JCB.91.3.227S

26. Keilin D. On cytochrome, a respiratory pigment, common to animals, yeast, and higher plants. *Proceedings of the Royal Society B: Biological Sciences.* 1925;98(690): 312–339. doi: https://doi.org/10.1098/rspb.1925.0039

27. Cori CF. Embden and the glycolytic pathway. *Trends in Biochemical Sciences.* 1983;8(7):257–259. doi: https://doi.org/10.1016/0968-0004(83)90353-5

28. Fiske CH, Subbarow Y. Phosphorus compounds of muscle and liver. *Science (1979).* 1929;70(1816):381–382. doi: https://doi.org/10.1126/science.70.1816.381-a

29. Lipmann F. Metabolic generation and utilization of phosphate bond energy. In: *Advances in Enzymology and Related Areas of Molecular Biology, Volume 1.* John Wiley & Sons, Ltd; 1941:99–162. doi: https://doi.org/10.1002/9780470122464.ch4

30. Kornberg H. Krebs and his trinity of cycles. *Nature Reviews. Molecular Cell Biology.* 2000;1(3):225–228. doi: https://doi.org/10.1038/35043073

31. Krebs H. The history of the tricarboxylic acid cycle. *Perspectives in Biology and Medicine.* 1970;14(1):154–172. doi: https://doi.org/10.1353/pbm.1970.0001

32. Krebs H, Johnson W. The role of citric acid in intermediate metabolism in animal tissues. *FEBS Letters.* 1980;117: K2–K10. doi: https://doi.org/10.1016/0014-5793(80)80564-3

33. Lipmann F. A phosphorylated oxidation product of pyruvic acid. *Journal of Biological Chemistry.* 1940;134(1):463–464. doi: https://doi.org/10.1016/S0021-9258(18)73290-0

34. Lipmann F. Enzymatic synthesis of acetyl phosphate. *Journal of Biological Chemistry.* 1944;155(1):55–70. doi: https://doi.org/10.1016/S0021-9258(18)43172-9

35. Harris JS, Klein JR. Acetylation of sulfanilamide by liver tissue in vitro. *Experimental Biology and Medicine.* 1938;38(1):78–80. doi: https://doi.org/10.3181/00379727-38-9746P

36. Lipmann F. Acetylation of sulfanilamide by liver homogenates and extracts. *Journal of Biological Chemistry.* 1945;160(1):173–190. doi:10.1016/S0021-9258(18)43110-9

37. Lipmann F, Kaplan NO, Novelli GD, Tuttle LC, Guirard BM. Coenzyme for acetylation, a pantothenic acid derivative. *Journal of Biological Chemistry.* 1947;167(3):869–870. doi: https://doi.org/10.1016/S0021-9258(17)30973-0

38. Lipmann F, Kaplan NO, Novelli GD, Tuttle LC, Guirard BM. Isolation of coenzyme A. *Journal of Biological Chemistry.* 1950;186(1):235–243. doi: https://doi.org/10.1016/S0021-9258(18)56309-2

39. Novelli GD, Lipmann F. The involvement of coenzyme a in acetate oxidation in yeast. *Journal of Biological Chemistry.* 1947;171(2):833–834. doi: https://doi.org/10.1016/S0021-9258(17)41096-9

40. Olson RE, Kaplan NO. The effect of pantothenic acid deficiency upon the coenzyme a content and pyruvate utilization of rat and duck tissues. *Journal of Biological Chemistry.* 1948;175(2):515–530. doi: https://doi.org/10.1016/S0021-9258(18)57172-6

41. Kaplan NO, Lipmann F. The assay and distribution of coenzyme A. *The Journal of Biological Chemistry.* 1948;174(1):37–44. doi: https://doi.org/10.1016/S0021-9258(18)57372-5

42. Korkes S, Stern JR, Gunsalus IC, Ochoa S. Enzymatic synthesis of citrate from pyruvate and oxalacetate. *Nature.* 1950;166(4219):439–440. doi: https://doi.org/10.1038/166439b0

43. Novelli GD, Lipmann F. The catalytic function of coenzyme a in citric acid synthesis. *Journal of Biological Chemistry.* 1950;182(1):213–228. doi: https://doi.org/10.1016/S0021-9258(18)56541-8

44. Stern JR, Ochoa S. Enzymatic synthesis of citric acid. I. Synthesis with soluble enzymes. *The Journal of Biological Chemistry.* 1951;191(1):161–172. doi: https://doi.org/10.1016/S0021-9258(18)50963-7

45. Stern JR, Ochoa S. Enzymatic synthesis of citric acid by condensation of acetate and oxalacetate. *The Journal of Biological Chemistry.* 1949;179(1):491. doi: https://doi.org/10.1016/S0021-9258(18)56860-5

46. Lipmann FA. On chemistry and function of coenzyme A. *Bacteriological Reviews.* 1953;17(1):1–16. doi: https://doi.org/10.1128/br.17.1.1-16.1953

47. Strecker HJ, Wood HG, Krampitz LO. Fixation of formic acid in pyruvate by a reaction not utilizing acetyl phosphate. *Journal of Biological Chemistry.* 1950;182(2):525–540. doi: https://doi.org/10.1016/S0021-9258(18)56487-5

48. Utter MF, Lipmann F, Werkman CH. Reversibility of the phosphoroclastic split of pyruvate. *Journal of Biological Chemistry.* 1945;158(2):521–531. doi: https://doi.org/10.1016/S0021-9258(18)43159-6

49. Chantrenne H, Lipmann F. Coenzyme A dependence and acetyl donor function of the pyruvate-formate exchange system. *Journal of Biological Chemistry.* 1950;187(2):757–767. doi: https://doi.org/10.1016/S0021-9258(18)56222-0

50. Stadtman ER, Novelli GD, Lipmann F. Coenzyme A function in and acetyl transfer by the phosphotransacetylase system. *Journal of Biological Chemistry.* 1951;191(1):365–376. doi: https://doi.org/10.1016/S0021-9258(18)50986-8

51. Ochoa S, Stern JR, Schneider MC. Enzymatic synthesis of citric acid. II. Crystalline condensing enzyme. *Journal of Biological Chemistry.* 1951;193(2):691–702. doi: https://doi.org/10.1016/S0021-9258(18)50926-1

52. Stern JR, Shapiro B, Stadtman ER, Ochoa S. Enzymatic synthesis of citric acid: III. Reversibility and mechanism. *Journal of Biological Chemistry.* 1951;193(2):703–720. doi: https://doi.org/10.1016/S0021-9258(18)50927-3

6 Protein and DNA Structure

We saw in Chapter 3 how Thomas Graham came to propose that proteins were colloids; that is, small molecules which non-specifically aggregated into larger complexes. Although hangers-on of the colloidal theory remained well into the 20th century, this hypothesis slowly crumbled until the early 20th century, when enough evidence emerged such that it was considered disproven by most scientists. One of the first reports to provide direct evidence against the colloidal theory came in 1902 by the American chemist Thomas Osborne, who was working on seed proteins at the Connecticut Agricultural Research Station. At the time, the main method used to estimate the molecular weight of proteins was by analysis of the number of sulfur atoms in a given protein since this element was known to be present in only one amino acid: cysteine (methionine also contains sulfur, but it was not discovered until 1922). Thus, since the average sulfur content in amino acids was known, the molecular weight of a protein could be estimated based on the number of these atoms present in a particular protein. However, the treatment of proteins for sulfur determination yielded erroneous results due to partial splitting off of sulfur from the protein or to possible reactions of sulfur with various atoms in the treatment solution. Thus, Osborne used a newly developed sulfur extraction method to more accurately determine the presence of sulfur in proteins, and determined that the minimum molecular weight for a protein should be at least 15,000; although, he stressed that these could only be estimates *since the methods of analysis preclude great accuracy.*[1]

Another important step toward disproving the colloidal theory of proteins came from an unlikely source: a proponent of the colloidal theory.[2] Theodor Svedberg was born in Sweden and inherited his love of nature from his father, who often took him out on excursions. An autodidact from the start, he gave public demonstrations of wireless telegraphy while still in grammar school, after he taught himself to build a Marconi-transmitter and a Tesla-transformer when given the opportunity to study by himself in the school's physical and chemical laboratories. Shortly after graduating from the University of Uppsala and reading two newly published manuscripts on colloids, he was convinced that this branch of chemistry would *ultimately help to explain the processes of living matter.*[2] In 1912, initiated by several faculty members and supported by none other than Svante Arrhenius, a chair was created for Svedberg at Uppsala University, where his main area of focus would be that of the physicochemical properties of colloids. After experimenting with sedimentation of small colloids in a gravity field—that is, letting things settle—Svedberg realized that using stronger centrifugal forces might allow better separation of the particles, stating that *[t]he lack of a reliable method for the determination*

of the molecular weights of substances possessing a very complicated structure has been a serious obstacle in the progress of our knowledge of the chemistry of the proteins.[3] Around the same time, he was offered an invitation to work at the University of Wisconsin for an eight-month visit where he experimented with centrifugal fields to separate proteins. Svedberg returned to Uppsala electrified and convinced he could build a better centrifuge that would allow him to precisely determine the molecular weights of proteins.

The main reason sedimentation rates could not be determined in early centrifuges was because the instruments generated so much heat that they warmed the samples to the extent that they created convection forces which pushed the particles down the sides of the tube independently from the centrifugal forces; therefore, these forces had to be eliminated. With his student, Herman Rinde, Svedberg redesigned the centrifuge such that it addressed this and other issues and renamed it the ultracentrifuge (Figure 6.1).[4] In order to keep the temperature inside the cell (where the sample was held) constant, they designed an apparatus with a constant flow of hydrogen gas around the rotor, which better decreased friction compared to air and, therefore, decreased the heat generated. In addition, they found that the majority of heat produced came from the bearings surrounding the rotor shaft, so they designed one which allowed cold water to be circulated to cool it down. However, this did not keep it sufficiently cool, so they also had to go one step further and isolated the rotor from the bearing using a screen of copper, in addition to spraying oil on the bearing/toothed wheel assembly to further decrease friction.

Finally, to ensure that the speed of the centrifuge remained constant, which was required for precise sedimentation measurements, they inserted a special clutch between the rotating toothed wheel (attached to the motor) and the shaft which efficiently translated the rotational energy to the rotor.[4] Observations of sedimentation rates were conducted through a window on top of the cell, where the sample was illuminated by a horizontal light redirected vertically via a prism mounted below the cell. Since this window was located on top of the centrifuge, samples had to be monitored in a separate room above the centrifuge and through an opening in the ceiling (Figure 6.1b). Pictures were also taken using a camera mounted above the cell window for more precise measurements of the average size and distribution of particles. With these modifications in place, centrifugation speeds of up to 5000 g were possible and Svedberg and Rinde were able to precisely measure the sedimentation rate of gold, which allowed them to calculate the average size of these particles in colloids down to a diameter of about 5 nm.[4]

Having shown the validity of their new instrument, Svedberg and Fåhraeus tackled the problem of determining

DOI: 10.1201/9781003379058-6

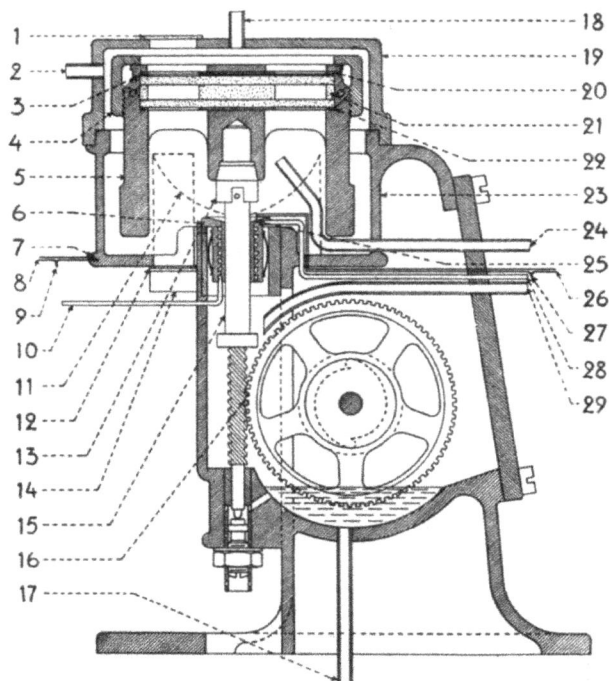

(a)

1, Upper window; 2, Hydrogen outlet; 3, Ebonite plate; 4, Lid of rotor; 5, Rotor; 6, Water-cooled spring bearing; 7–9, Thermocouple; 10, Water outlet; 11, Copper screen; 12, Lower window; 13, Carrying cone; 14, Reflecting prism; 15, Shaft of rotor; 16, Toothed wheel; 17, Oil outlet; 18, Hydrogen inlet; 19, Lid; 20, Rubber plate; 21, Cell; 22, Rubber plate; 23, Casing of centrifuge; 24, Hydrogen inlet; 25–27, Thermocouple; 28, Water inlet; 29, Oil inlet.

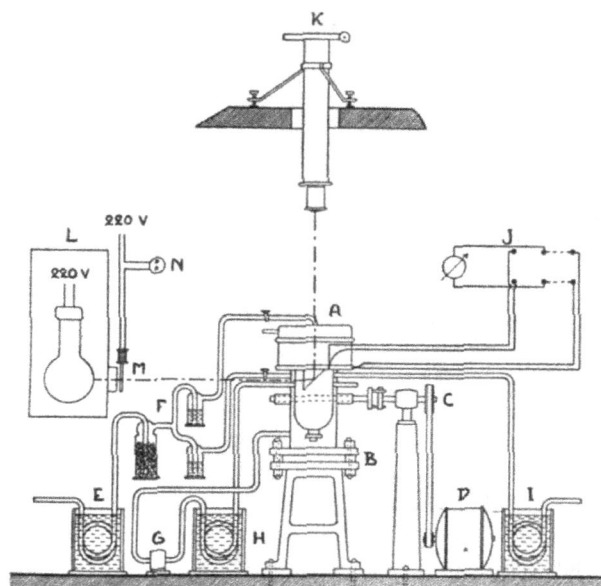

(b)

A, Centrifuge; B, Stand; C, Pulley; D, Motor; E, Cooling spiral for hydrogen current; F, Valves; G, Oil circulation pump; H, Cooling spiral for oil current; I, Cooling spiral for water current; J, Galvanometer and key; K, Camera; L, Lamp; M, Shutter; N, Switch for shutter.

FIGURE 6.1 Svedberg's new ultracentrifuge. (a) The sample was held in (21) and the rotor assembly was rotated by a shaft (14) connected to a toothed wheel (16). Heat was kept to a minimum by flowing hydrogen gas around the rotor (24) and by running cold water around the bearings (28). In addition, the rotor was further isolated from the bearings using a copper shield (11). Finally, the toothed wheel (16) was kept well lubricated by a steady stream of oil entering through an inlet (29). (b) Centrifuged samples were visualized using a lamp, l, and imaged by a camera, k, mounted above the ultracentrifuge.[4]

the molecular weight of proteins, for which they used hemoglobin, the protein in red blood cells to which iron binds. The molecular weight of this protein was suspected to be 16,700 at a minimum, based on the content of iron in the molecule (each molecule can bind up to four atoms of iron), but several investigations using various assays also reported a wide variety of molecular weights, from a fraction thereof to multiples. At this time, Svedberg was still convinced proteins were colloidal substances; therefore, he expected them to separate into multiple bands of low molecular weight when subjected to centrifugal forces. However, to his surprise, his and his colleague's experiments indicated that the molecular weight of hemoglobin was 4 times 16,700.[3] To confirm that their results were correct, the authors repeated their experiment using several different conditions: varying the solution in which hemoglobin was centrifuged, varying the concentration of hemoglobin within these solutions, varying the length of the sample column, the speed of the centrifuge, and even running the samples uninterrupted for up to up to one week; however, all experiments resulted in

a similar molecular weight for the protein. Nonetheless, not yet quite ready to do away with the colloidal theory of proteins, Svedberg concluded that:

> The technique of the method is not yet refined enough to permit us to decide whether all the molecules in a hemoglobin solution are composed of four groups of the weight 16,700 or whether there is an equilibrium between molecules in various states of aggregation- an equilibrium which might be regarded as dependent upon concentration, salt content, etc. There is a circumstance, however, strongly in favor of the view that at least in one and the same solution there is practically only one kind of hemoglobin molecule. It is the fact that in none of the series of determinations hitherto made has there been observed any systematic variation of molecular weight with distance from the center of rotation. If there were molecules of different mass present in the same solution the values of the molecular weight calculated from measurements of the central part of the centrifuged solution should be lower than those calculated from measurements of the peripheral part. This is not the case.[3]

Svedberg's laboratory subsequently further improved the centrifuge, which enabled them to more precisely elucidate the molecular weight of a great number of proteins. Some of these even registered in the millions! These experiments showed with little doubt that proteins were large, stable molecules and not colloids, and refuted a paradigm that existed for nearly a century.

* * * * *

By the turn of the 20th century, Fischer had reported that amino acids were linked together via peptide bonds; however, many remained skeptical as late as the mid-century. An important contribution to the resolution of this controversy (and to that of the structure of DNA, as we will see later in this chapter) was the discovery of X-ray diffraction by Luau, Friedrich, and Knipping in 1912.[5] In Chapter 1, we learned about Crook's and Thomson's development of the cathode ray tube, whose walls fluoresced when a current was discharged under vacuum. Around the same time, Philipp Lenard discovered that these rays could pass through a thin sheet of aluminum foil placed over a hole made in the glass at the back of the tube. As such, he found that the rays travelled a few inches in the air beyond the tube before being degraded at an exponential rate, which he determined by observing the fluorescence emitted on a screen at various distances beyond the tube.[6] In 1895, Wilhelm Conrad Röntgen reported the results of experiments on this phenomenon, one of which showed that brass weights enclosed in a wooden box appeared on an image without any sign of the box; importantly, he determined that this phenomenon could also be used to make an image of internal bodily structures, such as bones, on a screen; not surprisingly, the medical community quickly embraced and made good use of this new technology. Since Röntgen could not determine what these rays were, he called them X-rays, a term that has persisted.

The next breakthrough in protein science occurred in 1912, when Luau, Friedrich, and Knipping described how crystals could diffract X-rays and produce a pattern that was dependent on the molecular structure of the crystal (Figure 6.2). This phenomenon was different from that used in radiography as Röntgen had done, as it used a photographic plate to register the diffraction of X-ray waves by the atoms in the crystal sample in question, upon which

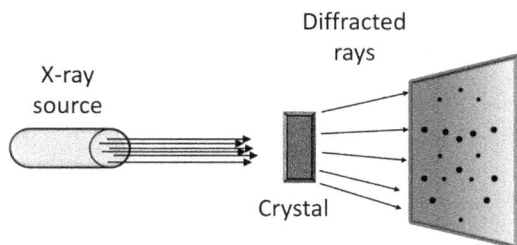

FIGURE 6.2 X-ray diffraction at a glance. An X-ray beam is directed toward a sample, which deflects the beam in a pattern that is dependent on the distribution of atoms within the crystal sample.

argument is based as to the form and nature of the bodies which diffracted the waves.[7] Astbury explains:

The object of using X-rays instead of, or rather in addition to, visible light is, of course, to extend the range of vision, i.e. to permit objects which are far too small to be seen by the human eye, even with the aid of the most powerful microscope, to be viewed at least mentally. Just as ripples are seriously distorted by pebbles which would be far too small to disturb the form of ocean waves, so X-rays, some ten thousand times shorter in wavelength than visible light, are seriously distorted by atoms and molecules which are far too small to disturb the main form of waves of visible light. Things are "seen" through the waves which are thrown back from the body looked at, and only such irregularities can be distinguished as are not much smaller than the wave-length of the waves used, because irregularities which are negligible compared with the waves make negligible impressions on them. The most carefully machined and polished surface is quite rough from the molecular point of view, even though by ordinary light it may appear so perfect as to be "optically flat"; but by using smaller and smaller wave-lengths, smaller and smaller irregularities can be detected until, finally, in the "light" of the X-rays, after making the appropriate calculations, the shapes of the atoms and molecules themselves can be distinguished.

The crystalline condition may be fairly called the "natural" state of solid matter, on account of the almost universal tendency of molecules, on solidification, to settle down, not in irregular heaps, but in regular and often highly symmetrical aggregates which are simply three dimensional, or space-, patterns analogous in every way to the familiar flat, or two-dimensional, patterns of the textile industries. This is a very fortunate circumstance in the application of the principles of X-ray analysis, for it means that if the form and dimensions of the molecular pattern underlying the crystal architecture of the substance which gives rise to that pattern can be deduced from the X-ray diffraction picture, as theoretically should be possible, then conclusions can be drawn as to the form and dimensions of the molecules themselves, because the *dimensions of a pattern are the expression of the dimensions of the units from which it is built.*[7] [emphasis in original]

Luau, Friedrich, and Knippling's paper was followed the next year by W.L. Bragg, in England, who used X-ray diffraction technology to solve the first crystal structures, that of NaCl and KCl.[5] Images of the first X-ray diffraction pattern of proteins (silk) were published soon after but the purity of the samples was not yet good enough to deduce any definite structure; however, it was enough to conclude that there was some sort of molecular order in this material.[5] It would not be until the 1920s before William Thomas Astbury made significant advances toward elucidating a structure for proteins.

Astbury was born in England in 1898, where he studied chemistry, physics, and mathematics.[8] Though his studies were interrupted by the war, he completed them upon his return—taking up mineralogy and crystallography—and he was recommended to W.L. Bragg by one of his professors. By this time, images of several simple textile fibers, such as silk and ramie, had been acquired with satisfactory results; however, most other fibers produced relatively blurry images. In 1926, Bragg was preparing a lecture at the Royal Institution, whose topic he decided would be "The imperfect Crystallization of Common Things", the discussion of which included that of proteins. Thus, he asked Astbury to prepare wool and hair fiber samples and to take images of their X-ray diffractions; this began a life-long passion in Astbury for the study of fibers.[5] In 1928, Braggs recommended Astbury to the University of Leeds as a lecturer in Textile Physics, which he reluctantly accepted since he wanted to stay with Braggs; nonetheless, he soon published a short paper—the first on the structure of proteins—describing that human hair and sheep's wool (which are made up of the protein, keratin) were *almost certainly built up in a rather imperfect manner of molecular chains [...] running roughly parallel to the fibre axis.*[9] Furthermore, these proteins seemed to have two separate molecular structures: a native, unstretched form, which he called α-keratin, and a stretched form, β-keratin, which could be extended up to three times their length when exposed to water or steam. Surprisingly, Astbury also found that these forms were reversible, stating that *[h]air has the striking property of always returning to its original length if wetted after extension.*[10]

From 1931 to 1935, Astbury and his colleagues published a series of papers studying over one hundred X-ray photographs from *a wide range of material under as great a variety of conditions*, and were able to more specifically describe the molecular structure of keratin.[10–12] However, they observed that in some of the images they obtained, *much of the center is obscured by* [a] *halo of high 'spacing'*, [which clearly] *arises from the scales of the hairs.*[10] Therefore, they developed a method for stripping the cuticles and scales from hair by using the strands *as the string of a wire bow, of the 'cello pattern', and drawing it rapidly backwards and forwards through coarsely powdered glass under a slight pressure* (Figure 6.3a).[10] They also had to secure the hairs tightly when stretching them but found that neither clamps nor cement could hold the hairs for long periods of time. Therefore, they built an instrument that allowed them to use both methods at once (Figure 6.3b). We can see from the image of the simple molecules of silk and cellulose, that the *crystalline aggregate formed by the fibre molecules are long and thin, and lie with* [its] *long* [axis] *roughly parallel to the fibre axis*, and that the *molecules are straight, fully extended line-patterns* (Figure 6.4 a and b).[7] However, as the complexity of the molecule increased, so too did the requirement for increased purity of the crystallized sample. As such, cotton or wool fibers produced much blurrier images, although the general parallel nature of the molecules could still be observed (Figure 6.4 c - e). Also noticed was the difference in patterns produced by unstretched and stretched wool molecules, which indicated that the fiber was present in two separate forms (Figure 6.4, d vs. e).

From these images, Astbury found that the distance between the intramolecular pattern in α-keratin was similar to the distance between the hexagonal glucose residues in

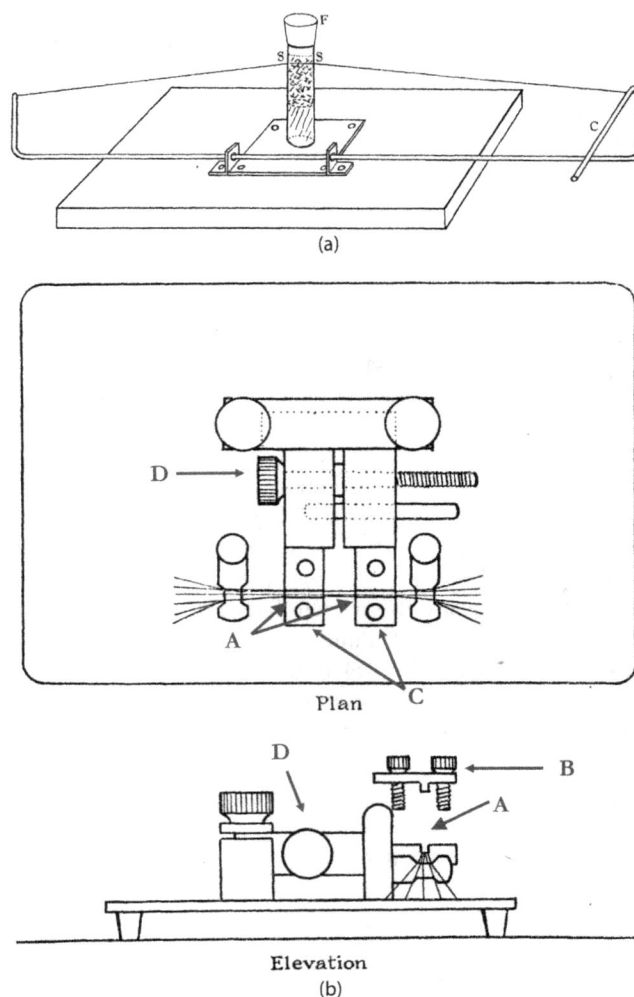

FIGURE 6.3 (a) Device used to strip scales of hair. A strand of hair was wrapped around a cylinder of *coarsely powdered glass under slight pressure* and fixed to a handle. The handle was moved back and forth to peel off the hair's *scale-bearing protective sheath.* (b) Device used to stretch hair strands. A bundle of hair was secured to the apparatus by cementing it in two grooves (A) and screwing in using clamps (B). The hair was then stretched by separating the two arms (C) holding the hair by way of a screw (D). (Figures adapted from Ref. [10].)

cellulose, whereas the distance in amino acids in β-keratin was slightly shorter than the *fully extended* pattern of silk, which indicated to him that *unstretched wool must be built of the same chains* [as stretched wool] *in a folded state, i.e., the mechanism of its unusual elasticity is simply that of a molecular spring* (Figure 6.5).[7] Another aspect of keratin discovered by Astbury in these papers involved molecular interactions: he found that the amino acid side chains in native, unstretched wool and hair (and to a lesser extent, in the stretched forms), had considerable interactions between the sulfur atoms of cysteine residues (called disulfide bridges), or by salt linkages between basic and acid residues (Figure 6.6). Importantly, Astbury's experiments provided additional support for Fischer's polypeptide hypothesis (i.e., that proteins were formed from the union of a series of amino acids linked

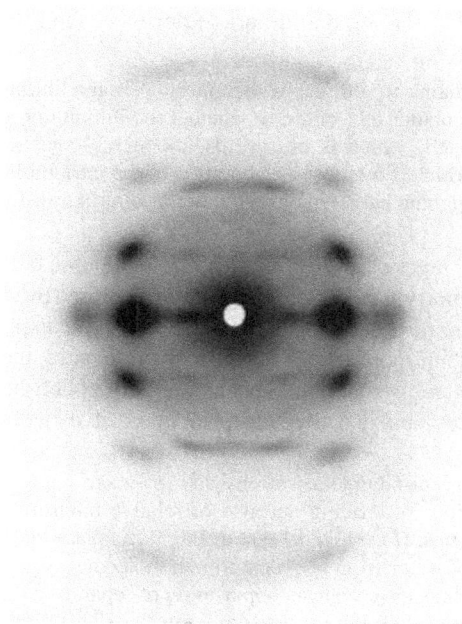

Natural silk fibre (fibroin)
(a)

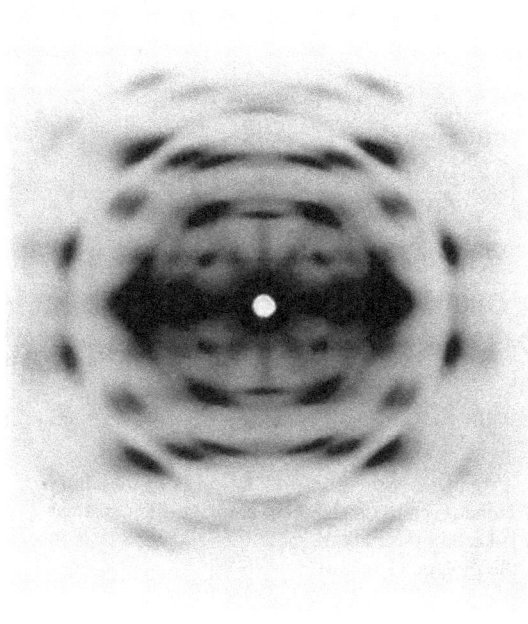

Native cellulose fibre (ramie)
(b)

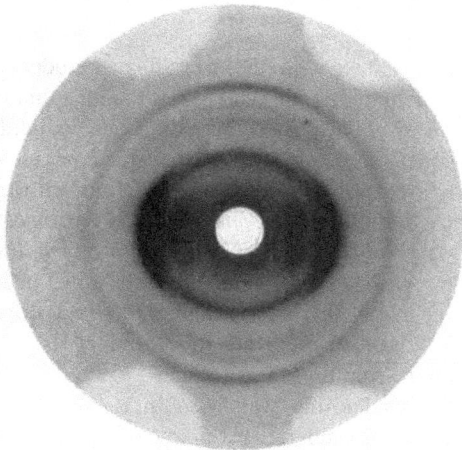

Cotton fibre
(c)

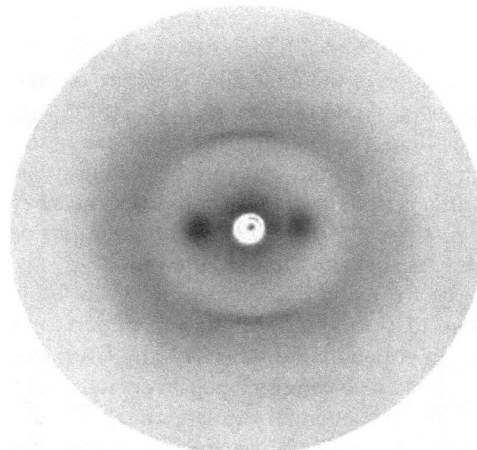

Unstretched wool fibre (α-keratin)
(d)

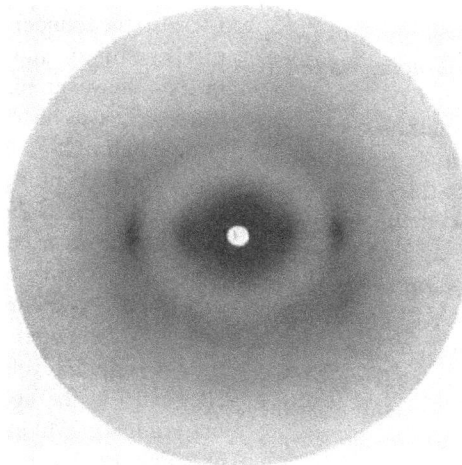

Stretched wool (β-keratin)
(e)

FIGURE 6.4 The X-ray diffraction patterns of stretched and unstretched fibers. Simple fabrics, such as silk (a) or cellulose (b), have well-defined patterns, whereas more intricate fabrics (cotton (c) or wool (d, e)) produce blurrier images requiring better protein preparations to obtain clear images.[7] (Adapted with permission from John Wiley and Sons. Astbury WT. The X-ray interpretation of fibre structure. Journal of the Society of Dyers and Colourists. Copyright © 1933 Society of Dyers and Colourists.)

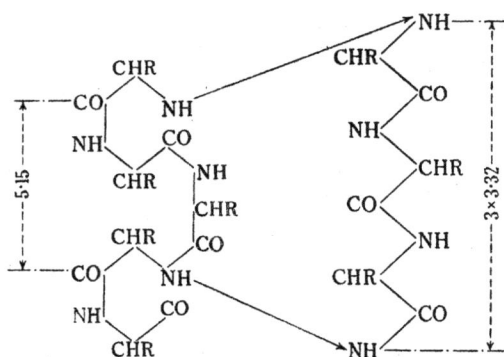

FIGURE 6.5 Proposed structures of keratin by Astbury. α-keratin (unstretched), characterized by a spring-like arrangement (left) and β-keratin (stretched), characterized by a straight arrangement of the molecules (right).[10]

by peptide bonds), since the images clearly indicated that the amino acids were connected to one another.[8]

As mentioned above, the extent to which a diffraction pattern can be analyzed depends in great part on the purity of the crystalized sample. And although Astbury made great strides in the elucidation of the molecular structure of proteins, further advances would require much better crystallization of proteins. John Desmond Bernal proposed, in the opening of his first paper on the structure of proteins, that *a knowledge of the crystal structure of the amino acids is essential for the interpretation of the x-ray photographs of animal materials.*[5] As such, he acquired preliminary X-ray diffraction patterns for crystalized amino acids in 1931, which allowed him to move on to the analysis of pepsin, a globular enzyme, in collaboration with Dorothy Crowfoot Hodgkin.[5,13] Bernal had received some pepsin crystals from a colleague in Svedberg's laboratory in Uppsala, but could see nothing but a *vague blackening* when he took them out of the solution and mounted them, a phenomenon which was found to be a result of the action of air on the crystal. This indicated to them that there was *complete alteration of the crystal* and suggested that this was why *previous workers have obtained negative results with proteins, so far as crystalline pattern is concerned.*[13]

As a possible remedy to this issue, the authors thought of drawing the crystals in their *mother liquor* to prevent any exposure to air. In doing so, they successfully maintained the structure of the crystal and produced the first clear images of proteins. In their report, Bernal and Crowfoot described pepsin molecules as *dense globular bodies, perhaps joined together by valency bridges, but in any event separated by relatively large spaces which contain water.*[13] In addition, their calculations confirmed Svedberg's large molecular weight values obtained by sedimentation in the centrifuge. In an article published back to back to Bernal and Crowfoot's, Astbury and Lomax stated that:

> we are left now with the paradox that the pepsin molecule is both globular and also a real, or potential, polypeptide chain system, and the immediate question is whether the

chains are formed by metamorphosis and linking-up of the globular molecules, or whether the initial unit is the chain itself, which is afterwards folded in some neat manner which is merely an elaboration of the intra-molecular folding that has been observed in the keratin transformation.[14]

Images of insulin, excelsin, lactoglobulin, hemoglobin, and chymotrypsin protein crystals were acquired by various groups in the next half decade, and this discipline exploded thereafter with more than a dozen new proteins by the late 1940s.[5] In 1938, Bernal, Fankuchen, and Peritz published a paper on hemoglobin and chymotrypsin, in which they observed that:

> the dried crystals of chymotrypsin show not only alterations of spacing but also of relative intensities of reflections. If we assume that drying takes place by the removal of water from between protein molecules, studies of these changes provide an opportunity of separating the effects of inter- and intra-molecular scattering. This may make possible the direct Fourier analysis of the molecular structure, once complete sets of reflections are available in different states of hydration.[5]

Thus, it was Max Perutz, a former student of Astbury and a student of Bernal, who tested the idea that protein crystals were essentially rigid, but could move slightly, in a relative manner, depending on their water content. His success in this area was two-fold: he shed light *on the behaviour of water between protein molecules in various states of humidity and on the possibility of limiting from the observations the overall shape of hemoglobin.*[5] Using this new knowledge using X-ray diffraction, Linus Pauling, Robert Corey, and H.R. Branson reported, in 1951, that they had solved two protein structures: a spiral and a pleated sheet structure.[15]

Linus Pauling was born in Portland, Oregon in 1901 and was considered by many the greatest chemist of the 20th century, with significant contributions in quantum mechanics, crystallography, mineralogy, structural chemistry, anesthesia, immunology, medicine, and evolution.[16] He was also the founder of modern molecular biology and structural chemistry, and received the Nobel Prize for chemistry in 1954; however, his contributions were not only limited to the sciences, as he took a stand against nuclear testing in the second half of his life and had a significant role in establishing a treaty banning atmospheric tests, actions for which he received the Nobel Peace Prize in 1962.[16] Pauling is the only person to have won two unshared Nobel Prizes. He was eager to learn very early in life, having read the bible and Darwin's Theory of Evolution by the age of nine, and he has been quoted as saying that his favorite book at that time was the Encyclopedia Britannica. He decided to study chemistry at the age of 13 and left school in 1917, without graduating, to attend Oregon Agricultural College as a chemical engineering major, where he was offered a full-time teaching position at the age of 19! After reading papers by Lewis and Langmuir concerning the electronic structure of molecules, Pauling found a *strong desire to understand the physical and chemical properties of substances*

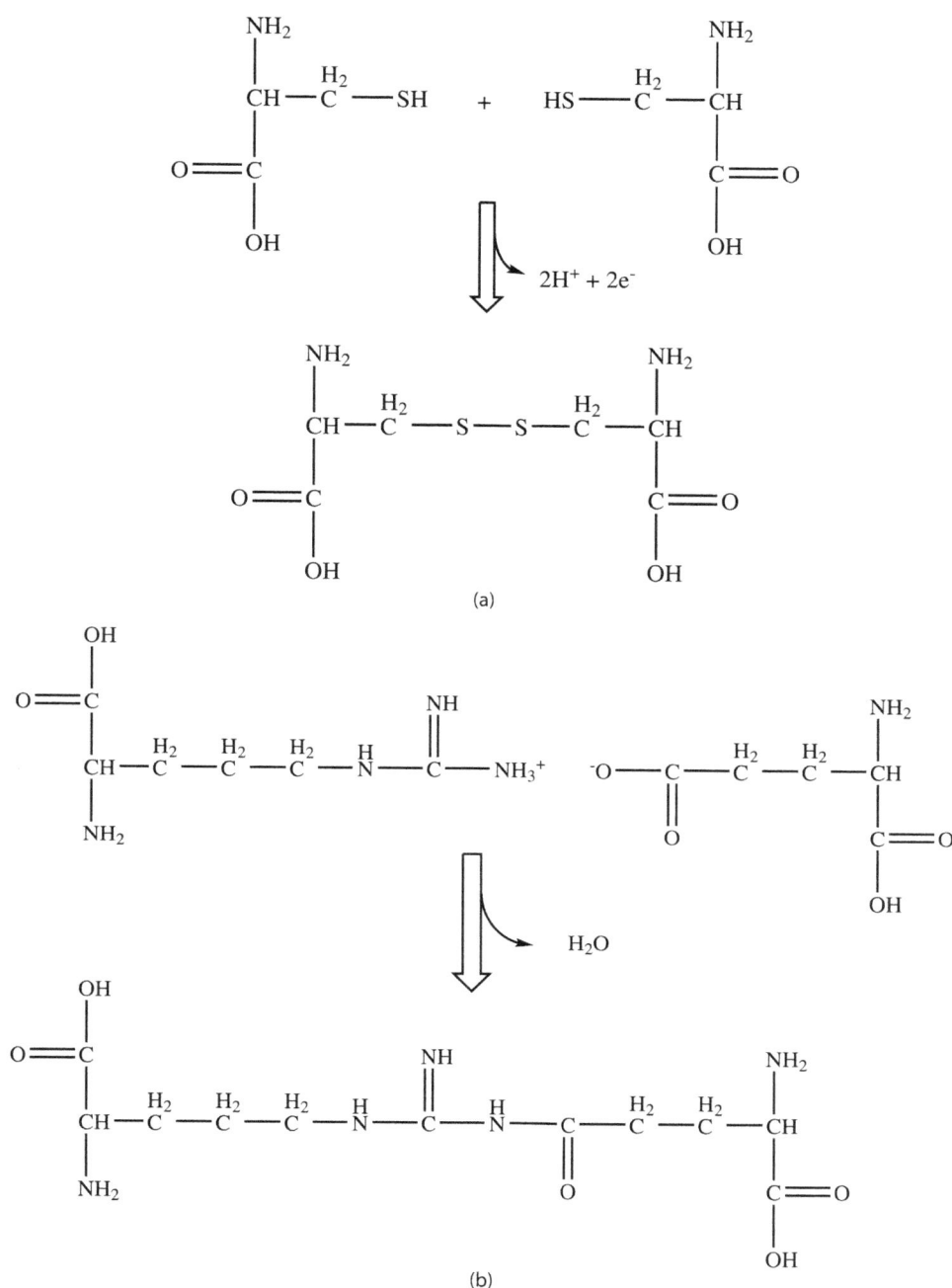

FIGURE 6.6 Interactions between amino acid side chains. (a) A disulfide bridge (S–S); (b) a salt bridge between the positively charged –NH₃ on the arginine and the negatively charged –O on the glutamic acid.

in relation to the structure of the atoms and molecules of which they are composed.[16]

Pauling started his Ph.D. on the determination of crystal structures by X-ray diffraction at a new school in Pasadena, the California Institute of Technology, in 1922, where he remained for over 40 years. He first started thinking about proteins in the early 1930s, and in collaboration with a visiting protein chemist from the Rockefeller Institute of Medical Research, Alfred Mirsky, began investigating the different properties of native versus denatured proteins (such as the change in appearance of egg white after it has

been cooked).[17] In 1936, they published a paper describing their theory of protein denaturing, in which they attribute the different properties of individual proteins *to their uniquely defined configurations*, and therefore, denatured proteins were *characterized by the absence of a uniquely defined configuration.*[18] Protein denaturization was proposed to come about as a result of exposure to excess heat, alkali, acids, alcohol, etc.—conditions which had been shown to result in a loss of protein function and immunological specificity, decreased solubility leading to coagulation, and changes in certain chemical properties, such as availability of sulfhydryl, disulfide, and phenol groups. They also concluded that one of the major forces holding proteins together was the hydrogen bond. Therefore, the breaking of hydrogen bonds in a protein was proposed to be one of the initial consequences of exposure of proteins to a denaturing agent, leaving amino acid side chains free to associate with other side chains in a manner that is not native to the protein, and which in turn leads to a loss of the protein's unique configuration and therefore, loss of its specific properties.[18] For example, alcohols such as ethanol (C_2H_5OH) denature proteins by forming hydrogen bonds with amino acid side chains, effectively competing against the native hydrogen bonds in proteins; in contrast, acids supply protons to electronegative atoms, whereas bases remove protons required for hydrogen bonding. This explained why denatured proteins coagulate, since the exposed side chains were now available to form nonspecific hydrogen bonds between amino acids from a variety of different proteins, forming a precipitate called a coagulum.

These ideas about protein denaturation led Pauling to ponder the precise nature of protein structure and folding, a topic to which he would devote 15 years.[17] In 1937, he first attempted to decipher the structure of α-keratin, on which Astbury had done much work using X-ray diffraction, with a plan to *use my knowledge of structural chemistry to predict the dimensions and other properties of a polypeptide chain, and then to examine possible conformations of the chain*.[17] However, he eventually reached the conclusion that a piece of the puzzle was still missing and stopped working on the problem. That same year, Robert Corey arrived at Caltech as a research fellow *interested in the problem of determining the structure of proteins*. As such, Pauling asked him to start with amino acids and small peptides, for which the precise structure had not yet been determined, reasoning that this information would be instrumental in elucidating the structure of proteins. After one year of research, Corey had succeeded in obtaining structures for glycine and alanine, but his progress was interrupted by the war. Nearly ten years later, Corey had managed to determine the structure of a dozen amino acids and some simple peptides with high accuracy, so Pauling decided to turn his attention back to α-keratin. Using the information he and others had gathered over the previous 15 years, Pauling's efforts resulted in the proposal of two structures for the polypeptide chain: the alpha helix and the pleated sheet.

He first started with the assumption that all amino acids were structurally equivalent with respect to the folding of the polypeptide chain even though they were asymmetrical molecules. Inspired by a theorem of mathematics, which states that *the conversion from one asymmetrical object into an equivalent asymmetric object* could be accomplished by one rotation about an axis followed by a translation along that axis, he set out to determine if any of these transformations resulted in hydrogen bonding between amino acid residues.[19] Limiting the number of possibilities was the fact that the α-carbon of an amino acid was known to be in resonance with its bonded nitrogen and oxygen, meaning that the C–N bond

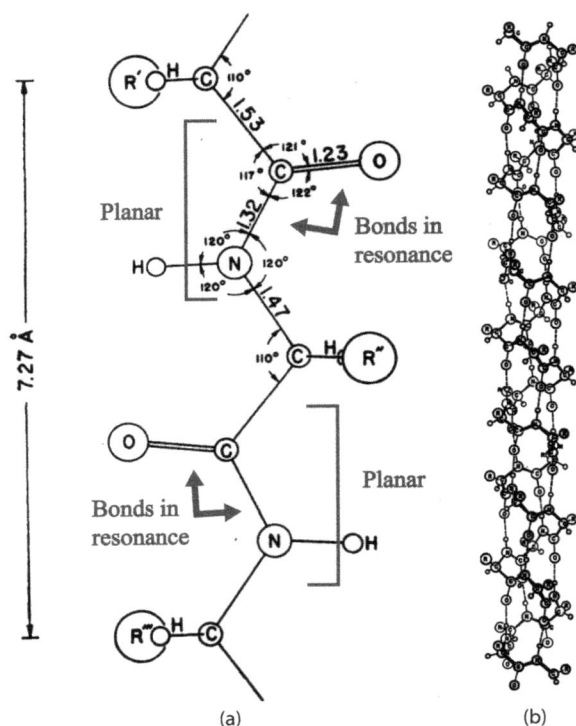

(a)   (b)

FIGURE 6.7 (a) A section of a protein with three amino acids (side chains R′, R″, and R‴), which shows bond angles, length between amino acids, and the C–N and C–O bonds in resonance; (b) the helical shape of a protein.[19] (Figures reprinted by permission from the Linus Pauling and the structure of proteins, *OSU Libraries Special Collections & Archives Research Center* (1951 p. 8).)

could not be rotated, and therefore, that the configuration of this section of a peptide bond was planar (Figure 6.7a). To help him visualize the various possible transformations, he sketched a protein chain on a piece of paper and attempted to fold it in a manner in which hydrogen bonds formed between the residues and found only five possible configurations.[17] After further consideration of these structures, he concluded that the only configuration that agreed with the known data was that of the helix (Figure 6.7b), which he suggested was a major component of α-keratin and other similar fibrous proteins, as well as some globular proteins.[19]

Next, he thought about the extended conformation of fibrin and β-keratin, which was assumed to form lateral hydrogen bonds with residues from a chain running in the opposite direction (Figure 6.8a).[20] However, Pauling and Corey reported that another configuration was possible, that of hydrogen bonds forming with residues in chains running in the same direction, which they termed the pleated sheet (Figure 6.8b). In this configuration, the hydrogen bonds between the two chains are perpendicular to the plane of the sheet, *directing their carbonyl groups in one direction and their imino groups in the other direction, and all the chains are oriented in the same way*.[20] They also found evidence that this type of configuration was common in stretched muscle, stretched hair, feather keratin, and a number of other fibrous proteins. Here, the resonance characteristic of the α-carbon

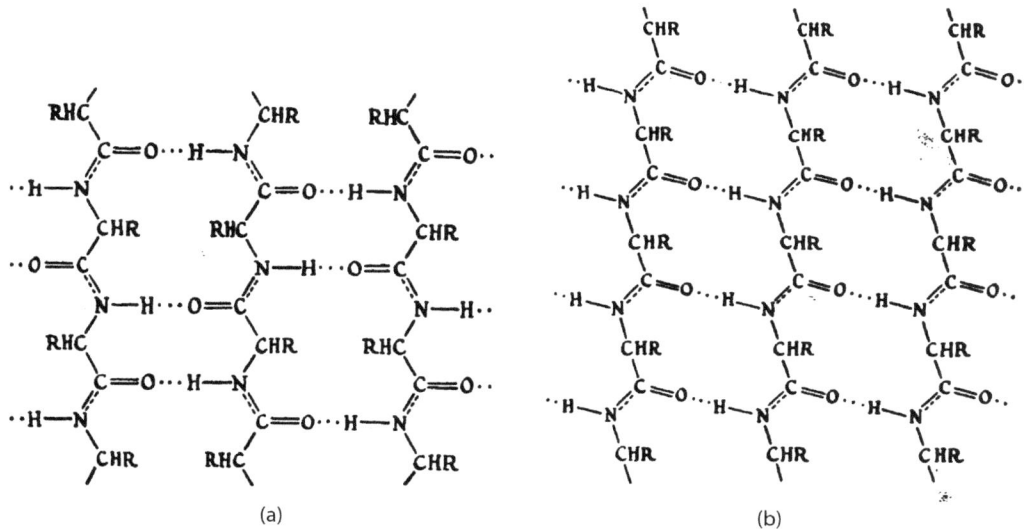

FIGURE 6.8 Structure of hydrogen bonding in proteins. (a) β-sheet in which alternate chains oriented in opposite directions form hydrogen bonds; (b) β-sheet in which all chains are in the same direction—the staggered arrangement of the hydrogen bonds forms a "pleated sheet" structure.[20] (Reprinted by permission from the Linus Pauling and the structure of proteins, OSU Libraries Special Collections & Archives Research Center (1951 p. 11).)

in amino acids contributed significantly to the property of these proteins since it allowed a certain amount deformation, and thus explained the stretchability of these molecules.

* * * * *

Despite the great advances in nucleic acid research in the early part of the 20th century, little was known about the structure and function of nucleic acids, which were still mostly thought of as small molecules with little biological importance since they were composed of only four subunits: ribonucleic acid (RNA) was composed of adenine, uridine, guanine, and cytosine; whereas deoxyribonucleic acid (DNA) was composed of adenine, thymine, guanine, and cytosine.[21,22] In fact, up until the 1940s, proteins were generally regarded to be the carriers of heritable information given the potential diversity of information that could be encoded from 21 amino acids versus that of nucleic acids with its 4 bases. However, some interesting discoveries came about in the 1920s and 1930s that inched our knowledge forward in terms of the importance of nucleic acids.

As mentioned at the end of Chapter 4, RNA was thought to be a component of yeast and plant cells, whereas DNA was a component of mammalian cells. However, in 1924, Robert Feulgen and Heinrich Rossenbeck reported the use of a procedure which specifically stained DNA (now called the Feulgen stain), which resulted in the surprising demonstration that DNA was present in the nucleus of all cells, regardless of their origin.[23] In contrast, it was later determined that RNA was mainly located in the cytoplasm, though some could also be found in the nucleus, and it became clear that the two kinds of nucleic acids were indeed equally found in all species. However, the functional differences between these nucleic acids, as well as how their subunits were connected, remained elusive. Nonetheless,

results suggested that DNA was somehow linked to cell division, whereas RNA was linked to protein synthesis since it was found to be present in large amounts in cells where protein synthesis was abundant (this will be further discussed in Chapters 7 and 8). In the 1940s, Oswald Avery and colleagues followed up on some results by Fred Griffith published in the 1920s, which would lead to the elucidation of the cellular role of DNA.

Griffith was born in England in 1877 and graduated in medicine at Liverpool in 1901.[24] He eventually worked for the government in the Pathological Laboratory of the Ministry of Health with his brother, Scott, during World War I until it was rebranded as the Emergency Public Health Laboratory Service. Sadly, Griffith and his brother were killed after receiving a direct hit from an air raid in 1941. In the 1920s, Griffith was investigating a trend in pneumonia where, over a period of five years, he noticed a significant decrease in the incidence of type II pneumococci and a corresponding increase in type IV.[25] He also found *a particularly striking* instance of a single pneumonia patient harboring 4 different types of pneumococci: one type I and three type IVs. To understand this phenomenon better, and *In the course [...] of other inquiries and of the routine examination of sputum* from pneumonia patients, he inoculated mice with samples of patient's bacteria, which, not surprisingly, usually resulted in the mice's death within a few days. However, upon examination of the types of bacteria obtained from the deceased mice, he found several instances in which mice injected with one type of pneumococcus would harbor other types.

Pneumococci virulence can easily be distinguished by two distinct phenotypes: smooth bacteria (S form) are virulent and grow in colonies with a smooth appearance when plated on blood agar plates; rough bacteria (R form) are

Killed S pneumococci	Living R pneumococci	No. of mouse	Result		Type of culture obtained from mouse
Pn. 85, Group IV, steamed 20 mins. Dose = deposit of 60 c.c. of broth culture	R 4, Type II. Dose = 0·25 c.c. of blood broth culture	405	Died	4 days	S colonies, Type II
		406	Killed	7 ,,	None
		407	,,	7 ,,	R colonies
		408	Died	4 ,,	S colonies, Type II
Pn. 160, Group IV, as above	R 4, Type II as above	409	Killed	7 days	S colonies, Type II
		410	Died	4 ,,	,, ,,
		411	,,	4 ,,	,, ,,
		412	,,	3 ,,	,, ,,
II B, Group IV, as above	R 4, Type II as above	413	Died	3 days	S colonies, Type II
		414	,,	2 ,,	,, ,,
		415	,,	3 ,,	,, ,,
		416	Killed	7 ,,	R colonies
None	R 4, Type II. Doses = 0·75, 1·0, 1·0 c.c. of blood broth culture	462	Killed	19 days	None
		463	,,	19 ,,	,,
		464	,,	19 ,,	,,

FIGURE 6.9 Results of one of Griffith's experiments showing that avirulent bacteria can be transformed to the virulent form. Heat-killed virulent bacteria were inoculated alone or with live avirulent bacteria into mice. Cultures from dead or killed mice were analyzed for the presence and type of bacteria. **Killed S. pneumococci**: Type of S bacteria that was heat-killed before injection; **Living S. pneumococci**: type of living bacteria that was co-injected with the killed bacteria. **Results** indicate whether the mouse died the infection or was killed to assess the bacterial culture present, as well as the number of days following injection at which this took place. Note how mice #405, #408, and #410–415 all died from an S-type bacterial infections even though they were injected with avirulent, R-type bacteria.[25]

nonvirulent and grow in colonies with a rough appearance. Thus, Griffith found, in one instance, a mouse which succumbed to disease from rough bacteria which *possessed a combination of rough cultural characteristics and virulence which had not previously been noted.* He offered two hypotheses to explain these odd results: (1) less invasive strains (types III and IV) were already present in the mouse at the time of inoculation of the more virulent strains (types I and II) and (2) the inoculated strain somehow transformed into another strain via an unknown process. About the possibility of the latter, Griffith admitted, *[o]n a balance of probabilities interchangeability of type seems a no more unlikely hypothesis than multiple infection with four or five different and unalterable serological varieties of pneumococci.*[25] Further experimentation into this phenomenon showed that S type of bacteria could indeed be "converted" to R type, and that these "new" R bacteria could revert back to the S type when inoculated into mice.

Thus, to test the requirements for the transformation of bacteria, Griffith carefully crafted a set of experiments in which he inoculated mice with one type of bacteria (S or R) along with heat-killed bacteria of the other type. Interestingly, he found that an S form (virulent) of pneumococcus that had been heat killed could transform the R form into a virulent strain when inoculated together in mice (Figure 6.9). Importantly, he found the temperature of the heat treatment was critical: temperatures too low or too high would prevent transformation. To explain these fascinating results, he turned to the most likely candidate of the times—proteins. It was known at the time that virulent pneumococci produced a soluble substance, assumed to be composed of proteins, that was specific for the various types of pneumococci. Consistent with this idea, Griffith found that when a virulent type transformed to the avirulent

type, the ability to secrete this substance was lost. In addition, it was assumed that the ineffectiveness of excessive heat treatment was due to the destruction of the protein. Thus, Griffith proposed that there was an *S antigen* that was secreted by virulent bacteria that could be transferred to nonvirulent types, stating that:

By S substance I mean that specific protein structure of the virulent pneumococcus which enables it to manufacture a specific soluble carbohydrate. This protein seems to be necessary as material which enables the R form to build up the specific protein structure of the S form.[25]

Unfortunately, elucidation of this "transforming principle" would have to wait until 1944, when Oswald T. Avery, Colin Macleod, and Maclyn McCarty published their seminal paper on transformation.[26] By this time, Griffith's experiments had been replicated several times by different researchers, including conditions in which the transformation was successfully performed *in vitro*, and Griffith's conclusions were widely accepted. A variety of alternate hypotheses were suggested, in addition to Griffith's, to account for the phenomenon, including a genetic explanation, in which the "transforming principle" was said to be similar to a gene, and the capsular protein generated as a result of transformation was a gene product. Concerning the latter, Avery *et al.* stated that [if] *this transformation is described as a genetic mutation–and it is difficult to avoid so describing it–we are dealing with authentic cases of induction of specific mutations by specific treatments.*[26]

Avery et al.'s paper began by describing the "reaction system"—that is, nutrient broth, presence of serum, and the bacterial strain used—required for the efficient and consistent transformation of pneumococci, followed by the method of extraction and isolation of the transforming principle.

In addition, they described manners in which the transforming principle could be inactivated and steps to be taken to prevent this from occurring (e.g., storage conditions, temperature during purification, etc.). Next, the authors described the results of various tests they performed to identify the purified substance: tests that identified peptide bonds and those using proteases came back negative, which indicated the absence of proteins in the isolate, whereas tests that reported the presence DNA were positive. In addition, analysis of the chemical contents of their samples showed that the ratio of nitrogen to phosphorus was within range of that expected for DNA, and DNase, an enzyme that degrades DNA, was found to inactivate the transforming principle. At the time, DNA had not been shown to be present in pneumococci, so the suggestion that it may be the transforming principle was all the more surprising. As such, the authors stated that:

> It is evident, therefore, that not only is the capsular material reproduced in successive generations but that the primary factor, which controls the occurrence and specificity of capsular development, is also reduplicated in the daughter cells. The induced changes are not temporary modifications but are permanent alterations which persist provided the cultural conditions are favorable for the maintenance of capsule formation.[26]

Although it would be more than a decade before the central dogma of biology, that DNA makes RNA and RNA makes proteins, would be put forth by Francis Crick (see Chapter 7), Avery was treading closely to it by the spirit of the following statement concerning bacterial transformation:

> Equally striking is the fact that the substance evoking the reaction and the capsular substance produced in response to it are chemically distinct, each belonging to a wholly different class of chemical compounds […] Thus, it is evident that the inducing substance and the substance produced in turn are chemically distinct and biologically specific in their action and that both are requisite in determining the type specificity of the cell of which they form a part.[26]

The authors concluded the paper by stating that if,

> the biologically active substance isolated in highly purified form as the sodium salt of desoxyribonucleic acid actually proves to be the transforming principle, […] then nucleic acids of this type must be regarded not merely as structurally important but as functionally active in determining the biochemical activities and specific characteristics of pneumococcal cells. Assuming that the sodium desoxyribonucleate and the active principle are one and the same substance, then the transformation described represents a change that is chemically induced and specifically directed by a known chemical compound. If the results of the present study on the chemical nature of the transforming principle are confirmed, then nucleic acids must be regarded as possessing biological specificity the chemical basis of which is as yet undetermined.[26]

* * * * *

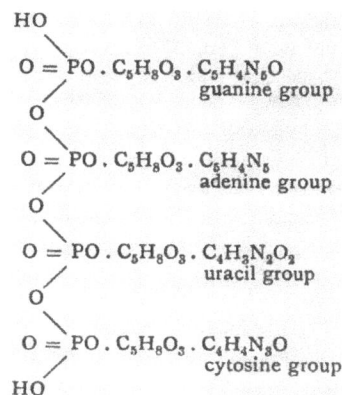

FIGURE 6.10 Structure of DNA according to Levene's tetranucleotide theory. Levene envisioned DNA to be a molecule composed of the four nucleotides connected end-to-end via phosphate bonds, in a constant and repeating order.[30]

Erwin Chargaff was inspired by Avery's manuscript the moment it was published and was among the first biochemists to reorganize his laboratory in order to study differences in the chemical make-up of nucleic acids.[27] Chargaff was an Austro-Hungarian analytical chemist who fled Europe early in the rise of the Nazi movement and subsequently relocated, in 1935, to the recently founded Department of Biochemistry of the College of Physicians and Surgeons of Columbia University in New York. Up until the 1940s, Levene's tetranucleotide theory—suggested for yeast nucleic acids in 1917—had been the accepted theory about the structure of both deoxyribonucleic and ribonucleic acids.[22,28] It stated that nucleic acids were formed by the end-to-end joining of each of the four nucleotides in a determined and invariable sequence (Figure 6.10), and, according to Gulland, this was supported by several lines of evidence: (1) both types of nucleic acids were essentially the same, so conclusions about one could automatically be applied to the other; (2) quantitative studies in yeast showed an equal amount of purines and pyrimidines; (3) the molecular weight of yeast nucleic acid was determined to be exactly that of the sum of the four nucleotides; (4) enzyme studies supported the tetranucleotide theory.[29]

However, evidence against this hypothesis had slowly come to light, starting with the fact that nucleic acids were shown to be much bigger molecules than previously thought; in fact, they were bigger than any other known molecule, which led to the suggestion that each of the tetranucleotide molecule assembled into a large molecule, a polytetranucleotide, which consisted of end-to-end joining of tetranucleotides.[21] However, in hindsight, Gulland admitted that the tetranucleotide theory would never have gained traction if the true molecular weight of the molecule had been discovered earlier, stating:

> It should be realised that the conception of a molecule composed of polymerised tetranucleotides has grown from a mental superposition of the later demonstrations of high

molecular weights on to the older ideas of a simple molecule containing one of each of the four appropriate nucleotides; had the true molecular sizes been realised earlier, it
is doubtful whether the conception would have gained such
firm hold as is apparently the case.

At present the facts tend to favour the hypothesis of molecules composed of polymerised tetranucleotides in the cases
of yeast ribonucleic acid and some examples, at any rate,
of deoxypentose nucleic acids, but sound evidence is non-
existent either to support the existence of a uniform tetranucleotide composed of one of each of the appropriate nucleotides or to indicate a regular sequence of nucleotides. Turning
to pentose nucleic acids in general, there are contrary indications to a uniform content of the four nucleotides.[29]

It was in the midst of these controversies that Chargaff
began studying nucleic acids and supposed, in 1944, following his reading of Avery's seminal paper on bacterial
transformation, that the *genetically significant differences
determined by DNA might be reflected in analytically
detectable differences in the content and order of the DNA
bases, i.e., the purines and pyrimidines.*[27] In the introduction of the first of several papers published by his laboratory
between 1948 and 1952, the authors stated that it is

hard to conceive of two macromolecular substances *[DNA
and RNA]*, synthesized in as different cellular systems as
calf thymus and yeast, as being identical, although, in the
absence of immunological or more refined chemical methods, a strict decision may not yet be possible.[31]

They went on to describe their improved method for isolation of intact DNA in yeast by first crushing them in a
buffer solution, noting that cold temperatures were required
throughout the process to avoid degradation.[31] The solution
was then centrifuged, resulting in three layers: two solid
layers consisting of a bottom layer containing intact cells
and an upper layer consisting of cell debris, and an opal-
escent top layer. The latter was removed and the top solid
layer was separated from the bottom layer, kept in a refrig-
erator for 72 hours, and centrifuged again. Ice cold ethanol
was then added to the viscous supernatant, which resulted
in the precipitation of "white threads" that could be wound
on a glass rod and separated. Analysis of this substance
revealed the isolation of pure DNA.

Chargaff's refinement of nucleic acid isolation allowed
him to quantify the nucleotides present in nucleic acids
more precisely. However, to do so, Vischer and Chargaff
first had to improve paper chromatography—a technique
whereby molecules were separated by passing a solution
of those molecules through filter paper—to allow better
separation of nucleotides.[32] This involved testing numerous
solvents to determine which would be the most adequate
carrier of nucleotides, while still allowing the best separation. To that end, they passed known mixtures of nucleotides
dissolved in various solutions through filter paper and dried
the paper in an oven. The purines were then fixed using
a mercuric nitrate solution, whereas the pyrimidines were

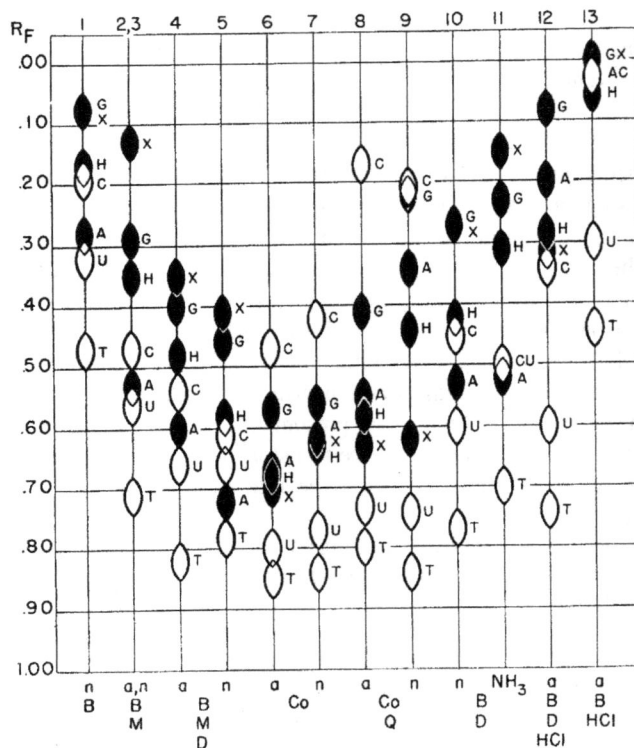

FIGURE 6.11 Separation of nucleotides in various solvents.
Known quantities of the indicated nucleotides were passed
through filter paper by various solvents and labeled as mercuric salts. A, adenine; G, guanine; H, hypoxanthine; X, xanthine;
U uracil; C, cytosine; T, thymine. Letters at the bottom represent
various solvents: a, acid; n, neutral; B, n-butanol; M, morpholine;
D, diethylene glycol; Co, collidine; Q, quinoline.[32]

fixed in a mercuric acetate solution, both of which resulted
in the formation of a mercuric complex. To visualize the
nucleotide bands, the filter paper was passed through a solution of ammonium sulfide, which resulted in *well defined
black spots of mercuric sulfide* where ever nucleotides were
present (Figure 6.11).[32] To quantify the amount of nucleotide, each band was cut out, allowed to air dry, and the
nucleotides were extracted from the filter paper using HCl
and subjected to ultraviolet spectroscopy (it was known at
the time that DNA absorbs light in the ultraviolet range in a
manner that was proportional to concentration).

Thus, having shown a proof of concept for their new technique, Vischer and Chargaff investigated the *distribution of
nitrogenous constituents of pentose nucleic acids in yeast
and pig pancreas.*[33] They also used their new technique
to study the decomposition of pyrimidines upon the harsh
treatment that were required to study them. Uracil, cytosine,
and thymine are molecules formed from the same basic
molecule and differing only by addition of one molecular
group (see Chapter 4, Figure 4.7) and cytosine was known
be converted to uracil quite easily. Therefore, it was conceivable that the harsh treatment used in the study of nucleic
acids might result in this alteration and for this reason, the
presence of uracil in nucleic acids was put into question by

many.[33] As such, Vischer and Chargaff subjected pyrimidines to a variety of chemical treatments and by using paper chromatography through which to run the products, found that using concentrated formic acid instead of strong mineral acids to extract these nucleotides preserved cytosine and prevented its conversion to uracil. Using this milder hydrolysis method, they showed that uracil was indeed present in RNA and not just a conversion product of cytosine; however, they found no thymine in RNA. Importantly, although no conclusive values could be given for the ratio of nucleotides, this study did conclusively show that nucleotides were not present in equal amounts as dictated by Levene's tetranucleotide theory and set in motion the demise of a paradigm that had existed for nearly half a century! These findings were followed up the next year in a paper studying the composition of nucleic acids in the spleen, in which they confirmed that the sugar was that of thymus DNA, that is, 2-desoxyribose.[34] Chargaff and his colleagues also investigated the composition of DNA in samples obtained from various sources using this technique and found that:

> The desoxypentose nucleic acids extracted from different species thus appear to be different substances or mixtures of closely related substances of a composition constant for different organs of the same species and characteristic of the species.
>
> It is, however, noteworthy—whether this is more than accidental, cannot yet be said—that in all desoxypentose nucleic acids examined thus far the molar ratios of total purines to total pyrimidines, and also of adenine to thymine and of guanine to cytosine, were not far from 1.[35]

As explained in the next section, this last observation would provide a crucial clue to Crick and Watson in elucidating the structure of DNA and in proposing a mechanism for DNA replication. Finally, Chargaff emphasized the fact that given the complexity now known concerning the composition of DNA, *there will be no objection to the statement that, as far as chemical possibilities go, [DNA] could very well serve as one of the agents, or possibly as the agent, concerned with the transmission of inherited properties.*[35]

* * * * *

Francis Crick and James Watson are without a doubt two names synonymous with DNA. The events surrounding their discovery of the structure of DNA in 1953, including its controversies, have been extensively documented and, therefore, will not be explored here in detail. However, its discovery is a significant part of the history of cell biology and this story would be incomplete without it. I will therefore summarize the essential elements of this time period, without going into the intricate details of the events. Francis Crick was born in Northampton, England, where he obtained a B.Sc. from the University College, London and, like many others, had his Ph.D. interrupted by the war.[36] During his Ph.D.—whose thesis

on "The viscosity of water at high pressures" he described as *unimaginably dull*—a bomb fell on his laboratory and put an end to his studies in physics.[37] At the age of 31, though he knew very little about biology or organic chemistry, he decided to change his career to the study of biology and started a Ph.D. in the laboratory of Max Perutz studying the structure of proteins. He obtained his Ph.D. in 1954, but two years into his degree, in 1951, a fateful encounter with James Watson would change his life.

James Dewey Watson was an American scientist born in Chicago in 1928. He then received an experimental scholarship from the University of Chicago, which allowed him to start his B.Sc. at the age of 15.[38] Watson was drawn to zoology due to a childhood interest in bird-watching, but soon developed an interest in genetics. He pursued the latter for his Ph.D. at Illinois University in the laboratory of Salvador Luria, an eminent scientist in the field of bacterial viruses, where he studied the effects of X-rays on bacteriophage multiplication. He met Maurice Wilkins at a symposium in 1951, where he saw X-ray diffraction patterns of DNA for the first time and was so excited that he went back to Illinois wishing to change the direction of his research. This was made possible by Luria, who arranged for Watson to work in England at the Cavendish laboratory with John Kendrew. It was there that Crick and Watson met for the first time and discovered their common interest in the structure of DNA. At this time, numerous scientists had made a number important discoveries on the molecular nature of DNA, but none could unify these findings to suggest a plausible molecular structure for DNA. Importantly, you may remember from Chapter 4 that previous work had identified a phosphate–sugar–base arrangement. However, the final piece of the puzzle was the X-ray crystallography work of Rosalind Franklin, who was born in London in 1920. There, she entered Newham College in 1938 to pursue studies in the natural sciences and obtained her Ph.D. in 1945, after which she worked in Paris doing research on the structure of coal using X-ray crystallography.[36] In 1950, she obtained a 3-year fellowship at King's College London, where she was to study the X-ray patterns of denatured proteins but this project was changed—unbeknownst to her—to the structure of DNA prior to her arrival when Maurice Wilkins received some good DNA samples.

Astbury had been the first to obtain X-ray diffraction images of DNA in 1936 and had noted that the molecule was capable of stretching 250% and that the *spacing along the fibre axis [...] is almost identical with that of a fully extended polypeptide chain system*, such as β-keratin and β-myosin.[39] However, he erroneously suggested that its structure was composed of *a close succession of flat or flattish nucleotides standing out perpendicularly to the long axis of the molecule to form a relatively rigid structure.*[39]

Crick and Watson both started working on the structure of DNA as side projects, to be indulged outside of their laboratories and in their own time. In fact, Crick's mentor, William Bragg, was not altogether happy with his student's extracurricular hobbies and strongly discouraged him from pursuing such avenues. This did not deter the budding scientists and the pair nevertheless started building DNA models. However, Watson, due to his lack of knowledge in crystallography, misunderstood part of one of Franklin's talks and thought there were eight molecules of water in a unit cell per DNA chain, instead of the reported eight per nucleotide, underestimating the water content by a factor of 11! This led them to suggest a three stranded helix structure for DNA with the sugar and phosphate backbone on the inside and the bases on the outside. After their model was completed, they invited Wilkins and Franklin to inspect the structure and Franklin immediately spotted the mistake. In addition, she did not yet believe that the data supported a helical structure for DNA and thought that the phosphates should be on the outside. News of this meeting reached Crick's mentor, Bragg, who at once instructed Crick and Watson to stop working on the structure of DNA and to leave this area of research to Wilkins and Franklin at King's College. Crick and Watson thus returned to their respective research areas.

However, the duo's fervor for elucidating the structure of DNA was reignited in 1952 when Linus Pauling sent a letter to his son at the Cavendish laboratory (who shared an office with Crick and Watson) announcing the eminent publication of his proposed structure for DNA. Alarmed at this news, they at once asked for an advance copy of the manuscript. When they received it, they immediately saw an error in the proposed triple helix structure: it was held together by hydrogen bonds which resulted in the neutralization of the charges on the phosphate groups, noting that *Pauling's nucleic acid, in a sense, was not an acid at all.*[36] Around the same time, Watson stopped by Wilkin's office at King's College for a visit, and the latter proceeded to show him an X-ray crystallography image of DNA that Franklin had recently taken (Figure 6.12). As soon as Watson saw the image, he was immediately convinced of the double helical structure of the molecule. The next day, he asked Bragg if he and Crick could be allowed to resume work on the structure of DNA. Bragg agreed.

Crick and Watson first decided to move the sugar/phosphate backbone to the outside of the structure (Figure 6.13a), as previously suggested by Franklin, which helped align the atomic distances in their model with the known crystallographic data. They also realized that the two strands must run opposite to one another given that the structure had a two-fold axis of symmetry. The model they first produced was one in which identical nucleotides were paired between the two chains. However, this required that distances between the backbones be varied to accommodate the different sizes of the nucleotides—T and C being single-ringed molecule, whereas A and G are the bigger, double-ringed, molecules—and thus, this was clearly wrong and had to be altered. Following much deliberation, Watson tried pairing a single-ringed with a double-ringed nucleotide—that is, adenine with thymine and guanine with cytosine (Figure 6.13 b and c)—and found that the

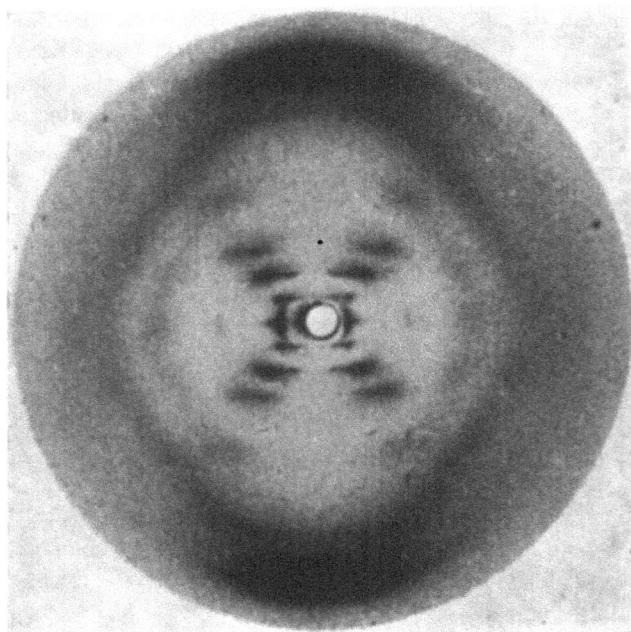

FIGURE 6.12 Franklin's famous X-ray image of DNA that inspired Crick and Watson.[40] Reprinted by permission from Springer Nature Customer Service Centre GmbH: Springer Nature, Nature, Franklin RE, Gosling RG. Molecular configuration in sodium thymonucleate. (Copyright © 1953.)

ensuing structure resolved all objections expressed about previous models. In fact, the molecule was well stabilized by two hydrogen bonds between each base, and since the distance between nucleotides would be constant, any nucleotide could be inserted at any position in the chain without altering its physical properties, which explained Chargaff's previous observations that nucleic acids were found to have equal ratios of A to T and C to G. When Watson showed Crick his model later that day, they knew they had solved the structure of DNA.

On April 25, 1953, Crick and Watson's paper on the structure of DNA was published (Figure 6.13d), in which they also noted that this model did not apply to RNA, *as the extra oxygen atom would make too close a van der Waals contact.* They concluded the manuscript by making the prescient observation that it *has not escaped our notice that the specific pairing we have postulated immediately suggests a possible copying mechanism for the genetic material.*[41] In a follow-up paper, they expanded on this last point, stating that if the sequence of one strand was known, the complementary strand could be written given that adenine only pairs with thymine and guanine with cytosine. In addition, they correctly suggested that *prior to duplication, the hydrogen bonds are broken, and the two chains unwind and separate. Each chain then acts as a template for the formation on to itself of a new companion chain.*[42]

* * * * *

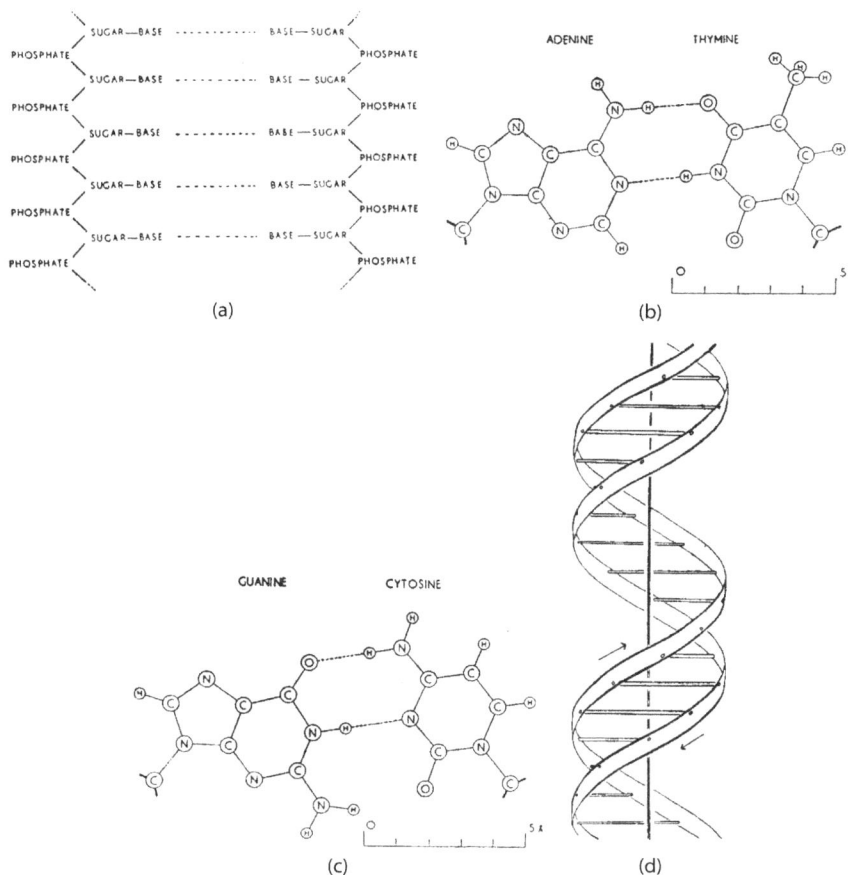

FIGURE 6.13 Crick and Watson's models of the structure of DNA and its hydrogen bonds. (a) Diagram representing the phosphate–sugar–base arrangement of each DNA strand, as well as the hydrogen bonds, as represented by the dotted line. (b) and (c) The hydrogen bonds of AT and GC pairs. (d) The Crick–Watson model presented in their landmark paper.[41,42] (Figure parts (a) and (d) reprinted by permission from Springer Nature Customer Service Centre GmbH: Springer Nature, Nature, Watson JD, Crick FHC. Genetical implications of the structure of deoxyribonucleic acid, Copyright © 1953. Figure parts (b) and (c) reprinted by permission from Springer Nature Customer Service Centre GmbH: Springer Nature, Nature, Watson JD, Crick FHC. Molecular structure of nucleic acids: A structure for deoxyribose nucleic acid, Copyright © 1953.)

REFERENCES

1. Osborne TB. Sulphur in protein bodies. *Journal of American Chemical Society.* 1902;24(2):140–167. doi:10.1021/ja02016a003

2. Claesson S, Pedersen KO. The Svedberg, 1884–1971. *Biographical Memoirs of Fellows of the Royal Society.* 1972;18:594–627. doi:10.1098/rsbm.1972.0022

3. Svedberg T, Fåhraeus R. A new method for the determination of the molecular weight of the proteins. *Journal of American Chemical Society.* 1926;48(2):430–438. doi:10.1021/ja01413a019

4. Svedberg T, Rinde H. The ultra-centrifuge, a new instrument for the determination of size and distribution of size of particle in amicroscopic colloids. *Journal of American Chemical Society.* 1924;46(12):2677–2693. doi:10.1021/ja01677a011

5. Hodgkin DC. Crystallographic measurements and the structure of protein molecules as they are. *Annals of the New York Academy of Sciences.* 1979;325(1 The Origins o):121–148. doi:10.1111/j.1749-6632.1979.tb14132.x

6. Ewald PP. *Fifty Years of X-Ray Diffraction.* Ewald PP, ed. Springer; 1962. doi:10.1007/978-1-4615-9961-6

7. Astbury WT. The X-ray interpretation of fibre structure. *Journal of the Society of Dyers and Colourists.* 1933;49(6):168–180. doi:10.1111/j.1478-4408.1933.tb01756.x

8. Bernal JD. William Thomas Astbury, 1898–1961. *Biographical Memoirs of Fellows of the Royal Society.* 1963;9:1–35. doi:10.1098/rsbm.1963.0001

9. Astbury WT, Woods HJ. The X-ray interpretation of the structure and elastic properties of hair keratin. *Nature.* 1930;126(3189):913–914. doi:10.1038/126913b0

10. Astbury WT, Street A. X-ray studies of the structure of hair, wool, and related fibres. I. General. *Philosophical Transactions of the Royal Society A: Mathematical, Physical and Engineering Sciences.* 1932;230(681-693):75–101. doi:10.1098/rsta.1932.0003

11. Astbury WT, Woods HJ. X-ray studies of the structure of hair, wool, and related fibres. II. The molecular structure and elastic properties of hair keratin. *Philosophical Transactions of the Royal Society A: Mathematical, Physical and Engineering Sciences.* 1933;232(707-720):333–394. doi:10.1098/rsta.1934.0010

12. Astbury WT, Sisson WA. X-ray studies of the structure of hair, wool, and related fibres—III: The configuration of the keratin molecule and its orientation in the biological

cell. *Proceedings of the Royal Society of London. Series A Mathematical and Physical Sciences.* 1935;150(871): 533–551. doi:10.1098/rspa.1935.0121

13. Bernal JD, Crowfoot D. X-ray photographs of crystalline pepsin. *Nature.* 1934;133(3369):794–795. doi:10.1038/133794b0

14. Astbury WT, Lomax R. X-ray photographs of crystalline pepsin. *Nature.* 1934;133(3369):795–795. doi:10.1038/133795a0

15. Pauling L, Corey RB. Two hydrogen-bonded spiral configurations of the polypeptide chain. *Journal of American Chemical Society.* 1950;72(11):5349–5349. doi:10.1021/ja01167a545

16. Dunitz JD. Linus Carl Pauling, 1901–1994. *National Academy of Sciences of the United States of America Biographical Memoirs.* Published online 1997:221–261. Accessed March 30, 2019. http://www.nasonline.org/publications/biographical-memoirs/memoir-pdfs/pauling-linus.pdf

17. Pauling L. Manuscript Notes and Typescripts; "The Discovery of the Alpha Helix". September 1982. Special Collections & Archives Research Center, Oregon State University Libraries: Published 2015. Accessed March 30, 2019. http://scarc.library.oregonstate.edu/coll/pauling/proteins/notes/1982a2.10.html

18. Mirsky AE, Pauling L. On the structure of native, denatured, and coagulated proteins. *Proceedings of National Academy of Sciences U S A.* 1936;22(7):439–447. doi:10.1073/pnas.22.7.439

19. Pauling L, Corey RB, Branson HR. The structure of proteins; two hydrogen-bonded helical configurations of the polypeptide chain. *Proceedings of National Academy of Sciences U S A.* 1951;37(4):205–211. Accessed November 2, 2022. Oregon State University Special Collections and Archives Research Center, Corvallis, Oregon. (1951p.8)

20. Pauling L, Corey RB. The pleated sheet, a new layer configuration of polypeptide chains. *Proceedings of National Academy of Sciences U S A.* 1951;37(5):251–256. Accessed November 2, 2022. Oregon State University Special Collections and Archives Research Center, Corvallis, Oregon. (1951p.11)

21. Gulland JM, Barker GR, Jordan DO. The chemistry of the nucleic acids and nucleoproteins. *Annual Review of Biochemistry.* 1945;14(1):175–206. doi:10.1146/annurev.bi.14.070145.001135

22. Cohen JS, Portugal FH. The search for the chemical structure of DNA. *Connecticut Medicine.* 1974;38(10):551–557. Accessed March 30, 2019. https://profiles.nlm.nih.gov/ps/access/CCAAHW.pdf

23. Brachet J. Ribonucleic acids and the synthesis of cellular proteins. *Nature.* 1960;186(4720):194–199. doi:10.1038/186194a0

24. Downie AW. Pneumococcal transformation-a backward view fourth Griffith memorial lecture. *Journal of General Microbiology.* 1972;73(1):1–11. doi:10.1099/00221287-73-1-1

25. Griffith F. The significance of pneumococcal types. *The Journal of Hygiene (London).* 1928;27(2):113–159. doi:10.1017/S0022172400031879

26. Avery OT, Macleod CM, McCarty M. Studies on the chemical nature of the substance inducing transformation of pneumococcal types: Induction of transformation by a desoxyribonucleic acid fraction isolated from pneumococcus type III. *Journal of Experimental Medicine.* 1944;79(2):137–158. doi:10.1084/jem.79.2.137

27. Cohen SS, Lehman IR. Erwin Chargaff, 1905–2002. *National Academy of Sciences of the United States of America Biographical Memoirs.* Published online 2010: 3–15. http://www.nasonline.org/publications/biographical-memoirs/memoir-pdfs/chargaff-erwin.pdf

28. Levene PA, Bass LW. *Nucleic Acids.* The Chemical Catalog Company, Inc.; 1931. Accessed November 7, 2022. https://babel.hathitrust.org/cgi/pt?id=uc1.b4165245&view=1up&seq=5

29. Gulland JM. Some aspects of the chemistry of nucleotides. *Journal of the Chemical Society (Resumed).* 1944;0(0):208. doi:10.1039/jr9440000208

30. Jones W. *Nucleic Acids: Their Chemical Properties and Physiological Conduct.* Longmans, Green, and Co.; 1914. Accessed November 7, 2022. https://archive.org/details/nucleicacidsthei00jonerich/page/n7/mode/2up

31. Chargaff E, Zamenhof S. The isolation of highly polymerized desoxypentosenucleic acid from yeast cells. *Journal of Biological Chemistry.* 1948;173(1):327–335. doi:10.1016/S0021-9258(18)35588-1

32. Vischer E, Chargaff E. The separation and quantitative estimation of purines and pyrimidines in minute amounts. *Journal of Biological Chemistry.* 1948;176(2):703–714. doi:10.1016/S0021-9258(19)52686-2

33. Vischer E, Chargaff E. The composition of the pentose nucleic acids of yeast and pancreas. *Journal of Biological Chemistry.* 1948;176(2):715–734. doi:10.1016/S0021-9258(19)52687-4

34. Chargaff E, Vischer E, Doniger R, Green C, Misani F. The composition of the desoxypentose nucleic acids of thymus and spleen. *Journal of Biological Chemistry.* 1949;177(1):405–416. doi:10.1016/S0021-9258(18)57098-8

35. Chargaff E. Chemical specificity of nucleic acids and mechanism of their enzymatic degradation. *Experientia.* 1950;6(6):201–209. doi:10.1007/BF02173653

36. Squires GI. The discovery of the structure of DNA. *Contemporary Physics.* 2003;44(4):289–305. doi:10.1080/0010751031000135959

37. Rich A, Stevens CF. Francis Crick (1916–2004). *Nature.* 2004;430(7002):845–847. doi:10.1038/430845a

38. James W. *Biographical.* NobelPrize.org. Nobel Media AB © The Nobel Foundation 1962. https://www.nobelprize.org/prizes/medicine/1962/watson/biographical/

39. Astbury WT, Bell FO. X-ray study of thymonucleic acid. *Nature.* 1938;141(3573):747–748. doi:10.1038/141747b0

40. Franklin RE, Gosling RG. Molecular configuration in sodium thymonucleate. *Nature.* 1953;171(4356):740–741. doi:10.1038/171740a0

41. Watson JD, Crick FHC. Molecular structure of nucleic acids: A structure for deoxyribose nucleic acid. *Nature.* 1953;171(4356):737–738. doi:10.1038/171737a0

42. Watson JD, Crick FHC. Genetical implications of the structure of deoxyribonucleic acid. *Nature.* 1953;171(4361): 964–967. doi:10.1038/171964b0

7 Protein Synthesis Part I
Localization of Protein Translation

By the early part of the 20th century, Fischer had demonstrated that proteins were composed of amino acids connected by peptide bonds; Pauling had determined two protein structures, the α-helix and β-sheet, and had shown that proteins were held together by hydrogen bonds and disulfide bridges; and the many enzymes known to play pivotal roles in cellular processes had been confirmed to be proteins. In addition, Mirsky and Pauling characterized the denaturation of proteins as the reversible loss of their properties brought about by *the absence of a uniquely defined configuration.*[1] However, a technique for the precise determination of the sequence of amino acids in a protein had yet to be elucidated. By the end of the 1940s, the sequences of only a few natural di- or tripeptides were known and these were usually elucidated by proteolysis using various enzymes or via synthesis. For example, glutathione was first isolated in yeast in 1921 and it was subsequently determined to be a tripeptide composed of two glutamic acids and one cysteine.[2] In 1930, the experiments by several groups enabled the elucidation of the exact sequence of these amino acids, which was accomplished following the analysis of peptide fragments generated by proteolysis with chemicals or enzymes (the sequence of the native protein can be elucidated by finding overlapping regions of the various fragments).[2] Although this method of trial and error and subsequent clever deduction was useful for the determination of amino acid sequences of small peptides, it was not appropriate for larger peptides or proteins since too many fragments would be generated to allow reconstruction by trial and error.

As such, an important technological advancement leading to the development of protein sequencing technology was the improvement, in 1941, of partition chromatography by Martin and Synge. Before this time, amino acids were separated by adding a drop of solution containing a mixture of amino acids—called the mobile phase—overtop an immiscible solution—the stationary phase—and letting them separate due to the effects of gravity and to each molecule's particular affinity for the stationary phase. Thus, molecules with greater affinity for the stationary phase were less mobile while those with less affinity moved progressively farther away from the starting point. However, the limitations of this technique included the fact that it often resulted in the movement of samples upward on the paper and only small amounts of samples could be tested to prevent diffusion of the sample as it was being dispensed. Martin and Synge realized that using a solid material, such as silica, might act as a support for the stationary phase and prevent any movement of the particles

in a direction other than the flow of the mobile phase. As such, a gelatinous mixture of silica and water was added to a tube, to which chloroform was added, and the mixture was allowed to settle as the chloroform drained at the bottom of the column.[3] After topping up the column with fresh chloroform, the amino acid samples, also dissolved in chloroform, were gently added to the top of the column and allowed to settle atop the silica. Separation of the molecules in the sample was initiated by addition of fresh chloroform to the top of the column, after which each amino acid migrated at a characteristic rate within the column depending on their affinity for the silica. As such, the relative distance travelled by each protein could be calculated and compared in terms of R values, that is, the distance travelled by an amino acid from the starting point. In 1943, Gordon, Martin, and Synge used their new technique to determine the full complement of amino acids in wool and gelatin, though the sequence thereof remained unknown.[4]

In 1945, Fred Sanger published the first of several papers which elucidated the sequence of the insulin protein using a technique he developed, whereby he identified the free amino terminal amino acids of insulin using a colorimetric reagent. Sanger was born into a Quaker family in Gloucestershire, England in 1913 where his father was the local doctor.[5] Influenced by his father's interests, he studied biology at Cambridge, where he became interested in the emerging field of biochemistry, reasoning that it offered a scientific basis to understand medical problems. He was offered a position at Cambridge soon after obtaining his doctorate degree in 1943 and chose to devote his efforts to the chemistry of insulin. At the time, detection of terminal amino acids was achieved by treating proteins with a reagent that reacted with free amino groups and yielded products that were insensitive to acid hydrolysis. Therefore, only the modified, terminal amino acid remained intact after acid hydrolysis of the protein. Experiments using these procedures consistently resulted in lysine derivative products and therefore, it was generally believed that lysine was the terminal amino acid of all proteins.[6] However, by the mid-1930s it was becoming clear that this could not be the case: assessment of the free amino nitrogen in several proteins, including insulin, showed that it was greater than could be accounted for by the presence of lysine alone. In addition, Jensen and Evans found a reagent that protected only phenylalanine from degradation and determined that it was a terminal amino acid in insulin.

In an effort to elucidate the true terminal amino acids of insulin, Sanger sought a reagent that could label all free amino groups equally well, since this would enable him to

DOI: 10.1201/9781003379058-7

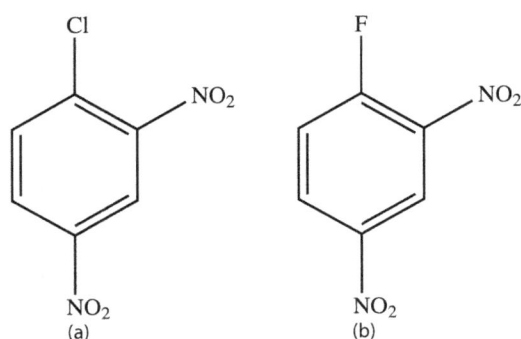

FIGURE 7.1 Fluorescent molecules used by Sanger to label terminal amino acids. (a) 2,4-dinitrochlorobenzene; (b) 2,4-dinitrofluorobenzene.

identify them following proteolysis of the protein. He was aware that 2,4-dinitrochlorobenzene (DNCB) had been used to label terminal amino acids but this reagent could not be developed into a reliable labeling reagent since it required heat for the reaction to occur, a condition under which proteins might be hydrolyzed. However, Sanger was convinced that DNCB would be ideal for his studies since the molecule was naturally of a bright yellow color and this would prevent having to add a coloring reagent to his sample mixture during chromatography. Fortunately, he found that the reaction could proceed at room temperature if the chloro group in DNCB was substituted for a fluoro group, resulting in 2,4-dinitrofluorobenzene (DNFB; Figure 7.1).[6] Thus, Sanger labeled the amino terminal amino acids of insulin with DNFB, completely hydrolyzed the protein with hydrochloric acid, and subjected his sample to partition chromatography as described by Martin and Synge. He then determined the identity of the labeled amino acids by comparing their R values to the R values of DNFB-labeled, synthetic amino acids that he ran along with his samples. As such, Sanger found that there were six free amino groups in insulin: 2x glycines, 2x lysines, and 2x phenylalanines; however, based on their R values, the lysine residues were determined not to be terminal amino acids, but were labeled simply because this amino acid contains an amino residue in its side chain. As a result, these data indicated that insulin was composed of a total of four separate chains, two of which had phenylalanine at their amino termini, whereas the other two had glycine.[6] In two subsequent papers, Sanger confirmed that the insulin chains were linked via disulfide bridges. To this end, he used performic acid to disrupt these bonds, which resulted in four separate chains, two of which contained glycine as the terminal amino acids (Chain A) while the other two contained phenylalanine, as well as the internal lysine residues (chain B).[7,8]

Despite these successes using partition chromatography, this technique also had its drawbacks. The major one was that some amino acids were adsorbed by the silica, which resulted in the loss of amino acids and required some considerations to be taken in the analysis. In addition, Consden, Gordon, and Martin found that the samples often moved too

rapidly through the column, which prevented proper separation, and therefore, reduce resolution.[9] Thus, the authors improved upon the capillary analysis technique of Schönbein and Goppelsroeder, a technique similar to that of partition chromatography but in which cellulose filter paper was used as a stationary phase instead of silica; this improvement was instrumental in the determination of the amino acid sequence of insulin by Sanger. As such, Consden, Gorden, and Martin built an apparatus consisting of a glass trough—*opened along its length by grinding* and which contained a suitable solvent (i.e., the mobile phase)—placed within one end of a stoneware drain pipe.[9] After drawing a line near one end of a strip of filter paper, at least 1.5 cm wide and 20–56 cm long, onto which a small drop of a sample of amino acids was deposited, the end of the paper with the sample was immersed in the glass trough and held in place using a microscope slide.[9] As the solvent gradually moved through the paper, the amino acids were carried along with the solvent and separated based on their affinity for that solvent. The paper was removed after 6–24 hours and dried in an oven. The authors tested a number of reagents to visualize the separated amino acids on the paper and found that spraying the paper with ninhydrin could color amino acids present in mere microgram quantities.[10]

They also noted that cellulose, being a two-dimensional medium, could allow for samples to be run in two directions by using a different solvent in each direction, resulting in separation by two characteristics. Thus, a drop of sample could be added to one corner of a 18 × 22.5 cm sheet of cellulose and the first solvent added to separate the amino acids in one direction on the paper; after drying the paper as in one-dimensional chromatography, the second solvent was added to separate the amino acids in the perpendicular direction.[9] In either case, an R value (called R_f for paper chromatography) was used to describe the distance of each sample from the origin. Therefore, using this technique, Consden et al. succeeded in efficiently separating all amino acids and, in fact, found it to be *especially convenient, in that it shows at a glance information that can be gained otherwise only as the result of numerous experiments.*[9]

In a subsequent paper, the same authors described a method by which the amino acid or peptide spots separated on the filter paper could be extracted and used for downstream processing.[11] Here, samples were hydrolyzed for a short time to produce peptides which were separated on paper as described above, along with samples of known amino acids or peptides (used as a standard), and the lane containing the standards was cut and stained with ninhydrin to identify the positions of the known samples. The "standards" strip was then used as a guide for cutting the unknown samples at appropriate positions, each with a taper at one end. To extract the samples from the paper, the ends opposite to the tapered edge was held against wet filter paper, the water of which was transferred to and crept down the strip towards the tapered end, carrying along with it the sample contained on the paper. When the samples reached the tapered end, they ran along the taper and into a capillary tube that was positioned at the tip to accept the solution that dripped off of the paper.[11] The capillary tube containing the samples could then be subjected to further hydrolysis and separated once again in the manner to determine the amino acids contained in the samples.

Another technique that was instrumental to Sanger was that of Synge's ionophoretic fractionation method.[12] In turn, Synge and his team developed their technique by making improvements to Theorell and Åkeson's "diaphragm cell" procedure to separate bases, neutral substances, and strong acids all at once. The cell itself was constructed from acrylic sheets but each compartment was separated using different materials: the outermost compartments were separated from the middle compartments by cellophane, and the middle compartments from each other by sheepskin parchment (Figure 7.2). The cathode compartment was filled with basic solution of 1% NH_3; the sample to be analyzed was added to the next compartment; a mildly acidic solution of 1% acetic acid was added to the third compartment; finally, the anode compartment was filled with highly acidic solution of 0.1N H_2SO_4. 200V was applied across the terminals to begin the separation of the amino acids in the sample, which, in time, migrated to one of the four compartments based on their charge: basic amino acids migrated towards the cathode; neutral amino acids remained in the original sample compartment; acidic residues migrated to the third compartment; and strong acids, that is, non-organic compounds, migrated to the anode.

Consden et al. combined these techniques with Sanger's end terminal DNFB labeling and succeeded in identifying the sequence of amino acids in the short cyclopeptide

gramicidin S., while Sanger used these techniques to sequence the remainder of the insulin protein. To this end, he first oxidized the protein to separate the four chains in an effort to *reduce the complexity of the peptide mixture dealt with*.[13] He then labeled the free amino groups using DNFB and subjected the protein to partial hydrolysis to produce short peptides which were subsequently separated by paper chromatography. From these experiments, the sequences of the last few terminal amino acids on each insulin chain, as well as the amino acids surrounding the internal lysine residue of chain B, were determined. Surprisingly, they found that the amino acids isolated during paper chromatography were present in equimolar concentrations and noted that:

> [...] it is extremely unlikely that two different proteins would give the same result, probably less likely than that they would appear homogeneous by any of the known physicochemical methods of determining protein purity [...]
>
> It is concluded that the insulin molecule is built up of two pairs of very similar, if not identical, polypeptide chains.[13]

Following these successes, Sanger turned his attention to deciphering the entire insulin protein, which was accomplished in two pairs of papers published in 1951 and 1953. He first started with chain B, which he partially hydrolyzed to obtain small peptides that were separated using a variety of columns, charcoal, and ionophoresis, in an effort to decrease the complexity of the samples in each fraction (Figure 7.3). After separating the samples using paper chromatography, he extracted the spots, labeled the N-termini with DNFB, subjected the peptides to complete hydrolysis in HCl, and once again, separated the amino acids using paper chromatography.[14] However, their initial results using acid hydrolysis resulted in fragments too small to be adequately analyzed. Therefore, they reasoned that using the proteolytic enzymes pepsin, chymotrypsin, and trypsin, might cleave the protein into bigger fragments and increase

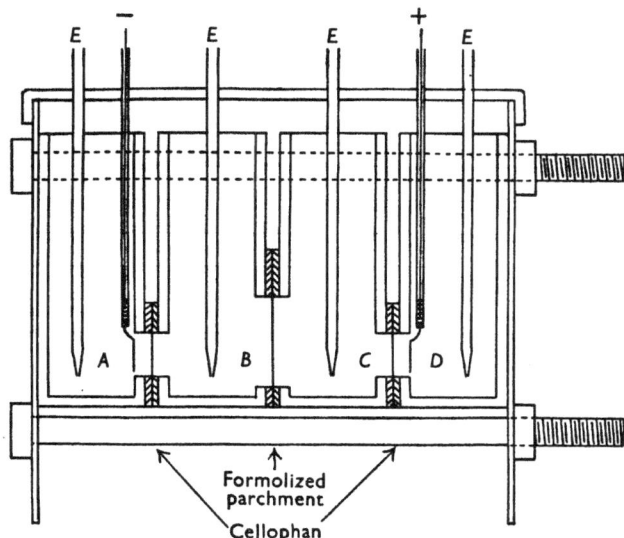

FIGURE 7.2 A side view of Synge's diaphragm cell assembly used in his ionophoretic technique. The compartments were separated as shown in the diagram and the cathode (A) was filled with NH_3, while the anode was filled with H_2SO_4 (D); Compartment C contained acetic acid. The sample to be analyzed was placed in a solution at pH6 (B) and 200 V were applied across the terminals (E). (Figure adapted from Ref [12].)

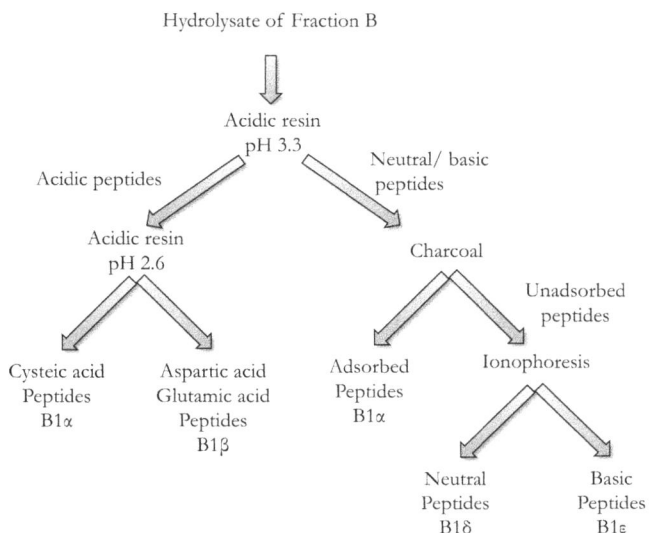

FIGURE 7.3 Scheme used by Sanger & Tuppy to separate hydrolyzed peptides of insulin chain B.[14]

the resolving power of their experiment.[15] Indeed, doing so enabled them to determine, for the first time, the complete sequence of a large protein (although in retrospect, insulin is, in fact, a rather small protein). Two years later, they published a similar pair of papers describing the sequence of insulin chain A, providing a full complement of the amino acid sequence of insulin.[16,17] In addition, these papers represented the first study in which a large protein was analyzed following enzymatic digestion, and as such, confirmed and/or shed light on the mechanism of action of trypsin (which splits carboxyl groups of arginine and lysine), chymotrypsin (which attacks carboxyl groups of phenylalanine and tyrosine), and pepsin (which splits bonds adjacent to tyrosine, phenylalanine, and leucine).[15]

The impact of Sanger's papers was immediately recognized as ground-breaking for several reasons: (1) it encouraged other laboratories to attempt sequencing other proteins; (2) it gave hope that the 3D structure of insulin could soon be determined now that the complete amino acid sequence was known; (3) it finally confirmed, without a doubt, that amino acids were linked via peptide bonds and that a protein could contain more than one peptide chain; (4) proteins were now confirmed to be composed of unique amino acid sequences with no repeating patterns; (5) the action of proteolytic enzymes, which had not been studied in full-length proteins, was confirmed to be specific for particular amino acids; (6) this validated the use of synthetic peptides to determine the specificity of action of novel proteolytic enzymes.[18] It should also be noted that du Vigneaud accomplished the sequencing of two small hormones, oxytocin and vasopressin, at nearly the same time as Sanger and using methods developed by Sanger and Consden et al.[18]

* * * * *

Although an efficient method for protein sequencing was well underway in the early part of the 1950s, very little was known about how proteins were synthesized within a cell. Fischer's hypothesis of the peptide bond was well accepted by this time and the structure of DNA had recently been elucidated along with the proposal of an elegant mechanism for replication. And although chemical differences between DNA and RNA were known, a functional role for RNA had yet to be elucidated. Suggestion of an important role for RNA in protein synthesis can be traced back to 1941, when Brachet and Caspersson independently noted that cells which produce vast quantities of proteins, such as pancreas cells or silk-producing cells of the silkworm, also contained very high quantities of RNA; in contrast, cells that were metabolically active but produced little protein, such as those of the kidneys or heart, had correspondingly low quantities of RNA.[19] In addition, experiments had shown that conditions which increased or decreased protein synthesis led to a concomitant increase or decrease in RNA concentrations, respectively.

In the late 1930s and early in the 1940s, Albert Claude at the Rockefeller Institute for Medical Research was using differential centrifugation to fractionate normal and tumor cells into their various components. As such, Claude broke cells open and resuspended the lysates in neutral water, after which he centrifuged the samples at high speeds to separate the nucleus from the cytoplasm. In the latter fraction, he identified *numerous highly refringent bodies of extremely small size* [50 to 200 nm in diameter], *which may be at rest or in active Brownian* [i.e., random] *movement.*[20] He showed that these particles were *complex formations in which a nucleoprotein of the ribose type occurs in association with a definite proportion of lipids, especially phospholipids.* These *small particles* were present in all cell types studied and accounted for a major part of cytoplasmic RNA, though the size and quantity were variable. Based on these characteristics, these particles were initially thought to be mitochondria, but their small size precluded this conclusion since mitochondria were known to be much greater than 200 nm. In addition, fractionation at higher speeds resulted in separation of mitochondria from this new structure.[20] Therefore, after careful consideration of his data, Claude concluded that:

the evidence, so far, indicates that the mass of the small particles does not derive from the grossly visible elements of the cell but constitutes a hitherto unrecognized particulate component of protoplasm, more or less evenly distributed in the fundamental substance [...] In order to differentiate the small particles from the other, already identified elements of the cell, it may be convenient in the future to refer to this new component under a descriptive name which would be specific. For this purpose the term microsome [small body] appears to be the most appropriate.[20]

One technological advancement that contributed to the elucidation of the role of RNA in cells was the development of the electron microscope. Discovery of the basic principle of electron microscopy occurred in Germany in 1926 when H. Busch studied the trajectories of particles in electric and magnetic fields and showed that these fields could act as a lens on an electron beam, that is, it could concentrate or diverge the beam (similar to what Crookes had found to occur in cathode-ray tubes).[21,22] Other German scientists used this principle a decade later and developed the electron microscope. The need for such an instrument arose due to the development of bacteriological research, since these organisms are much smaller than the 200 nm resolution limit of light microscopes. Resolving power—that is, the smallest distance two points can be from each other while still being seen as two separate points—is approximately equal to half the minimum wavelength used; therefore, since the shortest wavelength the eye can see is about 400 nm (seen as violet), the resolution of light microscopy is limited to about 200 nm. Electron microscopy increases this resolving power by using electrons, whose wave characteristics could theoretically provide unlimited resolution, but in practical terms, increased it to about 0.1 nm, sufficiently small to see bacteria and viruses.[21] However, one limitation of the electron microscope was that the instrument needed to be in a vacuum since this was the only environment where electrons travelled without hindrance, that is, without bumping into gas molecules. For this reason, a double chamber, akin to what can be found on a submarine, was required to load samples: the sample was first deposited in a lateral compartment separate from the main body of the microscope and not under vacuum. After closing and sealing this compartment, a vacuum was created, the separation between the two chambers removed, and the sample was positioned into the main body for imaging.[21]

The electron beam was directed through a number of electric coils that created the appropriate magnetic field to concentrate the beam on the sample and magnify the resulting signal, which was then projected on a fluorescent screen for visualization.[21] Deflection of electrons by the sample is dependent on its thickness and density, so the final image is a representation of the different densities and thicknesses of the sample. As such, very thin samples—of the order of 0.1 μm—were required to prevent artifacts due to the thickness of the samples or from the mounting media.[23] Paradoxically, these extremely thin slices had to be made unusually strong since they would be imaged *in vacuo*. To address the first issue—that of thickness—two methods were available at the time: the first was to fractionate cells until particles small enough to be imaged were obtained.[23] This provided information on the types of particles present in the sample but gave no information as to the localization of the particles within the cell. The second involved an instrument borrowed from light microscopy preparations, the microtome, which cut samples into very thin slices. However, microtomes of the times were not adequate to cut such thin slices and needed to be significantly improved, a task that would not be achieved until the 1950s. The other issue—that of maintaining the structural integrity of the particles in the sample when they were exposed to a vacuum—could be achieved by coating the sample with heavy metals—namely gold and later, osmium tetroxide—which, as an added benefit, also increased the density of the samples and therefore, increased the image's contrast.[23]

By the end the 1940s, electron microscopy images of sufficient quality were beginning to be produced and George Palade took advantage of these new developments to study cells in a way that was not previously possible. Thus, images obtained by Palade revealed a *'lace-like' network of slightly higher density than the rest of the cytoplasm*, which was suggested to consist of vesicles and tubules interconnected in a continuous network.[24] Since the structure was restricted to the inner region of the cytoplasm, it was named endoplasmic reticulum (endoplasmic—inside the cell; reticulum—network; ER) by Keith Porter in 1953. The same year, Palade presented electron microscopy images of a wide range of cell types, which all contained a new structure visible in the cytoplasm: it was round in shape, 10–15 nm in size, and seemed to have a particular affinity for the lower part of the outside of the ER.[25] He also found that it was present in all of the more

than 40 cell types he studied, though its presence was variable depending on the cell type: erythroblasts and undifferentiated cells of rapidly growing epithelia have poorly developed ER with *a large population of small, dense particles, most of them freely scattered in the cytoplasm*, whereas in the mature leukocytes and other well-differentiated cells, the particles seem to be *preferentially associated with the membrane of the endoplasmic reticulum*.[25] Given these observations, and since these particles had also been isolated in bacteria which do not have endoplasmic reticula, Palade proposed that the microsomes identified by Claude were *derived mainly from the endoplasmic reticulum and that their characteristically high RNA content is due to* [the] *associated small particles* that Palade identified.[26]

To confirm or disprove this hypothesis, Palade and Siekevitz studied samples derived from rat liver and attempted to tease apart the various morphological and biochemical components of microsomes.[26] As such, they used a portion of rat livers for in situ sample preparations for electron microscopy and homogenized the rest to isolate the microsomes by fractionation. Samples of the various fractions of homogenate, including those of microsome fractions, were fixed and embedded for electron microscopy, while others were used to examine the biochemical make-up in terms of protein, RNA, and phospholipid content. As expected, the ER in in situ samples was studded with small, dense particles which seemed to be arranged preferentially in circles or in double rows (Figure 7.4a, b).[26] Importantly, most of these structures could readily be identified in tissue homogenates prior to fractionation (Figure 7.4c), indicating that homogenization of the tissue did not alter the morphology of the constituents to any significant degree other than to fragment the ER into *independent vesicles, tubules, and cisternae*. In addition, the microsomal fraction contained tubules or cisternae of ER elements with dense particles attached in double rows or circles, indicating that they are *fragments of the [ER] derived, to a large extent, from the rough surfaced parts of the network [...] and not artefacts introduced by tissue homogenization*.[26] The authors also noted that two major components of the microsome—the membranous structure of the ER and the particles that are attached to them—could easily be distinguished in the microsomal fraction and concluded that these were *fundamentally distinct structures* and supported Claude and Porter's hypothesis that microsomes were preformed structures of the cytoplasm derived from the ER.

Biochemically, the authors found that microsomes contained large amounts of RNA, proteins, and phospholipids, with a protein: RNA ratio of about 1. They also treated the microsome fraction with various chemicals to determine the stability of the samples under different conditions. As such, they found that the RNA and protein components could be separated using versene (a reagent that chelates divalent cations), which indicated that the association between protein and RNA in microsomes required divalent cations. They also treated microsomes with deoxycholate (a detergent used to solubilize membranes) and found that

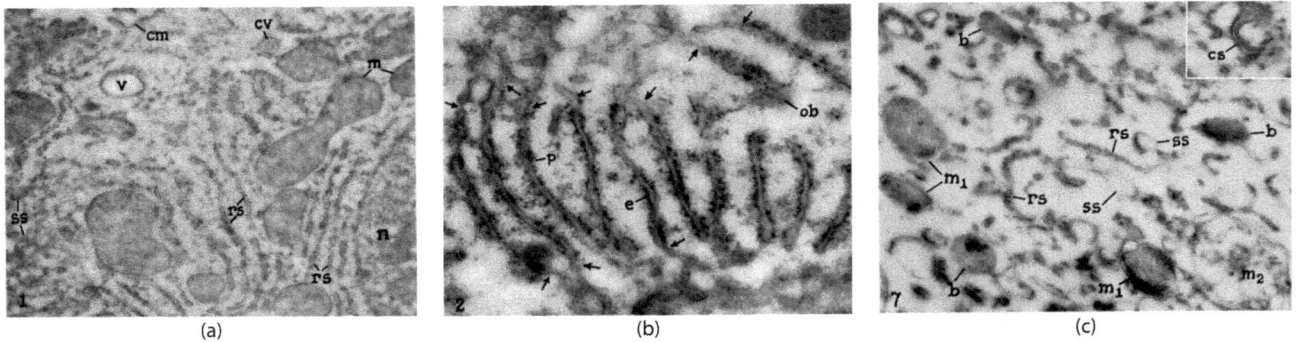

FIGURE 7.4 Electron microscopy images of liver samples. Low (a) and high (b) magnification of in situ preparation of a liver cell showing part of a nucleus (n), mitochondria profiles (m), and smooth-surfaced (ss) and rough-surfaced (rs) endoplasmic reticula (ER). p points to dense particles which characterize the rough ER, shown here as ten elongated profiles, (one which is marked "e"). (c) Electron microscopy image of a cross-section of a homogenized liver sample. Shown are intact mitochondria (m₁), rough (rs) and smooth (ss) surfaced endoplasmic reticula. (Figure adapted from Ref. [26].)

upon centrifugation all that remained in the pellet were small particles rich in RNA and protein, similar to those attached to the outer surface of the ER or microsomes. Taken together, these data supported the idea that microsomes were membranous fragments derived from rough-surfaced ER, to which small, dense particles, rich in RNA and protein, were attached, and once again supported Palade's hypothesis that the dense particles in microsomes were structures separate from the ER. These dense particles rich in RNA were renamed "ribosomes" in 1958 to replace the various names that could be found the literature to describe it, such as "microsomal particle", "ribonucleo-protein of the microsomal fraction", and "Palade granule".[27]

* * * * *

The first experimental indication that RNA was involved in protein synthesis came in the early 1950s, when Tore Hultin from the Wenner-Gren Institute for Experimental Biology in Stockholm wanted to determine if any particular cell fraction would precede the others during protein synthesis. Thus, he injected ¹⁵N-labeled glycine into newborn chicks and, after a period of time ranging from 10 minutes to 5 hours, determined the degree of incorporation of ¹⁵N in various fractions—nuclear, mitochondrial, microsomal, and cytoplasmic—isolated from the liver.[28] As such, Hutlin reported that ¹⁵N was incorporated much earlier and at a more rapid rate into the microsomal fraction than any other fraction, which agreed with the higher content of RNA in this fraction compared to the others. He also found an even higher degree of ¹⁵N incorporation in the cytoplasmic fraction, but attributed this to contamination from the microsomal fraction (we will see later that this was not a correct conclusion). As such, he noted that it,

> is of special interest that the present experiments indicate that the ribonucleic acid is related not only to the trans-formation of polypeptide chains into proteins of specific structure, but also to the initial incorporation of amino acids into peptide chains [...]

Since the microsomes are especially rich in ribonucleic acid, and the production of phosphate-bound energy is centered in the mitochondria, the experiments indicate that the immediate availability of ribonucleic acid is more important for the formation of proteins than a close proximity to energy-producing enzyme systems.[28]

Although others would soon replicate Hutlin's findings both *in vitro* and *in vivo*, direct evidence of the role of RNA in protein synthesis was still lacking. Surprisingly, viruses would play a vital role in the elucidation of the mechanism of protein synthesis in general. Many viruses are essentially RNA molecules enclosed in a protein coat, but the role of each of these components was not known in the 1950s. As such, Gierer and Schramm removed the protein components from tobacco mosaic virus (TMV) and found that the RNA component alone was apt for infection of plant leaves, albeit at a lower efficiency.[29] The same year, Heinz Fraenkel-Conrat published a study in which he combined the protein fraction of one strain of TMV with the nucleic acid of a different strain of the same virus and found that the symptoms of the infection consistently resembled that of the strain from which the RNA was derived, while the nature of the protein coat had no influence on infection phenotype.[30] However, he found that only the anti-serum of the strain from which the proteins were derived could prevent an infection, whereas anti-serum obtained from other strains had no effect. In addition, analysis of the amino acid composition of the protein coat of progeny viruses revealed that it was identical to that of the virus from which the RNA was derived.

Another manner in which RNA was implicated in protein synthesis was the fact that the latter could be inhibited by ribonuclease (RNAse), an enzyme that specifically cleaves RNA. Vincent Allfrey and his colleagues incubated rat liver ribosomal fractions with RNAse and found that under these conditions, incorporation of ¹⁴C into proteins was about half as much as control conditions that were not treated with RNAse.[31] Later, Brachet used a similar idea to test the requirements of RNA on different cell functions of onion root tip and then extended his finding to animal cells.[32,33] As such, he treated living amoeba with RNAse for 90 minutes and found that mobility was quickly halted and the organism became round and eventually lysed. Surprisingly, he also found that addition of purified yeast RNA could partially, and in some cases, completely reverse the effects of RNAse treatment. In addition, he found that RNAse first accumulated in the cytoplasm and cleaved cytoplasmic RNA, then attacked ribosomal RNA, and finally, targeted nuclear RNA. Given the wealth of data accumulated in the early years of the 1950s, Brachet concluded that:

> [w]e can thus from now on consider the correlation between the ribonucleic acid content and protein synthesis as a well-established fact, and we can draw the conclusion that ribonucleic acid somehow takes a direct role in protein synthesis.[19]

* * * * *

In 1954, Paul C. Zamecnik and Elizabeth B. Keller performed experiments on the requirements for amino acid incorporation into proteins using microsomes and discovered that the process required energy in the form of ATP. As such, amino acid incorporation did not proceed once mitochondria were removed from the extract unless ATP or an ATP-regenerating system was included.[34] This was followed up the next year by Mahlon Hoagland, who incubated a *105,000 × g rat liver extract*—that is, the pellet of extracts centrifuged at 105,000 × g—with ATP and radiolabeled pyrophosphate (^{32}PP) for 7 minutes and found that addition of amino acids to the reaction mixture significantly increased ^{32}PP incorporation into ATP as assessed using a Geiger counter. The reaction produced no accumulation of ^{32}PP even though a pyrophosphatase (KF) was included to prevent cleavage of PP into orthophosphate, and the author found no loss of ATP to AMP, which suggested that ATP was not hydrolyzed (in other words, ATP was used up in the reaction).[35] As such, this suggested to him that the reaction in question was reversible and therefore, free ^{32}PP competed with non-labeled, endogenous PP resulting from the splitting of ATP during the reaction and subsequently regenerated with the radioactive pyrophosphate. In the next experiment, he used a reaction that had been developed by Lipmann and Tuttle a decade earlier to ameliorate the identification of acetyl phosphate intermediates, in which hydroxylamine (NH_2OH) reacted with the acyl part of a target molecule to form a stable hydroxamic acid product (this was the same acetylation trap used by Lipmann in the elucidation of the Krebs cycle, see Figure 5.13a).[36] This product could then be reacted with iron salts to give a bright purple color that could be quantified. Thus, Hoagland repeated his experiments but omitting KF (so pyrophosphatases remained active) and with the addition of hydroxylamine as an acetylation trap for activated amino acids. With this experiment, he found an increase in ATP hydrolysis (i.e., an increase in AMP), a concomitant

accumulation of two equivalents of orthophosphate, as well as the production of amino hydroxamic acid (Figure 7.5a).[35] This suggested to Hoagland that hydroxylamine took part in an irreversible reaction with ATP-activated amino acids whereby AMP was released and the PP was hydrolyzed to orthophosphate. However, since AMP was no longer associated with the enzyme, it could not react with PP to reform ATP in the reversible reaction that prevailed in the absence of hydroxylamine.

These results led Hoagland to propose a mechanism for amino acid activation whereby an enzyme specific for each amino acid rendered ATP labile at its AMP-pyrophosphoryl site, favoring an attack by the carboxyl oxygen of an amino acid and creating an amino acyl–AMP compound (equations 1 and 2):[35]

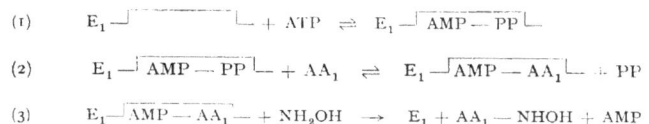

$$(1) \qquad E_1 \!\!-\!\!\overline{}\!\!-\!\!+ ATP \;\rightleftharpoons\; E_1 \!\!-\!\!\overline{AMP - PP}\!\!-$$

$$(2) \qquad E_1 \!\!-\!\!\overline{AMP - PP}\!\!-\!\!+ AA_1 \;\rightleftharpoons\; E_1 \!\!-\!\!\overline{AMP - AA_1}\!\!-\!\!+ PP$$

$$(3) \qquad E_1 \!\!-\!\!\overline{AMP - AA_1}\!\!-\!\!+ NH_2OH \;\rightarrow\; E_1 + AA_1 - NHOH + AMP$$

The combination of these two reactions does not result in any changes in ATP or PP since they are both reversible and accounted for the lack of ^{32}PP accumulation in the first set of experiments. However, addition of hydroxylamine (NH_2OH, equation 3) to the reaction resulted in the displacement of AMP from the enzyme complex by hydroxylamine, which reacted with the labilized amino acid. In this case, the accumulation of AMP and orthophosphate indicated that this reaction was not reversible.

The mechanism of amino acid activation by ATP was followed-up the next year by Hoagland, Keller, and Zamecnik. In those experiments, the authors did not use the 105,000 × g rat liver extracts but instead used the supernatant, which presumably contained the enzymes responsible for amino acid activation.[37] As such, they found that amino-hydroxamate formation was additive for each amino acid added to a reaction, which indicated that the amino acids and the enzymes in the reaction did not compete and suggested that each amino acid likely had its own dedicated enzyme for activation. In addition, they noted that hydroxamate formation in the presence of a mixture of amino acids reached a maximum when they used a 105,000 × g liver fraction supernatant that had been extracted using 35–40% saturated ammonium sulfate. However, after testing each amino acid separately under these conditions, they found that amino-hydroxamate was only formed in the presence of methionine. Therefore, they once again fractionated the 105,000 × g liver protein supernatant at different pH and tested each amino acid against the different fractions. Results indicated that each amino acid had an optimal pH range for amino-hydroxamate formation, supporting the idea that each amino acid was activated by a different enzyme whose activity was dependent on pH (Figure 7.5b).

Finally, the authors showed that protein synthesis halted in the presence of hydroxylamine, which confirmed their suspicion that protein synthesis required that amino acids

ATP LOSS, AND HYDROXAMATE APPEARANCE IN THE PRESENCE OF
AMINO ACIDS AND HYDROXYLAMINE (in μM/ml)

Addition	Hydroxamate formed	ATP lost	Pi formed
----	0	2.31	4.64
NH$_2$OH	0.34	1.39	2.78
AA	0	2.25	4.51
AA + NH$_2$OH	0.69	2.25	4.51
△ due to AA alone	0	0	0
△ due to AA in presence of NH$_2$OH	0.35	0.86	1.73

(a)

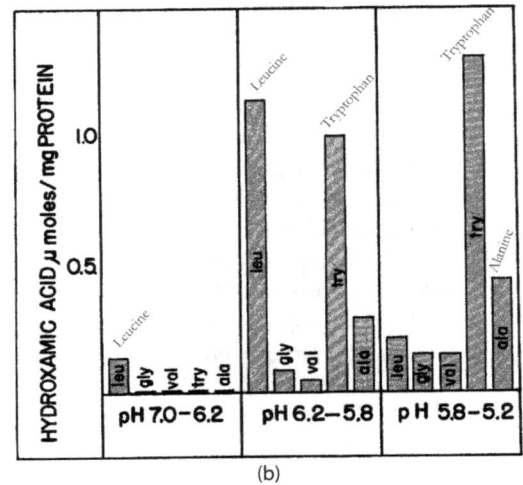

(b)

FIGURE 7.5 (a) Results of an experiment testing the requirements for amino acid activation. The presence of hydroxymate when both amino acids (AA) and hydroxylamine (NH$_2$OH) are added indicates that amino acids have been activated and have reacted with the hydroxylamine. The presence of hydroxymate in the absence of amino acids indicates that part of the hydroxylamine reacts nonspecifically. This value was subtracted from the results in the last line.[35] (b) Amino-hydroxamic acid formation using protein fractions extracted at different pH. Leucine is preferentially active between pH 5.8 and 6.2, whereas tryptophan and alanine are active from pH 5.2–5.8.[37]

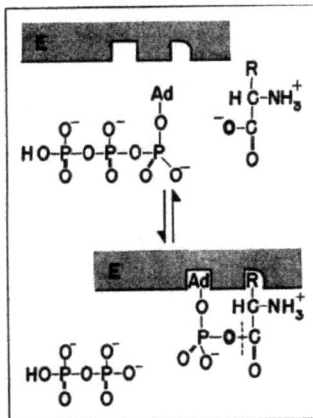

FIGURE 7.6 Mechanism for amino acid activation. An amino acid binds to a particular enzyme's catalytic site, along with ATP, and the enzyme catalyzes a chemical reaction which binds the two together, thereby activating the amino acid.[37]

go through this process of activation before they could be incorporated into proteins. Thus, these data further supported the idea that each amino acid bound to its own specific enzyme along with ATP to form an activated amino acid–AMP compound with the release of PP, and suggested that once activated, the amino acid was primed for incorporation into a growing polypeptide chain (Figure 7.6). Similar results and conclusions were obtained the same year by Paul Berg using baker's yeast and by Demoss and Novelli using a variety of bacteria,[38,39] indicating that this process was likely to be universal.

* * * * *

REFERENCES

1. Mirsky AE, Pauling L. On the structure of native, denatured, and coagulated proteins. *Proceedings of National Academy of Sciences U S A.* 1936;22(7):439–447. doi:10.1073/pnas.22.7.439
2. Harington CR, Mead TH. Synthesis of glutathione. *Biochemical Journal.* 1935;29(7):1602–1611. doi:10.1042/BJ0291602
3. Martin AJP, Synge RLM. A new form of chromatogram employing two liquid phases: A theory of chromatography. 2. Application to the micro-determination of the higher monoamino-acids in proteins. *Biochemical Journal.* 1941;35(12):1358. doi:10.1042/bj0351358
4. Gordon AH, Martin AJ, Synge RL. Partition chromatography in the study of protein constituents. *Biochemical Journal.* 1943;37(1):79–86. 10.1042/bj0370079
5. Berg P. Fred sanger: A memorial tribute. *Proceedings of the National Academy of Sciences.* 2014;111(3):883–884. doi:10.1073/PNAS.1323264111
6. Sanger F. The free amino groups of insulin. *Biochemical Journal.* 1945;39(5):507–515. doi:10.1042/bj0390507
7. Sanger F. Fractionation of oxidized insulin. *Biochemical Journal.* 1949;44(1):126–128. doi:10.1042/bj0440126
8. Sanger F. Oxidation of insulin by performic acid. *Nature.* 1947;160(4061):295–296. doi:10.1038/160295b0
9. Consden R, Gordon AH, Martin AJ. Qualitative analysis of proteins: A partition chromatographic method using paper. *Biochemical Journal.* 1944;38(3):224–232. doi:10.1042/bj0380224
10. Gordon AH. Electrophoresis and chromatography of amino acids and proteins. *Annals of New York Academy of Sciences.* 1979;325:95–106. doi:10.1111/j.1749-6632.1979.tb14130.x
11. Consden R, Gordon AH, Martin AJ. The identification of lower peptides in complex mixtures. *Biochemical Journal.* 1947;41(4):590–596. doi:10.1042/bj0410590

12. Synge RLM. Non-protein nitrogenous constituents of rye grass; Ionophoretic fractionation and isolation of a "bound amino-acid" fraction. *Biochemical Journal.* 1951; 49(5):642–650. doi:10.1042/bj0490642

13. Sanger F. The terminal peptides of insulin. *Biochemical Journal.* 1949;45(5):563–574. doi:10.1042/bj0450563

14. Sanger F, Tuppy H. The amino-acid sequence in the phenylalanyl chain of insulin. I. The identification of lower peptides from partial hydrolysates. *Biochemical Journal.* 1951;49(4):463–481. doi:10.1042/bj0490463

15. Sanger F, Tuppy H. The amino-acid sequence in the phenylalanyl chain of insulin. 2. The investigation of peptides from enzymic hydrolysates. *Biochemical Journal.* 1951; 49(4):481–490. doi:10.1042/bj0490481

16. Sanger F, Thompson EOP. The amino-acid sequence in the glycyl chain of insulin. II. The investigation of peptides from enzymic hydrolysates. *Biochemical Journal.* 1953;53(3):366. doi:10.1042/bj0530366

17. Sanger F, Thompson EOP. The amino-acid sequence in the glycyl chain of insulin. I. The identification of lower peptides from partial hydrolysates. *Biochemical Journal.* 1953;53(3):353–366. doi:10.1042/bj0530353

18. Smith EL. Amino acid sequences of proteins: The beginnings. *Annals of New York Academy of Sciences.* 1979;325(1 The Origins o):107–120. doi:10.1111/j.1749-6632.1979.tb14131.x

19. Brachet J. Ribonucleic acids and the synthesis of cellular proteins. *Nature.* 1960;186(4720):194–199. doi:10.1038/186194a0

20. Claude A. The constitution of protoplasm. *Science (1979).* 1943;97(2525):451–456. doi:10.1126/science.97.2525.451

21. Marton L. The electron microscope: A new tool for bacteriological research. *Journal of Bacteriology.* 1941;41(3): 397–413. doi:10.1128/jb.41.3.397-413.1941

22. Oatley CW. The early history of the scanning electron microscope. *Journal of Applied Physics.* 1982;53(2):R1–R13. doi:10.1063/1.331666

23. Palade GE. Albert Claude and the beginnings of biological electron microscopy. *Journal of Cell Biology.* 1971;50(1):5.

24. Palade GE. Microsomes and ribonucleoprotein particles. In: Roberts RB, ed. *Microsomal Particles and Protein Synthesis.* Pergamon Press; 1958:36.

25. Palade GE. A small particulate component of the cytoplasm. *Journal of Biophysical and Biochemical Cytology.* 1955;1(1):59–68. doi:10.1083/jcb.1.1.59

26. Palade GE, Siekevitz P. Liver microsomes: An integrated morphological and biochemical study. *Journal of Biophysical and Biochemical Cytology.* 1956;2(2):171–200. doi:10.1083/jcb.2.2.171

27. Roberts R. *Microsomal Particles and Protein Synthesis.* Pergamon Press; 1958. Accessed March 30, 2019. https://archive.org/details/microsomalpartic0000biop

28. Hultin T. Incorporation in vivo of 15N-labeled glycine into liver fractions of newly hatched chicks. *Experimental Cell Research.* 1950;1(3):376–381. doi:10.1016/0014-4827(50)90015-2

29. Gierer A, Schramm G. Infectivity of ribonucleic acid from tobacco mosaic virus. *Nature.* 1956;177(4511):702–703. doi:10.1038/177702a0

30. Fraenkel-Conrat H. The role of the nucleic acid in the reconstitution of active tobacco mosaic virus. *Journal of the American Chemical Society.* 1956;78(4):882–883. doi:10.1021/ja01585a055

31. Allfrey V, Daly MM, Mirsky AE. Synthesis of protein in the pancreas. II. The role of ribonucleoprotein in protein synthesis. *Journal of General Physiology.* 1953;37(2): 157–175. doi:10.1085/jgp.37.2.157

32. Brachet J. Effects of ribonuclease on the metabolism of living root-tip cells. *Nature.* 1954;174(4436):876–877. doi:10.1038/174876a0

33. Brachet J. Action of ribonuclease and ribonucleic acid on living amoebae. *Nature.* 1955;175(4463):851–853. doi:10.1038/175851a0

34. Zamecnik PC, Keller EB. Relation between phosphate energy donors and incorporation of labeled amino acids into proteins. *Journal of Biological Chemistry.* 1954;209(1): 337–354. doi:10.1016/S0021-9258(18)65561-9

35. Hoagland MB. An enzymic mechanism for amino acid activation in animal tissues. *Biochimica et Biophysica Acta.* 1955;16(2):288–289. doi:10.1016/0006-3002(55)90218-3

36. Lipmann F, Tuttle LC. A specific micromethod for the determination of acyl phosphates. *Journal of Biological Chemistry.* 1945;159(1):21–28. doi:10.1016/S0021-9258(19)51298-4

37. Hoagland MB, Keller EB, Zamecnik PC. Enzymatic carboxyl activation of amino acids. *Journal of Biological Chemistry.* 1956;218(1):345–358. doi:10.1016/S0021-9258(18)65898-3

38. Berg P. Acyl adenylates; the interaction of adenosine triphosphate and L-methionine. *Journal of Biological Chemistry.* 1956;222(2):1025–1034. doi:10.1016/S0021-9258(20)89959-1

39. Demoss JA, Novelli GD. An amino acid dependent exchange between 32P labeled inorganic pyrophosphate and ATP in microbial extracts. *Biochimica et Biophysica Acta.* 1956;22(1):49–61. doi:10.1016/0006-3002(56)90222-0

8 Protein Synthesis Part II
The Mechanism of Protein Translation

In 1957, Francis Crick gave what would become a most significant lecture on protein synthesis at University College London during a Society for Experimental Biology symposium on the "Biological Replication of Macromolecules". In the hour-long lecture, which is now referred to as the "Central Dogma" lecture and about which is said to have *permanently altered the logic of biology*,[1] Crick summarized the development in protein research with an emphasis on the mechanism of protein synthesis and makes clear that proteins' role in cells is at least of equal footing as that of DNA:

It is an essential feature of my argument that in biology proteins are uniquely important. They are not to be classed with polysaccharides, for example, which by comparison play a very minor role. Their nearest rivals are the nucleic acids. Watson said to me, a few years ago, 'The most significant thing about the nucleic acids is that we don't know what they do.' By contrast the most significant thing about proteins is that they can do almost anything. In animals proteins are used for structural purposes, but this is not their main role, and indeed in plants this job is usually done by polysaccharides. The main function of proteins is to act as enzymes. Almost all chemical reactions in living systems are catalysed by enzymes, and all known enzymes are proteins. It is at first sight paradoxical that it is probably easier for an organism to produce a new protein than to produce a new small molecule, since to produce a new small molecule one or more new proteins will be required in any case to catalyse the reactions.

I shall also argue that the main function of the genetic material is to control (not necessarily directly) the synthesis of proteins [...] Once the central and unique role of proteins is admitted there seems little point in genes doing anything else.

Biologists should not deceive themselves with the thought that some new class of biological molecules, of comparable importance to the proteins, remains to be discovered. This seems highly unlikely. In the protein molecule Nature has devised a unique instrument in which an underlying simplicity is used to express great subtlety and versatility; it is impossible to see molecular biology in proper perspective until this peculiar combination of virtues has been clearly grasped.[2]

Crick went on to discuss two topics which he called *Sequence Hypothesis* and the *Central Dogma*, the former of which was an idea which had been referred to in the past and was already widely accepted, in which it was proposed *that the specificity of a piece of nucleic acid is expressed solely by the sequence of its bases, and that this sequence is a (simple) code for the amino acid sequence of a particular protein.* However, the Central Dogma, which can be simply expressed as *DNA makes RNA, RNA makes proteins*, was a new concept altogether, and it stated that:

once 'information' has passed into protein it cannot get out again. In more detail, the transfer of information from nucleic acid to nucleic acid, or from nucleic acid to protein may be possible, but transfer from protein to protein, or from protein to nucleic acid is impossible. Information means here the precise determination of sequence, either of bases in the nucleic acid or of amino acid residues in the protein.[2]

This hypothesis assumed that there was an RNA template in the cytoplasm since the cytoplasm is where proteins are synthesized and suggested that its synthesis was under the control of DNA. Crick also suggested that the simplest model for this hypothesis (though wrong) was one in which each ribosomal RNA had a different sequence of RNA nucleotides (based on the gene from which it was transcribed), which gave rise to proteins with different amino acid sequences. Furthermore, in contrast to the belief at the time that all the proteins present in microsomes were undergoing synthesis, he suggested that most of the proteins present were, in fact, structural—that is, they formed the scaffold upon which synthesis could occur—and only a small fraction was actively being synthesized.

At the time, the best hypothesis for the manner in which RNA aided in protein synthesis was that the RNA template assumed a different configuration to form the 20 different "cavities" into which the different amino acid side chains could bind. In contrast, Crick proposed the "Adaptor Hypothesis" in his lecture, in which *the amino acid is carried to the template by an 'adaptor' molecule, and that the adaptor is the part which actually fits on to the RNA. In its simplest form one would require twenty adaptors, one for each amino acid.* He went on to suggest that proteins, amino sugars, or even nucleotides were contenders to be these adaptor molecules; however, he felt that the latter was the likeliest candidate since it offered a ready mechanism for association: it would use the same base-pairing mechanism found in DNA, that is, via hydrogen bonding. Crick extended this thought by suggesting that the template might consist of a single RNA chain:

Each adaptor molecule containing, say, a di- or trinucleotide would each be joined to its own amino acid by a special enzyme. These molecules would then diffuse to the microsomal particles and attach to the proper place on the bases of the RNA by base-pairing, so that they would then be in a position for polymerization to take place [...]

DOI: 10.1201/9781003379058-8

Thus one is led to suppose that after the activating step, discovered by Hoagland [...] some other more specific step is needed before the amino acid can reach the template.[2]

He also stressed the fact that in the Central Dogma, at least two types of RNA were required: the first, the *template RNA*, would be synthesized in the nucleus using DNA as a template and *carries the information for sequentialization*; the second, the *soluble RNA*, which would associate with amino acids and enable their incorporation into proteins.

Finally, Crick discussed the *coding problem*, that is, the challenge to explain how the sequence of 4 bases can "code" for the sequence of 20 amino acids. George Gamow had previously suggested the possibility that three bases might code for each amino acid, that adjacent triplets of bases might overlap, and that more than one triplet of nucleotides might stand for a particular amino acid. The "triplet code" aspect seemed likely since a doublet code would only produce $4 \times 4 = 16$ permutations—not enough for all 20 amino acids—whereas the triplet code would allow for more than enough permutations ($4 \times 4 \times 4 = 64$). However, the main point of contention with his scheme was his support for an overlapping code (Figure 8.1). Specifically, it was noted that the space between each nucleotide is approximately equal to that of the distance between amino acids in a polypeptide chain, which would not leave much room for all the molecules to fit during protein synthesis. Otherwise, this issue was left rather untouched during his talk. As we will see in the following sections, many of the ideas brought forward by Crick in this lecture have been shown to be fantastic bits of insight.

* * * * *

In 1958, the year after Crick's seminal lecture, his predictions already began to be proved correct. That year, Hoagland, Zamecnik, and their colleagues published a paper describing the requirement of a new kind of RNA, which they named soluble RNA (later renamed transfer RNA, tRNA). They started from the knowledge gained from previous reports, including those of Caspersson and Brachet, that RNA was required for protein synthesis, and sought to determine if this might be related to the relatively unexplored stage between amino acid activation and their incorporation into proteins. Recall from the last chapter that Zamecnik's group had found that incorporation of amino acids into proteins required the presence of microsomes, an enzyme contained in the $105,000 \times$ g fraction,

ATP, a nucleotide regenerating system, and amino acids, and resulted in the binding of AMP to amino acids, as mediated by the enzyme fraction. In addition, incubation of the protein synthesis mixture with hydroxylamine was found to halt protein synthesis, thereby confirming that amino acid activation was a requirement for this process. In 1957, a preliminary report published by the same group revealed that cytoplasmic RNA could bind amino acids and transfer them to microsomal proteins.[3]

This new step in the process of amino acid incorporation into proteins was further explored the next year, when they found that binding of amino acids to RNA was dependent on ATP concentration (the manner of which, they noted, was similar to that which they observed for the activation of amino acids), as well as a pH 5 supernatant fraction (Figure 8.2).[4] The authors also found that the transfer of amino acids to RNA was sensitive to ribonuclease (which degrades RNA)—a reagent that did not affect ATP-PP exchange or amino acid hydroxamate formation—which indicated that binding of amino acids to RNA came after these steps. Further evidence supporting the idea that transfer occurred following binding to RNA was provided by the fact that the rate of transfer of amino acids to proteins when protein-bound RNA was added to a complete system was similar to that of the upper part of the curve when soluble amino acids were added, which indicated to them that the process was the similar but took longer to proceed using free amino acids (Figure 8.3a).[4] As with amino acid activation, the extent of binding of amino acids to RNA was dependent on the number and concentration of amino acids added to the reaction. They also returned to experiments using microsomes to further determine the requirements for the transfer of RNA-bound amino acids to proteins. As such, they determined that the transfer was essentially complete between about 5–10 minutes, and therefore, was very efficient (Figure 8.3b). Finally, analysis of cellular RNA content indicated that this type of RNA accounted for only 2% of total cellular RNA. From these experiments, the authors concluded that:

> The product formed, an RNA or ribonucleoprotein to which amino acids are apparently covalently linked, is capable of interacting with enzymatic components of the activating enzyme preparation and with microsomes to effect the transfer of the amino acid to peptide linkage in protein. It is therefore suggested that this particular RNA fraction

	Overlapping	Partial Overlapping	Non-overlapping
mRNA	A U G G C A A G U G C C A	A U G G C A A G U G C C A	A U G G C A A G U G C C A
Codon 1	A U G	A U G	A U G
Codon 2	U G G	G G C	G C A
Codon 3	G G C	C A A	A G U

FIGURE 8.1 Possible schemes for the triplet code of "template" RNA. In an overlapping or partially overlapping code, all nucleotides in RNA can be used as the first nucleotide of a triplet, resulting in each nucleotide being part of more than one codon. In contrast, each nucleotide is only used once in a nonoverlapping code.

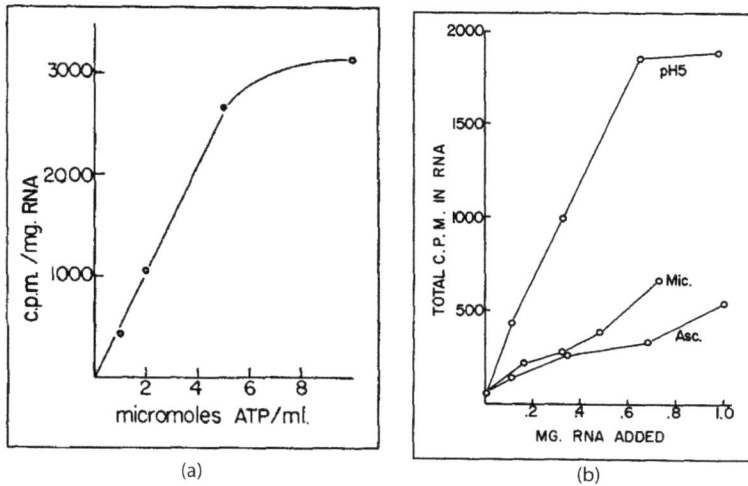

FIGURE 8.2 Binding of amino acids to RNA requires ATP and the pH 5 fraction. (a) ^{14}C-labeled amino acids were incubated in a protein synthesis reaction mixture with different concentrations of ATP and radioactivity was assessed in the isolated RNA; (b) ^{14}C-labeled amino acids were incubated in a protein synthesis mixture as indicated above, but with the added RNA supplied from either the pH 5 fraction (pH5), the microsomal fraction (mic.), or whole cells (Asc.).[4]

functions as an intermediate carrier of amino acids in protein synthesis […]

Since the amino acid activation reaction is insensitive to ribonuclease and since an activating enzyme has been isolated relatively free from RNA, it is still necessary to invoke an initial enzymatic activation reaction as originally postulated, followed by a transfer of amino acid to linkage on RNA.[4]

Therefore, these data, in line with Crick's "adaptor hypothesis", suggested that:

pH 5 RNA molecules, each charged with amino acids in characteristic sequence, polymerize in microsomes (in specific order determined by the complementary structure of microsomal RNA) to higher molecular weight units with

resultant configurational changes which permit peptide condensation between contiguous amino acids.[4]

* * * * *

The next chapter in the history of protein synthesis requires us to take a step back and look at the development of phage research in the first part of the 20th century. In 1915, Frederick Twort observed a phenomenon in growing bacteria that was independently discovered and more thoroughly investigated by the French-Canadian, Félix d'Herelle, working at the Institue Pasteur in 1917. d'Herelle was investigating a bacterial disease that affected locusts in Mexico when he noticed that a bacterial culture contained colonies that seemed to have an *indented irregular*

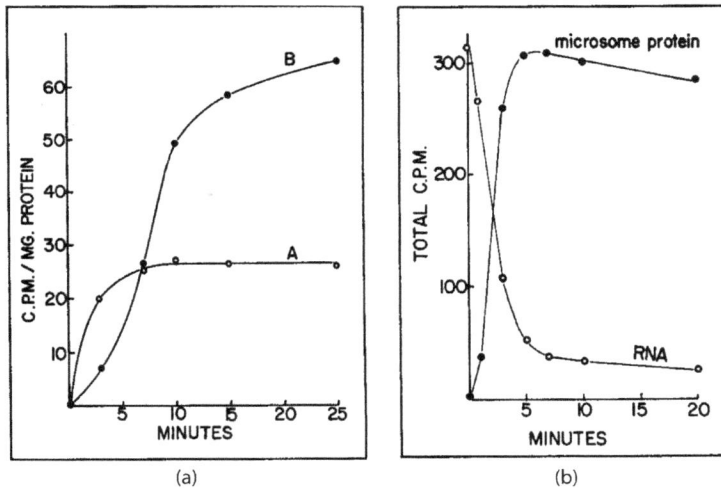

FIGURE 8.3 Amino acid transfer to microsomal proteins is rapid and efficient. (a) RNA-bound ^{14}C-leucine (A) or soluble ^{14}C-leucine (B) was added to a complete protein synthesis system and amino acid transfer to proteins was assessed as reported by radioactivity in protein. (b) ^{14}C-leucine-bound RNA was added to a complete protein synthesis system and radioactivity of either microsomes or RNA assessed.[4]

contour [...] sufficiently pronounced to arouse my curiosity.[5] Though he initially hypothesized that the *true pathogenic agent* [was] *necessarily an ultramicroscopic organism*, he soon realized that this was not the case, but rather, that both the bacterium and an *invisible agent* were both required for infection. Furthermore, in experiments performed in an attempt to reconcile this last hypothesis *with the observed facts in some of the human diseases, particularly bacillary dysentery and typhoid fever*, he found that fecal samples that were obtained from a diseased individual and filtered to remove the bacteria, subsequently added to fresh bacteria, and inoculated into animals, sometimes induced symptoms in the latter comparable to those seen in humans.

One day, he realized that the bacterial samples which produced these symptoms were always collected when the disease was in convalescence. Therefore, he added several drops of filtered fecal samples that were collected daily from a patient who had been admitted at the Pasteur Hospital with severe dysentery to bacterial cultures and incubated the mixture overnight. Although most of the cultures grew as normal, he found that one of the fecal samples completely inhibited the growth of the bacterial culture—as suspected, this sample had been collected when the disease was in convalescence. Next, he added to this "sterile" culture an amount of fresh bacteria with a *marked turbidity*—which indicated the presence of great number of bacteria—and found that *all of the* [bacteria] *originally present had been dissolved* after about 10 hours' incubation. Surprisingly, after filtering the "sterile" culture and repeating the process with fresh bacteria, he found that the *disappearance of* [the bacteria] *was effected with greater and greater rapidity*, which suggested that *the principle [...] behaved like a filtrable organism, parasite of bacteria*. He gave this "principle" the name of bacteriophage (bacteria-eating) and suggested that it was a *virus pathogenic for the bacterium*.

It took a number of years for d'Herelle's ideas about a bacteria-infecting virus to be proven correct, and the emergence of electron microscopy allowed Luria (Watson's Ph.D. mentor), Delbrück, and Anderson to study the interaction of bacteriophages with their bacterial hosts during the virus' various life cycles.[6] As such, they found that these viruses were "sperm-shaped" particles that could attach to the surface of bacteria and cause them to lyse. Indeed, one of their images showed a bacterium, *one end of which has burst open and has liberated a flood of material in which several hundred* [virus] *particles [...] are visible. Along with the virus particles, a granular material has come out from the bacterium.* Furthermore, from the images they obtained in experiments where bacteria were incubated with two different viruses, they concluded that [the] *structure of the visible particles is specific for each strain and a bacterium liberates the particles which are characteristic for the virus which has acted upon it.*[6] It would eventually be determined that viruses are composed of a protein structure which houses a DNA or RNA molecule. An infection is caused when a virus attaches its tail to a cell and inserts

its genetic material, which uses the host cell machinery to replicate itself and self-assemble into new viruses, at which point they burst out of the cell, lysing it in the process, and these new viruses repeat the process in new cells.

This brings us back to our discussion on protein synthesis. By 1956, there were contradicting evidence as to whether or not there was a high turnover of RNA in phage-infected *E. coli*. Therefore, Elliot Volkin and Lazarus Astrachan developed a protocol that could more reliably assess the turnover of RNA following phage infection. They infected *E. coli* with bacteriophage and, after an adequate amount of time had elapsed for phage adsorption on the bacteria, added $^{32}PO_4$ to the culture to label nascent polynucleotides and then isolated the RNA to determine the amount that was radiolabeled.[7] The results showed that RNA synthesis was constant in uninfected cells and RNA itself was relatively stable, with a turnover of about 1 hour. In contrast, they found that in infected cells, addition of $^{32}PO_4$ for 6 minutes—called a pulse—followed by addition of excess of unlabeled PO_4—called a chase—resulted in a significant increase in RNA synthesis due to the rapid turnover of this newly synthesized RNA, which was determined to be about 10 minutes (Figure 8.4).[8,9] Surprisingly, RNA analysis indicated that the percent composition of nucleotides from newly synthesized RNA was in agreement with the virus' DNA rather than the host DNA. Thus, these data indicated that following phage infection, host RNA

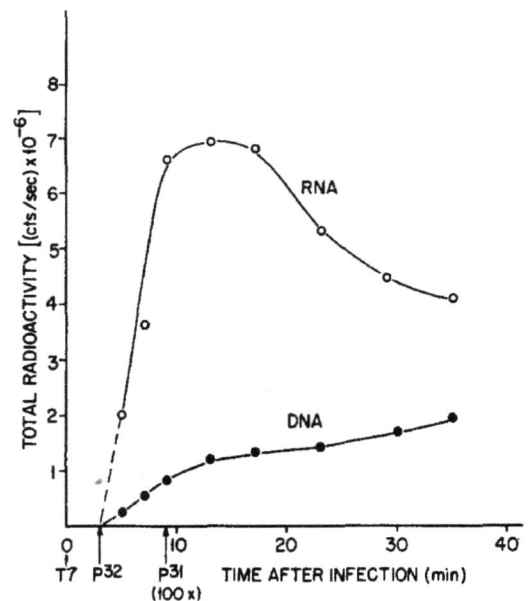

FIGURE 8.4 RNA turnover experiment. Bacteria were infected with T7 phages and $^{32}PO_4$ was added 3 minutes after infection, and 100× excess PO_4 was subsequently added after 6 minutes. The steady rate of incorporation of PO_4 into DNA indicates that the decrease in RNA observed is not due to the instability of the radioactive phosphate.[8] (Reprinted from Virology 6(2), Volkin E, Astrachan L, Countryman JL. Metabolism of RNA phosphorus in Escherichia coli infected with bacteriophage T7, 545–555. Copyright © 1958, with permission from Elsevier.)

was rapidly degraded and viral RNA was synthesized. As Volkin and Astrachan noted, the *similarity between the RNA and DNA synthesized after infection suggests a hitherto unsuspected connection between RNA synthesis and DNA synthesis.*[8] In addition, such a quick turnover in RNA synthesis suggested that this type RNA was unstable, as opposed to that of tRNA or ribosomal RNA, which had previously been shown to be relatively stable.

In previous studies using subcellular fractionation, Volkin and Astrachan had found that newly synthesized RNA following phage infection preferentially sedimented with a cell fraction containing the smallest proportion of RNA.[10] In 1960, Nomura, Hall, and Spiegelman extended these findings by performing similar experiments in an attempt to further qualify where the newly synthesized RNA was localized. As such, the authors infected *E. coli* with bacteriophage and incubated the culture with $^{32}PO_4$ for 2 minutes before stopping the assimilation of phosphorus with NaN_3.[11] Cells were then crushed with alumina and centrifuged into various fraction according to the scheme in Figure 8.5a. Following the quantitation of RNA in each fraction, the authors determined that newly synthesized RNA (i.e., radioactive RNA) was mainly present in the ribosomal fraction of cells. They also found that it sedimented at a slower rate than typical ribosomal RNA, which suggested that it may be a different type of RNA.[11] In addition, they determined that the ^{32}P-labeled RNA had a different electrophoretic mobility from the non-labeled RNA, suggesting that there must be a difference in nucleotide composition between the samples (Figure 8.5b). Subsequent analysis of sample's nucleotide content confirmed this and supported the ideas of Volkin and Astrachan that following phage infection viral RNA was synthesized.

At this time, the RNA template for protein synthesis was thought to be a stable molecule similar to DNA in its nucleotide composition and permanently attached to ribosomes, such that there would be one ribosome/RNA structure for each gene coded by DNA. However, the experimental evidence against this mechanism, as summarized by Jacob & Monod in 1961, did not support this hypothesis; as such, the following characteristics were suggested for the RNA template: it should be a polynucleotide; it should have a wide variety of molecular weights; its base composition should reflect that of DNA; it should be found associated with ribosomes; and it should have a high rate of turnover.[12] They also noted that the new species of RNA discovered by Volkin and Astrachan (and others since then who studied other organisms) met these specifications and named it messenger RNA (mRNA). Thus, the authors proposed the following mechanism for protein synthesis:

> The molecular structure of proteins is determined by specific elements, the structural genes. These act by forming a cytoplasmic "transcript" of themselves, the structural messenger, which in turn synthesizes the protein. The synthesis of the messenger by the structural gene is a sequential replicative process, which can be initiated only at certain points on the DNA strand [...][12]

The same year, two groups expanded on Volkin and Astrachan's and Nomura et al.'s findings, publishing back-to-back papers describing an unstable RNA species that seemed to be the intermediate carrier of information between genes and proteins as suggested by Jacob and Monod. Brenner, Jacob, and Meselson began their paper by proposing three hypotheses that could explain how RNA synthesis in *E. coli* switched from bacterial to viral following an infection.[13] Model I was the *classical model* in which *the bacterial machinery is switched off, and new ribosomes are then synthesized by the phage genes.* Model II assumed that *proteins are assembled directly on the DNA; the new RNA is a special molecule which enters old ribosomes and destroys their capacity for protein synthesis.* Model III supported the mRNA hypothesis with the following consequences upon phage infection: (1) *to switch off the synthesis of new ribosomes*; (2) *to substitute phage messenger RNA for bacterial messenger RNA* [...] [which] *can occur quickly only if messenger RNA in unstable.*[13]

To fully understand the results of their experiments, we need to have a look at additional properties that had been elucidated about ribosomes. Since ribosomes were discovered and mostly studied by ultracentrifugation, their names take the form of their rate of sedimentation, in Svedberg units (S). As such, the bacterial ribosome complex has a sedimentation rate of 70S and is composed of 50S and 30S subunits (eukaryotic cells have slightly different ribosomes: an 80S complex composed of 60S and 40S subunits—many antibiotics take advantage of these differences by inactivating bacterial ribosomes while sparing human ribosomes).[14] It was also found that the concentration of divalent cations present in the solution was important for the separation of ribosomes into their various subunits. Specifically, low concentrations of Mg^{2+} resulted in the dissociation of the ribosome complex into its 30S and 50S subunits, whereas higher concentrations were required for the subunits to remain associated into a 70S complex; at even higher concentrations, two 70S ribosomes aggregated into one 100S ribosome.

Thus, to distinguish between the group's three proposed models, Brenner, Jacob, and Meselson first pulsed bacteria with ^{14}C-uracil during a phage infection to label newly synthesized RNA, and then transferred the bacteria to media containing ^{13}C-uracil to stop labeling of RNA. Fractionation by centrifugation then allowed the authors to determine if newly synthesized RNA following phage infection could be isolated with ribosomes. Consistent with Nomura et al.'s results, the authors found that the bulk of RNA synthesis after phage infection was associated with ribosomes since the peak of radiolabeled RNA overlapped with what was determined to be the 70S ribosome fraction (Figure 8.6).[13] Further analysis also revealed that labeled RNA within bacteria quickly decreased after transfer to non-labeled media, indicating that this RNA species was highly unstable. The authors used a similar method to determine that no new ribosomes were synthesized from either the host or the phage. As a whole, these data effectively eliminated model I.

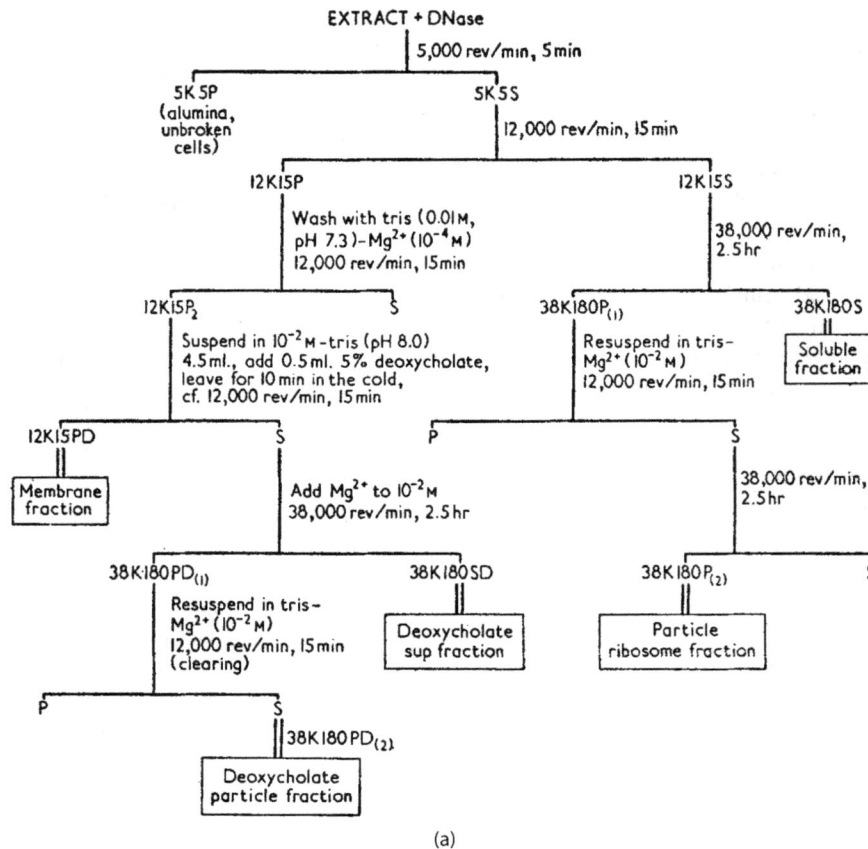

(a)

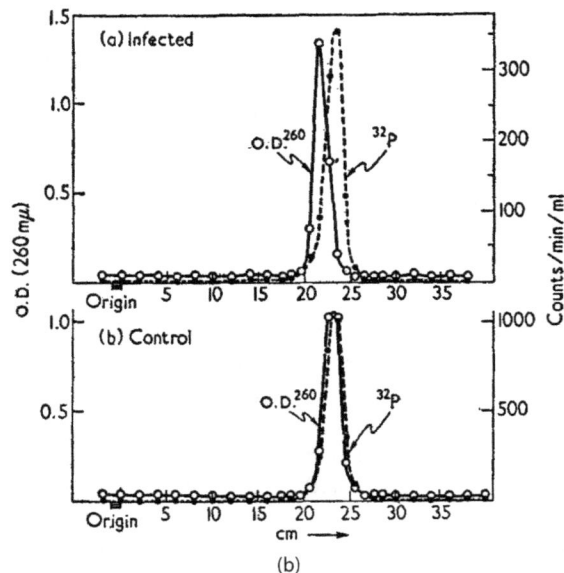

(b)

FIGURE 8.5 (a) Fractionation scheme for Nomura et al.'s experiments; (b) shift in mobility in newly synthesized, ^{32}P-labeled RNA in bacteriophage-infected *E. coli*. ^{32}P was added to uninfected cells (bottom) or 5 minutes after phage infection (top) and assimilation was allowed to proceed for 7 minutes before stopping the reaction. Ribosomal RNA was extracted and run on a starch electrophoresis column, where the profiles of ultraviolet absorption (O.D.260, original RNA) and radioactivity (^{32}P, newly synthesized RNA) were assessed.[11] (Reprinted from J Mol. Biol. 2(5), Nomura M, Hall BD, Spiegelman S. Characterization of RNA synthesized in *Escherichia coli* after bacteriophage T2 infection, 306–326, with permission from Elsevier, Copyright © 1960.)

To address the other two models, the authors needed to determine whether or not viral protein synthesis occurred on bacterial ribosomes. As such, they grew bacteria in ^{15}N-media to label all pre-existing proteins—including ribosomes—and infected them with bacteriophage; the bacteria were then incubated with ^{14}N (to wash out the ^{15}N, thus, any new ribosomes would not be radioactive) and ^{35}SO$_4$ media for two minutes (which labeled all newly synthesized proteins), and finally,

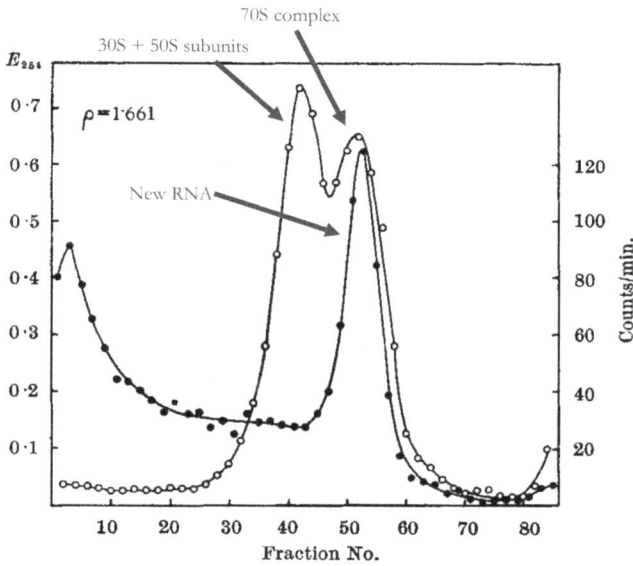

FIGURE 8.6 Newly synthesized RNA is found in the 70S ribosomal fraction. *E. coli* were infected with phage and fed ^{14}C-uracil for 2 minutes before ribosomes were isolated and their profile (open circle) as well as that of the attached RNA (closed circles), was assessed.[13] (Adapted by permission from Springer Nature Customer Service Centre GmbH: Springer Nature, Nature, Brenner S, Jacob F, Meselson M. An unstable intermediate carrying information from genes to ribosomes for protein synthesis, Copyright © 1961.)

transferred the bacteria to media containing excess, non-labeled SO_4. and methionine—in essence, they conducted a pulse-chase experiment for two parameters.[13] The authors then centrifuged the reactions to purify the ribosomes, and assessed if the newly synthesized, ^{35}S-labeled, viral proteins were found in existing, ^{15}N-labeled ribosomes, or if they were found in newly synthesized, ^{14}N-labeled, viral ribosomes. Results showed that the peak of ^{35}S proteins coincided only with the 70S complex of the ^{15}N-labeled ribosome, indicating that viral proteins were synthesized on existing ribosomes (Figure 8.7a). In addition, labeling was lost following addition of excess $^{32}SO_4$ and ^{32}S-methionine—the kinetics of which were consistent with the passage time of proteins on ribosome—which confirmed that labeling was specific for protein synthesis (Figure 8.7b). As such, the authors concluded that:

> most, if not all, protein synthesis in the infected cell occurs in ribosome. The experiment also shows that pre-existing ribosomes are used for synthesis and that no new ribosomes containing stable sulphur-35 are synthesized. This result effectively eliminates model II […]
>
> It is a prediction of the hypothesis that the messenger RNA should be of simple copy of the gene and its nucleotide composition should correspond to that of the DNA […]
>
> One last point deserves emphasis. Although the details of the mechanism of information transfer by messenger are not clear, the experiments with phage-infected cells show unequivocally that information for protein synthesis cannot be encoded in the chemical sequence of the ribosomal

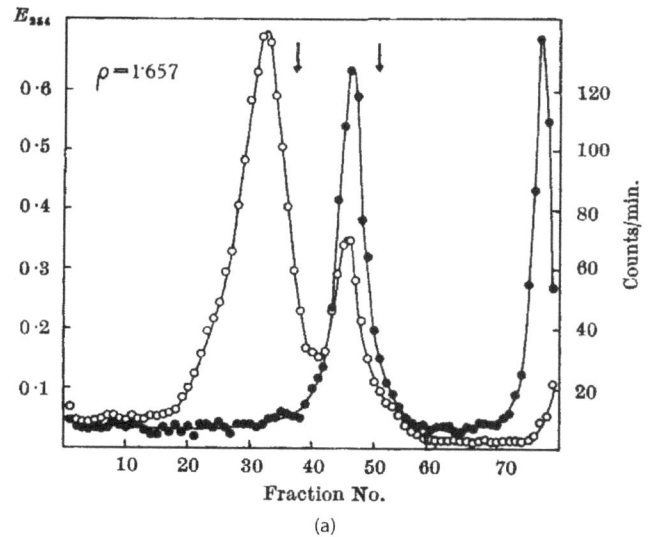

(a)

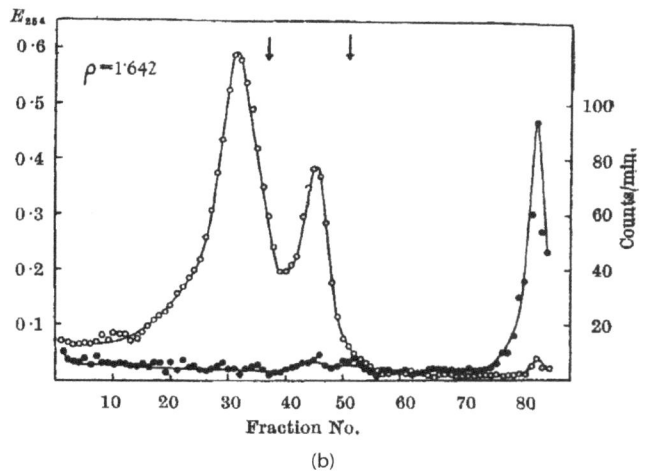

(b)

FIGURE 8.7 Viruses use the host's ribosome to translate its RNA. ^{15}N-labeled *E. coli* were infected with bacteriophage and then transferred to non-radioactive ^{14}N media. ^{35}S was then added for 2 minutes, after which half the solution was removed for analysis (a) and the other was transferred to non-radioactive media and grown for additional 8 minutes (b). Ribosomes were then extracted and assessed for ultraviolet absorption (preexisting proteins, ^{15}N ribosomes, open circle) and radioactivity (newly synthesized viral proteins, $^{35}SO_4$-labeled proteins, closed circle). The arrows indicate the expected positions of ^{14}N ribosomes.[13] (Reprinted by permission from Springer Nature Customer Service Centre GmbH: Springer Nature, Nature, Brenner S, Jacob F, Meselson M. An unstable intermediate carrying information from genes to ribosomes for protein synthesis, Copyright © 1961.)

RNA. Ribosomes are non-specialized structures which synthesize, at a given time, the protein dictated by the messenger they happen to contain.[13]

Although these experiments elegantly showed that mRNA was the template for protein synthesis following phage infection, this was not shown for normal, uninfected cell. However, James Watson and his group were also using isotope-labeled molecules to study the distribution of newly synthesized RNA, proteins, and ribosomes in uninfected bacteria. When each group heard about the other's research, they agreed that their papers should be published back-to-back. For their part, Watson's group's first indication that mRNA might be different from typical ribosomal RNA was that *traces [...] of ribonuclease degrade* [the former] *under conditions where RNA in ribosomes is untouched.* The authors also reasoned that since:

> ribosomal and transfer RNA comprise at least 95 per cent of *E. coli* RNA [...] *[mRNA]*, if present, can amount to at most only several per cent of the total RNA. Now, if the messenger were stable, only a corresponding fraction of newly synthesized RNA could be *[mRNA]*; the great majority of new RNA being the metabolically stable ribosomal and *[tRNA]*'s. If, however, *[mRNA]* is turning over [...] then a much larger fraction of newly made RNA must be messenger.[15]

The authors tested their hypothesis by exposing *E. coli* to 10–20 second pulses of ^{14}C-uracil and quickly chilled the cells on ice while stopping the incorporation of radioactive RNA by addition of NaN_3. They then extracted the cell contents by crushing the cells with alumina and isolated the RNA by centrifugation.[15] As such, they found, as Brenner et al. had found, that newly synthesized, ^{14}C-uracil RNA in infected bacteria sedimented with 70S ribosomes and similar results were obtained in uninfected *E. coli*. They also found that in low Mg^{2+} (conditions under which ribosomes dissociate into their 30S and 50S subunits), *the majority of the radioactivity is not associated with ribosomes but moves at a slower 14-16s peak* (Figure 8.8). Together, these

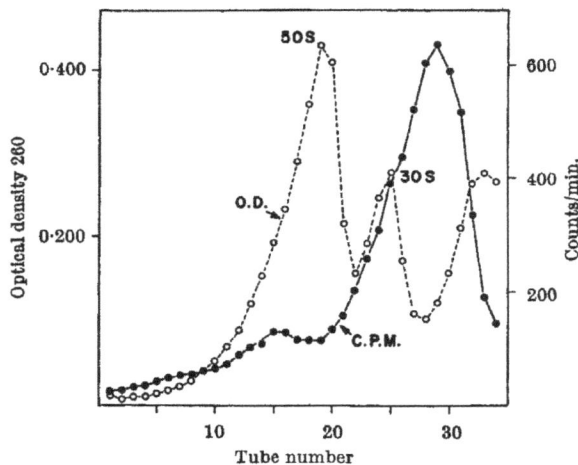

FIGURE 8.8 Sedimentation of newly synthesized RNA in uninfected bacteria. RNA from uninfected cells was labeled with ^{14}C-uracil for 20 seconds and ribosomes were isolated using centrifugation in low Mg^{2+} concentrations, in which 30S and 50S subunits dominate. Ultraviolet absorption (ribosomes, open circles) and radioactivity (RNA, closed circles) were then assessed.[15] (Reprinted by permission from Springer Nature Customer Service Centre GmbH: Springer Nature, Nature, Gros F, Hiatt H, Gilbert W, Kurland CG, Risebrough RW, Watson JD. Unstable ribonucleic acid revealed by pulse labelling of *Escherichia coli*, Copyright © 1961.)

groups' findings supported Jacob and Astrachan's messenger RNA hypothesis and indicated that mRNA is relatively unstable, even in uninfected cells, and associates with 70S ribosomes to initiate protein synthesis.

Although some studies had implied that mRNA was derived from DNA, Hall and Spiegelman proved it conclusively in 1961 using hybridization techniques recently developed. Marmur, Doty, et al. had previously found that double-stranded DNA which was denatured into single-stranded molecules by the action of heat could reassemble into their native, double-stranded form if subjected to slow cooling.[16,17] Surprisingly, Alexander Rich found that the same phenomenon could occur even if one strand was DNA and the other RNA.[18] This indicated that as long as two nucleic acid sequences were complimentary, they could combine and be held together by hydrogen bonds between the bases—as dictated by the Watson–Crick base pairing model—regardless of the type of nucleic acid.

Hall and Spiegelman used this information to test the hypothesis that newly synthesized RNA following phage infection was indeed derived from phage DNA, reasoning that *formation of a double-stranded hybrid during a slow cooling of a mixture of two types of polynucleotide strands can be accepted as evidence for complementarity of the input strands.*[19] Thus, the authors radio-labeled phage RNA with ^{32}P and phage DNA with ^{3}H, and began by confirming that they could clearly separate single-stranded DNA and RNA from double-stranded molecules by centrifugation. As such, they mixed these three nucleic acid species at room temperature and centrifuged them in a CsCl gradient for 5 days, the results confirming that each species clearly separated with little overlap (Figure 8.9a). The authors stated that any *distortion of the distribution of H³-DNA or P³²-RNA [...] which leads to regions of overlap between H³ and P³² would be indicative of such interactions.*[19]

Thus, when they mixed RNA with heat-denatured DNA and allowed the mixture to slowly cool down stepwise from 65 to 26°C in a water bath over about 30 hours, they found that a new ^{32}P-RNA band appeared *approximately centered on the band of H³-DNA* and concluded that [this] *new P³²-containing band must contain an RNA-DNA hybrid having approximately the same density as denatured T2-DNA* (Figure 8.9b).[19] They also determined that the DNA molecule had to be denatured to the single-stranded state for hybridization to occur, supporting the idea that hybridization between DNA and RNA was taking place. Finally, they tested the extent of hybridization of RNA and DNA from heterologous sources and found that none of these mixtures resulted in any significant degree of hybridization, confirming that the DNA and RNA sequences must be significantly identical for hybridization to occur. With these experiments, the authors concluded that:

> The fact that T2-RNA and DNA do satisfy the specificity requirement *[for hybridization]* must reflect a correspondence in structure between the two. Structural specificity of this order in single polynucleotide strands can only reside

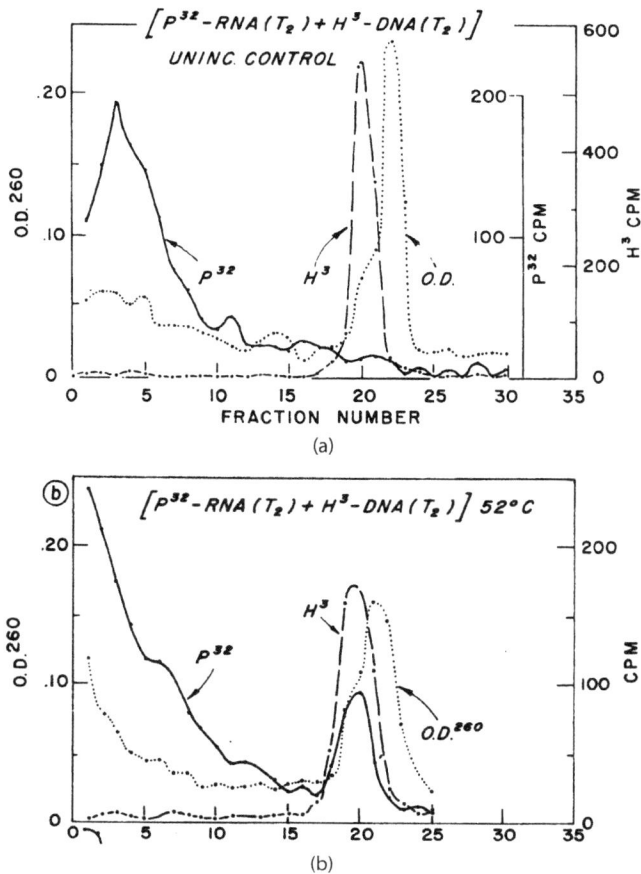

FIGURE 8.9 (a) Single-stranded DNA, double stranded DNA, and single-stranded RNA separate well by centrifugation in a CsCl gradient. (a) Unlabeled, double-stranded DNA (dotted line), ^{3}H-labeled, single-stranded DNA (broken line), and ^{32}P-labeled RNA (solid line) were mixed together and separated by centrifugation. The different fractions were assessed for radioactivity and ultraviolet absorption. (b) RNA hybridizes with single-stranded DNA. ^{32}P-labeled RNA and ^{3}H-labeled single-stranded DNA were mixed and cooled down at 52°C before being separated by centrifugation and assessed for ultraviolet absorbance and radioactivity.[19] (Reprinted from Hall BD, Spiegelman S. Sequence complementarity of T2-DNA and T2-specific RNA. Proc Natl Acad Sci U S A 1961;47(2):137.)

in definite sequences of nucleotides. We conclude that the most likely interrelationship of the nucleotide sequences of T2-DNA and RNA is one which is complementary in terms of the scheme of hydrogen bonding proposed by Watson and Crick [...]

It seems more likely that the RNA molecules directly concerned with specifying protein synthesis in normal cells would have a base ratio corresponding to DNA and would possess other properties analogous to those found for T2-specific RNA. Its principle characteristics may be summarized as follows: (1) a weak linkage with the ribosome fractions since it can be broken by dialysis against *[low]* Mg^{++}, (2) an active metabolic turnover, (3) an average sedimentation coefficient of about 8S, (4) a base composition which is closely analogous to its homologous DNA [...], and (5) a sequence complementary to its homologous DNA [...]

The demonstration of sequence complementarity between homologous DNA and RNA is happily consistent with an attractively simple mechanism of informational RNA synthesis in which a single strand of DNA acts as a template for the polymerization of a complementary RNA strand.[19]

The same year, Spiegelman published results from experiments addressing issues of occurrence of natural RNA/DNA hybrids, as well as the universality of these hybrids. In a first paper, Spiegelman, Hall, and Strock confirmed the *existence of native DNA-RNA hybrids in T2-infected E. coli* [which] *is consistent with the assumption that DNA serves as a template for the synthesis of complementary informational RNA.*[20] However, they added the caveat that this *view would require the existence of a DNA-dependent enzymatic mechanism for synthesizing polyribonucleotide;* an enzyme with these characteristics, called RNA polymerase, had been discovered the previous year by Severo Ochoa.

Later the same year, Hayashi and Spiegelman addressed the issue of universality of this mechanism when they investigated whether these observations would hold in uninfected cells. However, they noted that one advantage of the infection model was that *the synthesis of the ribosomal RNA components is suppressed [...] [and this] advantage is not generally present in uninfected cells, which, consequently, complicates the search for normal informational* [i.e., messenger] *RNA.*[21] Therefore, their approach was to determine if ribosomal RNA synthesis was suppressed when cells transitioned from fast to slow growth (other studies had suggested that this was the case). Thus, the authors were encouraged *to look more carefully into such transitions as pertinent to the purposes we had in mind.* They reasoned that:

Since the bulk of the RNA is ribosomal, cells growing at higher rates possess more ribosomes. Consider then the situation when one subjects a culture to a "step-down" transition by transferring cells from a rich to a synthetic medium. The growth rate is decreased by a factor of about 2. More important, at the moment they are introduced into synthetic medium and for some time thereafter, the cells have more ribosomes than they can usefully employ. From the viewpoint of selective advantage, it is perhaps not surprising to find that such step-down transitions are accompanied by a dramatic cessation of net RNA synthesis. Nevertheless, protein synthesis proceeds for a while at near normal rates [...] It seemed probable that the remaining residue of RNA synthesis would be restricted to the variety immediately necessary for the fabrication of new protein molecules, i.e., the normal informational RNA for which we were searching.[21]

Thus, the authors first examined the nucleotide ratio between DNA and newly synthesized RNA during a "step-down" transition by pulsing normally growing and "transitioning" bacteria with ^{32}P for 10 or 20 minutes to label newly synthesized RNA and DNA; they then isolated each component and their nucleic acid content was analyzed.[21] Results revealed that the *'step-down' culture is extremely*

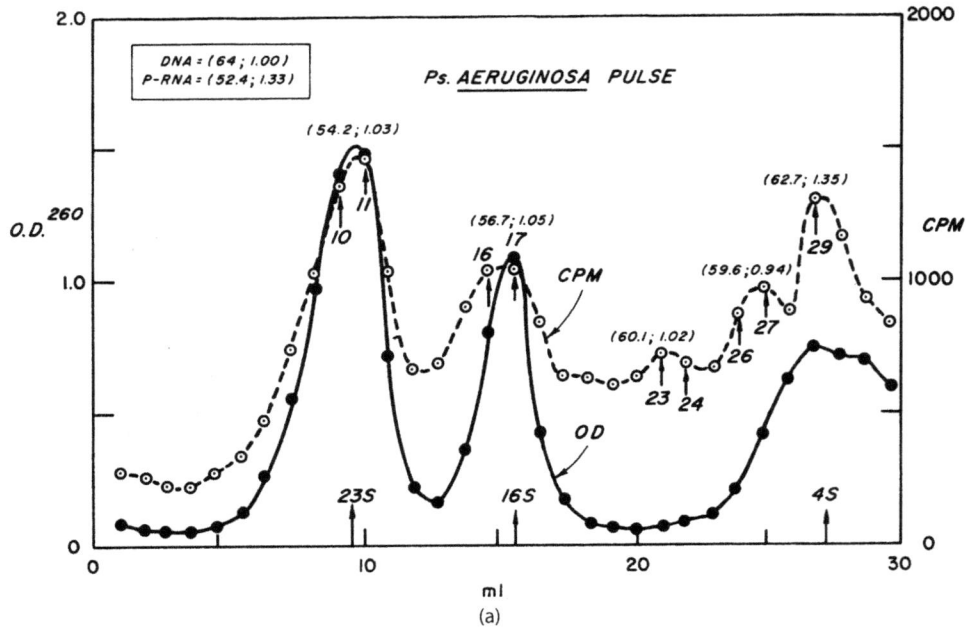

FIGURE 8.10   Ratio of bases in RNA compared to DNA in *E. coli*. Bacteria growing in either regular (control) or enriched (step-down) media were washed and resuspended in regular media for 20 minutes, followed by a 10- or 20-minute pulse in ^{32}P. RNA was isolated from each condition and its optical density (preexisting RNA; nonradioactive; closed circles) and radioactivity (newly synthesized RNA; open circles) profile were assessed (a). Some samples analyzed (indicated in the graph by the numbered arrows) were further analyzed for nucleotide composition and compared to bulk RNA and DNA (b). Note the similarity between total DNA and step-down nucleotide content.[21] (Reprinted from Hayashi M, Spiegelman S. The selective synthesis of informational RNA in bacteria. Proc Natl Acad Sci U S A. 1961;47(10):1564–1580.)

heterogeneous, ranging in size all the way from above the 23S region down to 4S (Figure 8.10a). In addition, the authors found that the base ratio of RNA in control cultures resembled that of total RNA in several organisms, whereas step-down cultures had a base ratio similar to DNA, supporting the hypothesis that mRNA was derived from DNA (Figure 8.10b). Hyahashi and Spiegelman also performed a hybridization experiment to determine if the sequence of RNA synthesized after step-down was similar to DNA. As such, the authors pulsed cultures to label RNA during a step-down transition, isolated and purified it, collected samples from several fractions, performed a hybridization experiment with either homologous or heterologous single-stranded DNA by cooling from 55 to 28°C in 25 hours, and separated the nucleic acids by CsCl gradient centrifugation to determine the extent of hybridization. As such, they found that *excellent hybridization occurs* between RNA and homologous DNA (Figure 8.11), whereas there was *no*

suggestion of any detectable mating between RNA and heterologous DNA.[21] In the discussion of the paper, the authors rejoiced at their decision to explore the use of step-down cultures as a tool for their studies:

It is apparent from the data presented that the choice of cultures in "step-down" transition was a happy one. They obviously provide almost ideal experimental material for the study of non-ribosomal RNA synthesis. In particular, they permitted the ready demonstration of RNA molecules in non-infected cells which satisfy the complementarity criterion established with the T2-E. coli complex [...]

It would appear from the results summarized here that the synthesis of polyribonucleotide strands complementary to homologous DNA is a generalized feature of normal and virus-infected bacterial cells. With respect to the genetic transcription mechanism, the present data are consistent with the conclusions drawn from our previous experiments with T2-infected cells. It would appear that complementary

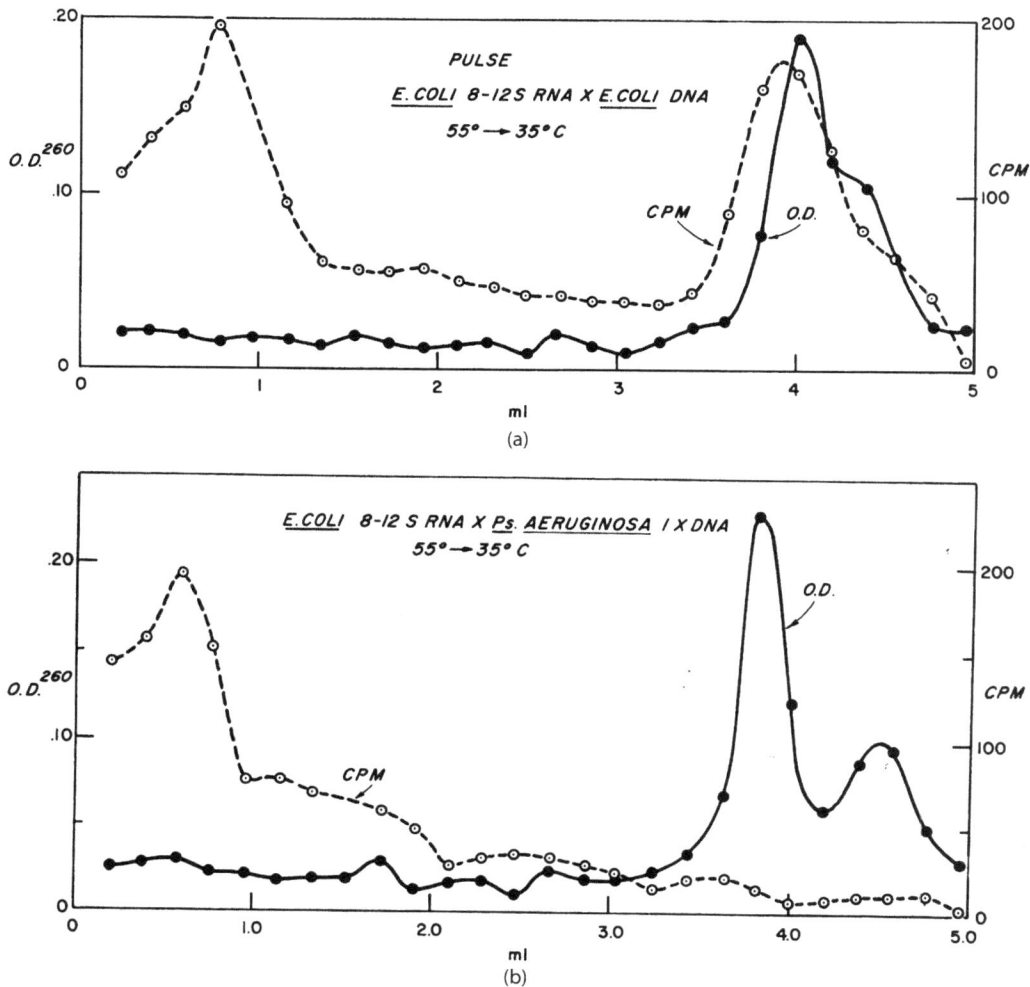

FIGURE 8.11 mRNA sequences are similar to that of DNA. A mixture of H³-RNA from *E. coli* "step-down" cultures were cooled with single stranded DNA from homologous (i.e., same organism, therefore, same sequence; (a), open circles) or heterologous (i.e., different organism, therefore, different sequence; (b), open circles) sources and separated by density gradient centrifugation. Double-stranded DNA was included as a control (closed circles). Note that the radioactive peak in (a) aligns well with the double-stranded DNA peak, suggesting that DNA/RNA hybridization occurred.[21] (Reprinted from Hayashi M, Spiegelman S. The selective synthesis of informational RNA in bacteria. Proc Natl Acad Sci U S A. 1961;47(10):1564–1580.)

RNA strands are the intermediaries between DNA and the protein-synthesizing apparatus.[21]

* * * * *

Astoundingly, Crick's "Central Dogma" for protein synthesis, that is, that DNA makes RNA and RNA makes proteins, had been reasonably confirmed by 1961, only a short 3 years after it was proposed. Now the looming question that remained concerned the way in which the genetic code converted mRNA into amino acids. Key to elucidation of this mechanism was the discovery, by Severo Ochoa and colleagues in 1960, of an enzyme that catalyzes the synthesis of polynucleotides, called polynucleotide phosphorylase. This allowed researchers to make synthetic mRNA to test the incorporation of amino acids into proteins in a controlled manner. As such, Marshall Nirenberg and his team synthesized specific RNA sequences in an effort to test the requirements of protein synthesis in bacteria.

To do so, they incubated a variety of RNA polynucleotides with radioactive amino acids, isolated the resulting proteins by centrifugation, and assessed the presence of radioactivity in the synthesized peptide. Thus, in a first experiment, they found that of all polynucleotides, only polyuridylic acid resulted in the synthesis of peptides composed solely of phenylalanine (Figure 8.12a).[22] In the converse experiment, the addition of polyuridylic acid to a mixture of amino acids resulted in the synthesis of proteins only when phenylalanine was present (Figure 8.12b). The same year, Ochoa's group published their findings using similar methods, and confirmed Niremberg's results concerning the sole incorporation of phenylalanine from polyU RNA.[23] The authors also tested heteropolynucleic acids on protein synthesis and found that although incorporation of phenylalanine into a

POLYNUCLEOTIDE SPECIFICITY FOR PHENYLALANINE INCORPORATION

Experiment no.	Additions	Counts/min/mg protein
1	None	44
	+ 10 μg Polyuridylic acid	39, 800
	+ 10 μg Polyadenylic acid	50
	+ 10 μg Polycytidylic acid	38
	+ 10 μg Polyinosinic acid	57
	+ 10 μg Polyadenylic-uridylic acid (2/1 ratio)	53
	+ 10 μg Polyuridylic acid + 20 μg polyadenylic acid	60
	Deproteinized at zero time	17
2	None	75
	+ 10 μg UMP	81
	+ 10 μg UDP	77
	+ 10 μg UTP	72
	Deproteinized at zero time	6

(a)

SPECIFICITY OF AMINO ACID INCORPORATION STIMULATED BY POLYURIDYLIC ACID

Experiment no.	C^{14}-amino acids present	Additions	Counts/min/mg protein
1	Phenylalanine	Deproteinized at zero time	25
		None	68
		+ 10 μg polyuridylic acid	38,300
2	Glycine, alanine, serine, aspartic acid, glutamic acid	Deproteinized at zero time	17
		None	20
		+ 10 μg polyuridylic acid	33
3	Leucine, isoleucine, threonine, methionine, arginine, histidine, lysine, tyrosine, tryptophan, proline, valine	Deproteinized at zero time	73
		None	276
		+ 10 μg polyuridylic acid	899
4	S^{35}-cysteine	Deproteinized at zero time	6
		None	95
		+ 10 μg polyuridylic acid	113

(b)

FIGURE 8.12 Polyuridine RNA codes for phenylalanine incorporation into proteins. (a) mRNA composed solely of the indicated nucleotides was added to a reaction mixture for protein synthesis, which contained only radioactive phenylalanine, and radioactivity was assessed; (b) mRNA composed solely of uridine was added to reaction mixtures for protein synthesis, along with the indicated radioactive amino acids, and radioactivity was assessed.[22] (Reprinted by permission from Myrna Milgram Weissman. Nirenberg MW, Matthaei JH. The dependence of cell-free protein synthesis in *E. coli* upon naturally occurring or synthetic polyribonucleotides. Proc Natl Acad Sci U S A. 1961;47(10):1588–1602. doi:10.1073/PNAS.47.10.1588.)

polypeptide chain could also be induced by polyUC and polyUA (nucleic acids composed of random combinations of U and C or U and A nucleotides, respectively), the former could also incorporate serine, leucine and proline, whereas the latter also induced the incorporation of tyrosine and isoleucine (Figure 8.13a). Interestingly, a mixture of polyUC in a ratio of 1:5 was ineffective in incorporating any of the amino acids tested, whereas the reverse mixture—poly UC in a 5:1 ratio—incorporated several. Thus, based on results using nucleotides mixed in different ratios, the authors suggested that proline was likely synthesized from a triplet code using 1U and 2C (and CCC), serine and leucine from 2U and 1C, and tyrosine and isoleucine from 2U and 1A. In subsequent papers, Ochoa and his group synthesized polynucleic acids from the combination of two or three nucleotides in known ratios and determined the relative incorporation of each amino acid from these different polynucleotides, which allowed them to deduce the three nucleotides required for incorporation of each amino acid (Figure 8.13b).[24]

However, these experiments did not reveal the actual sequence of nucleotides in the triplet code. In addition, it did not confirm nor disprove that the genetic code was a triplet; indeed, this was an assumption derived from logical arguments. Nonetheless, one thing that was clear from these data was that a particular amino acid could be coded for by more than one triplet code. Interestingly, a connection was made between the proposed triplet codes and the consequences of mutations on protein synthesis:

It is of interest in this connection that, in a nitrous acid mutant of tobacco mosaic virus described by Tsugita and Fraenkel-Conrat, a proline residue was replaced by leucine. As the nitrous acid effect is due to deamination of C to U, replacement of C by U would be in line with the code letters suggested for these amino acids. Our results would also be compatible with the fact that proline:phenylalanine ratio of wild cucumber and tobacco mosaic virus protein varies in the same direction as the C:U ratio of their respective nucleic acids.[23]

A different system was developed by Niremberg and Leder in 1964 which would allow to precisely define the sequence of nucleotides in each triplet code that codes for

Amino acid	Polynucleotide													
	U-rich							A-rich					C-rich	
	U	UA (5:1)	UC (5:1)	UG (5:1)	UAC (6:1:1)	UCG (6:1:1)	UAG (6:1:1)	A	AU (5:1)	AC (5:1)	AG (5:1)	ACG (4:1:1)	C	CI[a] (5:1)
Ala	–	–	–	–	–	3	–	–	–	–	–	6	–	22
Arg	–	–	–	–	–	3	–	–	–	–	13	27	–	19
Asn	–	7	–	–	7	–	5	–	28	30	–	17	–	–
Asp	–	–	–	–	–	–	3	–	–	–	–	4	–	–
Cys	–	–	–	20	–	25	32	–	–	–	–	–	–	–
Glu	–	–	–	–	–	–	2	–	–	–	30	21	–	–
Gln	–	–	–	–	*	–	*	–	–	44	6	30	–	–
Gly	–	–	–	4	–	3	–	–	–	–	5	3	–	5
His	–	–	–	–	3	–	–	–	–	9	–	13	–	–
Ile	–	20	–	–	16	–	32	–	20	–	–	–	–	–
Leu	–	14	20	13	25	25	27	–	3	–	–	–	–	–
Lys	–	3	–	–	2	–	–	100	100	100	100	100	–	–
Met	–	–	–	–	–	–	4	–	–	–	–	–	–	–
Phe	100	100	100	100	100	100	100	–	–	–	–	–	–	–
Pro	–	–	8	–	3	3	–	–	–	5	–	9	100	100
Ser	–	–	25	–	25	26	–	–	–	–	–	8	–	–
Thr	–	–	–	–	9	–	–	–	–	23	–	24	–	–
Try	–	–	–	5	–	4	*	–	–	–	–	–	–	–
Tyr	–	25	–	–	25	–	20	–	3	–	–	–	–	–
Val	–	–	–	20	–	20	25	–	–	–	–	–	–	–

(a)

Distribution of amino acids among triplets

Triplet series	Triplets	Amino acids
1[a]	AAA	Lys
2[a]	UUU	Phe
3[a]	CCC	Pro
4	GGG	–
5[a]	AUU UAU UUA	Tyr, leu, ile
6	AGG GAG GGA	–, gly, –
7[a]	ACC CAC CCA	His, pro, thr
8[a]	UAA AUA AAU	Asn, lys, ile
9	UGG GUG GGU	Try, gly, –
10[a]	UCC CUC CCU	Ser, pro, leu
11	GAA AGA AAG	Arg, –, glu
12[a]	GUU UGU UUG	Cys, leu, val
13	GCC, CGC, CCG	Arg, –, ala
14[a]	CAA ACA AAC	Asn, thr, gln
15[a]	CUU UCU UUC	Ser, phe, leu
16	CGG GCG GGC	–, gly, –
17	AUG AGU UAG UGA GAU GUA	Glu, met, –, –, –, asp
18[a]	AUC ACU UAC UCA CAU CUA	His, tyr, gln, thr, ile, asn
19	AGC ACG GAC GCA CAG CGA	–, ser, –, asp, ala, –
20	UGC UCG GUC GCU CUG CGU	–, –, arg, –, ala, –

(b)

FIGURE 8.13 (a) The indicated ratios of RNA polynucleotides were added to a mixture and the amino acid incorporated into the resulting protein was assessed. (b) The three-nucleotide code for incorporation of each amino acid into proteins proposed by Ochoa in 1964. Note that the specific order of nucleotides could not be determined.[24] (Reprinted by permission from Springer Nature Customer Service Centre GmbH: Springer Nature, Bull N Y Acad Med., Ochoa S. The chemical basis of heredity: the genetic code, Copyright © 1964.)

each amino acid. The genesis of this assay occurred when Niremberg wondered how long an mRNA would have to be for tRNA to associate with ribosomes.[25] To this end, the authors developed a new approach for isolating ribonucleotides: they passed a mixture of ribosomes, tRNA, mRNA, and a complex of all three components (as would occur during protein synthesis) through filter paper and found that only the latter was retained on the filter paper, while the individual components simply washed through.[25,26] They confirmed the validity of this system by using polyU mRNA—which was known to stimulate incorporation of phenylalanine into proteins—noting that the *rapidity of this assay, compared to others which depend upon the centrifugation of ribosomes, has greatly simplified this study.*[26]

Next the authors determined the effects of different lengths of polyU mRNA on tRNA binding to ribosomes and found that mRNAs required at least three nucleotides for binding of tRNAs. In addition, only multiples of three nucleotides increased the binding of tRNA to ribosome, providing, for the first time, direct proof that the genetic code was composed of nonoverlapping triplet nucleotides. The next step was clearly to synthesize all 64 possible combinations of triplets in the genetic code and test them against each of the 20 amino acids, which took Nirenberg and a team of dozens of scientists one year to accomplish, such that the genetic code was completely deciphered by 1966 (Figure 8.14).[25] In later studies, researchers found that the AUG codon, which codes for methionine, initiates the protein synthesis process, meaning that all proteins begin with this amino acid, whereas the codons UAA, UAG, and UGA do not code for any amino acids, but rather, signals the end of the protein, at which point the ribosome dissociates and is released from the mRNA, along with the completed protein. Additional studies comparing the genetic code of bacteria, invertebrates, and mammals, found that this code was universal, and therefore, the process of protein synthesis was essentially the same in all organisms. As Marshall Nirenberg stated in his review of the discovery of the genetic code, *These results had a profound philosophical impact on me because they indicate that all forms of life on this planet use essentially the same language.*[25]

Although practically nothing was known about the complex mechanism of protein synthesis at the beginning of the 1950s, this process and the genetic code were elucidated less than 20 years later. This phenomenal feat was accomplished by scores of scientists in several laboratories and confirmed that DNA was the site of storage of genetic information; however, that is the extent of its function. To be of any use, this molecular code needs to be transcribed to messenger RNA, a heterogeneous polyribonucleotide with a base sequence complimentary to the section of DNA from which it is derived, and which in turn is used as a template for protein synthesis on ribosomes (Figure 8.15a). In the previous chapters, we discussed that proteins are

composed of a series of amino acids connected via peptide bonds. Additional research revealed that these amino acids needed to be activated by the transfer of energy in the form of ATP before they could be incorporated into a soluble complex called transfer RNA, which acts as an adaptor molecule between amino acids and mRNA. Thus, a sequence of three nucleotides on the aminoacyl-tRNAs complex forms hydrogen bonds with a complementary sequence on the mRNA, which is in a complex with a ribosome; the amino acid is then transferred from the tRNA to the growing peptide chain (Figure 8.15b). The initiation of protein synthesis was determined to begin after ribosomes bind to a recognition sequence near the beginning of the mRNA and "scans" the mRNA—that is, it glides along the mRNA—until it encounters an AUG codon, which stimulates the incorporation of methionine as the

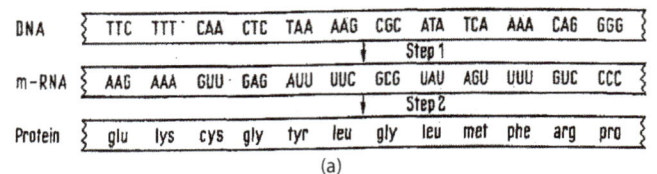

FIGURE 8.15 (a) A two-step transcription of the genetic code, as proposed by Ochoa in 1964. DNA sequences are transcribed into mRNA, which contain a sequence which is complimentary to the DNA sequence. The mRNA is then translated into proteins by ribosomes, which read the sequence three nucleotides at a time to specify incorporation of each amino acid.[24] (b) Illustrated summary of protein synthesis. mRNA binds to ribosomes and a tRNA, containing methionine and whose nucleotide triplet is complimentary to the triplet read by the ribosome, binds to the ribosome. The next triplet is read and the appropriate tRNA binds. This time, the ribosome catalyzes the binding of the two amino acids via a peptide bond. The ribosome then ejects the first tRNA (which is now detached from its amino acid) and reads the third triplet and attaches it to the second amino acid. This continues until the ribosome reads one of the three stop codons, which catalyzes the attachment of water and the release of the protein.[27] (Figure (a) reprinted by permission from Springer Nature Customer Service Centre GmbH: Springer Nature, Bull N Y Acad Med., Ochoa S. The chemical basis of heredity: The genetic code, Copyright © 1964.)

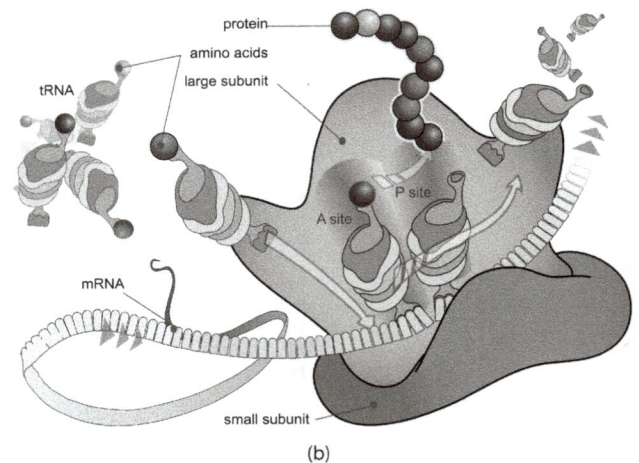

FIGURE 8.14 The genetic code as it is known today.

first amino acid of all proteins; synthesis ends when the ribosome encounters one of the three termination codons, at which point the ribosome complex disassembles and releases the mRNA and the protein into the cytosol or the inside of the ER.

* * * * *

REFERENCES

1. Judson HF. *The Eighth Day of Creation: Makers of the Revolution in Biology.* Commemorat. CSHL Press; 1996.
2. Crick FH. On protein synthesis. *Symposia of the Society for Experimental Biology.* 1958;12:138–163. In copyright. Wellcome Collection.
3. Hoagland MB, Zamecnik PC, Stephenson ML. Intermediate reactions in protein biosynthesis. *Biochimica Biophysica Acta.* 1957;24(1):215–216. doi:10.1016/0006-3002(57)90175-0
4. Hoagland MB, Stephenson ML, Scott JF, Hecht LI, Zamecnik PC. A soluble ribonucleic acid intermediate in protein synthesis. *Journal of Biological Chemistry.* 1958;231(1):241–257. doi:10.1016/S0021-9258(19)77302-5
5. D'Hérelle F. *The Bacteriophage and Its Behavior.* Williams & Wilkins Company; 1926. Accessed November 7, 2022. https://archive.org/details/thebacteriophage0000unse
6. Luria SE, Delbrück M, Anderson TF. Electron microscope studies of bacterial viruses. *Journal of Bacteriology.* 1943;46(1):57–77. doi:10.1128/jb.46.1.57-77.1943
7. Volkin E, Astrachan L. Phosphorus incorporation in *Escherichia coli* ribonucleic acid after infection with bacteriophage T2. *Virology.* 1956;2(2):149–161. doi:10.1016/0042-6822(56)90016-2
8. Volkin E, Astrachan L, Countryman JL. Metabolism of RNA phosphorus in *Escherichia coli* infected with bacteriophage T7. *Virology.* 1958;6(2):545–555. doi:10.1016/0042-6822(58)90101-6
9. Volkin E, Astrachan L. RNA metabolism in T2-infected *Escherichia coli.* In: McElroy WD, Glass B, eds. *A Symposium on the Chemical Basis of Heredity.* Johns Hopkins P:Oxford; 1957:686–695.
10. Volkin E, Astrachan L. Intracellular distribution of labeled ribonucleic acid after phage infection of Escherichia coli. *Virology.* 1956;2(4):433–437. doi:10.1016/0042-6822(56)90001-0
11. Nomura M, Hall BD, Spiegelman S. Characterization of RNA synthesized in *Escherichia coli* after bacteriophage T2 infection. *Journal of Molecular Biology.* 1960;2(5):306–326. doi:10.1016/S0022-2836(60)80027-7
12. Jacob F, Monod J. Genetic regulatory mechanisms in the synthesis of proteins. *Journal of Molecular Biology.* 1961;3:318–356. doi:10.1016/S0022-2836(61)80072-7
13. Brenner S, Jacob F, Meselson M. An unstable intermediate carrying information from genes to ribosomes for protein synthesis. *Nature.* 1961;190(4776):576–581. doi:10.1038/190576a0
14. Attardi G. The mechanism of protein synthesis. *Annual Review of Microbiology.* 1967;21(1):383–416. doi:10.1146/annurev.mi.21.100167.002123
15. Gros F, Hiatt H, Gilbert W, Kurland CG, Risebrough RW, Watson JD. Unstable ribonucleic acid revealed by pulse labelling of *Escherichia coli. Nature.* 1961;190(4776):581–585. doi:10.1038/190581a0
16. Marmur J, Lane D. Strand separation and specific recombination in deoxyribonucleic acids: Biological studies. *Proceedings of National Academy of Sciences U S A.* 1960;46(4):453–461. doi:10.1073/pnas.46.4.453
17. Doty P, Marmur J, Eigner J, Schildkraut C. Strand separation and specific recombination in deoxyribonucleic acids: Physical chemical studies. *Proceedings of National Academy of Sciences U S A.* 1960;46(4):461–476. doi:10.1073/pnas.46.4.461
18. Rich A. A hybrid helix containing both deoxyribose and ribose polynucleotides and its relation to the transfer of information between the nucleic acids. *Proceedings of National Academy of Sciences U S A.* 1960;46(8):1044–1053. doi:10.1073/pnas.46.8.1044
19. Hall BD, Spiegelman S. Sequence complementarity of T2-DNA and T2-specific RNA. *Proceedings of National Academy of Sciences U S A.* 1961;47(2):137. doi:10.1073/pnas.47.2.137
20. Spiegelman S, Hall BD, Storck R. The occurrence of natural DNA-RNA complexes in *E. coli* infected with T2. *Proceedings of National Academy of Sciences U S A.* 1961;47(8):1135–1141. doi:10.1073/pnas.47.8.1135
21. Hayashi M, Spiegelman S. The selective synthesis of informational RNA in bacteria. *Proceedings of National Academy of Sciences U S A.* 1961;47(10):1564–1580. doi:10.1073/pnas.47.10.1564
22. Nirenberg MW, Matthaei JH. The dependence of cell-free protein synthesis in *E. coli* upon naturally occurring or synthetic polyribonucleotides. *Proceedings of National Academy of Sciences U S A.* 1961;47(10):1588–1602. doi:10.1073/PNAS.47.10.1588. Marshall Nirenberg died in 2010. The paper is contributed to this volume by his wife Myrna Milgram Weissman October 2022.
23. Lengyel P, Speyer JF, Ochoa S. Synthetic polynucleotides and the amino acid code. *Proceedings of National Academy of Sciences U S A.* 1961;47(12):1936–1942. doi:10.1073/pnas.47.12.1936
24. Ochoa S. The chemical basis of heredity: The genetic code. *Bulletin of the New York Academy of Medicine.* 1964;40(5):387–411. doi:10.1007/BF02151242
25. Nirenberg M. Historical review: Deciphering the genetic code—A personal account. *Trends in Biochemical Sciences.* 2004;29(1):46–54. doi:10.1016/j.tibs.2003.11.009
26. Nirenberg M, Leder P. RNA codewords and protein synthesis. The effect of trinucleotides upon the binding of sRNA to ribosomes. *Science (1979).* 1964;145(3639):1399–1407. doi:10.1126/science.145.3639.1399
27. File: Ribosome mRNA translation en.svg—Wikipedia. Accessed November 1, 2022. https://en.wikipedia.org/wiki/File:Ribosome_mRNA_translation_en.svg

9 The Energy of Cells Part II
Oxidative Phosphorylation

We saw in Chapter 5 that foodstuff initially gets broken down in the glycolytic pathway, quickly yielding a small burst of energy usable by cells. During anaerobic respiration, which includes the process of fermentation, this pathway results in the accumulation of lactic acid in mammals or ethanol and CO_2 in yeast and other fermenting organisms. However, if oxygen is available, the end-product of glycolysis, the three-carbon molecule, pyruvic acid, is converted to a high-energy compound, acetyl-CoA, which then combines with oxaloacetate to form citric acid, the first molecule of the Krebs cycle. The Krebs cycle is a series of redox reactions that create $FADH_2$ and NADH as by-products and its end-product, oxaloacetate, is recycled to form citrate in a new cycle. The end result of glycolysis and the Krebs cycle is the net synthesis of ATP, whose terminal phosphates can be used as a signaling tag (discussed in Chapters 12 and 13) or as a high-energy bond to enable cellular reactions. In this chapter, we discuss the experiments which allowed the elucidation of the mechanism of oxidative phosphorylation, that is, how $FADH_2$ and NADH donate electrons to the electron transfer chain of mitochondria—the powerhouse of the cell and the main site of ATP synthesis—which creates a H^+ gradient across the mitochondrial membrane that fuels the synthesis of ATP.

In 1932, Otto Warburg and W. Christian identified a new class of molecules involved in the transfer of hydrogen ions. Warburg had always been suspicious about the reactions occurring in Thunberg tubes and questioned whether they were physiologically relevant.[1] However, a visit to a colleague's laboratory in the U.S. was to change his mind. A colleague performed some experiments on red blood cells for him which showed that methylene blue was reduced in the metabolism of glucose to CO_2 and pyruvate. Warburg was intrigued by this phenomenon and, therefore, began an investigation of the chemistry of methylene blue.[1] As such, he found that incubation of ATP, glucose-6-phosphate (the first intermediate of glycolysis), and methylene blue with yeast cell extracts led to an O_2 uptake, and dialysis of the extracts revealed that two components were required for this hydrogen-transferring process: a heat-labile enzyme and a heat-stable co-enzyme.[1] The enzyme component was further fractionated into two components, one of which had a yellow color when oxidized and became clear when reduced—this molecule was appropriately named "yellow enzyme".[2] This substance, which Warburg and Christian succeeded in isolating a few years later, was later called lumiflavin and was determined to be part of a greater family of molecules called flavoproteins which are used as carriers of hydrogen (Figure 9.1).

In 1934, Warburg isolated another important flavoprotein molecule: the famed co-enzyme initially discovered by Harden and Young in 1906 which up until now, could not be purely isolated or fully identified. However, its successful isolation indicated why this component was so elusive: only 4.8 g was isolated from 100 L of horse erythrocytes![1] It was determined to contain nitrogen, phosphorus, pentose, and adenine, but also another component new to coenzyme fractions, nicotinamide, which was later shown to be the hydrogen acceptor group of the molecule. This compound contained two phosphates, but another similar molecule was purified the next year which contained three phosphates. These molecules were named diphospho- and triphospho- pyridine nucleotides (DPN and TPN) but were changed in the 1960's to nicotinamide adenine dinucleotide and nicotinamide adenine dinucleotide phosphate, respectively (NAD^+ and NADP) (Figure 9.2).[1] Although the structural identity of these molecules was now known, their precise role in respiration remained to be elucidated. Warburg found that incubation of glucose-6-phosphate with NADP and a protein isolated from red blood cells led to a transfer of two hydrogens from glucose-6-phosphate to NADP, which were then transferred to yellow enzyme. This type of reversible hydrogen transfer was a new idea at the time, and it would be almost twenty years before the concept was confirmed by identifying the precise hydrogen acceptor on the pyridine ring of nicotinamide.[1]

In 1931, Nilsson had suggested that phosphorylation was somehow coupled to electron transfer but offered little evidence to support such ideas.[3] His claims had come from previous experiments which indicated that more phosphate was esterified (phosphate linked to ATP rather than present in the soluble form) during the Krebs cycle than could be accounted for by substrate-level phosphorylation. In addition, several reactions in the Krebs cycle were known to be dependent on NAD^+ as a cofactor, such as the oxidation of isocitrate to oxalosuccinate or the oxidation of malate to oxaloacetate, but these had not yet been directly demonstrated to result in phosphate esterification. Severo Ochoa and his group had reported the failure of such an attempt, but Friedkin and Lehninger reasoned that:

> These failures to observe coupled esterification, if the experimental conditions were valid, would force the conclusion that the pyridine nucleotide-catalyzed oxidations of the Krebs cycle cause no esterification of phosphate and that the remaining three oxidative steps of the cycle [...] must account for all the esterification of phosphate observed during complete oxidation of pyruvate. As a

DOI: 10.1201/9781003379058-9

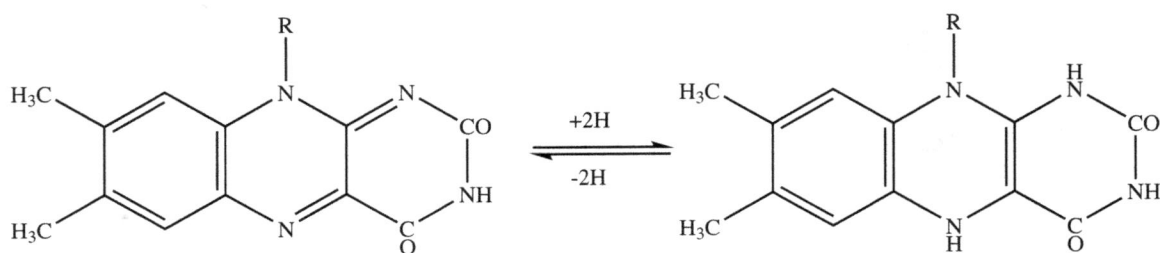

FIGURE 9.1 Redox reaction of flavins.

corollary of this conclusion each of these three oxidations must then be accompanied by the esterification of as many as 4 or 5 molecules of inorganic phosphate, a yield which approaches complete thermodynamic efficiency of the conversion of energy yielded in these oxidations into phosphate bond energy.

It appeared to us that the negative results reported by Ochoa in his study of phosphate esterification coupled to the pyridine-catalyzed oxidations might possibly have been a reflection of some conditions of his experiments which were adverse to the demonstration of phosphorylation by measurement of net phosphate uptake.[4]

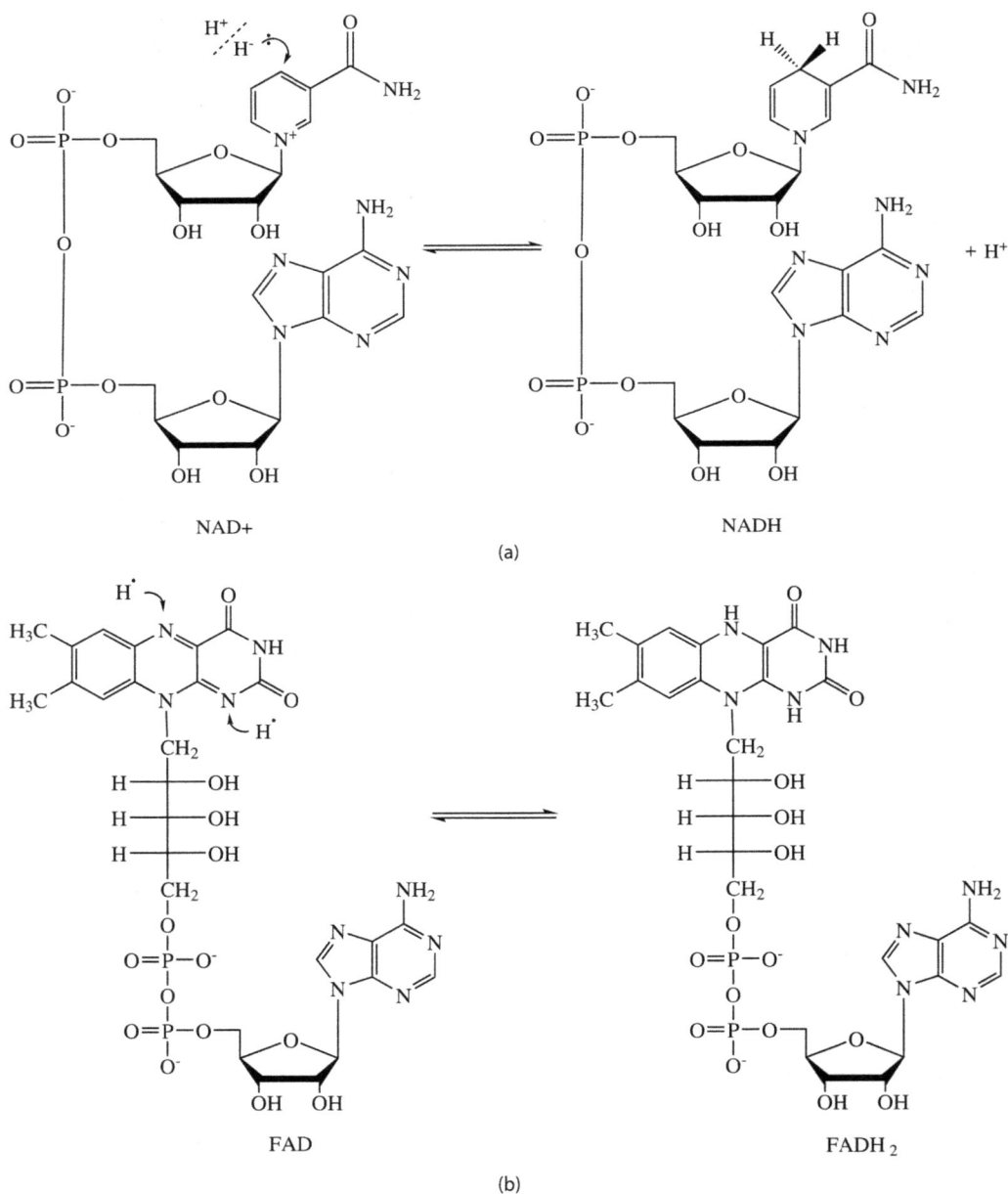

FIGURE 9.2 Examples of important hydrogen carriers which take part in redox reactions: NADH (a) and $FADH_2$ (b).

As such, Friedkin and Lehninger investigated the possibility that phosphorylation was directly linked to redox reactions by measuring the regeneration of ATP in the reaction:

The demonstration of coupled phosphorylation in these preparations by measuring net uptake of inorganic phosphate has not been possible, since the necessary trans-phosphorylases required to cause phosphorylation of acceptors such as glucose, creatine, etc., by adenosine triphosphate (ATP) have been removed in the washing procedure. However, it is readily possible to demonstrate the maintenance of the esterified phosphate of ATP coupled to oxidation. In the absence of oxidizable substrate, ATP is quickly dephosphorylated by phosphatases *[resulting in an increase in P_i]*; in the presence of active oxidation of added substrates of the Krebs cycle the level of esterified phosphate may be maintained at or somewhat below the starting level. The use of inorganic phosphate labeled with P^{32} *[would provide]* proof that such maintenance is the dynamic resultant of dephosphorylation and oxidation-coupled rephosphorylation, since after oxidation has taken place a large part of the P^{32} *[would be]* found in the esterified P fraction.[4]

Thus, using rat liver homogenates, the authors determined that phosphorylation was indeed coupled to electron transport between either purified NADH or NADH produced from redox reactions of the Krebs cycle and oxygen, since addition of increasing concentrations of NADH resulted in a concomitant increase in oxygen uptake and ^{32}P esterification (Figure 9.3), though they also noted that excessively high concentrations resulted in the loss of activity.[4] They also found that cytochrome *c*—a mitochondrial resident protein shown to be involved in redox reactions—was required for this process, but attempts to localize the esterification of P_i to ADP during the transport of electrons through cytochrome *c* were not successful. Finally, the authors tested a number of compounds for their ability to act as a phosphate acceptor and found that ADP (and ATP when the system allows for continuous hydrolysis of ATP)

was the only primary phosphate acceptor in this system.[5] To sum up, these experiments showed that NADH generated during the Krebs cycle donated electrons to cytochrome c in mitochondria, which in turn, donated them to oxygen, resulting in the esterification of P_i to ADP at a location that was yet to be determined.

In the mid-1940s, with the advent of electron microscopy, structural studies of mitochondria enabled the elucidation of a direct link between mitochondria and oxidative phosphorylation, pioneered by Albert Claude and George Palade. Claude was instrumental in determining the adequate conditions for cellular fractionation using differential centrifugation and the characterization of isolated organelles.[6] As such, he isolated mitochondria and determined that succinoxidase and cytochrome oxidase (enzyme complexes involved in respiration) were exclusively located in mitochondria, consistent with the abovementioned involvement of cytochrome *c* in respiration.[6] By the 1950s, Palade had developed electron microscopy preparation techniques well enough to obtain high-definition images of mitochondria, from which he defined these structures as such:

Two spaces or chambers are outlined by the mitochondrial membranes, an outer chamber contained between the two membranes, and an inner chamber bounded by the inner membrane. The inner chamber is penetrated and, in most cases, incompletely partitioned by laminated structures which are anchored with their bases in the inner membrane and terminated in a free margin after projecting more or less deeply inside the mitochondrion.[7]

* * * * *

Although a role for redox reactions in the esterification of phosphate to form ATP was well established and the site for these reactions was confirmed to be mitochondria, the precise mechanism for the synthesis of ATP had yet to be determined. However, from the early 1950s to the late 1970s, four hypotheses would take center stage in an attempt

Substrate	Oxygen uptake	Inorganic P	Esterified P	Incorporation of P^{32} into esterified P
	c.mm.	γ	γ	*per cent*
0 time		100	302	0.088
None	3	240	158	3.25
Malate	15	158	240	24.3
Fumarate		158	236	24.2
Oxalacetate	15	206	196	10.1
cis-Aconitate		182	212	16.3
Citrate		164	232	20.0
α-Ketoglutarate	13	214	196	12.0
Succinate	54	120	284	37.4
Pyruvate		185	211	13.8
" + malate	26	115	287	39.2

FIGURE 9.3 Redox-coupled phosphorylation. Various Krebs cycle intermediates were added to a reaction mixture and esterification of $^{32}P_i$ was assessed (last column).[4]

to explain it. The first of these, the chemical hypothesis, was proposed in 1953 by Edward Slater and was primarily derived from the known mechanism of substrate-level phosphorylation during glycolysis and the Krebs cycle, that is, synthesis of ATP during oxidative phosphorylation was proposed to occur by phosphorylation of ADP in enzymatic reactions via an intermediate:[8]

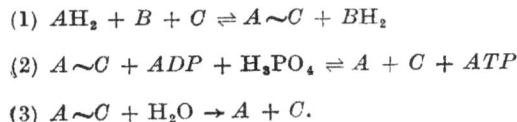

(1) $AH_2 + B + C \rightleftharpoons A{\sim}C + BH_2$

(2) $A{\sim}C + ADP + H_3PO_4 \rightleftharpoons A + C + ATP$

(3) $A{\sim}C + H_2O \rightarrow A + C.$

Although this hypothesis was well supported by the results available at the time, years of failed attempts at finding the elusive intermediate would ensue and even Slater himself eventually admitted that his proposal lacked *usefulness or credibility* and that his intermediate probably did not exist. Nonetheless, it was the only hypothesis considered to have any merit for many years.[9]

In 1961, RJP Williams proposed a mechanism by which protons in the mitochondrial membrane could be used as a driving force for ADP phosphorylation. However, the same year, Peter Mitchell advanced a similar hypothesis which gained more traction. Stating a number of shortcomings of the chemical hypothesis—that is, failure to find the high-energy intermediate; unexplained mitochondrial swelling; unexplained association of oxidative phosphorylation with mitochondria—Mitchell's hypothesis proposed that a hydrogen gradient between the inside and the outside of mitochondria (i.e., on either side of the membrane), generated from electrons donated by enzymatic reactions of the Krebs cycle, could provide the energy required for phosphorylation of ADP to form ATP.[10] The first important feature of the "chemiosmotic hypothesis", as it was known, was the suggestion that the plasma membrane of mitochondria served as a partial barrier to H^+ and OH^- ions and contained a reversible complex which Mitchell called the

"ATPase system". The system was assumed to be anisotropic, that is, H^+ ions could only enter the system from one side and OH^- ions from the other, but it was also equipped with OH^- and H^+ "translocation systems" that shuttled these components to their respective sides, thereby creating an excess of H^+ on the outside and an excess of OH^- on the inside of mitochondria (Figure 9.4). As Mitchell explained,

> According to the chemi-osmotic coupling hypothesis, the differential of the electro-chemical activity of the hydrogen and hydroxyl ions across the membrane, generated by electron transport, causes the specific translocation of hydroxyl and hydrogen ions from the active centre of the so-called ATPase system, thus effectively dehydrating ADP + P.
>
> [...] The underlying thesis of the hypothesis put forward here is that if the processes that we call metabolism and transport represent events in a sequence, not only can metabolism be the cause of transport, but also transport can be the cause of metabolism. Thus, we might be inclined to recognize that transport and metabolism, as usually understood by biochemists, may be conceived advantageously as different aspects of one and the same process of vectorial metabolism.[10]

And thus, the hypothesis proposed that the synthesis of ATP depends on the flow of protons from one side of the membrane to the other via an ATPase system.

In 1963, Paul Boyer added to the long list of proposed high-energy intermediates—in line with Slater's chemical hypothesis—and provided evidence that phosphohistidine might be a candidate; however, their claimed molecule would suffer the same fate as the other proposed molecules when he realized the next year that was not an intermediate in oxidative phosphorylation, but part of an enzyme involved in the Krebs cycle.[11] He then started thinking about conformational changes in enzymes and proposed, in 1965, an early version of his "conformational hypothesis" to explain oxidative phosphorylation: conformational changes in the ATPase—that is, changes in the molecule's

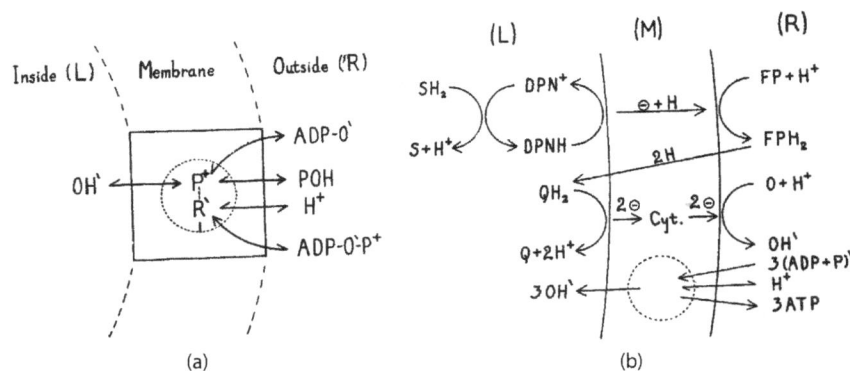

FIGURE 9.4 Diagrams of Mitchell's chemiosmotic mechanism. (a) The reversible ATPase enzyme was proposed to be embedded in the membrane and powered by a H^+ gradient on either side of it: high concentrations of OH^- on the inside of the membrane and of H^+ on the outside, coupled to the esterification of P_i to ADP to produce ATP. (b) Protons are transferred in the mitochondrial membrane from one electron transport molecule to the next, until they reach the ATPase.[10] (Reprinted by permission from Springer Nature Customer Service Centre GmbH: Springer Nature, Nature, Mitchell P. Coupling of phosphorylation to electron and hydrogen transfer by a chemiosmotic type of mechanism, Copyright © 1961.)

3D structure—could provide the energy required for phosphate transfer to ADP.[12,13] In time, Boyer would merge Mitchell's chemiosmotic hypothesis with his own to make a more complete version of his hypothesis which would prove to be accurate.[9] In the next section, I will present the main evidence in support of the mechanism of ATP formation during oxidative phosphorylation.

* * * * *

In 1960, Efraim Racker and his group published the first of a long series of papers concerning the characterization of an enzyme containing ATPase activity, the first of which reported the successful purification of the catalytic site of the enzyme—which they called coupling factor 1 (F_1)—from beef heart mitochondria.[14] As such, the authors found that the supernatant fraction of lysates from disrupted mitochondria (which contained F_1) was responsible for most of the ATPase activity and determined that only the terminal P_i of ATP was hydrolyzed in this system (Figure 9.5). Surprisingly, they also found that excess ADP was inhibitory on ATPase activity, and importantly, that the activity gradually increased above pH 5.0 with maximal ATPase activity between pH 8.5–9.3.[14]

In an accompanying paper, they described results which supported the idea that ATPase was capable of operating in the reverse direction. Here, the authors found that inclusion of highly purified ATPase in a system apt for metabolizing components of the Krebs cycle resulted in uptake of O_2 and P_i, whereas the omission of ATPase resulted only in O_2 uptake without concomitant esterification (i.e., P_i uptake, Figure 9.6). They also tested whether the apparent redox-linked esterification in their system was due to ADP phosphorylation and found a significant increase in $^{32}P_i$-ATP exchange in the presence of their purified enzyme. Thus, these results indicated that a single enzyme was responsible for the reversible reaction of the hydrolysis of ATP into ADP and P_i and for the synthesis of ATP from ADP and P_i.[15]

In the 1960s, the study of oxidative phosphorylation was approached mainly by using either intact cells or purified mitochondria and testing their ability to metabolize different substrates under various conditions, which revealed much about the kinetics and sequence of events leading to sugar metabolism. Another strategy used was the isolation

Coupling factor	O_2 up-take	P_i up-take	P:O	ATPase	P_i^{32}-ATP
μg	μatoms	μmoles		μmoles ATP^{32} disappeared	c.p.m./ μmole ATP
0	6.3	0.1	0.02	3.6	90
5	6.2	0.1	0.02	3.7	190
10	5.5	1.0	0.18	3.7	440
20	5.5	1.5	0.27	3.4	740
40	5.5	2.0	0.36	3.9	1030
20 + DNP	5.5	0.1	0.02	4.3	50
20*				3.4	0

* Particulate fraction omitted.

FIGURE 9.6 Coupling factor increases P_i esterification and $^{32}P_i$-ATP exchange. Coupling factor was added to a reaction mixture and various parameters were assessed. DNP was used to inhibit oxidative phosphorylation, and therefore, the P_i-ATP exchange. Note how P_i uptake and $^{32}P_i$-ATP exchange increases proportionately to the amount of coupling factor added.[15]

of various mitochondrial fractions to test the manner in which they interacted in an attempt to reveal their structural arrangement, a strategy that led to the discovery of numerous components of the system. However, it was determined that this manner of investigation also led to disruption and inactivation of the components under study. Thus, a third strategy was adopted by D.E. Griffith's group which combined elements of both of these ideas to determine the best conditions to isolate each of the different complexes of the ETC *such that the functional requirements for over-all electron transfer are optimally satisfied*, and then incubated the isolated complexes in various combinations to determine which substrate these combinations could metabolize and the extent to which electron transfer occurred.[16] The individual complexes isolated and used for recombination are shown in Figure 9.7.

Thus, the authors found that when complex I and complex III were combined in a ratio up to 1.5, cytochrome *c* was maximally reduced by NADH, whereas minimal or no activity

Additions	Initial ATP	Final			
		ATP	ADP	AMP	P_i
	μmoles	μmoles			
None........................	6.2	4.1	2.2	0	2.2
Dinitrophenol (5×10^{-4} M).......	6.2	3.0	3.1	0	2.9

FIGURE 9.5 Only the terminal phosphate in ATP is cleaved by ATPase. Concentrations of ATP, ADP, AMP, and P_i were assessed following ATP cleavage by ATPase. DNP was used here to inhibit oxidative phosphorylation. The fact that no AMP was present in the final solution indicated that only ATP was hydrolyzed.[14]

Complex I—DPNH-coenzyme Q reductase, which catalyzes the reaction

$$DPNH + Q + H^+ \rightarrow DPN^+ + QH_2 \qquad (1)$$

Complex II—Succinic-coenzyme Q reductase, which catalyzes the reaction

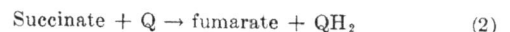

$$Succinate + Q \rightarrow fumarate + QH_2 \qquad (2)$$

Complex III—QH_2-cytochrome *c* reductase, which catalyzes the reaction

$$QH_2 + 2 \text{ ferricytochrome } c \rightarrow Q + 2 \text{ ferrocytochrome } c \qquad (3)$$

Complex IV—Cytochrome *c* oxidase, which catalyzes the reaction

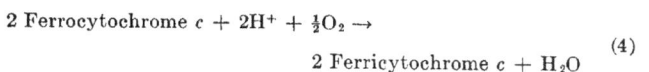

$$2 \text{ Ferrocytochrome } c + 2H^+ + \tfrac{1}{2}O_2 \rightarrow$$
$$2 \text{ Ferricytochrome } c + H_2O \qquad (4)$$

FIGURE 9.7 Complexes of the ETC isolated by Hatefi et al. in their experiments and the reactions they catalyze.[16]

was detected for the individual complexes (Figure 9.8a). In contrast, the combination of complex II and III resulted in significant reducing activity by succinic acid. In addition, they found that mixing of the complexes had to be performed at high concentrations and then diluted for the experiment in question, from which the authors concluded that *interaction between the two particles resulting in reconstituted activity takes place in the prescribed proportions* [... and] *results from a particle-particle interaction.* When the three complexes were combined, cytochrome *c* could be reduced by either NADH or succinate and the authors confirmed that the activity observed was due to a transport of electrons from NADH to complex I, through complex II, and finally, to cytochrome *c* through complex III.[16] Furthermore, the combination of all four complexes with cytochrome *c* resulted in an increase in O$_2$ (Figure 9.8b), though the authors noted that activity was progressively lost with time due to the known ease of dissociation of complex IV. Therefore, these experiments indicated that complexes of the ETC needed to be arranged in a specific order to function, and the authors suggested the sequence in Figure 9.8c as the proper arrangement.[16]

At the same time, future Venezuelan politician, Humberto Fernández-Morán, and his colleagues reported the isolation of a mitochondrial fraction that contained the complete electron transfer chain capable of transporting electrons from NADH and succinate to oxygen, with or without the enzyme responsible for catalyzing the phosphorylation of ADP (Racker's F$_1$ complex). The authors used sucrose gradient centrifugation to isolate two analogous fractions from the mitochondrial particulate fraction, one which contained *an essentially intact electron transfer chain divested of the primary dehydrogenating enzyme complexes* [elementary particle, EP], and another with *the capacity for both electron transfer and oxidative phosphorylation* [ETP$_H$].[17] As expected, they determined that the ETP$_H$ was composed of flavins, cytochromes, iron, copper, coenzyme Q, and phospholipids, in addition to a variety of proteins. However, they found that the conditions of the isolation process played a significant role in the ratio of the individual components isolated; thus, the condition used was one in which *both the preservation of the stoichiometry among the functional components found in the mitochondrion (or in the EPT$_H$* [sic]) *and the preservation of activity were the yardsticks of suitability.*[18]

The group's research also led to the capture of the first clear images of the ATPase in mitochondria. Electron microscopy preparations up to that point revealed:

merely a general structural framework consisting essentially of lipoprotein. Thus, in the mitochondrial membranes only certain stereotyped, uniform features of the 'osmium-stained' smooth membranes are seen, with no indications of the specific enzymic complexes and other constituents.[17]

Enzyme preparation	Inhibitor	Specific activity	
		Succinic-cytochrome *c*	DPNH-cyto-chrome *c*
DPNH-coenzyme Q reductase (I)		0.0	8.9
Succinic-coenzyme Q reductase (II)		0.0	0.0
QH$_2$-cytochrome *c* reductase (III)		0.0	0.0
I + III		0.0	57.4
II + III		48.4	0.0
I + II + III		56.6	42.9
I + II + III	Antimycin A	<4.0	0.0
I + II + III	Amytal	52.0	0.0
I + II + III	Thenoyltrifluoro-acetone	0.0	

(a)

Enzyme preparation	Specific activity	
	Succinic oxidase	DPNH oxidase
DPNH-coenzyme Q reductase (I) + cytochrome *c*	0	0
Succinic-coenzyme Q reductase (II) + cytochrome *c*	0	0
QH$_2$-cytochrome *c* reductase (III) + cytochrome *c*	0	0
Cytochrome oxidase (IV) + cytochrome *c*	0	<1
I + II + III + cytochrome *c*	3.1	2.9
I + II + III + IV + *cytochrome c*	28.7	14.5
I + II + III + IV + *cytochrome c* + cyanide	0	0
I + II + III + *IV* + *cytochrome c* + antimycin A	0	0
I + II + III + IV + *cytochrome c* + Amytal	25.7	0
I + II + III + IV + *cytochrome c* + thenoyltrifluoroacetone	0.7	

(b)

(c)

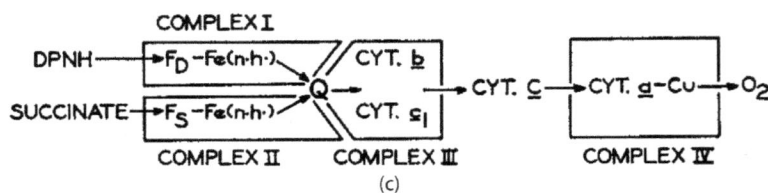

FIGURE 9.8 Hatefi et al.'s reconstitution of the electron transfer chain. Isolated complexes were incubated in various combinations and amount of reduced cytochrome c (a) or oxygen consumption (b) was assessed. Antimycin, amytal, and thenoyltrifluoroacetone are inhibitors of oxidative phosphorylation. Italicized components were premixed, followed by addition of the other components. (c) The sequential arrangement of ETC components as proposed by Hatefi et al.[16]

To that end, the authors used newly developed staining techniques and took advantage of significant improvements in the resolution of electron microscopes to increase the quality of images of the ultrastructures of mitochondria. These advances allowed, for the first time, the visualization of *a characteristic polyhedral or round structural unit, 80 to 100 Å in diameter, as a basic component of mitochondrial membranes* [... the] *vast numbers of these particles—as well as their regularity and periodicity—are the most remarkable features* (Figure 9.9).[17] They found that these particles, calculated to be in numbers of approximately 10,000 to 100,000 per mitochondrion, were associated with the external mitochondrial envelope and consisted of a head piece, a stalk, and a base piece which was attached to the cristae.[17] Finally, they observed that the ETP and the ETP_H fractions were basically the same except that the ETP fraction (which was incapable of phosphorylating ADP) contained mitochondria with only one membrane, whereas the fully active fraction contained the expected double membrane.[17]

During the course of their studies, Kagawa and Racker found that the F_1-ATPase (i.e., the coupling factor) they had isolated and characterized was not sensitive to oligomycin treatment, a substance that was known to inhibit oxidative phosphorylation and ATPase activity. Therefore, they

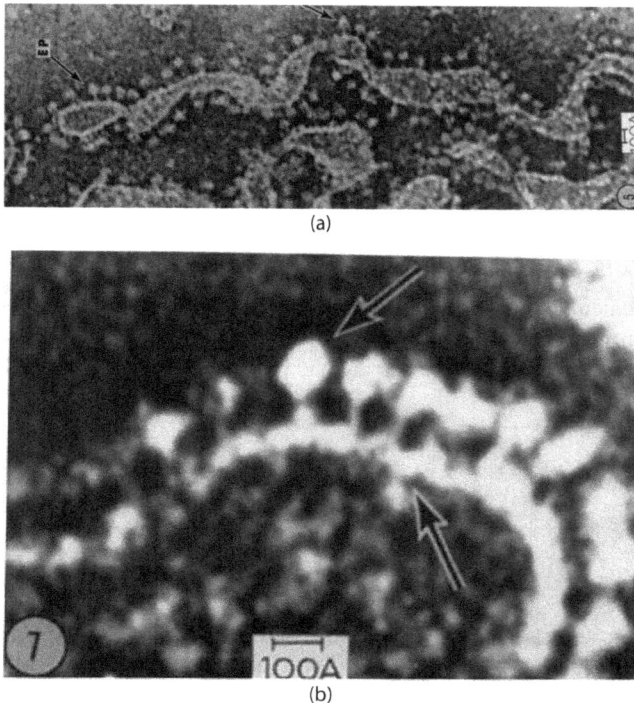

(a)

(b)

FIGURE 9.9 First clear images of the ATPase in mitochondria showing the head piece, the stalk, and the base piece (elementary particles, EP).[17] (Used with Permission of the Rockefeller Institute Press, from Fernandez-Moran H, Oda T, Blair PV, Green DE. A macromolecular repeating unit of mitochondrial structure and function. Correlated electron microscopic and biochemical studies of isolated mitochondria and submitochondrial particles of beef heart muscle. J Cell Biol. 22:63–100.; Permission conveyed through Copyright Clearance Center, Inc. Copyright © 1964.)

searched for another subunit that might mediate this sensitivity. As such, they determined that treating mitochondrial particles sequentially with trypsin, urea, and again with trypsin, efficiently isolated a subunit from ATPase, which they called F_o (for oligomycin-sensitive).[19] The authors found that addition of purified F_o to soluble F_1 preparations restored oligomycin sensitivity to F_1 activity. Furthermore, the authors discovered that the activity conferred by F_o, in addition to its ability to bind to F_1, was particularly sensitive to phospholipase treatment, indicating that phospholipids were essential to the stability of the complex. Based on Fernández-Morán's images of mitochondria and other studies from Racker's group, it was determined that the F_1 subunit was the head piece and stalk of ATPase, where the F_o-subunit was the membrane-bound element.

* * * * *

In the mid1960s, concrete proof for Mitchell's chemiosmotic theory had started to surface. At the time, Jagendorf and Uribe were studying chloroplasts, a plant organelle that converts light into sugars, which can then be used to generate ATP in mitochondria. Jagendorf had previously reported an increase in hydrogen ion uptake in chloroplasts following exposure to light, which they proposed could be interpreted by Mitchell's chemiosmotic theory, i.e., that light induced a translocation of H^+ to the inner space of the organelle's double membrane. The authors reasoned that if this was correct, *then the same high-energy condition should be formed artificially, entirely in the dark and without electron transport, by loading the inner space of the grana disk membranes with protons.*[20]

Thus, the authors exposed chloroplasts to an acid bath followed by an increase to pH 8, concomitant with addition of ADP and P_i to allow phosphorylation to occur, and indeed, found a significant production of ATP (Figure 9.10). They also found that the nature of the acid supplied had a great impact on the degree to which phosphorylation occurred, with succinate generating more ATP than glutamate. In addition, they determined that the phenomenon was reversible in that significant ATP synthesis resulted if chloroplasts were charged to pH 4, discharged to pH 6.5, and again decreased to pH 4 before transferring to pH8.[20] The activity of the system was also shown to be dependent on the pH of the acid and alkaline stages, with an increase in activity observed the more acidic the acid stage was and the more basic the alkaline stage was, though the latter had maximum activity between pH 8.4 and 9.0. Finally, the absolute difference between the pH of the acid and alkaline stages was shown to be critical, the greater difference yielding more ATP. These results therefore supported the chemiosmotic theory in that a proton gradient on either side of a membrane enabled ATP synthesis. The authors explained:

When placed in acid, protons would be expected to penetrate the interior space depending on the external pH. Presumably, cations would be simultaneously displaced to preserve electrical neutrality. Added organic acids could penetrate in the undissociated form (without the need to

REVERSIBILITY OF INDUCTION OF X_E BY ACID

Acid	pH in Successive Minutes			ATP
	1	2	3	
Glutamic (5 mM)	4.0	—	—	15.6*
	4.0	6.5	—	0.0
	4.0	6.5	4.0	13.7
Succinic (5 mM)	4.0	—	—	49.7
	4.0	6.5	—	2.8
	4.0	6.5	4.0	33.6

FIGURE 9.10 Induction of ATP formation in chloroplasts is reversible. Chloroplasts were incubated successively in pH 4, 6.5, and 4, as indicated (columns 1-3), before being transferred to a reaction mixture at pH 8 and assessed for ATP formation.[20] (Reprinted by permission from Jean W Jagendorf; Jagendorf AT, Uribe E. ATP formation caused by acid-base transition of spinach chloroplasts. Proc Natl Acad Sci U S A. 1966;55(1):170–177.)

expel cations) and provide an internal reservoir of dissociable protons, thereby accounting for the higher yields. On raising the external pH to 8 or more, a proton gradient would be present from inside to outside, and this in itself would have thermodynamic potential. The height of the potential would depend on the ratio of pH's inside and outside, and this could account for the requirement of a distinct pH difference between the acid stage and the phosphorylation reaction. Indeed, the alkaline displacement of the phosphorylation reaction pH curve due to raising the acid stage pH [...] was one of the predictions of this model, and the experiments were performed in order to test it. The actual yield of ATP at any pH [...] probably reflects a balance between the pH dependence of the relevant chemical reactions, and the thermodynamic effect which is a matter of the change in pH from the previous acid stage.

Given a pH differential across the grana disk membrane, its potential might be translated into ATP formation via an anisotropic, membrane-bound, reversible ATPase as discussed by Mitchell.[20]

Peter Mitchell's group confirmed these findings in mitochondria the same year showing that incubation of rat liver mitochondria first in an alkaline bath and then in an acid bath resulted in ATP synthesis, noting that:

From the point of view of the chemiosmotic hypothesis, it is especially interesting that our observations indicate that

ATP synthesis in rat liver mitochondria is associated with the passage of protons inwards, as predicted, whereas the synthesis of ATP in chloroplasts is associated with the passage of protons outwards. This observation is in keeping with other evidence [...] which implies that the polarity of the ATPase system is opposite to that in the lamellae membrane of chloroplast.[21]

Shortly after, it was discovered that a particular type of bacterium, *H. halobium*, harbored a cell membrane protein, bacteriorhodopsin (also known as purple membrane), whose function was similar to photoreceptors in plants, that is, it converted light into the release of protons. As such, Oesterhelt and Stoeckenius found that illumination of the bacteria caused a 0.1–0.2 decrease in pH which returned to normal basal levels following transfer to darkness, a cycle which could be repeated indefinitely (Figure 9.11a).[22] They also found that uncouplers of oxidative phosphorylation increased the rate at which pH returned to basal levels upon incubation in the dark and abolished the subsequent decrease in pH when illuminated again, which implied to the authors *that illumination generates a pH and/or electrical gradient across the cell membrane.* However, they calculated that the number of protons released during illumination could not be accounted for simply by the release of protons from the membrane and reasoned, as was suggested

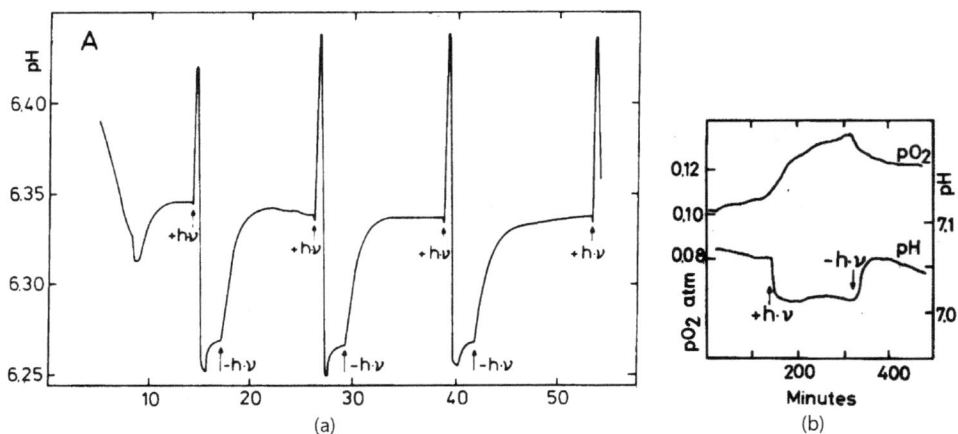

FIGURE 9.11 Exposure to light results in a decrease in pH and an increase in oxygen consumption. (a) Purple membranes were illuminated (+hν) or placed in the dark (−hν) and pH was assessed. (b) Same as in (a) but pH and oxygen were assessed. Notice that changes in pO$_2$ inversely reflect the presence or absence of light.[22] (From Oesterhelt D, Stoeckenius W. Functions of a new photoreceptor membrane. Proc Natl Acad Sci U S A. 1973;70(10):2853–2857.)

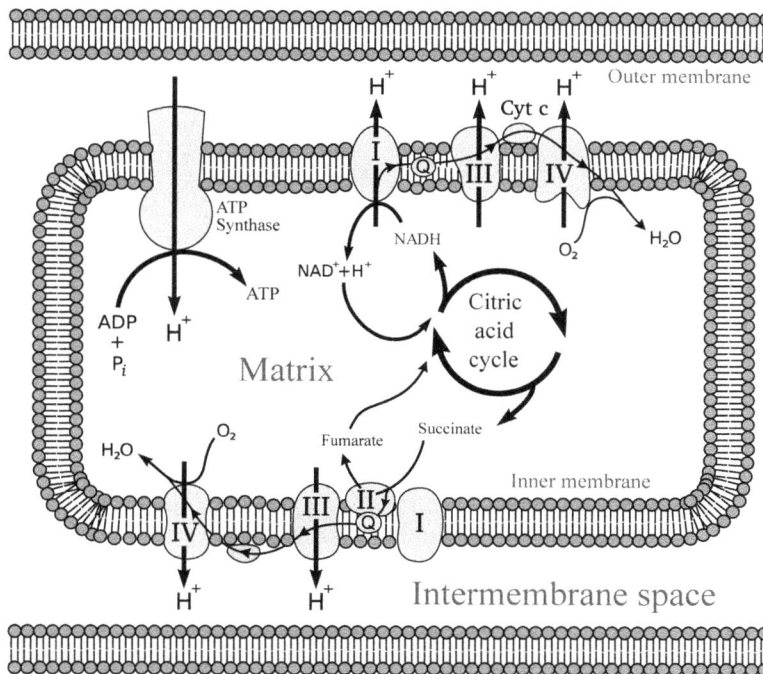

FIGURE 9.12 The electron transfer chain. The four complexes of the ETC are labeled I through IV. NADH donates its electrons to complex I, whereas $FADH_2$ donates electrons to complex II. Electrons are transferred from one complex to the next via oxidation/reduction reactions, terminating with their transfer to oxygen—the final electron acceptor—in complex IV to form water.[24]

by Mitchell's chemiosmotic hypothesis, that illumination of *H. halobium* induced a *net outward proton translocation from the cell*. Thus, they proposed that the bacteria:

through cyclic light-induced conformational changes of its bacteriorhodopsin, directly converts absorbed light energy into a proton gradient [...] across the membrane analogous to observations in other prokaryotic cells, mitochondria, and chloroplasts.[22]

Consistent with this, the authors found a negative correlation between decreased pH following illumination and O_2 consumption and ATP synthesis (Figure 9.11b).[22]

The next year, Racker and Stoeckenius replicated these findings using purified bacteriorhodopsin inserted into phospholipid vesicles. As such, they confirmed that bacteriorhodopsin alone was sufficient for the observed increase in pH following illumination, a phenomenon that was inhibited in the presence of uncouplers of oxidative phosphorylation, which indicated once again that illumination resulted in a movement of protons across the membrane in native bacteria that was enabled by bacteriorhodopsin.[23] In addition, significant ATP formation resulted from illumination of vesicles containing both ATPase and bacteriorhodopsin. These experiments supported the chemiosmotic hypothesis (which was considered a theory after these crucial experiments) and confirmed that a proton gradient across a double phospholipid membrane—as shown in mitochondria and chloroplasts—was crucial to drive ATP formation.

＊ ＊ ＊ ＊ ＊

In this chapter, we discussed the transfer of electrons from molecules oxidized in the Krebs cycle to the electron transport chain, a mitochondrial structure composed of four independent protein complexes and cytochrome c. As such, NADH donates electrons to complex I and $FADH_2$ donates electrons to complex II, and these are transferred to complex II, complex III, cytochrome c, and complex IV, through a series of redox reactions (Figure 9.12). In the process, each electron transferred from one complex to the next induces the translocation of protons (H^+) from the inside of mitochondria to its intermembrane space (the area between the two lipid bilayers), thus generating a proton gradient across that membrane. In the last complex, the electrons are transferred to oxygen, along with two H^+, thereby generating water as a byproduct. In the next chapter, we will see how the cell uses this gradient to generate the vast majority of ATP required for all its energy needs.

＊ ＊ ＊ ＊ ＊

REFERENCES

1. Krebs HA. Otto Heinrich Warburg, 1883–1970. *Biographical Memoirs of Fellows of the Royal Society.* 1972;18:628–699. doi:10.1098/rsbm.1972.0023
2. Kalckar HM. The nature of energetic coupling in biological syntheses. *Chemistry Reviews.* 1941;28(1):71–178. doi:10.1021/cr60089a002
3. Needham DM. Adenosinetriphosphate as fuel and as phosphate-carrier. In: *Machina Carnis: The Biochemistry of Muscular Contraction in Its Historical Development.* Cambridge University Press; 1971:782.

4. Friedkin M, Lehninger AL. Esterification of inorganic phosphate coupled to electron transport between dihydro-diphosphopyridine nucleotide and oxygen. I. *Journal of Biological Chemistry*. 1949;178(2):611–623. doi:10.1016/S0021-9258(18)56879-4

5. Lehninger AL. Esterification of inorganic phosphate coupled to electron transport between dihydrodiphosphopyridine nucleotide and oxygen. II. *Journal of Biological Chemistry*. 1949;178(2):625–644. doi:10.1016/S0021-9258(18)56880-0

6. Ernster L, Schatz G. Mitochondria: A historical review. *Journal of Cell Biology*. 1981;91(3 Pt 2):227s–255s. doi:10.1083/JCB.91.3.227S

7. Palade GE, Siekevitz P. Liver microsomes: An integrated morphological and biochemical study. *Journal of Biophysical and Biochemical Cytology*. 1956;2(2):171–200. doi:10.1083/jcb.2.2.171

8. Slater EC. Mechanism of phosphorylation in the respiratory chain. *Nature*. 1953;172(4387):975–978. doi:10.1038/172975a0

9. Prebble JN. Contrasting approaches to a biological problem: Paul Boyer, Peter Mitchell and the mechanism of the ATP synthase, 1961–1985. *Journal of History of Biology*. 2013;46(4):699–737. doi:10.1007/s10739-012-9343-7

10. Mitchell P. Coupling of phosphorylation to electron and hydrogen transfer by a chemi-osmotic type of mechanism. *Nature*. 1961;191(4784):144–148. doi:10.1038/191144a0

11. Allchin D. To err and win a Nobel prize: Paul Boyer, ATP synthase and the emergence of bioenergetics. *Journal of History of Biology*. 2002;35(1):149–172. doi:10.1023/A:1014583721788

12. Boyer PD. Carboxyl activation as a possible common reaction in substrate-level and oxidative phosphorylation and in muscle contraction. In: King T, Mason Howard, Morrison M, eds. *Oxidases and Related Redox Systems*. John Wiley & Sons; 1964:994–1008.

13. Boyer PD, Chance B, Ernster L, Mitchell P, Racker E, Slater EC. Oxidative phosphorylation and photophosphorylation. *Annual Reviews of Biochemistry*. 1977;46(1):955–966. doi:10.1146/annurev.bi.46.070177.004515

14. Pullman ME, Penefsky HS, Datta A, Racker E. Partial resolution of the enzymes catalyzing oxidative phosphorylation. I. Purification and properties of soluble dinitrophenol-stimulated adenosine triphosphatase. *Journal of Biological Chemistry*. 1960;235:3322–3329.

15. Penefsky HS, Pullman ME, Datta A, Racker E. Partial resolution of the enzymes catalyzing oxidative phosphorylation. II. Participation of a soluble adenosine tolphosphatase in oxidative phosphorylation. *Journal of Biological Chemistry*. 1960;235:3330–3336. Accessed March 30, 2019. http://www.ncbi.nlm.nih.gov/pubmed/13734097

16. Hatefi Y, Haavik AG, Fowler LR, Griffiths DE. Studies on the electron transfer system. XLII. Reconstitution of the electron transfer system. *Journal of Biological Chemistry*. 1962;237:2661–2669.

17. Fernandez-Moran H, Oda T, Blair P, Green DE. A macromolecular repeating unit of mitochondrial structure and function. Correlated electron microscopic and biochemical studies of isolated mitochondria and submitochondrial particles of beef heart muscle. *Journal of Cell Biology*. 1964;22:63–100. doi:10.1083/jcb.22.1.63

18. Blair P, Oda T, Green DE, Silver DR, Fernandez-Moran H, Merk FB. Studies on the electron transfer system. LIV. Isolation of the unit of electron transfer. *Biochemistry*. 1963;2(4):756–764. doi:10.1021/bi00904a023

19. Racker E. A mitochondrial factor conferring oligomycin sensitivity on soluble mitochondrial ATPase. *Biochemical and Biophysical Research and Communications*. 1963;10:435–439. doi:10.1016/0006-291X(63)90375-9

20. Jagendorf AT, Uribe E. ATP formation caused by acid-base transition of spinach chloroplasts. *Proceedings of National Academy of Sciences U S A*. 1966;55(1):170–177. doi:10.1073/pnas.55.1.170

21. Reid RA, Moyle J, Mitchell P. Synthesis of adenosine triphosphate by a protonmotive force in rat liver mitochondria. *Nature*. 1966;212(5059):257–258. doi:10.1038/212257a0

22. Oesterhelt D, Stoeckenius W. Functions of a new photoreceptor membrane. *Proceedings of National Academy of Sciences U S A*. 1973;70(10):2853–2857. doi:10.1073/pnas.70.10.2853

23. Racker E, Stoeckenius W. Reconstitution of purple membrane vesicles catalyzing light-driven proton uptake and adenosine triphosphate formation. *Journal of Biological Chemistry*. 1974;249(2):662–663.

24. File: Mitochondrial electron transport chain—Etc4.svg—Wikimedia Commons. Accessed November 1, 2022. https://commons.wikimedia.org/wiki/File:Mitochondrial_electron_transport_chain%E2%80%94Etc4.svg

10 The Energy of Cells Part III
The Mechanism of ATP Synthesis

This chapter continues from the last chapter, in that we now focus in on the molecular mechanism by which a phosphate group is added to ADP to produce ATP, which is one of the most intricate mechanisms in this book due to the complexity of the enzyme involved, the ATPase. The elucidation of this phenomenon came in a large part from the researches in the Boyer laboratory over about 25 years, but mostly between the mid-1960s and late 1970s. Boyer first drew inspiration from the studies of Mildred Cohn in 1953, in which he used the newly developed technique of isotopic labeling of molecules—in this case, he used ^{18}O-labeled phosphate—to follow the *path of the phosphate group through a series of reactions in which the phosphorus leaves no trace.*[1] The author's rationale for the experiments (Figure 10.1),

> depends upon the fact that the phosphate group does not necessarily proceed intact through a sequence of phosphorylation, transphosphorylation, and dephosphorylation reactions, but may lose one or more of its original oxygen atoms […] If inorganic phosphate labeled with O^{18} is taken up in organic linkage by the formation of a carbon-oxygen bond […] the oxygen bridging the carbon and phosphorus becomes labeled with O^{18}. Should the organic phosphate now be cleaved by the rupture of the phosphorus-oxygen bond as in phosphatase reactions, the organic moiety remaining would contain O^{18}. Moreover, if inorganic phosphate were formed in such a reaction, one of the four labeled oxygens would have been replaced by normal oxygen from the water […][1]

Thus, he hoped to detect unknown reactions involved in oxidative phosphorylation by *following the loss of O^{18} from inorganic phosphate.* Initial experiments using this technique had shown that only *10 per cent of the O^{18} initially present in the inorganic phosphate remained after 1 hour of reaction* in a mixture fit for oxidative phosphorylation. As such, the author realized that:

> […] the inorganic phosphate must have passed through many reaction cycles. If one envisages a sequence of reactions in which inorganic phosphate is taken up in organic linkage and then split hydrolytically with an O-P cleavage, 75 per cent of the labeled O^{18} will be retained in the inorganic phosphate formed after one cycle. Should this inorganic phosphate now react in another cycle, again 75 per cent of the remaining labeled O^{18} would be retained; that is, 56 per cent of the initial O^{18} […][1]

To account for this loss, Cohn suggested that only water could serve as the source of this exchange since *there is no other source of* [unlabeled oxygen] *in the system sufficiently large to account for the large amount introduced into the inorganic phosphate.* In addition, he also used a variety of oxidation substrates and electron acceptors in an attempt to localize the turnover of phosphate but all reagents used resulted in a significant loss of ^{18}O from phosphate. However, he did find that reagents that uncoupled redox reactions from phosphorylation (i.e., allowed redox reactions to proceed but blocked the phosphorylation event), such as 2,4-dinitrophenol (DNP), significantly prevented the *phosphate turnover reaction observed by the disappearance of O^{18}.* Therefore, the author suggested that *the requirements for the phosphate turnover reaction seem to parallel the requirements for phosphorylation* rather than those required for the oxidation of substrates. Cohen also emphasized the fact that *this rapid phosphate turnover reaction does not occur during phosphorylation accompanying the substrate level oxidation* [i.e., glycolysis], but only in *those processes which are sensitive to 2,4-dinitrophenol.*[1] Finally, one of the explanations provided by Cohn for the phenomenon he observed was one in which water participated either directly or indirectly in a reversible reaction, and suggested it,

> may be the mechanism which operates in the as yet unknown transphosphorylating systems involved in phosphorylation associated with the electron transport system […]
>
> In a reversible system, any exchange with water of a compound containing O^{18} would ultimately be reflected in a loss of O^{18} from inorganic phosphate.[1]

The next year, Boyer and his group, seeking a high-energy intermediate in oxidative phosphorylation, explored Cohen's reaction further; specifically, they were intrigued by the fact that his and others' results suggested that oxidative phosphorylation might be a reversible process. As such, Boyer, Falcone, and Harrison labeled P_i with ^{18}O or ^{32}P and added it to purified liver mitochondria in the absence of substrate (which inhibited oxidative phosphorylation), and assessed the extent of $^{32}P_i$ and ^{18}O transferred to ATP and water, respectively.[2] The authors found that although there was no net uptake of oxygen in the reaction, there was nevertheless a rapid exchange of the phosphate in P_i with ATP and of oxygen in P_i with water (Figure 10.2). In addition, analysis of the rate of exchange in these reactions indicated that between 15 and 20 oxygens were exchanged for every phosphate exchanged with ATP, making it highly unlikely that the exchanges occurred as a result of a simple reversal of the reaction $P_i + ADP \rightleftharpoons ATP + H_2O$.[2,3]

DOI: 10.1201/9781003379058-10

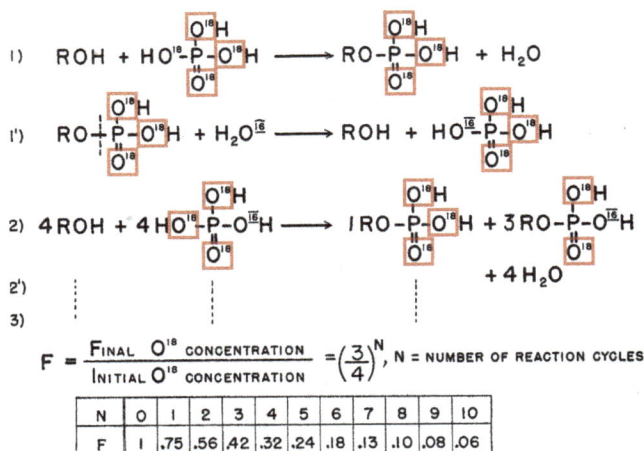

$$F = \frac{\text{FINAL O}^{18} \text{ CONCENTRATION}}{\text{INITIAL O}^{18} \text{ CONCENTRATION}} = \left(\frac{3}{4}\right)^N, \quad N = \text{NUMBER OF REACTION CYCLES}$$

N	0	1	2	3	4	5	6	7	8	9	10
F	1	.75	.56	.42	.32	.24	.18	.13	.10	.08	.06

FIGURE 10.1 Cohn's scheme to follow ^{18}O-labeled phosphate over a series of reactions. A phosphorylation/dephosphorylation reaction with phosphate containing four ^{18}O oxygens would be diluted by ¼ as a result of a first reaction (1 and 1′, ^{18}O shown in red boxes; R, any organic molecule). In the next exchange, there would be a three in four chance that an ^{18}O would take part in an exchange rather than the single, unlabeled, ^{16}O, the former further diluting the ^{18}O content of phosphate (2). Therefore, the average total isotope-labeled phosphate would progressively get diluted as the number of exchanges increases. The table at the bottom shows the statistical number of ^{18}O in PO$_4$ (F) after the indicated number of reaction cycles (N). (Figure adapted from ref. [1].)

Like Cohn before them, Boyer, Luchsinger, and Falcone found that addition of low concentrations of the inhibitor of oxidative phosphorylation, DNP—which they suggested *dissociates electron transport from the initial inorganic phosphate uptake*—resulted in the inhibition of both P$_i$-ATP and phosphate oxygen exchanges, both of which could be partially rescued by addition of ATP.[2,3] As such, the authors suggested that:

Present information definitely points to the explanation that the rapid exchange of P$_i$ oxygen with water oxygen as described herein results from the reversal of one, or more than one, of the phosphate uptake reactions of oxidative phosphorylation. Based on the validity of this explanation, the conclusion may be reached that the phosphate uptake reaction or reactions concerned are in a dynamic state in both the presence and the absence of net oxidative phosphorylation.[3]

In addition, Boyer reasoned that:

[if] the exchanges [...] are independent of oxidation and reduction reactions, then the over-all process of ATP formation can reasonably be divided into at least two parts, an oxidation-reduction part and a phosphorylation part that involves loss of P$_i$ oxygen atoms to water. Such a distinct separation of both phosphorylation and oxygen loss from oxidation and reduction would have considerable mechanistic implication.[4]

Importantly, their experiments supported the suggestion that the ATPase responsible for ATP hydrolysis to ADP and

EXCHANGE REACTIONS CATALYSED BY LIVER MITOCHONDRIA IN THE ABSENCE OF ADDED SUBSTRATE

Oxygen uptake	<0·05	μM
Inorganic phosphate (phosphorus-32) – ATP exchange		
Total phosphorus-32 added	3,670	c.p.m.
Total phosphorus-32 found in ATP	820	c.p.m.
Amount of reaction, ^{32}P$i \rightleftharpoons$ ATP	14	μmoles
Inorganic phosphate – oxygen-18 exchange		
Initial atom % excess oxygen-18 in inorganic phosphate	0·278	
Final atom % excess oxygen-18 in inorganic phosphate	0·063	
Amount of reaction, Pi – ^{18}O $\rightleftharpoons$ HOH	280	μ atoms

FIGURE 10.2 Oxygen is readily exchanged between P$_i$ and H$_2$O. ^{32}P-phosphate or ^{18}O-phosphate were added to a liver mitochondria reaction mixture without substrate and the amount of ^{32}P in ATP and ^{18}O in water were then assessed. Note the 20-fold higher exchange of ^{18}O-P$_i$ $\rightleftharpoons$ H$_2$O compared to ^{32}P $\rightleftharpoons$ ATP exchange.[2] (Adapted by permission from Springer Nature Customer Service Centre GmbH: Springer Nature, Nature, Boyer, PD, Falcone, AB, and Harrison, WH. Reversal and mechanism of oxidative phosphorylation. Copyright © 1954.)

P$_i$ could also operate in the reverse reaction, thus acting as an ATP synthase to synthesize ATP from the condensation of ADP and P$_i$. Boyer and his group continued to explore these exchange reactions a decade later, noting that:

The over-all process of oxidative phosphorylation [...] differs in an essential feature from 'substrate level' phosphorylation in that oxygen from P$_i$ appears in water instead of in one of the substrates. A dynamic equilibrium state of the overall process will thus give P$_i$ $\rightleftharpoons$ ATP, ADP $\rightleftharpoons$ ATP, P$_i$ $\rightleftharpoons$ HOH, and ATP $\rightleftharpoons$ HOH exchanges. Abundant evidence shows that these exchanges as observed with mitochondria and mitochondrial particles occur as partial reactions of oxidative phosphorylation. All the exchanges, and particularly the P$_i$ $\rightleftharpoons$ ATP, ADP $\rightleftharpoons$ ATP exchanges, may involve the phosphorylation reaction(s). The P$_i$ $\rightleftharpoons$ HOH, and ATP $\rightleftharpoons$ HOH exchanges give a means of probing at the water-forming reaction or reactions.[4]

As such, the authors used isotopic labeling to assess the rate of exchange of these reactions. Results of these experiments confirmed the previously noted partial inhibition of P$_i$ $\rightleftharpoons$ ATP and P$_i$ $\rightleftharpoons$ HOH exchanges in mitochondria using known inhibitors of oxidation-reduction reactions (cyanide, amytal, and antimycin).[4] Surprisingly, they also found that ATP $\rightleftharpoons$ HOH exchanges, which had been shown to only occur during oxidative phosphorylation (the other two exchanges were known to occur as a result of other reactions), were largely unaffected by the inhibitors, providing extra support for the idea that the exchange reactions were independent of the redox reactions of the ETC. As the authors explained,

If reduction or oxidation of electron carriers were a required concomitant step for the [...] exchange reactions, pronounced inhibition of these exchanges should be observed by conditions that prevent dynamic oxidation-reduction [...]

The experiments reported herein confirm and considerably extend earlier results that blocking of electron transport at various steps of the respiratory chain in presence of substrate gives only moderately weak inhibition of the P$_i$ $\rightleftharpoons$ ATP

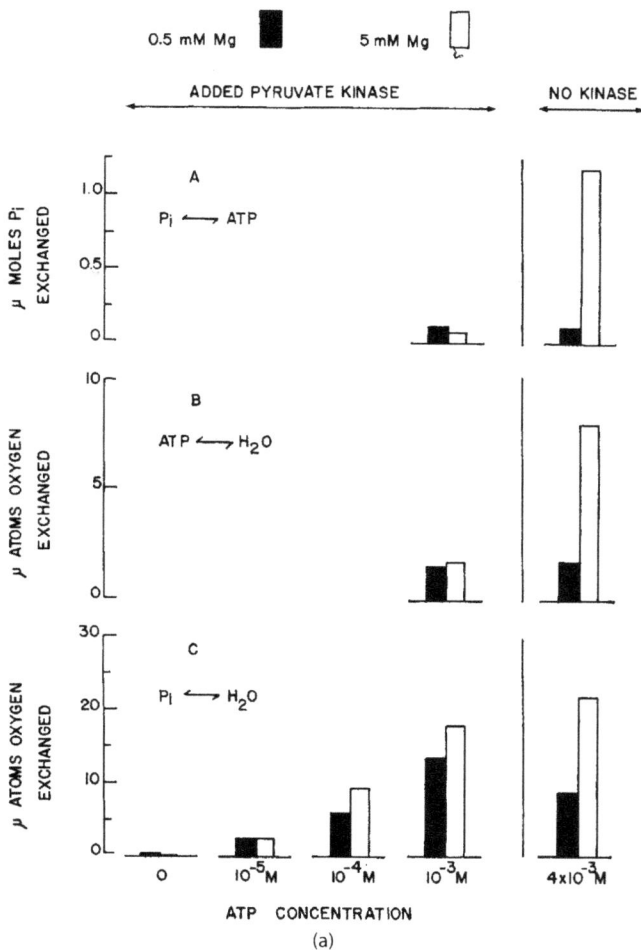

(a)

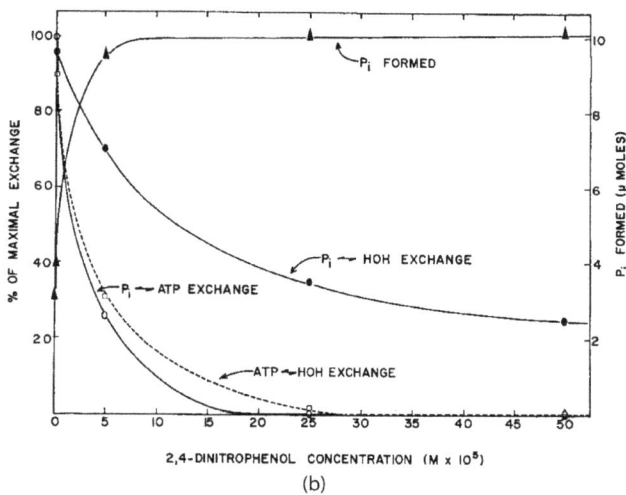

2,4-DINITROPHENOL CONCENTRATION (M x 10⁵)

(b)

P_i concentration		Incubation time	ATP cleaved	Oxygen from water appearing in P_i	Oxygen incorporated into P_i by exchange
Initial	Final				
mM		*min*	*μmoles/min*	*μatoms/min*	*μatoms/min*
0.1	0.3	2	1.02	1.33	0.31
1.0	1.6	5	1.18	2.14	0.96
10.0	10.6	10	—ᵃ	4.38	3.4

ᵃ Not conveniently measurable because of high initial P_i concentration.

(c)

and the $P_i \rightleftharpoons$ HOH exchanges. The additional finding that the ATP $\rightleftharpoons$ HOH exchange is likewise unrelated to capacity for oxidation and reduction is particularly significant.[4]

The authors also discussed various hypotheses that explained the results and concluded that those—including their own—which involved a high-energy intermediate in the synthesis of ATP were highly unlikely and proposed that a *reversible conformation change involving the protein of the respiratory enzyme* would better explain the reported data.[4] As such, they pointed out that the:

Absence of any phosphorylated intermediates would mean that the first covalent bond formed by either adenosine diphosphate or inorganic orthophosphate in oxidative phosphorylation is that in ATP [...] Lack of a phosphorylated intermediate would also mean that the $P_i \rightleftharpoons$ HOH and ATP $\rightleftharpoons$ HOH exchanges, as well as the $P_i \rightleftharpoons$ ATP exchange, might occur only as a consequence of the dynamic reversal of ATP formation, and thus show a dependence on ADP.[5]

Thus, they tested this hypothesis by adding titrating amounts of ATP to a reaction mixture in the presence or absence of pyruvate kinase, the presence of which would favor ATP synthesis and ensure that ADP levels are kept at a minimum. Results showed that removal of ADP completely inhibited both the $P_i \rightleftharpoons$ ATP and the ATP $\rightleftharpoons$ HOH exchanges but only minimally inhibited the $P_i \rightleftharpoons$ HOH exchange (Figure 10.3a).[5] The authors also noted a similar effect on exchanges by different Mg^{2+} concentrations and by the inhibition profile of the oxidative phosphorylation inhibitor, DNP (Figure 10.3b), which indicated that *the ATP $\rightleftharpoons$ HOH and $P_i \rightleftharpoons$ ATP exchanges behaved similarly but in a manner different from the $P_i \rightleftharpoons$ HOH exchange.*[5] Mitchell, Hill, and Boyer also tested the effects of P_i concentration on exchange reactions and found that it significantly increased $P_i \rightleftharpoons$ HOH exchange (Figure 10.3c). As such, the authors concluded that:

The continuation of an appreciable $P_i \rightleftharpoons$ HOH exchange when the ADP level is markedly reduced by pyruvate kinase action has the important implication that this exchange may not require dynamic reversal of ATP formation from

FIGURE 10.3 Effect of ATP, ADP, DNP, and P_i concentrations on exchange reactions. (a) Submitochondrial particles were incubated with increasing concentrations of ATP in the absence or presence of pyruvate kinase, the presence of which ensured that ADP levels were kept to a minimum. As such, ATP formation was favored in the presence of the kinase, while in its absence, a mixture of ADP and ATP was present. The reaction mixture contained ³²P-phosphate and ¹⁸O-H_2O and either low (black bars) or high (white bars) Mg^{2+} concentrations. (b) Submitochondrial particles were added to a reaction mixture with increasing concentrations of DNP and the various exchanges were assessed. (c) Submitochondrial particles were incubated in a reaction mixture containing titrating concentrations of P_i and P_i oxygen exchanges with water was assessed. (Figures adapted from Ref. [5].)

ADP and P_i. In contrast, the marked inhibition of the $P_i \rightleftharpoons$ ATP and ATP $\rightleftharpoons$ HOH exchanges by ADP removal implies that these exchanges occur principally or solely as a result of dynamic reversal of ATP formation from ADP and P_i. The occurrence of the $P_i \rightleftharpoons$ ATP exchange as an accompaniment of the ATP $\rightleftharpoons$ HOH exchange favors this interpretation. An alternate possibility, namely, that the requirement of ADP for the ATP $\rightleftharpoons$ HOH exchange could reflect some type of protein conformation change accompanying ADP binding, cannot be eliminated by present data. Such conformation change, however, would not be expected to result in a complete dependence on ADP.

A requirement of ATP formation for the ATP $\rightleftharpoons$ HOH and $P_i \rightleftharpoons$ ATP exchanges to occur, even if protein conformation effects are not involved, does not mean, however, that the reaction must be concerted. For example, water oxygen could enter P_i by dynamic reversal of transient metaphosphate formation and only appear in ATP when ADP is present. Further, mechanisms can be visualized in which an oxygen from P_i passes through another substance before forming water but in which reversible BTP formation is also required for exchanges to be observed. The question whether the oxygen loss from P_i to form HOH in oxidative phosphorylation is direct or indirect thus assumes considerable importance.

[...the] effects of 2,4-dinitrophenol with particles are in harmony with earlier observations which showed that 2,4-dinitrophenol concentrations in excess of those required to uncouple oxidative phosphorylation by intact mitochondria are required to inhibit the $P_i \rightleftharpoons$ HOH exchange. Perhaps the most convincing evidence is the inability of excess pyruvate kinase, [...], and thus ADP removal, to depress the $P_i \rightleftharpoons$ HOH exchange when ATP is added to particles. This result [...] can best be explained by a prominent $P_i \rightleftharpoons$ HOH exchange associated with the cleavage of ATP by particles, independent from a reversible formation of ATP from ADP and P_i.[5]

Finally, Mitchell, Hill, and Boyer speculated that the energy requirement for oxidative phosphorylation could be satisfied by a proton gradient as suggested by Mitchell's chemiosmotic hypothesis:

The lack of demonstrable phosphorylated intermediates, the apparent independence of the phosphorylation and water loss reactions from the oxidation-reduction reactions, and the dependency of the ATP $\rightleftharpoons$ HOH exchange on ADP are pertinent to any consideration about the nature of a "high energy" compound of state involved in oxidative phosphorylation. Such results suggest that an energized state [...] might not represent a covalent bond or an unspecified type of combination of a respiratory component and unknown substance. Conceivably, such an energized state might be something as simple as localized proton production.[5]

After this, Boyer elected to put the ATP synthesis problem to the side for several years, although the thought of it never truly left his mind. He returned to the problem in the early 1970s, recalling that every *ATP molecule formed in net oxidative phosphorylation shows extensive incorporation of water oxygens into the γ-phosphoryl group* and

that any *mechanism proposed for oxidative phosphorylation must account for this oxygen exchange*.[6] However, most mechanisms, including his previous support for a high-energy intermediate, did not. Thus, his group resumed studying this issue, starting with the fact that the reasons for the insensitivity of $P_i \rightleftharpoons$ HOH exchanges by DNP was not yet known, and sought other uncouplers that might mediate the same effects. As such, they found that the uncoupler, S-13, could inhibit $P_i \rightleftharpoons$ ATP and ATP $\rightleftharpoons$ HOH exchanges whereas the $P_i \rightleftharpoons$ HOH exchanges proceeded largely unaffected, akin to the effects of DNP (Figure 10.4a).[6] Thus, the authors suggested that:

Such studies, together with the requirement of ADP for the $P_i \rightleftharpoons$ HOH exchange, suggest that the exchange in the presence of uncouplers could be explained by a continued dynamic reversal of ATP formation limited to the catalytic site. Lack of the $P_i \rightleftharpoons$ ATP and ATP $\rightleftharpoons$ HOH exchanges in the presence of uncouplers would be explained if there exists an energy requirement for the release of the bound ATP. This concept may be diagrammed simply as follows:

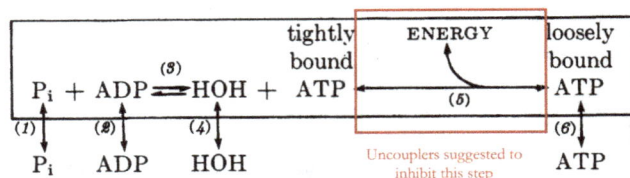

Steps occurring at the catalytic site are depicted within the [black] rectangle. Rapid interchange of water with the site in step 4 seems logical and has some support from the demonstration that the ^{18}O from labeled P_i readily appears in water of the medium. Dynamic reversal of steps 1 and 3 in the presence of uncouplers would thus suffice for the $P_i \rightleftharpoons$ HOH exchange. Inhibition by uncouplers of an energy-driven release of ATP by way of steps 5 and 6 would not inhibit the $P_i \rightleftharpoons$ HOH exchange but would block the $P_i \rightleftharpoons$ ATP and ATP $\rightleftharpoons$ HOH exchanges.[6]

Boyer also explained that:

ADP and P_i can bind and be converted into a tightly bound ATP. The water formed freely interchanges with medium water. Reversal of this reaction results in the incorporation of one water oxygen atom into the bound P_i. If the P_i can tumble freely at the catalytic site, when bound ATP is again formed there are three chances out of four that it will contain a water oxygen atom. Various exchanges of phosphate oxygen atoms with water oxygen atoms are measurable [...] The oxygen exchanges thus provide sensitive probes of reaction steps that otherwise might be hidden.[7]

In addition, Boyer, Cross, and Momson acknowledge that:

we had to overcome a limitation in our own thinking, namely the supposition that in net oxidative phosphorylation, as in substrate-level phosphorylation, use of energy was limited to events occurring before or during but not after the formation of each ATP molecule.[6]

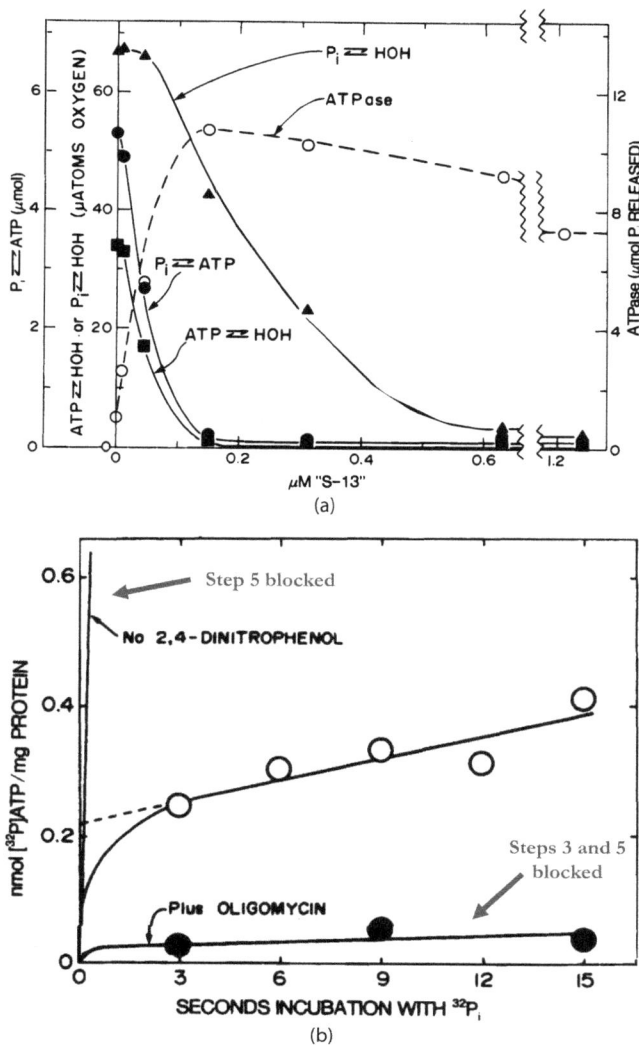

FIGURE 10.4 (a) The uncoupler, S-13, does not inhibit ATPase or the $P_i \rightleftharpoons$ HOH exchange. Rat liver mitochondria were incubated in a reaction mixture containing $^{32}P_i$ and ^{18}O-H_2O, without or with the indicated concentrations of the uncoupler, S-13. Note that the $P_i \rightleftharpoons$ HOH reaction was minimally affected at an S-13 concentration which inhibits the $P_i \rightleftharpoons$ ATP and ATP $\rightleftharpoons$ HOH reactions by 50%, and only inhibited by 35% when the others are completely inhibited. (b) Oligomycin, an inhibitor of oxidative phosphorylation, inhibits $^{32}P_i$ incorporation into ATP. Submitochondrial particles were incubated in a reaction mixture containing $^{32}P_i$ for the indicated times and incorporation into ATP was assessed. Addition of DNP allowed the assessment of the $p_i \rightleftharpoons$ ATP exchange only, whereas addition of oligomycin blocked all exchanges of P_i with ATP. (Figures adapted from Ref. [6].)

As such, the authors reasoned that if the above mechanism was correct, there should be *a small amount of* [ATPase-] *bound ATP [...] that rapidly equilibrates with P_i but not with ATP of the medium* during oxidative phosphorylation in the presence of uncouplers.[6] Furthermore, the proposed mechanism could explain previous findings by another group, whereby a *small 'jump' in total amount of* [^{32}P]*ATP present occurred upon addition of ADP to mitochondria in the*

presence of $^{32}P_i$, substrate, and O_2, which they suggested *could have resulted from a shift in equilibrium of step 3* in the above mechanism. Support for this reasoning was provided from an experiment in which they treated mitochondrial particles with $^{32}P_i$ in the presence of DNP (to stop oxidative phosphorylation)—conditions which ensured that the incorporation of ^{32}P into ATP occurred only as a result of rapid reversible exchange with ATP (step 3 above) and not of ATP synthesis (step 5 above)—and assessed the amount of *rapidly labeled, bound ATP*. Indeed, results showed there was a quick appearance of bound ^{32}P-ATP in the presence of DNP, whereas exchange was completely inhibited by the potent inhibitor, oligomycin (which, in contrast to other inhibitors, had been shown to also block the $P_i \rightleftharpoons$ HOH exchange), thereby confirming that the appearance of ^{32}P-ATP was as a result of a reversible exchange with H_2O (Figure 10.4b).

Boyer's group had also previously shown that concentrations of P_i above 10 mM increased the $P_i \rightleftharpoons$ HOH exchange and reasoned that if the bound ATP species was an intermediate in this exchange, addition of excess $^{32}P_i$ would lead to a concomitant increase in ATPase-bound ^{32}P-ATP. Indeed, they showed that addition of 50 mM $^{32}P_i$ led to a four-fold increase in bound ^{32}P-ATP. Thus, together, these findings indicated *that a prominent function of energy in oxidative phosphorylation is to cause release of preformed ATP*, since no energy was included in any of these studies, yet significant exchanges still took place.[6] However, how ATPase could be reversed to enable synthesis of ATP without the use of energy had yet to be determined; but in term of mechanism, Boyer suggested that:

[the] simplest, and on the basis of present information the most likely, way to bring about release of a preformed ATP from a tight, noncovalent binding site is by protein conformational change [...]

A conformationally interlinked matrix of proteins within the membrane provides an attractive hypothesis for capture, transmission, and use of energy in oxidative and photosynthetic phosphorylation, active transport, and energy-linked reduction.[6]

In 1977, Boyer's group published back-to-back papers which further supported the claim that energy was required only to release ATP from ATPase and that *ATP synthesis may involve conformationally induced changes in ATP affinity*. In these papers, the authors updated their proposed mechanism and stressed the importance of differentiating between the various exchanges taking place:

In this figure *[Figure 10.5a]* reactants at the catalytic phosphorylation site in the membrane are depicted in the *[black]* rectangle. Because the oxygen exchanges are more rapid than other exchanges and because no localization of water has been detected, either no specific binding site for water exists or the water binding and dissociation is more rapid than other steps involved.

The total oxygen exchange reactions observed with substrates such as P_i and ATP can usefully be separated into two components, the medium and intermediate exchanges.

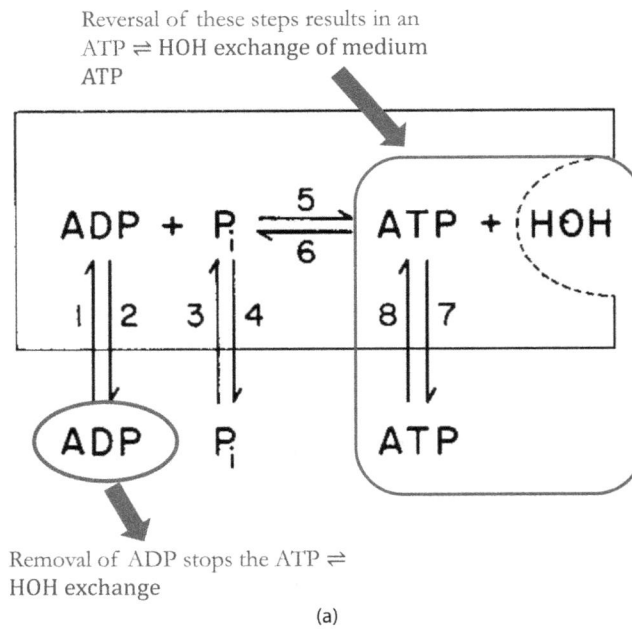

(a)

Required reaction steps for oxygen exchanges and the $P_i \rightleftharpoons ATP$ exchange of oxidative phosphorylation

Exchange reaction	Minimal required steps						
	ADP binding (Step 1)	ADP release (Step 2)	P_i binding (Step 3)	P_i release (Step 4)	Substrate interconversion (Steps 5 & 6)	ATP release (Step 7)	ATP binding (Step 8)
ADP $\rightleftharpoons$ ATP[a]	+	+			+	+	+
$P_i \rightleftharpoons$ ATP[b]			+	+	+	+	+
Medium $P_i \rightleftharpoons$ HOH			+	+	+		
Intermediate $P_i \rightleftharpoons$ HOH				+	+		+
Medium ATP $\rightleftharpoons$ HOH Intermediate[b]					+	+	+
ATP $\rightleftharpoons$ HOH			+		+	+	

[a] Requires presence of bound P_i derived either from medium P_i or from ATP cleavage.
[b] Requires presence of bound ADP derived either from medium ADP or from ATP cleavage.

(b)

Exchanges of phosphate oxygens with water oxygens catalyzed by ATP synthase.

Exchange	Measurement
Intermediate $P_i \rightleftharpoons$ HOH	Hydrolysis of γ-^{18}O-ATP and determination of ^{18}O in P_i formed
Intermediate ATP $\rightleftharpoons$ HOH	Synthesis of ATP from ^{18}O-P_i and determination of ^{18}O in ATP formed
Medium $P_i \rightleftharpoons$ HOH	Determination of loss of ^{18}O from ^{18}O-P_i when P_i binds, undergoes exchange, and returns to the reaction medium
Medium ATP $\rightleftharpoons$ HOH	Determination of loss of ^{18}O from γ-^{18}O-ATP when ATP binds, undergoes exchange, and returns to the reaction medium

(c)

FIGURE 10.5 Diagram and requirements for medium and intermediate exchange reactions. (a) The catalytic site is represented by the black rectangle. The dashed line represents the uncertainty for a specific binding site for water. (b) + indicates a requirement for the component in question.[8] (c) Summary of exchanges and their measurements.[7] (d) Chemical reactions of the various exchanges investigated by Boyer's group. (Figure parts (a and b) adapted from ref. [8]. Figure (c) used by permission from Paul D. Boyer—Nobel lecture Copyright © The Nobel Foundation 1997.) *(Continued)*

Intermediate P_i ⇌ H_2O exchange

Medium P_i ⇌ H_2O exchange

Intermediate ATP ⇌ H_2O exchange

Medium ATP ⇌ H_2O exchange

(d)

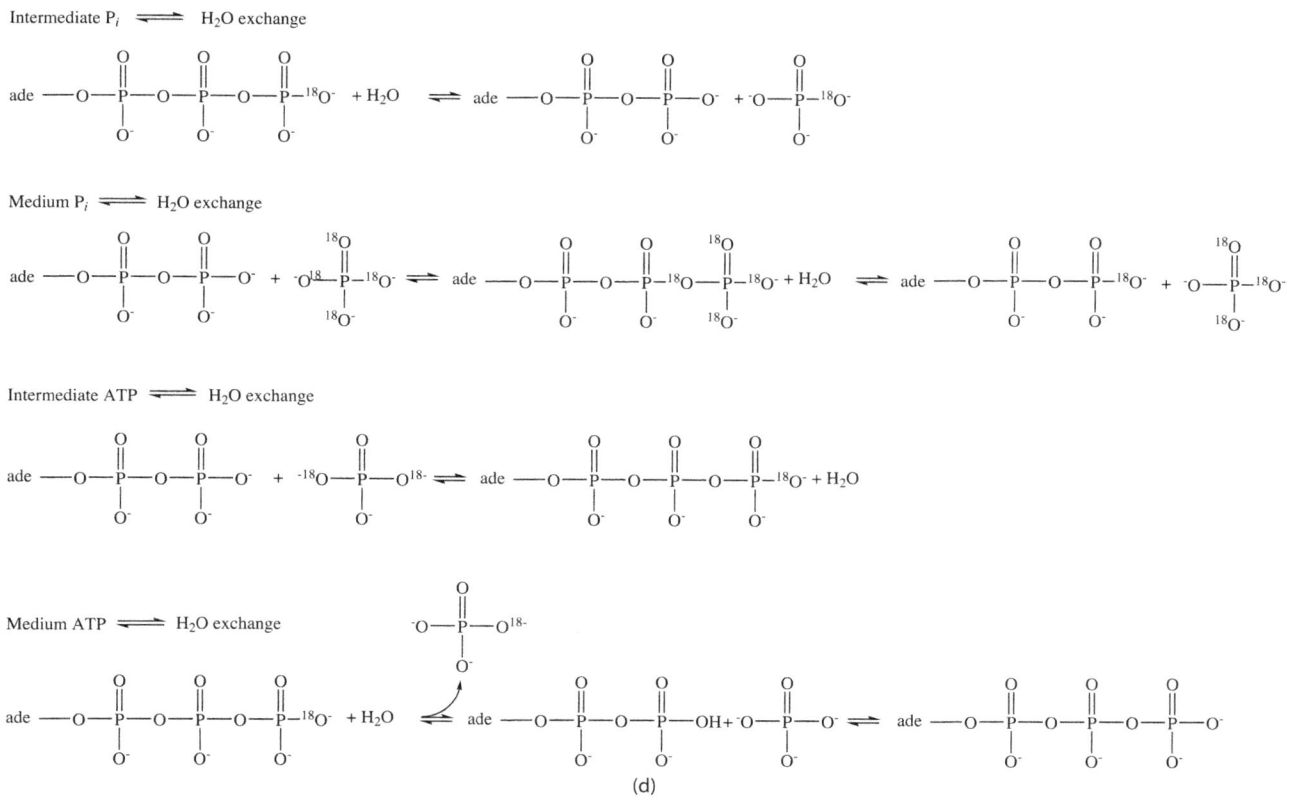

FIGURE 10.5 (*Continued*)

Medium exchange occurs when a reactant binds to a catalytic site, undergoes exchange, and is released as the same reactant [*these reactions are summarized in the table in Figure 10.5c; chemical reactions are depicted in Figure 10.5d*]. Intermediate exchange occurs when a reactant binds, undergoes exchanges and conversion to a product, and the product containing atoms incorporated by exchange is released. Intermediate exchange is defined as an exchange of the product released. For example, if ATP that binds and is subsequently released as ATP contains water oxygens, a medium ATP ⇌ HOH exchange has occurred. If ATP binds and the P_i formed from ATP and released to the medium contains more than the 1 oxygen required for ATP cleavage, an intermediate P_i ⇌ HOH exchange has occurred. Such definitions of exchange are independent of the mechanism of exchange.

In accord with [*the diagram in Figure 10.5a*], oxygen exchanges of ATP and P_i and the P_i ⇌ ATP and ADP ⇌ ATP exchanges require reaction steps as summarized in [*the table Figure 10.5b*].

In [*the diagram and the table, Figure 10.5 a and b*], only one bound species of each reactant is indicated. More than one species may exist, for example a loosely bound form not capable of undergoing covalent change, and a more tightly or competently bound form that undergoes substrate interconversion. The single binding steps of [*the diagram and the table*] include any intermediate species between free reactants and bound reactants that are catalytically competent.

[*Figure 10.5b*] illustrates well the power of the exchange measurements, as they require different steps of the overall reaction sequence and may thus reveal changes in those steps

not measurable by other approaches. Modifiers of the different steps of the reaction sequence given in [*Figure 10.5a*] may have different effects on the rates of the exchange reactions. This offers the possibility of demonstrating the steps affected by energy in ATP formation by measuring the effect of uncouplers on the exchange reactions in which case uncouplers are used to change the energy level in the submitochondrial particles.[8]

As such, the authors further tested the effects of uncouplers on the exchange reactions, reasoning that *uncouplers collapse an energized state of the mitochondria, probably the transmembrane electrochemical potential or pH gradient, that can serve to drive both oxidative phosphorylation and the exchange reactions.* In addition, previous results indicated that *intermediate P_i ⇌ HOH exchange was uncoupler-insensitive but the medium exchange was uncoupler sensitive.*[8] Thus, they used the uncoupler, S-13, to test *the relative sensitivities of the medium and intermediate exchanges further,* and confirmed that only the intermediate P_i ⇌ HOH exchange was insensitive to S-13 (Figure 10.6); the authors suggested that this phenomenon could be explained *if the reversible formation of bound ATP from bound ADP and P_i continues in presence of relatively high uncoupler concentrations.*[8] In conclusion, the authors explained that:

The continuation of the intermediate P_i ⇌ HOH exchange and the ATPase in the presence of uncouplers means that P_i bound to the catalytic site is still undergoing oxygen

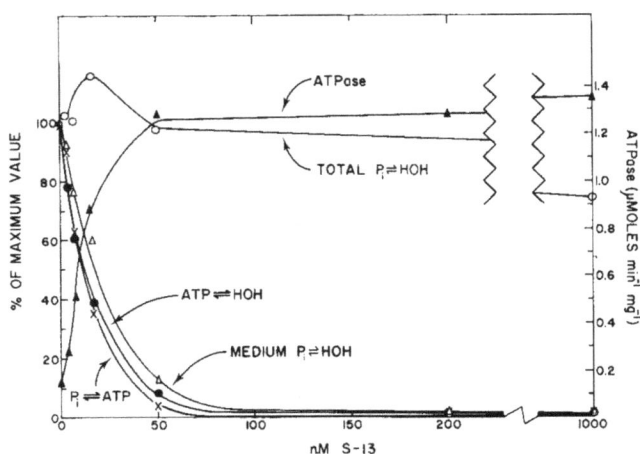

FIGURE 10.6 Effects of S-13 on intermediate $P_i \rightleftharpoons$ HOH exchange. Increasing concentrations of S-13 was added to a reaction mixture to test the sensitivities of the various exchanges. Note that medium $P_i \rightleftharpoons$ HOH exchanges are inhibited, while intermediate exchanges are not since total $P_i \rightleftharpoons$ HOH remains high.[9]

exchange and is still capable of being rapidly released to the medium. What is lost in the presence of uncoupler is the ability for medium P_i to be bound in the manner necessary for participation in the exchange reaction.

An important conclusion from our present data about the exchange of P_i oxygens with water is that the presence of Mg^{2+}, ADP, and P_i alone with submitochondrial particles but without oxidizable substrates does not allow the medium $P_i \rightleftharpoons$ HOH exchange to occur [...]

The continuation of a prominent intermediate exchange in the presence of the uncoupler S-13 at a 10-fold higher concentration than necessary to stop net oxidative phosphorylation indicates a remarkable insensitivity of the covalent bond forming and breaking step (ADP + $P_i \rightleftharpoons$ ATP + HOH) to uncouplers [...][8]

Their second paper addressed the fact that the mechanism they had suggested in their first paper did not adequately explain results of experiments in which ADP removal inhibited exchange reactions.[9] They also wanted to ascertain if the reported inhibition of $P_i \rightleftharpoons$ HOH exchange resulting from ADP removal had, in fact, only inhibited the medium $P_i \rightleftharpoons$ HOH, as had been the case with the uncoupler, S-13. As such, the authors first tested the effects of arsenate, which had previously been shown to inhibit oxidative phosphorylation by blocking P_i binding, in their system. In contrast to S-13, this compound significantly inhibited both medium and total $P_i \rightleftharpoons$ HOH exchanges (Figure 10.7 a, compare this to Figure 10.6), which indicated that the intermediate $P_i \rightleftharpoons$ HOH exchange was also inhibited (intermediate exchange = total exchange − medium exchange), though only partially so, as observed by the residual ATP $\rightleftharpoons$ HOH and $P_i \rightleftharpoons$ HOH activity.[9] As such, the authors suggested that:

The component of the ATP $\rightleftharpoons$ HOH exchange left in the presence of high arsenate concentrations can result from ATP binding, hydrolysis at the catalytic site without P_i

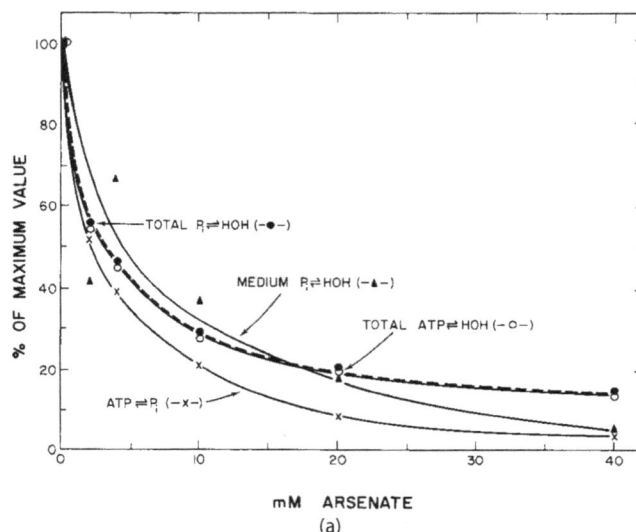

mM ARSENATE
(a)

Measure- ment	Observed catalytic rates	
	No ATP-regenerating system	With ATP-regenerating system
ATPase	250 nmol min⁻¹ mg⁻¹	750 nmol min⁻¹ mg⁻¹
$P_i \rightleftharpoons$ ATP	190 nmol min⁻¹ mg⁻¹	0.2 nmol min⁻¹ mg⁻¹
Total ATP $\rightleftharpoons$ HOH	509 natoms min⁻¹ mg⁻¹	<5 natoms min⁻¹ mg⁻¹
Medium P_i $\rightleftharpoons$ HOH	930 natoms min⁻¹ mg⁻¹	<70 natoms min⁻¹ mg⁻¹
Total $P_i \rightleftharpoons$ HOH	1850 natoms min⁻¹ mg⁻¹	375 natoms min⁻¹ mg⁻¹
Intermediate $P_i \rightleftharpoons$ HOH	920 natoms min⁻¹ mg⁻¹	>305 natoms min⁻¹ mg⁻¹

(b)

FIGURE 10.7 (a) Effects of arsenate on intermediate $P_i \rightleftharpoons$ HOH exchange. Submitochondrial particles were incubated with increasing concentrations of arsenate and the indicated exchanges were assessed. Note how total $P_i \rightleftharpoons$ HOH exchanges remain high while medium $P_i \rightleftharpoons$ HOH exchanges are completely inhibited. (b) The intermediate $P_i \rightleftharpoons$ HOH is minimally affected by low ADP concentrations. A reaction mixture was incubated with or without an ATP regenerating system and the various exchanges were assessed.[9]

release, re-formation of bound ATP, then ATP release. Such exchange corresponds to medium ATP $\rightleftharpoons$ HOH exchange.[9]

They also found that this residual ATP $\rightleftharpoons$ HOH exchange could be completely abolished if DNP was added, which indicated that *although ATP is still undergoing reversible hydrolysis at the catalytic site with resultant incorporation of water oxygen atoms into bound ATP, such bound ATP is not being released to the medium.*

In an attempt at deciphering the order of release and binding of ADP and P_i to the enzyme, the authors measured the rate of exchanges at different substrate (i.e., ATP, ADP, and P_i) concentrations, proposing that:

A compulsory order in which ADP leaves the catalytic site before P_i (or P_i binds before ADP) should result in an inhibition of the rate of the ATP $\rightleftharpoons$ P_i exchange when the ATP and ADP concentrations are increased at a constant ATP/ADP ratio. A compulsory order in which P_i leaves the catalytic site before ADP (or ADP binds before P_i) should result in an inhibition of the rate of the ATP $\rightleftharpoons$ ADP exchange when the ATP and P_i concentrations are increased at a constant ATP/P_i ratio.[9]

Their experiments showed that the $P_i \rightleftharpoons$ ATP and ADP $\rightleftharpoons$ ATP exchanges increased to a plateau when concentrations of ADP and ATP or ATP and P_i, respectively, were titrated in a constant ratio, which gave *strong evidence for a random binding of P_i and ADP to the catalytic site.* Next, they tested the effect of an ATP-regenerating system (which kept ADP levels to a minimum) on exchange reactions to determine the requirements for ADP on the various exchanges. Thus, the authors found that *the only exchange reaction not very strongly inhibited in the presence of an ATP-regenerating system is the intermediate $P_i \rightleftharpoons$ HOH exchange* (Figure 10.7b), which indicated that *the medium component of the total $P_i \rightleftharpoons$ HOH exchange is preferentially inhibited by ADP removal; the intermediate $P_i \rightleftharpoons$ HOH exchange continues at a rapid rate.*[9]

With these results in hand, the authors presented yet another revised model for the mechanism of ATP synthesis during oxidative phosphorylation, which involved an *alternating catalytic site scheme* that better accounted for their new results:

This figure depicts participation of two identical catalytic sites on the energy-transducing membrane; these sites function cooperatively through conformational interactions during ATP synthesis or hydrolysis. The model suggests that during oxidative phosphorylation the binding of ADP and P_i at one site is necessary for subsequent events leading to release of ATP from the second site. The sequence of events for ATP synthesis is regarded as the reversal of the sequence for ATP hydrolysis. The cycle of *[Figure 10.8]* would operate in the clockwise direction during ATP synthesis and in the counter-clockwise direction during ATP hydrolysis. The model depicted obviously does not show all separate binding and release steps that must occur or specify their possible order.[9]

The authors also noted that the scheme had the advantage that energy was solely required for the concomitant transition of P_i to a productive mode and the release of ATP from the catalytic site, as accomplished by:

[c]onverting ATP > E* $\cdot \substack{< \text{ADP} \\ \cdot \, P_i}$ to ATP$\cdot$E$^{*}\substack{< \text{ADP} \\ < P_i}$; that is both catalytic sites are modified in one conformational transition. We regard this as a particularly attractive feature of the model [...] *[In addition, the]* presence of a single type of catalytic site [...] and interconversion of these sites has appealing features of simplicity and symmetry. The model presented is the simplest we have been able to devise that accommodates present experimental information, needs only a single site of energy input, and one type of catalytic site.
The alternating catalytic site model gains support from a recent demonstration in our laboratory of a transitorily tightly bound ATP as an intermediate step in ATP synthesis by chloroplasts and indication that the total number of

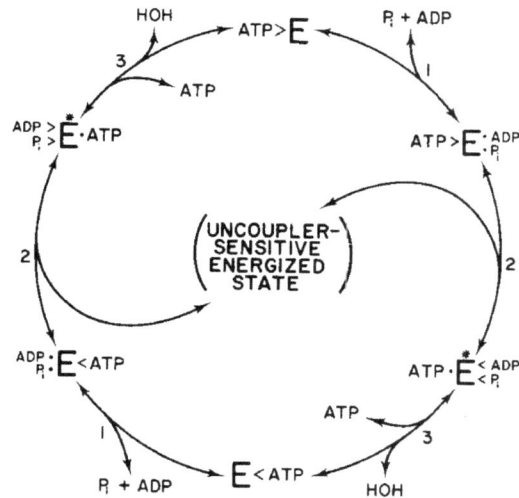

FIGURE 10.8 Diagram of the "alternating catalytic site" mechanism proposed by Kayalar, Rosing, and Boyer. Transitory, tightly bound substrates are indicated by < and >; loosely bound substrates are indicated by a dot; "E" is the form of the enzyme capable of converting ADP and P_i to ATP without energy, whereas E* is the energized form required to release ATP.[9]

catalytic sites participating at one time may be greater than one per CF$_1$-ATPase [...]

As noted briefly above, the model [...] accommodates the conclusion from our previous paper that energy derived from electron transfer reactions may be regarded as driving an energy-requiring conformational change that promotes binding of P_i and ADP in a productive mode. This change is recognized *[in the diagram...]* by conversion of the non-energized membrane, E, to an energized form, E*. Thus the designation ATP$\cdot$E$^{*}\substack{< \text{ADP} \\ < P_i}$ recognizes that this form has the capacity to make ATP at the catalytic site without additional energy input. It is tempting to suggest that the energized form, E*, represents a conformationally energized form of the ATPase complex in the membrane. However, present evidence requires only that the energized form be distinct from as well as derivable from the energized state that is dissipated by uncouplers. The state dissipated by uncouplers in mitochondria [...] likely represents a transmembrane potential or an intramembrane localized proton.

Finally, consideration is essential of a possible relation of our dual site model to the suggested presence of three α-β pairs in the F$_1$-ATPase. Firstly, presence of three rather than two α-β pairs is not firmly established. Secondly, an alternating three-site model, similar to the two-site model, could also accommodate our experimental findings, in particular, inhibiting effects of ADP and ATP removal on exchanges.[9]

Several years later, Boyer's group provided evidence for the three catalytic site mechanism by performing kinetics studies on the rate of binding and release of substrates over a range of substrate concentrations, however, they noted that *product release from one catalytic site is very slow until substrate binds at a second catalytic site.*[10] As such, the authors presented an updated version of their mechanism that better accommodated the new experimental evidence; as the authors explained:

> The properties of F_1-ATPase are explainable by occurrence of both a negative cooperativity of binding and a positive cooperativity of catalysis. Although at low ATP concentrations, substrate binding and product formation at one catalytic site per enzyme molecule has occurred, release of products from these catalytic sites is relatively slow. At higher substrate concentrations, additional ATP binding occurs and release of products from catalytic sites is now much more rapid. This behavior has been referred to in some previous publications from this laboratory as alternating site cooperativity.
>
> An additional important characteristic [...] is that each site is regarded as identical and participating in sequence. A depiction of this is given in [Figure 10.9]. Each catalytic site in turn goes through three major stages, ATP binding, an interconversion of tightly bound ATP to tightly bound ADP and P_i, and a release of ADP and P_i.[10]

By the late 1970s, Mitchell's chemiosmotic hypothesis, which stated that the energy used for oxidative phosphorylation was provided by a proton gradient across the mitochondrial membrane, had been sufficiently proven and was now largely considered a theory. However, the manner in which this energy was harnessed to drive oxidative phosphorylation was still under investigation. Boyer proposed

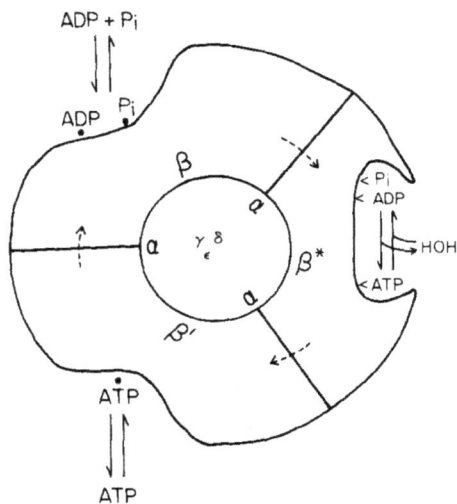

FIGURE 10.9 An updated version of Boyer's "alternating catalytic site" scheme. In this model, each catalytic binding site is proposed to be equal and capable of binding substrate, although only in a sequential manner. At any one time, one site will be bound by ATP, one by ADP and P_i, and one by an intermediate stage of ADP and P_i bound and in the process of covalent binding resulting in ATP formation. Dots represent loosely bound substrates, whereas < represents tightly bound substrate in transition. Note that rotation in clockwise direction induces ATP synthesis, whereas rotation in the anticlockwise direction results in ATP hydrolysis. In addition, product release from the catalytic site is increased by an increase in substrate concentration. See text and figure text for details.[10]

that ATPase might be an enzyme that rotates around a central axis, a consideration which led to the postulate of "rotational catalysis":

> To me, there seemed only one way that all catalytic sites could proceed sequentially and identically, with modulation by one or more single-copy, minor subunits. This was by rotational catalysis, in which large catalytic subunits moved rotationally around a smaller asymetric core [...] The internal core was likened to a cam shaft that modulated the conformation of the β-subunits.[7]

However, following failed attempts at proving this rotational mechanism, Boyer admitted that it *seemed apparent that an adequate evaluation of the possibility of rotational catalysis would need to await the knowledge of the 3-dimensional structure of the F1-ATPase.*[7]

With the advent of increased resolution of electron microscopes and X-ray crystallography, and better sample preparation techniques, excellent images of the ATPase were obtained starting in the 1990s, which enabled the elucidation of the precise mechanism of ATPase. These studies showed that ATPase was formed from a large number of subunits, with a rotating motor—the F_o subunit—embedded in the mitochondrial membrane, and an extracellular, catalytic domain, the F_1 subunit found in the mitochondrial lumen (Figure 10.10).[11] The F_1 subunit is composed of five subunits in the stoichiometry $\alpha_3\beta_3\delta_1\varepsilon_1\gamma_1$, where α- and β-domains form the catalytic sites of the enzyme and the γ domain forms a central stalk that connects the F_1 subunit to the F_o subunit. In addition,

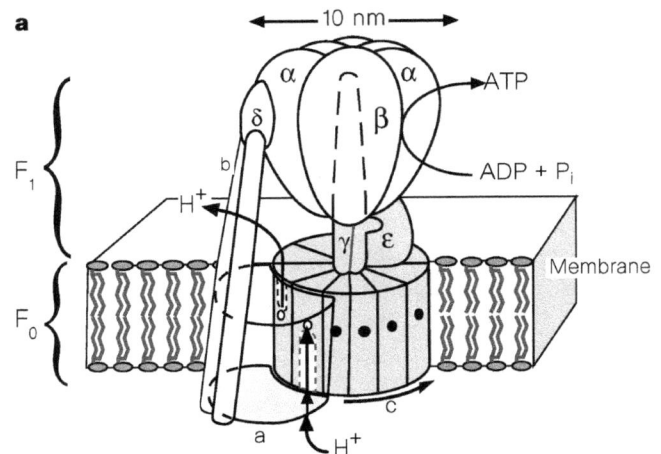

FIGURE 10.10 Diagram of the ATPase. The rotor of the enzyme consists of the c-ring, the ε-subunit, and the γ-subunit, and the stator consists of the a-, b-, δ-, and $\alpha_3\beta_3$-subunits. Protons enter F_o and bind one c-subunit, which cause a rotational force on the enzyme. Another channel releases H+ ions into the mitochondrion with produces water when it reacts with oxygen.[14] (Adapted with permission of Nature Research, from Elston, T, Wang, H and Oster, G. Energy transduction in ATP synthase. Nature 391:6666 Copyright © 1998; Permission conveyed through Copyright Clearance Center, Inc.)

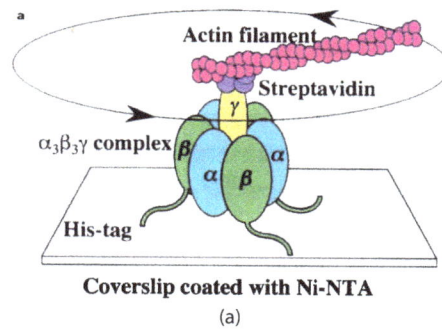

(a)

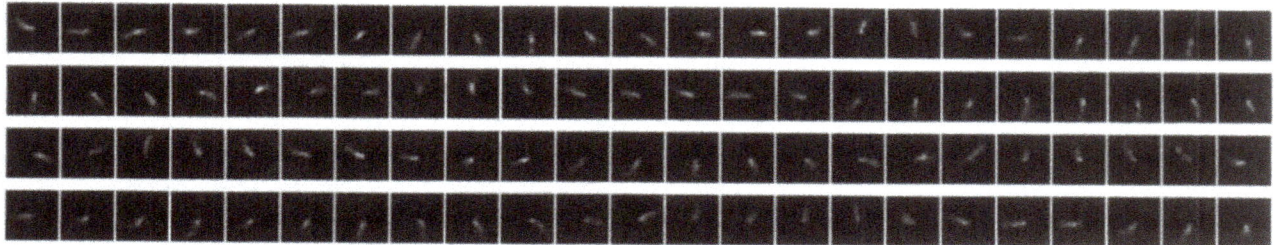

(b)

FIGURE 10.11   Visualization of ATPase rotation. ATPase was immobilized on a glass coverslip and a fluorescent protein was attached to the rotating shaft. Fluorescence microscopy was used to visualize the rotating enzyme upon addition of ATP.[13] (Adapted by permission from Springer Nature Customer Service Centre GmbH: Springer Nature, Nature, Noji, H, Yasuda, R, Yoshida, M and Kinosita, K. Direct observation of the rotation of F1-ATPase, Copyright © 1997.)

it was found that the enzyme is adorned with a peripheral stalk connecting the F_1 subunit to a nonrotating subunit of the F_o subunit and whose function appears to be to help the F_1 subunit resist the rotational torque of the F_o subunit. The F_o subunit is composed of more than a dozen proteins, of which the c subunits are arranged in the form of a cylinder and is the source of rotation of ATPase.

The asymmetry of the central stalk of the F_1 subunit requires that each catalytic β subunit adopt a different conformation, resulting in each subunit having a different binding affinity for P_i, ADP, and ATP. One such conformation, called the ADP-inhibited state, is common to two of the catalytic sites at any one time and can bind either ADP or ATP. The third catalytic site, distorted due to the curvature of the central stalk, cannot bind any nucleotides and is therefore known as the "empty" or "open" state. In contrast, the α-subunits remain closed at all times and bind both Mg^{2+} ions and a nucleotide which remain bound during the catalytic cycle.[12] The rotation of the F_o subunit during ATP hydrolysis is driven by the H^+ gradient across the mitochondrial membrane: a proton in the mitochondrial intermembrane space (the side distal to the F_1 complex) enters a channel in the α-subunit and binds to and neutralizes the charge of one c subunit, resulting in its tendency to move toward a more hydrophobic environment (on the inside of the membrane) (Figure 10.10).[12] This generates a rotational force on the protein, resulting in the release of a proton on the inside face of mitochondria through another channel in the α-subunit; two of these protons bind to reactive oxygen species (O^-) in mitochondria to produce water. In 1997, Noji et al. took a novel approach at observing the rotational

force of the ATPase by attaching a fluorescently labelled protein to the rotating shaft of the enzyme (Figure 10.11a).[13] Using microscopy, they found that addition of ATP resulted in the rotation of the fluorescent protein (Figure 10.11b).

* * * * *

REFERENCES

1. Cohn M. A study of oxidative phosphorylation with O18-labeled inorganic phosphate. *Journal of Biological Chemistry*. 1953;201(2):735–750. doi:10.1016/S0021-9258(18)66231-3
2. Boyer PD, Falcone AB, Harrison WH. Reversal and mechanism of oxidative phosphorylation. *Nature*. 1954;174(4426): 401–402. doi:10.1038/174401b0
3. Boyer PD, Luchsinger WW, Falcone AB. O18 and P32 exchange reactions of mitochondria in relation to oxidative phosphorylation. *Journal of Biological Chemistry*. 1956;223(1):405–421. doi:10.1016/S0021-9258(18)65150-6
4. Boyer PD, Bieber LL, Mitchell RA, Szabolcsi G. The apparent independence of the phosphorylation and water formation reactions from the oxidation reactions of oxidative phosphorylation. *Journal of Biological Chemistry*. 1966;241(22): 5384–5390. doi:10.1016/S0021-9258(18)96442-2
5. Mitchell RA, Hill RD, Boyer PD. Mechanistic implications of Mg++, adenine nucleotide, and inhibitor effects on energy-linked reactions of submitochondrial particles. *Journal of Biological Chemistry*. 1967;242(8):1793–1801. doi:10.1016/S0021-9258(18)96072-2
6. Boyer PD, Cross RL, Momsen W. A new concept for energy coupling in oxidative phosphorylation based on a molecular explanation of the oxygen exchange reactions. *Proceedings of National Academy of Sciences U S A*. 1973;70(10): 2837–2839. doi:10.1073/pnas.70.10.2837

7. Boyer PD. Paul D. Boyer: Nobel lecture: Energy, life, and ATP. NobelPrize.org. Nobel Media AB Copyright © The Nobel Foundation 1997.

8. Rosing J, Kayalar C, Boyer PD. Evidence for energy-dependent change in phosphate binding for mitochondrial oxidative phosphorylation based on measurements of medium and intermediate phosphate-water exchanges. *Journal of Biological Chemistry*. 1977;252(8):2478–2485.

9. Kayalar C, Rosing J, Boyer PD. An alternating site sequence for oxidative phosphorylation suggested by measurement of substrate binding patterns and exchange reaction inhibitions. *Journal of Biological Chemistry*. 1977;252(8): 2486–2491. doi:10.1016/S0021-9258(17)40484-4

10. Gresser MJ, Myers JA, Boyer PD. Catalytic site cooperativity of beef heart mitochondrial F1 adenosine triphosphatase. Correlations of initial velocity, bound intermediate, and oxygen exchange measurements with an alternating three-site model. *Journal of Biological Chemistry*. 1982;257(20):12030–12038.

11. Stock D, Gibbons C, Arechaga I, Leslie AG, Walker JE. The rotary mechanism of ATP synthase. *Current Opinion in Structural Biology*. 2000;10(6):672–679. doi:10.1016/S0959-440X(00)00147-0

12. Walker JE. The ATP synthase: The understood, the uncertain and the unknown. *Biochemical Society Transactions*. 2013;41(1):1–16. doi:10.1042/BST20110773

13. Noji H, Yasuda R, Yoshida M, Kinosita K. Direct observation of the rotation of F1-ATPase. *Nature*. 1997;386(6622): 299–302. doi:10.1038/386299a0

14. Elston T, Wang H, Oster G. Energy transduction in ATP synthase. *Nature*. 1998;391(6666):510–513. doi:10.1038/35185

11 Techniques

Several important techniques were developed between the 1960s and the 1990s without which, much of what we now know about how cells work would not be possible. Therefore, I have chosen to slightly forgo chronology, as used thus far, in favor of a chapter dedicated to these technological advances. I have decided to focus on a few techniques that are used in most, if not all laboratories today: sodium dodecyl sulfate polyacrylamide gel electrophoresis (SDS-PAGE) electrophoresis and western blotting, DNA cloning, the development of fluorescently tagged proteins, and finally, fluorescence-activated cell sorting (FACS). The first of these, SDS-PAGE electrophoresis was initially used to identify serum proteins, but was subsequently found to be useful for characterizing proteins in term of size. The development of antibodies and western blotting further enabled the direct detection of specific proteins following separation by SDS-PAGE. In addition, this technique allowed researchers to determine the state of interactions between proteins and helped to elucidate protein pathways.

DNA cloning ushered in an era of unprecedented control over the manipulation of DNA, allowing the overexpression of specific proteins of interest in cell lines and the purification of proteins of interest in bacteria. Further developments in this technique enabled the activation of specific genes at specific times or restricted to specific tissues, increasing a researcher's control over gene expression. This technique enabled scientists to determine the role of a given protein by introducing the gene which codes for it in cells that do not normally express the protein in question or to overexpress the protein product in cells that already have that particular gene. FACS, initially designed as a result of the need for a fast and efficient way to count blood cells in routine clinical work-ups, became an incredible tool to quickly assess a number of different parameters, such as the presence of specific cell markers (cell-surface proteins), to isolate a specific population of cells based on the expression of certain proteins, or for the determination of cell health following cytotoxic injury. Finally, the discovery of green fluorescent protein (GFP) in the 1970s led to the development of protein tagging with fluorescent proteins and allowed researchers to visualize the location of proteins in the cell.

* * * * *

We saw in Chapter 3 how Theodor Svedberg, in the early 20th century, invented the ultracentrifuge to better separate proteins and in the process, proved that proteins were not colloidal solutions but large, discrete molecules. This technique, however, was deficient in the fact that it could not separate proteins with similar characteristics, such as two proteins of similar size. Therefore, in 1925, Svedberg encouraged his pupil, Arne Tiselius, to continue the work

he had started on an electrophoretic method for the separation of proteins that might address this issue.[1] The theory of electrolysis—the movement of ions in a solution of electrolytes—was first described in 1834 by Michael Faraday, in a paper that also introduced the terms electrode, electrolyte, electrolysis, cation, and anion.[2] Simply put, immersion of two unreactive metals in a solution of ions will result in the flow of electrons if the metals are connected to an electrical source, such as a battery (Figure 11.1). For example, if two bars of platinum are immersed in an electrolyte solution and connected to a battery, an electrical field will be created that pushes electrons from the negative pole (the cathode), through the electrolyte solution, and to the positive pole (the anode); the rate at which electrons flow (electrons per second) is called the current.[3] In the process, H_2 gas is evolved from the cathode where electrons are released and electrons are picked-up at the anode with the evolution of O_2 gas. This continues as long as there are free electrons in the battery and ions in the solution. The force created by the electrical field that pushes electrons from one electrode to the other will also be exerted on any molecule that is placed in that same electrical field, the force being proportional to both the strength of the electrical field and the net charge of the molecule.

Thus, the movement of proteins in an electric field—initially known as "cataphoresis" but better known as "electrophoresis"—began in the early years of the 20th century by William B. Hardy and Leonor Michaelis.[1] Hardy had discovered that the movement of proteins in an electric field depended on the pH of the solution, and Michaelis used this information to investigate and characterize the properties of enzymes based on their movement in solutions of different pH values. Therefore, the choice of buffer for electrophoresis was critical and much time was dedicated to characterizing the movement of proteins in different buffers. As such, an adequate buffer should (1) be effective at maintaining pH values in all solutions; (2) act as a good conductor of current, that is, allow the movement of electrons; (3) should not react with the molecules desired to be separated.[3] The concentration of the buffer was also of importance since too low a concentration would not conduct electrons properly whereas too high would increase the current and therefore, also increase the heat generated (which, as we saw with the ultracentrifuge, creates unwanted convection forces).

Thus, Tiselius invented the "moving boundary method" for the separation of proteins as the subject of his thesis, an electrophoretic method which used ultraviolet light to follow the movement of proteins in a U-shaped tube (Figure 11.2a). However, like ultracentrifugation, this method suffered from the fact that similar proteins were only partially separated.[4] Another limitation was that the proteins often reached the

DOI: 10.1201/9781003379058-11

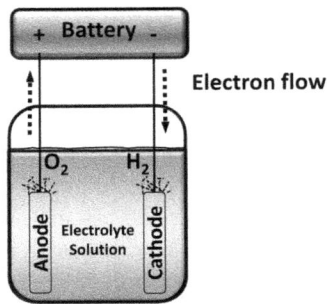

FIGURE 11.1 An electrolytic cell. When attached to a battery, electrons flow out of the cathode, through the electrolyte solution, and back to the anode.

end of the tube before they could be adequately separated. Thus, Tiselius subsequently modified the apparatus so as to create a *slow and uniform movement of the solution in the electrophoresis tube at an exactly known rate and in a direction opposite to the migration, by slowly lifting a cylindrical glass tube by clockwork out of the liquid in one electrode tube during the electrophoresis* (Figure 11.2b).[4] In addition, the circular tubes he was using in his initial apparatus were found to result in considerable thermal convection currents, a problem he much improved by using "flattened", rectangular tubes and a cold water bath to keep them cool. Tiselius described his electrophoresis apparatus as consisting of a

> central U-tube connected by thick rubber tubing with the large volume electrode vessels of which there were two pairs, one of about 2 litres capacity each for very difficult separations, the other of about 0.5 litre each. Reversible AgCl electrodes with saturated KCl solution was used as in the older apparatus, but of course the electrodes were given

much larger surface. The volume of KCl in each tube was 25-100 c.c.; it was allowed to form a layer at the bottom, running down through funnel tubes shortly before starting an experiment.

The U-tube, with an internal rectangular cross-section of 3 × 25 mm., was made of plain parallel glass plates, cemented together with acid-proof silicate cement; the curved walls of the U-shaped bottom part were cut from an ordinary cylindrical glass tube of the right diameter.[4]

To assess the migration of proteins, Tiselius adapted the Topler *schlieren* projection method, which used the diffraction pattern that results when light is passed through solutions of different concentrations.[3] Here, he used an objective placed as close to the tube as possible, where a light was shined through the tube and the image was captured by a camera; this method proved *much superior to the earlier light absorption method, especially when studying mixtures.*[4] An advantage of this method was that the concentration of the separated proteins could be assessed based on the refractive index of the particular band. As such, Tiselius showed the efficiency of his new apparatus by testing the migration of known protein solution, such as blood serum (Figure 11.2c).

However, as often happens in science, better methods for protein separation were soon developed. By the 1950s, zone electrophoresis—a technique whereby samples were applied to a supporting matrix at a given distance from electrodes such that during electrophoresis, proteins of different mobilities gradually separated from each other—was superior to the Tiselius apparatus in several ways: *freedom from quantitatively important boundary anomalies, the possibility of preparing electrophoretically discrete proteins, and adaptability to small quantities of material.*[3,5] However,

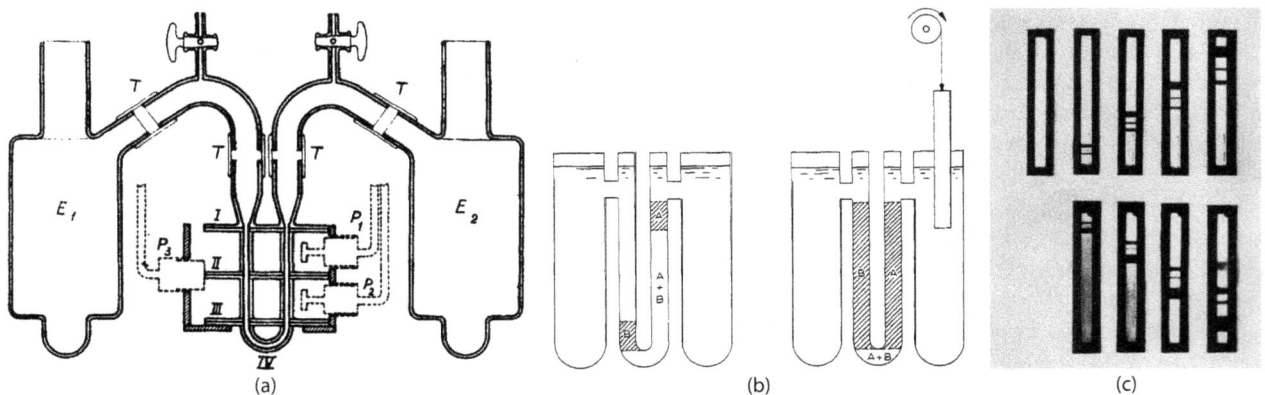

FIGURE 11.2 (a) Tiselius' first apparatus for electrophoretic analysis, where E_1 and E_2 were AgCl electrodes immersed in KCl solutions. Samples were added to the apparatus in the bottom section of the U-tube (IV), which was cut-off from the other sections by moving section III toward the left using the pneumatics system (P1-3). Current was then applied to separate the samples.[4] (b) Compensation arrangement. A plunger was inserted into one of the electrode vessels before the run was started and it was slowly removed as migration of the sample proceeded. This caused the movement of the boundaries in the opposite direction of its migration, allowing the sample to be run for a longer period of time and resulting in a better separation of the proteins within the sample.[1] (c) Results of the separation of serum using the Tiselius apparatus. Proteins moving the to the anode and cathode are represented on the top and bottom row, respectively.[4] (Figure parts (a, c) used with permission of Royal Society of Chemistry, from Tiselius A. A new apparatus for electrophoretic analysis of colloidal mixtures. Transactions of the Faraday Society 33(0), Copyright © 1937; Permission conveyed through Copyright Clearance Center, Inc. Figure part (b) used with permission from Tiselius A. Arne Tiselius: Nobel lecture; Copyright © The Nobel Foundation 1948.)

resolution was still superior in the Tiselius method compared to the two-zone electrophoresis methods available in the early days of the 1950's, one of which used filter paper as a supporting matrix whereas the other used starch grains (both were used primarily to reduce convection currents).[5,6] In 1955, Oliver Smithies attempted to develop an electrophoretic method that combined the advantages of these zone electrophoretic methods, while also hoping to increase their resolution.

As such, after trying a number of different supporting matrices, Smithies found that using a starch gel proved *to have a resolving power in many cases superior to that of the Tiselius method.*[5] This increase in resolving power was due to the sieving power of the pores within the gels, which *imposes an appreciable frictional resistance* to the movement of proteins since *the average pore size approaches the range of dimensions of proteins*, and resulted in their separation in *degrees proportional to their dimensions.*[6] Furthermore, an added advantage was that the gel concentration could be varied as needed, though Smithies found that gels between 10 and 15% provided satisfactory strength while not being too viscous when hot. Thus, the author prepared starch gels by boiling reagent grade soluble starch— or even potato starch in some cases—in borate buffer with *vigorous swirling [...] just short of boiling, when the starch grains are ruptured and a viscous homogeneous solution is obtained* and poured it into a plastic tray.[5]

The electrodes used were reversible Ag/AgCl coiled coils immersed in a concentrated NaCl solution with filter papers providing a bridge between the different compartments and the gel. The best method to introduce samples in the gel was determined to be to make a transverse cut in the gel, into which a piece of filter paper—with the protein samples adsorbed onto it—was then *introduced [...] and allowed to adhere to the undisplaced cut surface of the gel.* In this manner, *very slight differences between samples can be detected, as the components common to all samples give uninterrupted bands across the gel, while any additional (or absent) components show as bands only.*[5] An alternative to this was to cut out slots in the gel in the form of blocks into which samples mixed with starch in a partially settled state were inserted into the space. Following electrophoresis, the gel was removed from the plastic tray and cut in the horizontal plane, that is, parallel to the bottom of the gel, and the two fragments were separated and dyed using a variety of protein dyes to test their effectiveness for visualizing protein bands; amido- black 10B and bromophenol blue were found to be the most suited for this assay.

Thus, like Tiselius before him, Smithies used blood serum to verify the validity of his apparatus and, indeed, found the *classical patterns* that were expected for blood serum. However, the increase in resolution revealed the presence of additional serum components not previously identified, thus proving the effectiveness and superiority of his new apparatus. In fact, it was later shown that while the filter paper technique resolved serum proteins into 5–7 bands, the starch gel increased the number of detectable

proteins to between 20 and 30![6] However, despite the increase in resolution, starch gels still required a significant improvement: by the time two similar proteins were adequately separated, they had likely diffused to the point where band sharpness was less than optimal. Thus, Leonard Ornstein and Baruch J. Davis resolved this issue by developing a new method, disc electrophoresis—whose name comes from the fact that it was dependent on discontinuities in the support medium—to achieve an increase in resolution.[6,7] The authors were *stimulated* by Smithies' results using starch gels as a sieve to separate proteins but were also aware of the difficulties and variabilities in gel preparations.[7] They had previously used polyacrylamide gels as a means to embed tissues for sectioning and thought that they might be amenable to gel electrophoresis, since they were *thermostable, transparent, strong, relatively inert chemically, can be prepared with a large range of average pore sizes, and are non-ionic.*[7]

Davis also noted that since *these characteristics offer kinds of flexibility and versatility not easily attainable with starch gels at the present time, it was felt that polyacrylamide gel electrophoresis would provide a valuable complement to starch gel electrophoresis.*[6] Polyacrylamide is a molecule that forms long-chained polymers with pores that are somewhat malleable, such that a protein with enough momentum could effectively enlarge the pore and *slide through the network of linear chains.*[7] It is made from the combination of straight-chained molecules, acrylamide, and a cross-linking agent, bis-acrylamide, which combine to make a gel lattice of carbon-carbon-linked molecules, where the size of the pores is dependent on the ratio of acrylamide and bi-acrylamide (Figure 11.3).[6] Polymerization is activated by addition of ammonium persulfate (APS) and TEMED was used as a catalyst. Ornstein commented on his and Davis' initial experience in developing the gel:

> The first attempt, designed and executed by B. J. Davis, was so successful that it encouraged Davis and me through almost a year of successive failures until an understanding of mechanism and a reasonable degree of reproducibility began to be achieved.[7]

During the course of this optimization, the authors had noted that diffusion of the samples was a serious problem as it *continuously dilutes the separating fractions and blurs their boundaries*, and therefore, it was *desirable to reduce the running time to the minimum necessary to achieve a desired separation of the constituent ions.* As such, they reasoned that if a thinner starting zone was used—that is, a more compact area at the beginning of electrophoresis—the resolution would be increased since electrophoresis could be run longer without worrying about the diffusion of the sample.[7] To that end, they learned that in 1897, Kohlrausch had observed that, under specific conditions, the separation between two superimposed ions placed in an electric field could be maintained if the ion with the greater mobility was placed below the other.[7] However, Ornstein and Davis found that these conditions led to high convection forces

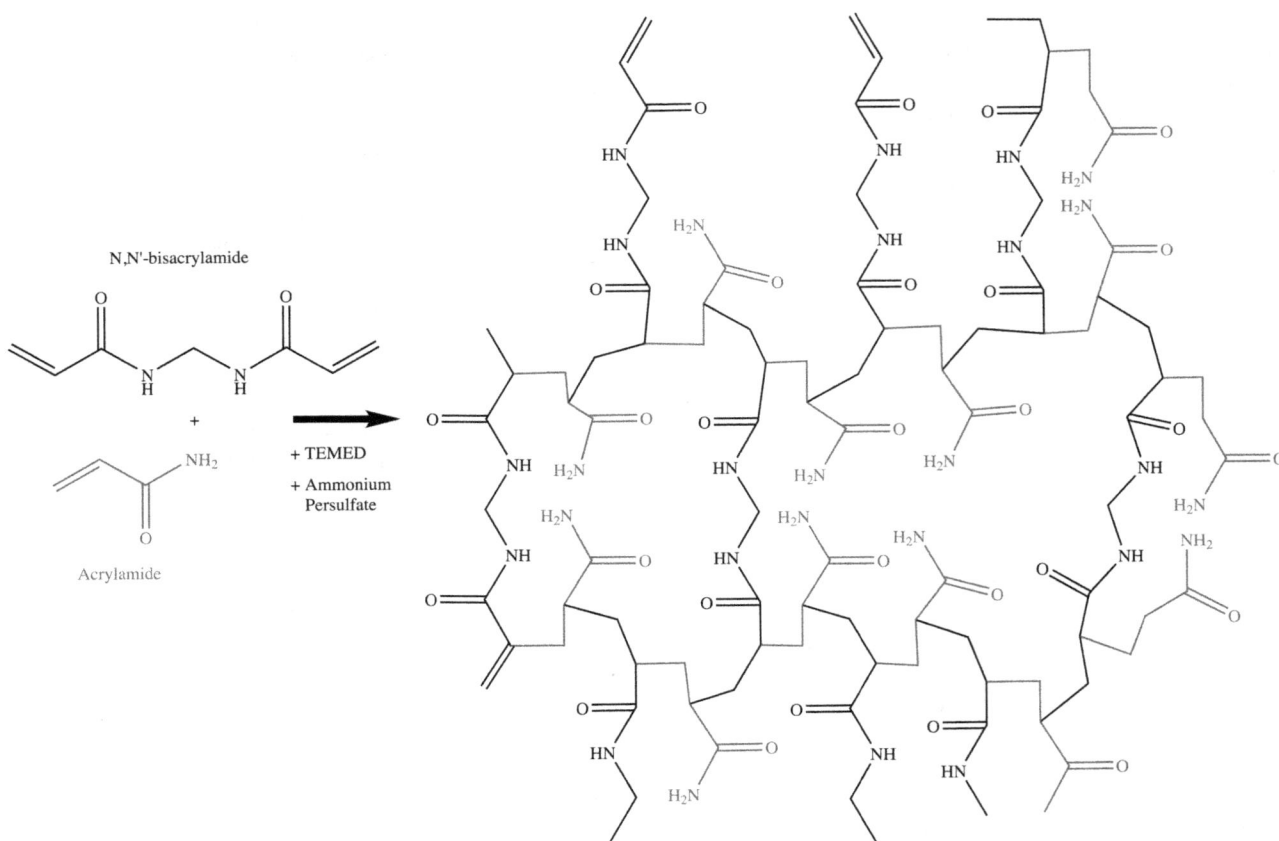

FIGURE 11.3 Polymerization of polyacrylamide. Cross-linking of acrylamide and bis-acrylamide is activated by APS and catalyzed by TEMED. This results in the production of a porous gel in which the size of the pores is dependent on the concentration and ratio of each chemical.

and a significant decrease in resolution; nonetheless, they determined that if a gel matrix was used to prevent the convection forces, then *given sufficient time, a sharp moving boundary will form and be maintained independent of the densities of the starting solutions and their initial concentrations, provided that the faster ions precede the slower.*[7] Thus, Ornstein and Davis found that by *stacking* a low-concentration/low pH gel on top of a higher concentration *separating* gel at a higher pH, *a sharp moving boundary could be formed that would produce a thin starting zone for the separation of proteins in the separating gel* (Figure 11.4):[7]

Above pH 8.0, most serum proteins have free mobilities in the range from −0.6 to −7.5 units. If the effective mobility of glycine, $m_a x_a$, were less than −0.6, the mobilities of the serum proteins would fall between that of the glycine and that of chloride [...]

If [...] a one milliliter mixture with mobilities ranging from −1.0 to −6.0 units is placed over the chloride solution, by the time the boundary has migrated one centimeter in the applied electric field, all the proteins will have concentrated into very thin discs, one stacked on top of the other in order of decreasing mobility, with the last followed immediately by glycine. We will call this process "steady-state stacking".

If the chloride boundary (and the following stack of discs) is permitted to pass into a region of smaller pore size

such that the mobility of the fastest protein drops below that of the glycine, the glycine will now overrun all the protein discs and run directly behind the chloride, and the proteins will now be in a uniform linear voltage gradient, each effectively in an extremely thin starting zone, and will migrate as in "ordinary zone electrophoresis".

Alternatively, if the boundary (and the following stack of discs) is permitted to pass into a region of higher pH, e.g., a pH of 9.8 (the pK_a, of glycine) [...] the glycine will now overrun all the protein discs and run directly behind the chloride, and the proteins will now be in a uniform linear voltage gradient, each effectively in an extremely thin starting zone, and will migrate as in "free electrophoresis".[7]

Chloride was chosen as the leading ion *since its mobility relative to [serum] proteins is high*, whereas glycine was chosen as the trailing ion *because its pK and mobility of glycinate are such that the effective mobility can be set at a value below that of the slowest known proteins of net negative charge in the pH range in which most of the serum proteins are anionic.*[6] The buffer also used a counter ion, that is, an *ion of sign opposite that of the slow, fast, and sample ions,* [serving] *as a buffer to set initially and then to regulate the various pH's behind the moving anion boundary.* In this case, the authors chose TRIS as its buffering capacity was adequate for this purpose and was *a relatively*

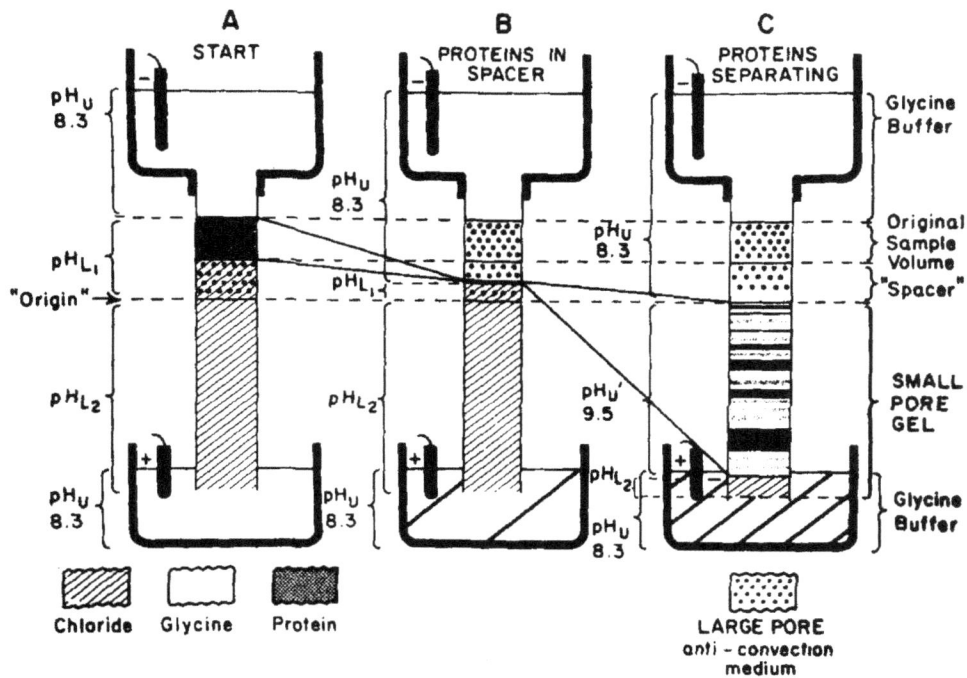

FIGURE 11.4 The stacking gel method of Ornstein and Davis. A gel with large pores at a lower pH (the stacking gel, dotted gel in the illustration) is stacked atop a gel with smaller pores at higher pH (the separating gel). The stacking gel condenses the proteins into a thin layer atop the separating gel, resulting in all proteins having the same starting point before being separated in the separating gel.[7] (Used with permission of John Wiley and Sons—Books, from Ornstein L. Disc electrophoresis. I. Background and theory. Ann N Y Acad Sci. 121, Copyright © 1964; Permission conveyed through Copyright Clearance Center, Inc.)

innocuous substance to most proteins.[6] In addition, *because of the special viscous properties of* [polyacrylamide gels], *proteins of equal free mobility but of appreciably different molecular weight (different diffusion constant) will migrate with markedly different mobilities and will easily be separated.*[7] As with Smithies and Tisselius before them, Ornstein and Davis used the well-characterized serum proteins as samples to test the effectiveness of their new method (Figure 11.5).

Electrophoresis using polyacrylamide was modified in 1967 by Shapiro, Viñuela, and Maizel by adding the stain Coomassie blue to the samples so that the movement of proteins could be visualized during electrophoresis as an indication of how far the proteins have migrated through the gel. More importantly, the authors found that adding the detergent, sodium dodecyl sulfate (SDS) and a reducing agent, β-mercaptoethanol (BME), to the samples significantly increased the resolution.[8] Using these modifications, the authors suggested that the technique could now potentially be used to *approximate the molecular weight of each component of a mixture of unknown polypeptides simply by running a suitable marker of known size in a second, parallel gel.* As to the mechanism, they suggested that:

> [...] SDS minimizes the native charge differences and that all proteins migrate as anions as the result of complex formation with SDS. The extensive disruption of hydrogen, hydrophobic, and disulfide linkages by SDS and *[BME]*

FIGURE 11.5 Separation of blood serum using Ornstein and Davis' electrophoretic method.[6] (Adapted with permission of John Wiley and Sons—Books, from Davis BJ. Disc electrophoresis. II. Method and application to human serum proteins. Ann N Y Acad Sci., 121, copyright © 1964; Permission conveyed through Copyright Clearance Center, Inc.)

results in the quantitative solubilization of many relatively insoluble proteins. These factors and the ease of the polyacrylamide technique strongly recommend it as the electrophoretic method of choice.[8]

Weber and Osborn found similar results the next year using *E. coli* proteins but wondered how reproducible and applicable these results were. Therefore, they began a series of experiments in search of an easier and quicker method to determine protein molecular weights accurately using SDS-PAGE.[9] Until that time, protein molecular weights could be determined by several methods, such as analysis of amino acid sequences, X-ray crystallography, or equilibrium centrifugation in a guanidine-HCl solution, which for the most part, yielded very reliable results. However, some of the drawbacks of these techniques included the *large amount of protein required* for processing and the fact that they were *very demanding experimentally.*[9] As such, the authors thought that SDS-PAGE might be used as a simple means by which to assess protein molecular weights.

Weber and Osborn used 40 different well-characterized proteins of various sizes and isoelectric points and applied them to a modified version of SDS-PAGE—using a phosphate-based gel buffer with a 10% acrylamide gel and stained the protein bands using Coomassie blue—and found that all proteins tested migrated strictly according to their molecular weights.[9] As an added benefit, they found that protein bands could be relatively easily extracted from the gel, purified, and used in downstream experiments. Thus, the authors proposed that:

> with enough markers an accuracy of about 10% in the determination of the molecular weight of an unknown protein falling in this molecular weight range may be possible [...]
>
> The good resolution and the fact that an estimate of the molecular weight can be obtained within a day, together with the small amount of protein needed, makes the method strongly competitive with others commonly employed.[9]

It was later shown that SDS coats proteins with a constant negative charge, and which allowed the separation of proteins based solely on their molecular weights. However, when Rosalin Pitt-Rivers and Ambesi Impiomba tried to elucidate the mechanism of action of SDS on protein, their results differed from what had previously been reported.

Therefore, they performed a series of experiments to determine the precise properties of SDS binding to proteins. As such, the authors soaked a variety of proteins in an SDS solution for 48 hours followed by dialysis to remove excess SDS, and the amount of SDS bound to the proteins was determined in a spectrophotometer using its extinction coefficient.[10] Results of this experiment showed that there was a significant difference in SDS binding between proteins that contained disulfide bridges and those that did not, the former reaching equilibrium in an average of about 8 days, while the latter reached equilibrium in an average of less than 5 days.[10] In addition, while the amount of SDS bound to proteins without disulfide bridges was consistent at about 1.4 g SDS/g of protein, this decreased to under 1 g SDS/g of protein in proteins with disulfide bridges; this could be brought up to 1.4 g SDS/g of protein by incubating the proteins with a reducing agent (such as BME) that disrupted these linkages. This indicated that disulfide bonds restrict the unfolding of proteins and therefore the ability for SDS to maximally bind to them, and thus supported the use of BME and heat to completely denature proteins.[10]

Later, Reynolds and Tanford studied the effects of SDS on proteins subjected to polyacrylamide electrophoresis to further elucidate its mechanism of action. The authors concluded that the *high level of binding of SDS to proteins and the constant binding ratio on a gram to gram basis [...] assure a constant charge per unit mass.*[11] Other studies found that the binding of SDS with proteins involved association of the hydrophobic tail of SDS with the hydrophobic amino acids in proteins, which results in the exposure of the negatively charged head group of SDS and gives proteins a net negative charge that is proportional to its molecular weight (Figure 11.6). Thus, proteins subjected to electrophoresis migrate in a manner that is strictly dependent on their molecular weight since their native charges are eliminated by the SDS.

Finally, in 1970, Uli Laemmli published a paper describing an improved method of disc electrophoresis to separate the structural components of T4 bacteriophage head proteins, which enabled the identification of a number of new proteins involved in the T4 head assembly.[13] However, very little detail was given about the reasoning behind the improvements or, indeed, the development of the technique itself (the author noted that these details would be included

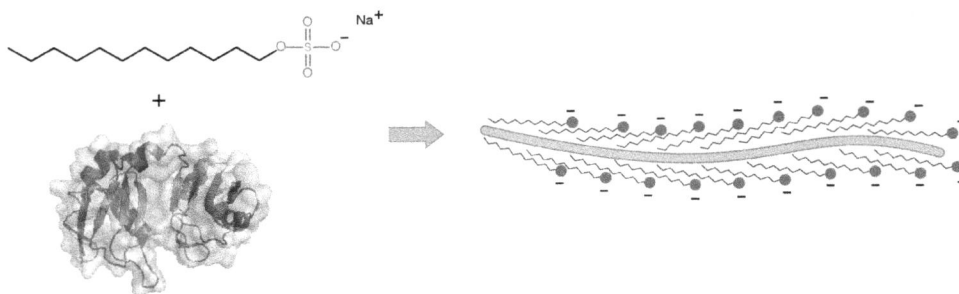

FIGURE 11.6 Mechanism of SDS binding to proteins. Heat denatures and unfolds proteins, allowing binding of the hydrophobic ends of SDS to proteins, thereby leaving their negative ends exposed.[12]

in a subsequent paper co-authored with Jacob Maizel, Jr., but this paper was never published, as far as I can tell).[14] Nonetheless, the methods section of this paper precisely described the composition of the gel and the buffers used: the stacking gel was 3% acrylamide in a Tris-Cl buffer at pH 6.8/0.1% SDS and the separating gel was 8% or 10% acrylamide in a Tris-Cl buffer at pH 8.8/0.1% SDS, with the addition of TEMED and APS; the electrophoresis buffer was a Tris-Cl/0.1% SDS buffer with glycine as the trailing ion; the sample buffer—the buffer in which the protein samples were diluted—was composed of Tris-Cl at pH 6.8 and 2% SDS, 10% glycerol, 5% BME as the reducing agent, and bromophenol blue as the dye to track the migration of the sample; finally, the samples were boiled to ensure the complete denaturation of the proteins. These methods have become standard for SDS-PAGE and the buffer in which samples are diluted has become known as Laemmli buffer.

* * * * *

The next major development in gel electrophoresis came from a DNA technique in which samples were transferred from a gel to a membrane for visualization, and its development significantly increased the usefulness of SDS-PAGE electrophoresis and completely changed the way it was used. By the mid-1970s, gel electrophoresis had been adapted for use with DNA, and as such, was used to confirm the fragments generated when cutting DNA during cloning experiments (see next section for an overview of this technique). These fragments were subsequently cut out of the gel, extracted, and sequenced to confirm their identity, a process that was time-consuming and also led to a decrease in resolution. Therefore, Edwin Southern sought a new technique that would make this process less of a burden on experiments. To that end, he found that DNA could readily and faithfully be transferred to and immobilized on nitrocellulose membranes if the membrane was laid on top of the gel: DNA diffused through the gel and adhered to the surface of the membrane in the presence of the right solution.[15] Although the time required to accomplish a complete transfer of a DNA sample to the membrane varied, overnight incubation was thought to be adequate in most cases. After the transfer, the identity of the DNA fragments was confirmed by using radiolabelled RNA molecules that were complimentary in nucleotide sequence and therefore bound to the DNA if that particular sequence was present. This technique revolutionized the field of DNA cloning and was named Southern blotting in honor of its developer. Later, a similar technique was developed to identify RNA transferred to nitrocellulose membranes after gel electrophoresis, a technique that was named Northern blot in homage to Southern.[16]

In 1979, two groups, Renart, Reiser, and Stark, and Towbin, Staehelin, and Gordon, independently adapted Southern's technique for the detection of proteins, mainly for the confirmation of antibody specificity following protein purification and for the discovery of new antibodies against specific targets.[16,17] For their part, Renart, Reiser,

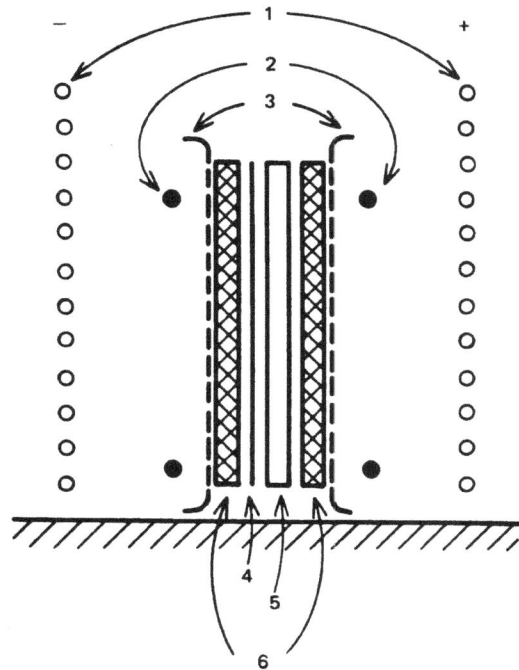

FIGURE 11.7 Towbin, Staehelin, and Gordon's protein transfer apparatus. A sheet of nitrocellulose (4) was laid on top of a polyacrylamide gel (5), and the whole was sandwiched between two scotch-bright pads (6) and held together using empty pipette tip trays and some rubber bands (3). The assembly was then placed between two electrodes (1) in a chamber containing acetic acid—with the nitrocellulose sheet facing the cathode—and a voltage was applied across the terminals.[16] (From Towbin H, Staehelin T, Gordon J. Electrophoretic transfer of proteins from polyacrylamide gels to nitrocellulose sheets: Procedure and some applications. Proc Natl Acad Sci U S A. 1979;76(9):4350–4354. Used with permission.)

and Stark used an activated diazobenzyloxymethyl (DBM) paper as the supporting matrix to immobilize the proteins and passively transferred the proteins out of the gel as Southern had done. In contrast, Gordon's group laid a nitrocellulose membrane on top of a gel and sandwiched them between Scotch-Bright pads supported on each side by *stiff plastic grid[s]* (e.g., disposable pipette trays), the whole being held together by a rubber band (Figure 11.7).[16] The sandwich was then placed in a buffer solution and a voltage was applied, which resulted in the movement of the proteins towards the anode (in the case where SDS was used during electrophoresis). In either case, both of these techniques resulted in the faithful transfer of proteins from the gel to the membrane, which could then be detected using antibodies. However, Towbin, Staehelin and Gordon found that the presence of SDS was detrimental to the transfer of proteins and opted instead to develop the technique using urea, a reagent not widely used in electrophoresis of proteins.

In 1981, Neal Burnette adapted Towbin's technique to be used with SDS and named it "Western blot", in keeping with *the established tradition of 'geographic' naming of* transfer techniques.[18] As such, Burnette used the idea of the

TIME OF TRANSFER (h)

0   1   4   12   22

GELS

BLOTS

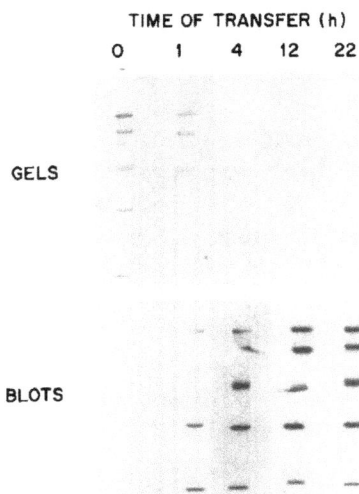

FIGURE 11.8 Burnette's Western blot using SDS-based buffers. A gel was cut into five lanes and each lane was transferred to a membrane under similar conditions for the indicated times.[18] (Reprinted from Anal Biochem. 112(2), Burnette WN. "Western blotting": Electrophoretic transfer of proteins from sodium dodecyl sulfate-polyacrylamide gels to unmodified nitrocellulose and radiographic detection with antibody and radioiodinated protein A, 195–203, Copyright © 1981, with permission from Elsevier.)

"Towbin sandwich" to transfer radiolabeled proteins from a gel to a membrane and found that resolution was superior to DBM transfer (Figure 11.8). However, he did confirm the inefficiency of protein transfer in the presence of SDS—as reported by Towbin, Staehelin, and Gordon—but found that it was partly due to their transfer time of 1 hour not being long enough. Indeed, Burnette determined that transfer of proteins in the presence of SDS was most efficient for smaller proteins, while larger proteins required extensive transfer times, up to 24 hours. Burnette also noted that the use of antibodies for the detection of nitrocellulose-adsorbed proteins led to high background since the antibodies could also bind to the membrane, being proteins themselves. As such, he incubated the membrane with a number of inert protein solutions prior to protein detection with antibodies, in an attempt to block the membrane sites not covered by the transferred proteins and found that a 5% solution of bovine serum albumin worked best (non-fat milk is also commonly used now). Finally, Burnette concluded that:

> The technique of Western blotting should prove valuable for the analysis of proteins fractionated on the basis of molecular weight in SDS gel electrophoresis. Use of unmodified nitrocellulose sheets and an electrophoretic mode of transfer provide speed and a simplicity of technique that, combined with essentially complete and quantitative protein transfer, is not achievable by other blotting methods. Coupled with antigen detection by antibody […,] the Western blot is a very sensitive method for visualizing specific proteins in complex antigenic mixtures.[18]

* * * * *

Another technique that had a significant impact on cell biology research is cloning. If you recall from Chapter 2, the exchange of genetic material between chromosomes was first observed in the early 20th century and was the basis for Morgan's Theory of Linkages, which explained why genes do not necessarily segregate according to Mendelian laws if they are far enough apart on the same chromosome. In the 1940s, a phenomenon was observed independently by Delbrück and Bailey and by Hershey and Rotman in which bacteria transformed simultaneously with two separate bacteriophage DNA (i.e., DNA was inserted into the bacteria) produced progeny phages containing fragments of DNA derived from each inserted DNA molecule.[19] In 1961, Meselson and Weigle investigated the mechanism by which DNA exchanges might occur, reasoning that it could occur in one of two ways: either by *de novo* synthesis, where the new DNA is synthesized using either parent as template, or by breakage and reassembly, in which the *recombinant sequence is formed by the association of DNA fragments from different parental lines.*[19]

To test these hypotheses, the authors transformed cells simultaneously with both unlabeled λ phage DNA and λ phage DNA labeled with heavy isotopes, and used CsCl density-gradient centrifugation to detect any changes in distribution of the heavy isotopes in the progeny DNA. As such, they found that a certain percentage of progeny phages were indeed composed of DNA that was partially isotopically labeled, indicating that DNA from both parents recombined to make the progeny's DNA and suggested *that recombination occurs by breakage of parental chromosomes followed by the reconstruction of genetically complete chromosomes from the fragments.*[19] In addition, the authors consistently found a class of recombinant DNA that contained more than 50% of the original parental DNA, which indicated that *recombination by chromosome breakage may occur without separation of the two subunits of the parental chromosome*, and therefore, that *chromosomes need not replicate in order to recombine*. The authors also confirmed Morgan's theory when they noted that the *probability of recombination between two loci is at least approximately proportional to the amount of DNA between them.*

In 1967, a number of groups independently discovered an enzyme that could stitch DNA fragments together.[20–23] For his part, Gellert used the newly discovered fact that linear phage λ was converted to a covalently linked circle shortly after infection of *E. coli*. In addition, phage λ DNA was known to possess termini with complimentary nucleotides that could close into circles via non-covalent hydrogen bonds, and Gellert proposed that this was an intermediate species in the formation of covalent circles. Thus, he found an *E. coli* extract with activity responsible for the transformation of linear phage DNA into covalent circles and determined that, indeed, the hydrogen-bonded circle was a required intermediate in the process.[20] The assay used was dependent on the fact that covalent *circles are most readily distinguished from other forms of λ DNA* [linear or hydrogen bonded] *by sedimentation at an alkaline pH sufficiently high to denature the DNA.* Since circularization of linear DNA had been reported in other organisms, Gellert suggested that this activity *may more plausibly be considered a general repair activity designed for sealing breaks within DNA molecules.*[20]

Weiss and Richardson also purified this DNA-sealing activity from *E. coli* extracts but used a different approach to report the ligation of DNA strands. The authors used pancreatic DNAse to generate single-stranded DNA breaks, dephosphorylated the terminal phosphates generated, and then rephosphorylated them with ^{32}P-labeled phosphate (Figure 11.9a).[21] In addition, they used the fact that phosphates engaged in a phosphate bond are not soluble in acid, whereas terminal phosphates are (Figure 11.9b). Thus, the authors added their bacterial extracts to this radiolabeled, nicked DNA and measured the extent of ligation by assessing the amount of acid-insoluble, ^{32}P-labeled DNA. Consistent with Gellert's results, Weiss and Richardson found an increase in radioactivity in the presence of their enzyme fraction, which indicated that the DNA break had been repaired, and named the purified enzyme *polynucleotide ligase.*

We also saw in Chapter 5 how Griffith transformed avirulent strains of bacteria into virulent strains by mixing heat-treated virulent strains with avirulent strains, and how Avery, Macleod, and McCarthy proved that the transforming principle was DNA and not proteins. By the 1960s,

it had been shown that a particular strain of *E. coli*, K12, was insensitive to transformation unless an intact "helper phage" was added either before or during transformation.[24] This helper phage DNA was therefore thought to make cells competent for infection, but the exact conditions under which this occurred were not known. In 1967, Mandel showed that *E. coli* that were competent for transformation had increased cell wall permeability and this was followed up in 1970 by Mandel & Higa, who investigated the effects of monovalent and divalent cations on this phenomenon. As such, the authors incubated *E. coli* in various concentrations of ice-cold $CaCl_2$ for 20 minutes before phage DNA was added to the cells, with or without helper phage, and found an *extremely rapid rise in competence* of *E. coli* up to a particular $CaCl_2$ concentration, and a slow decline in competence as the $CaCl_2$ concentration was further increased.[25] The authors also found that entry of DNA into cells was complete within 2 minutes, a time-frame that was much quicker than helper phage-assisted infection, and indeed, they showed that transformation proceeded without the need for helper phage under these conditions. The fact that $CaCl_2$ increased permeability of the cell wall was confirmed by the co-incubation of the membrane-impermeable, DNA-binding substance, actinomycin D, which decreases cell viability by inhibiting transcription. As such, the authors found only 25% of cells incubated in $CaCl_2$ survived in the

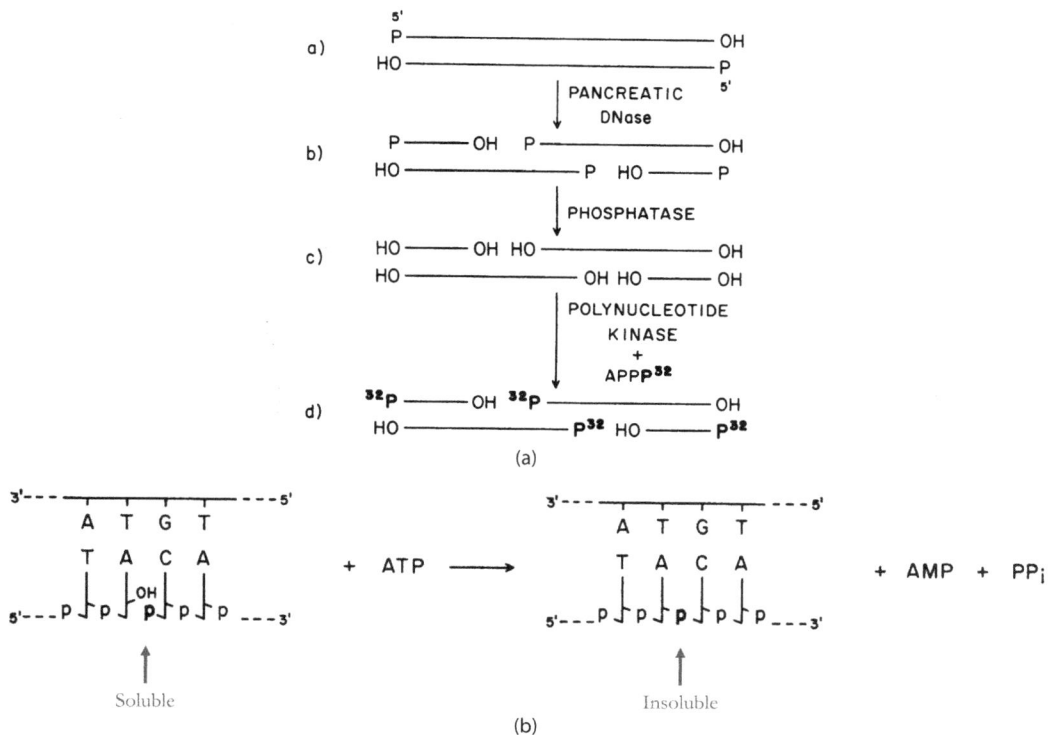

FIGURE 11.9 Weiss and Richardson's scheme to test the ability of enzymes to ligate DNA. (a) DNA was incubated with DNase to produce single-stranded nicks, followed by phosphatase to remove the exposed 5′ phosphates. Addition of a kinase and ^{32}P$_i$ allowed labeling of the DNA. (b) Incubation of radioactive, nicked DNA with polynucleotide ligase resulted in ligation of the nick as reported by an increase in radioactivity in acid-insoluble ^{32}P$_i$.[21] (Adapted from Weiss B, Richardson CC. Enzymatic breakage and joining of deoxyribonucleic acid, I. Repair of single-strand breaks in DNA by an enzyme system from *Escherichia coli* infected with T4 bacteriophage. Proc Natl Acad Sci U S A. 1967;57(4):1021–1028. Used with permission.)

presence of actinomycin D, compared to 100% survival in the absence of $CaCl_2$.

Later, it was shown that plasmids transformed into bacteria were replicated along with genomic DNA from one generation to the next, and importantly, the genes on the plasmids remained active. For example, if bacteria were transformed with a plasmid containing a particular drug resistance gene—for example, resistance to ampicillin— the transformed bacteria would express that protein product and become resistant to ampicillin.[26] As we will see later in this chapter, these discoveries were of great importance to the development of cloning.

* * * * *

During the 1950s, a phenomenon called "host-controlled modifications" was identified, whereby the DNA of certain bacteriophages and viruses was modified by the host during infection.[27] Such modifications—which were subsequently determined to not involve mutations—were initially identified by a decrease in the ability of bacteriophage λ to grow in a given E. coli host following their growth in a first E. coli strain; thus, phage growth was said to be "restricted" by the first strain. As such, Arber and Dussoix assessed host-controlled modifications in a variety of λ phages and E. coli hosts and found that certain variants of λ phage could only infect specific strains of E. coli. In addition, the authors determined that the DNA of phages that could not infect a particular E. coli host was quickly degraded by the host.[28] Therefore, based on experimental results, Arber suggested that a highly specific 'restriction enzyme' only initiates the degradation, for example, by cleavage of the DNA, and that these cleavage products are then subject to the action of less specific nucleases.[27] This phenomenon was proposed to provide bacteria with a mechanism whereby the expression or integration of foreign genetic elements can be prevented, akin to an immune system in eukaryotes.

In 1968, Meselson and Yuan isolated an enzyme responsible for the degradation of foreign DNA by host E. coli.[29] The fact that the enzyme could cleave DNA of circularized plasmids—which have no open ends—indicated that it was an endonuclease, that is, a nuclease that cleaves nucleotides within a DNA molecule (as oppose to an exonuclease, which only cleaves terminal nucleotides). In addition, the authors showed that the cleavage products of the restriction enzyme were duplexes containing little or no single stranded DNA, suggesting that the enzyme cleaved double-stranded DNA, although cleavage occurred one strand at a time. It was later shown that, although a specific sequence of DNA was required for the enzyme to bind, the actual cleavage site was random and removed from the recognition sequence, and that, paradoxically, the enzyme could modify DNA in a manner that prevented its restriction abilities. These and other aspects of this endonuclease would limit its usefulness in cloning, though, as we will soon see, other enzymes would provide more specific and consistent cleavage sites, characteristics which would be invaluable to making cloning a household name.

Two years later, back-to-back papers were published describing the isolation and characterization of the first restriction enzyme that enabled the wide-spread use of cloning. Smith and Wilcox discovered an enzyme in Hemophilus influenzae, which they called endonuclease R, that had similar properties to Meselson and Yuan's enzyme, in that it was an endonuclease that cleaved foreign, but not host, double-stranded DNA, but only at a few specific sites.[30] This aspect was most interesting to the authors and they suggested that it was likely that this recognition specificity resides in the base sequence of the sites, and proposed this to be at most six bases long based on the following reasoning:

> For phage T7 DNA we observed 40 to 45 breaks per molecule. Since T7 DNA is about 40,000 base pairs in length, the average fragment is about 1000 base pairs in length. For phage P22 DNA, the average fragment is approximately 1300 base pairs in length. To attain this degree of specificity a site would have to be five to six bases in length providing that the enzyme recognizes a completely unique sequence.[30]

Smith's follow-up paper confirmed this idea and identified the base sequence of the recognition site.[31] To accomplish this, the authors labeled T7 phage DNA with ^{33}P and digested it to completion using endonuclease R (Figure 11.10). Next, they dephosphorylated the 5′ terminal nucleotide of each cleavage product, rephosphorylated it using ^{32}P-labeled

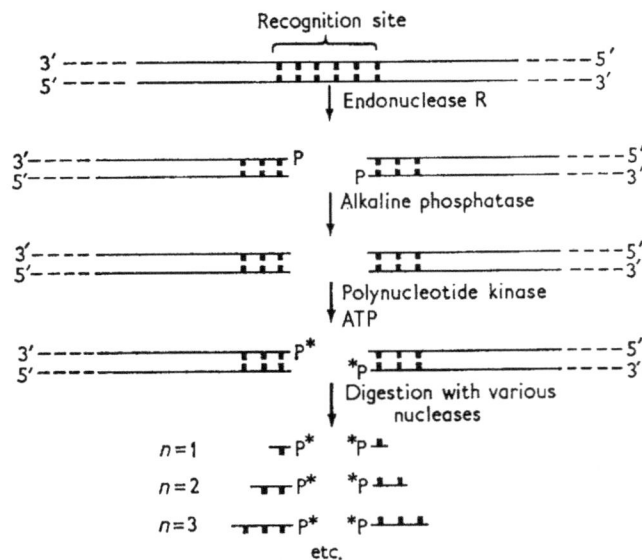

FIGURE 11.10 Scheme for the determination of the recognition site of endonuclease R. ^{33}P-labeled DNA was treated with endonuclease R and 5′ phosphates removed using phosphatase, followed by rephosphorylation with [32]P. The DNA was then completely digested using a variety of enzymes, followed by sequence analysis of the radioactive fragments.[31] (Reprinted from J Mol Chem. 51(2), Kelly TJ, Smith HO. A restriction enzyme from Hemophilus influenzae: II. Base sequence of the recognition site, 393–409, Copyright © 1970, with permission from Elsevier.)

phosphate, and digested the fragments further using nonspecific DNases; this resulted in a mixture of polynucleotides of different lengths. Analysis of these fragments was performed by electrophoresis on thin-layer cellulose followed by autoradiography. However, since the DNA was labeled with both ^{32}P and ^{33}P, the usual method of autoradiography was modified slightly, whereby radioactivity was assessed with and without a screen, the former of which reported only the higher-energy [32]P (since the lower-energy ^{33}P was absorbed by the screen), whereas the latter reported both labels.[31] As such, analysis of nucleotides contained in the various fragments revealed that terminal nucleotides only contained adenine and guanine (purines), and polynucleotide analysis revealed the presence of an AC sequence on the 3′ side of the cleavage site. In addition, it was found that the fourth nucleotide was irrelevant for the enzyme's recognition of the sequence. Thus, the authors concluded that endonuclease R cleaved DNA at a site 5′ to a PuAC sequence recognition site:

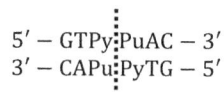

$$5' - GTPy\!\mid\!PuAC - 3'$$
$$3' - CAPu\!\mid\!PyTG - 5'$$

where Pu stands for purine, either A or G, and Py stands for pyrimidine, either T or C.[31] The authors noted the symmetry of the sequence and proposed that this was reflected in the enzyme itself:

> These considerations suggest the possibility [...] that recognition is accomplished by two identical subunits related by a 2-fold axis of symmetry. In this model each subunit recognizes the same sequence of three bases on opposite strands of the DNA duplex. From the standpoint of economy of genetic information it is much cheaper to specify two identical subunits each capable of recognizing three bases in a symmetrical sequence of six than to specify a larger protein capable of recognizing the entire sequence [...]
>
> In retrospect, it is probably not very surprising that endonuclease R recognizes a symmetrical sequence since the enzyme carries out an essentially symmetrical operation, namely the cleavage at equivalent points of two DNA strands of opposite polarity. This consideration suggests that symmetrical recognition sequences may be the rule for restriction enzymes in general.[31]

By 1973, more than twenty restriction enzymes had been discovered that could cleave DNA at specific sites within a recognition sequence. However, unlike endonuclease R, most of these would be found to cleave DNA in a staggered manner, leaving an overhang of a few base pairs on each strand. Since the list of restriction enzymes was steadily growing, Smith and Nathans proposed certain rules for the naming of restriction enzymes, whereby the first three letters indicated the bacterium in which the enzyme was discovered, the next letter an abbreviation of the strain of bacteria, and the number represents the order in which the enzyme was discovered.[32]

The introduction of these last few techniques culminated in 1972, with the introduction of DNA cloning by Paul Berg and his group, whose goal was to

> develop a method by which new, functionally defined segments of genetic information can be introduced into mammalian cells.
>
> Since purified SV40 DNA can also transform cells (although with reduced efficiency), it seemed possible that SV40 DNA molecules, into which a segment of functionally defined, nonviral DNA had been covalently integrated, could serve as vectors to transport and stabilize these nonviral DNA sequences in the cell genome.[33]

As such, Berg and his group combined restriction digest and ligation to accomplish the first part of their plan (an outline of this scheme is shown in Figure 11.11a). Earlier that year, Morrow and Berg had found that the newly discovered EcoRI endonuclease cut at a single site within the SV40 genome. Thus, Jackson, Symons, and Berg used this enzyme to linearize ^{3}H-labeled SV40 DNA, and then added a series of 50–100 adenine nucleotides to the 3′ termini of some of the DNA and thymidine to the others, using an enzyme called terminal transferase.[33] However, the efficiency of this reaction was very low, so they tried another method, suggested to them by Lobban and Keiser, which involved removing 30–50 nucleotides at each 5′ end before addition of the transferase, and the authors found that efficiency significantly increased.

Thus, the authors mixed the two constructs—that is, the one ending with adenines and the one ending with thymines—and found that they formed hydrogen-bonded circles, which, as expected, subsequently formed covalent circles following the addition of DNA polymerase and ligase to seal the DNA.[33] Covalent bonding of the fragments was confirmed using the usual method of density gradient centrifugation, which showed that the recombinant DNA was twice a heavy as the original DNA (Figure 11.11b). Jackson, Symons, and Berg thus concluded that *two unit-length linear SV40 molecules have been joined to form a covalently closed-circular dimer.*[33] Given this success, the authors next attempted to insert a fragment of another related genome, λdvgal, which contained a cluster of genes responsible for the metabolism of galactose in *E. coli*, called the galactose operon, which was not present in the SV40 genome. As such, they used the same procedure described above and found that, indeed, insertion was possible, and all the standard tests supported the conclusion that *the newly formed, covalently closed-circular DNA contains one SV40 DNA segment and one λdvgal DNA monomeric segment.*[33] In addition, the authors suggested that:

> The methods described in this report for the covalent joining of two SV40 molecules and for the insertion of a segment of DNA containing the galactose operon of *E. coli* into SV40 are general and offer an approach for covalently joining any two DNA molecules together [...]
>
> One important feature of this method, which is different from some other techniques that can be used to join

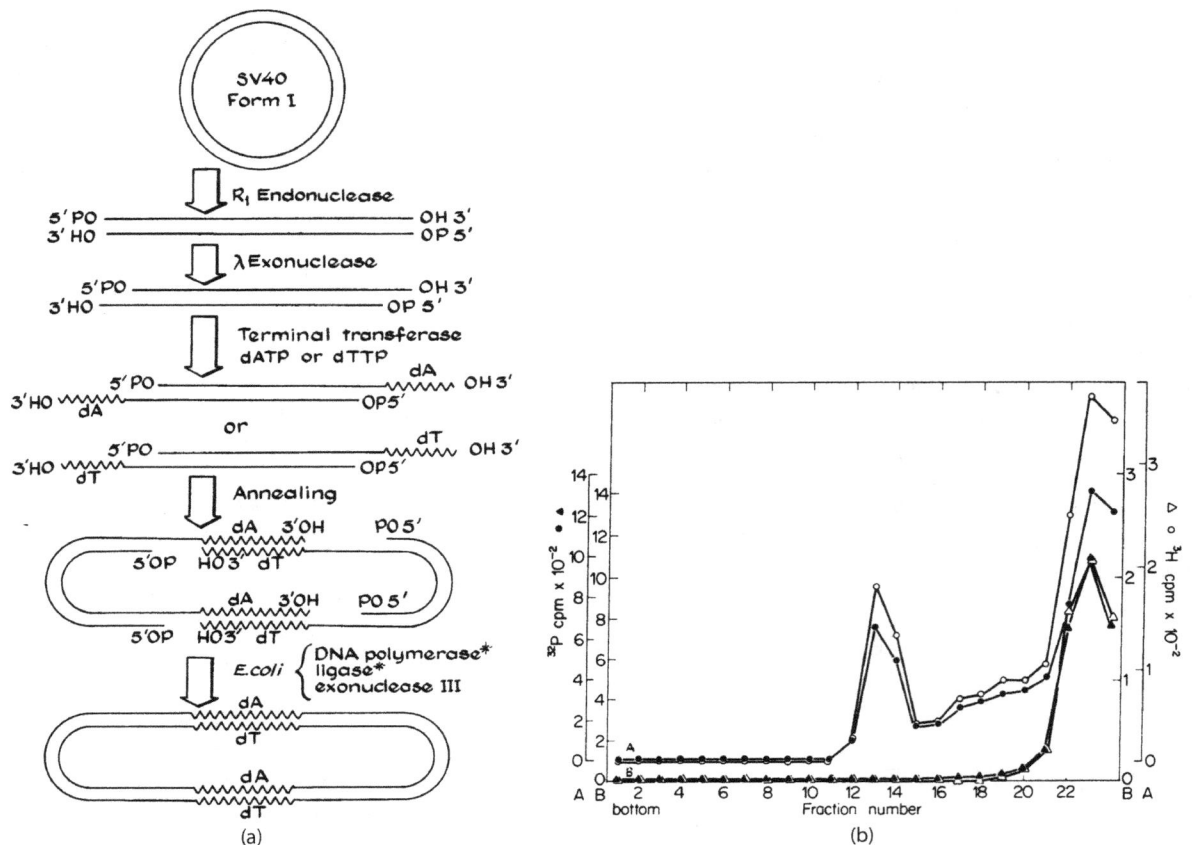

FIGURE 11.11   (a) Protocol for synthesis of SV40 plasmid dimer. ^{3}H-lableled SV40 plasmids were digested using EcoRI, which linearized the plasmid. The 5′ termini were further trimmed using λ-exonuclease and 50–100 radioactive adenines or thymines were added to these ends, thereby creating poly(dA) or poly(dT) termini. These were then incubated together such that the complimentary dA and dT sequences annealed and were held together via hydrogen bonds. The gaps were then filled-in by the addition of DNA polymerase and the fragments joined together by DNA ligase. (b) Samples were treated as described above and separated using density gradient centrifugation. Only in the presence polymerase, ligase, and exonuclease (A lines) do the samples circularize, as indicated by the peaks around fraction 13 (B lines are the same as A lines but without polymerase, ligase, and exonuclease). Lines with filled-in symbols used ^{3}H as a read-out, whereas lines with open symbols used ^{32}P.[33] (Reprinted from Jackson DA, Symons RH, Berg P. Biochemical method for inserting new genetic information into DNA of simian virus 40: Circular SV40 DNA molecules containing lambda phage genes and the galactose operon of *Escherichia coli*. Proc Natl Acad Sci U S A. 1972;69(10):2904–2909. Used with permission.)

unrelated DNA molecules to one another is that here the joining is directed by the homopolymeric tails on the DNA. In our protocol, molecule A and molecule B can only be joined to each other; all AA and BB intermolecular joinings and all A and B intramolecular joinings (circularizations) are prevented. The yield of the desired product is thus increased, and subsequent purification problems are greatly reduced [...]

Such hybrid DNA molecules and others like them can be tested for their capacity to transduce foreign DNA sequences into mammalian cells, and can be used to determine whether these new nonviral genes can be expressed in a novel environment [...][33]

The year after the publication of this paper, Cohen et al. engineered the first cloned genes that were transformed into *E. coli*. The authors noted that the new restriction endonucleases discovered *appeared to have great potential value for the construction of new plasmid species by joining DNA molecules from different sources* and succeeded

in extracting a gene from one species and inserting it into the genome from another species.[34] To that end, they used EcoRI to cleave the R6-5 plasmid, which contained several antibiotic-resistance genes, into twelve different fragments and separated them by agarose gel electrophoresis. These fragments were extracted from the gel, cloned into another plasmid which did not have any of these antibiotic-resistance genes, and transformed *E. coli*. After plating the transformed bacteria on plates with different antibiotics, the authors identified one clone that was resistant to kanamycin, neomycin, and sulfonamide, which indicated not only that genes could be clones from one organism to another, but that these genes were indeed functional in their new host.[34]

As the number of restriction enzymes increased over the years, so too did the number of different nucleotide-specific DNA cleavage sites, which also increased the possibilities for the insertion of specific genes at specific locations in a plasmid. In addition, the onset of plasmid manufacturing specifically designed for cloning made this process all the more simple. These plasmids were engineered with bacterial and eukaryotic antibiotic-resistance genes, as well as clusters of DNA—called "multiple cloning site"—that matched the recognition sequence of a variety of restriction enzymes. The former enabled the addition of antibiotics during the expansion of transformed bacteria, which ensured that only successfully transformed bacteria survived and expanded, whereas the latter simply provided a variety of options for cloning.

* * * * *

Despite the tremendous power now available in the hands of researchers as a result of the development of cloning, this technique was still a long and tedious process since a gene of interest needed to be removed from a genome and inserted into a plasmid. This required knowledge of the DNA sequence surrounding the gene of interest, information that was not available at the time and difficult to obtain. Starting in 1983 and culminating in the publication of a paper in 1987, Kary Mullis at Cetus Corporation designed a technique that could relatively easily accomplish such a task and, in the process, revolutionized the biological sciences! Up to this point, the only available method for the *de novo* synthesis of DNA molecules was the step-wise assembly of nucleotides into double-stranded DNA, which was still complicated and *often fraught with peril due to the inevitable indelicacy of chemical material.*[35] Thus, Mullis and Faloona developed a method, which they called "polymerase chain reaction" (PCR), which

involves the reciprocal interaction of two oligonucleotides and the DNA polymerase extension of products whose synthesis they prime, when they are hybridized to different

strands of a DNA template in a relative orientation such that their extension products overlap. This method consists of repetitive cycles of denaturation, hybridization, and polymerase extension and seems not a little boring until the realization occurs that this procedure is catalyzing the doubling in each cycle of the amount of the fragment defined by the positions of the 5′ ends of the two primers on the template DNA, that this fragment is therefore increasing in concentration exponentially, and that the process can be continued for many cycles and is inherently very specific […]

Several embodiments have been devised that enables one not only to extract a specific sequence from a complex template and amplify it, but also to increase the inherent specificity of this process by using nested primer sets, or to append sequence information to one or both ends of the sequence as it is being amplified, or to construct a sequence entirely from synthetic fragments.[35]

As such, the authors incubated DNA, the four deoxynucleoside triphosphates, and two PCR primers—sequences of nucleotides complementary to sections of DNA *which corresponds to the extremities of the fragment to be amplified*—for 2 minutes at 95°C to separate the DNA strands, and then lowered the temperature to 30°C to allow the primers to bind, or anneal, to the source DNA (Figure 11.12). Then DNA polymerase purified from *E. coli* was added and extension—that is, the addition of nucleotides to the ends of the primers using the complimentary strand as a template—was allowed to proceed for 2 minutes. This cycle was repeated up to 27 times by returning the temperature to 95°C to once again separate double-stranded DNA. As such, amplification of the target sequence occurs in an exponential manner since twice as many target sites are present with each subsequent cycle. In addition, the authors noted that

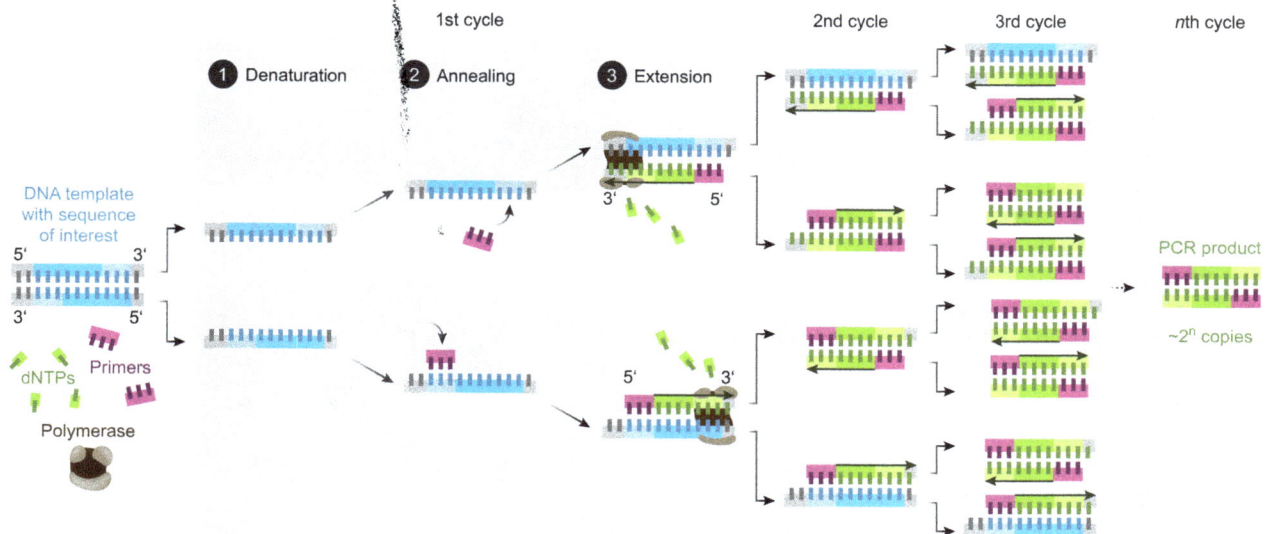

FIGURE 11.12 The PCR reaction. During the first cycle, primers bind to complimentary sequences on each DNA strand, which has been made single-stranded by heating, and DNA polymerase then fills in the gaps, creating two identical copies of the molecule. The DNA is then denatured again, bound by more primers, and once again the gaps are filled, resulting in four molecules. This is repeated for a number of cycles—usually about 30—thereby resulting in the creation of about 2^{30} (1 billion) molecules (assuming the reaction started with only one DNA molecule).[36]

the unique property of the targeted sequence for regenerating new primer sites with each cycle is intrinsic to the support of a chain reaction, and the improbability of it happening by chance accounts for the observed specificity of the overall amplification.[35]

In addition to amplification of a sequence of choice, Mullis et al. described a manner whereby a target sequence could be amplified with extra nucleotides at its extremities, thereby creating a PCR product which was different from its source DNA. This was accomplished by adding the desired extra nucleotides to the ends of the primers used in the PCR. Although these extra nucleotides would not bind to the target DNA in the first cycle, they would be part of the new DNA strand synthesized. Thus, from the second cycle on, there would be an increasing number of DNA molecules that contained those nucleotides so the primers would bind to those in their entirety. As such, the final PCR product would be DNA molecules homologous to the source target but with additional nucleotides flanking it. And since these nucleotides are specified by the primers, which in turn, are designed by the researcher, any sequence could be added; as such, Mullis and Faloona (and many other since) used this feature to *insert restriction site linkers onto amplified fragments of human genomic sequences to facilitate their cloning.* For example, the authors used their PCR method to rapidly and easily clone a gene from human DNA into a plasmid ready for bacterial transformation. They used the *olinucleotide primers and probes previously described for the diagnosis of sickle cell anemia,* and amplified the 110 base-pair fragment of β-hemoglobulin gene using PCR. In addition, they modified the primers such that they contained PstI and HindIII restriction site linkers on either ends to facilitate cloning. The PCR products and the target plasmid were digested using PstI and HindIII, ligated, and transformed into an appropriate *E. coli* plasmid host, which then started expressing the protein.

The above procedure was immediately recognized for its potential and researchers everywhere were now able to quickly and easily extract genes of unknown sequences, as long as a few nucleotides flanking the desired sequences were known. They could also clone genes in a manner that combined fragments—or even entire genes—from two or more different genes, such as to create new proteins that contained two or more protein sequences (or protein fragments). For example, we will learn in the next section about a protein—green fluorescent protein (GFP)—that fluoresces green when exposed to light. We will also see in Chapter 13 that this protein was cloned into a plasmid containing a gene of interest, such that this particular protein was now green when illuminated. This enabled Jennifer Lippincott-Schwartz to visualize her protein of choice as it moved throughout the cell. But first, let's talk about the discovery of GFP.

* * * * *

In 1955, Osamu Shimomura was a senior at Nagasaki University in Japan, where he worked as a teaching assistant.[37] His mentor had a particular interest in his education

and gave him the opportunity to work at Nagoya University as a visiting researcher. There, Shimomura first learned about tiny, egg-shaped, crustaceans called *cypridina*, which release two components of an enzymatic reaction, luciferin and luciferase, when they come upon predators. These components react together to produce a bright cloud of blue luminescence and allows the crustaceans to escape into the darkness of the surrounding water (Figure 11.13a).[37] In Nagoya, the budding scientist was asked to purify and crystallize these components, which he accomplished after about 10 months of

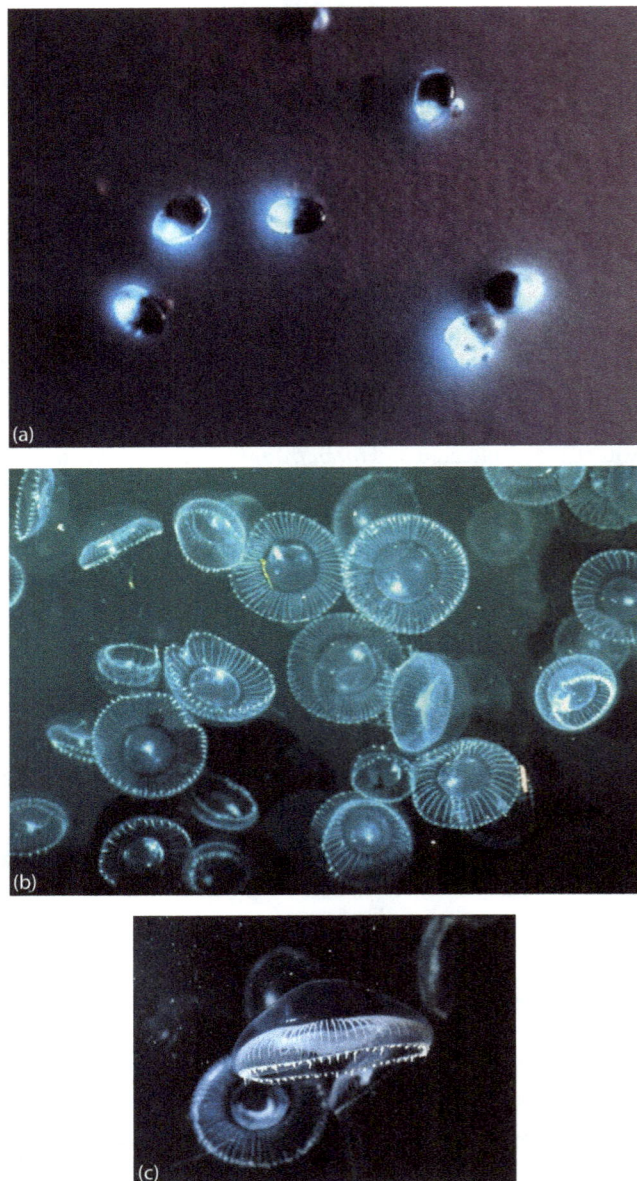

FIGURE 11.13 (a) Glowing cypridina from a luciferin–luciferase reaction. (b) and (c) *Aequorea victoria* jellyfish using aequorin and GFP to emit light. Jellyfish *aequorea* measure 7–10 cm in diameter, with a ring of light-emitting organs at the ends of their umbrella.[37] (Adapted with permission from John Wiley and Sons, Shimomura O. The discovery of aequorin and green fluorescent protein. J Microsc. Copyright © 2005 The Royal Microscopical Society.)

trial and error. In 1959, Shimomura received and accepted an invitation to join Frank Johnson's laboratory at Princeton University as a research associate. At the suggestion of his new mentor, he agreed to study the luminescence emitted by the jellyfish, *Aequorea* (Figure 11.13b and c).

The initial thought was that the luminescence was the product of a luciferin–luciferase reaction, but all efforts to purify these components failed. Nonetheless, at the dismay of his supervisor, Shimomura set-out to *isolate the luminescent substance regardless of what it might be*.[37] After trying everything he could think of, Shimomura was stumped and decided to take *several days soul searching, trying to imagine the reaction that might occur in luminescing jellyfish and searching for a way to extract the luminescent principle*. Then, Shimomura realized that an enzyme is likely responsible for the reaction, even if that enzyme is not luciferase, therefore, *the activity of this enzyme or protein can probably be altered by a pH change*.[37] Thus, he searched for pH conditions that would reversibly inactivate the enzyme and found that the "squeezate"—the turbid liquid that resulted from squeezing the rings of 20 or 30 jellyfish through gauze—*was luminous at pH 6 and pH 5, but not at pH 4*.[37]

Therefore, Shimomura lysed cells at pH 4, filtered the mixture to get rid of cell debris, and neutralized the solution to return it to a pH of 5–6; to his surprise, the solution instantly regained its luminescence, *indicating that* [he] *had succeeded in extracting the bioluminescence substance!*[37] Even more surprisingly, *the inside of the sink lit up with a bright blue flash* when he discarded his enzyme solution in the sink at the end of the experiment, which he reasoned had occurred due to the seawater from the *overflow from an aquarium*. Thus, he *mixed a small amount of seawater with the extract: it gave explosively strong luminescence*, which he deduced was due to the action of Ca^{2+}, which is readily available in seawater.[37] From there, Shimomura and his supervisor extracted the protein in question from 10,000 specimens purchased from local kids—*paying a penny per jellyfish*—which yielded about 1 mg of highly purified protein with a molecular weight of about 20 KDa and emitted blue light upon addition of Ca^{2+}; they named the protein "aequorin".[37,38] During the purification of aequorin and in a short footnote of their account in 1962, the authors noted that *a protein giving solutions that look slightly greenish in sunlight through*[sic] *only yellowish under tungsten lights, and exhibiting a very bright, greenish fluorescence in the ultraviolet of a Mineralite, has also been isolated from squeezates*.[38] However, virtually nothing was known about this protein—which was later named "Green Fluorescent Protein" (GFP)—in the more than ten years after its purification other than its fluorescence spectrum.[39]

At this point, I should take a few moments to explain the phenomenon of fluorescence: chromophores are molecules that absorb light and give a substance its color, that is, the colors that are *not* absorbed bounce off of the molecule and are captured by our eyes. Fluorophores are special types of chromophores which are "excited" by a specific range of light wavelengths and emit light at a longer wavelength. Furthermore, if two fluorophores are in sufficiently close proximity and the emission of one coincides with the excitation of the other, only the emission of second fluorophore will be seen since the emission of the first will be "captured" by the second; this is called Förster resonance energy transfer (FRET). Morin and Hastings suggested that FRET might occur between aequorin and GFP in jellyfish since aequorin was found to have an emission maximum at 470 nm—which is within GFP's excitation spectrum—and although both aequorin and GFP were present in Aequorea, these jellyfish only emitted green light.[37] As such, in 1974, Morise et al. purified and characterized GFP and confirmed that, indeed, the luminescence seen in Aequorea was the result of FRET between aequorin and GFP.[39] Soon thereafter, the chemical structure of GFP was determined (Figure 11.14) and Prasher et al. determined its nucleotide sequence and cloned the gene in 1992.[40–42] It was also determined that the fluorescence of GFP was due to amino acids 65–67 (Ser-Tyr-Gly), buried in the center of a β-barrel structure, which become fluorescent following a number of reactions with adjoining amino acids.[43]

Importantly, in 1994, two groups succeeded in transforming *E. coli* with a plasmid containing the GFP gene and showed that its expression resulted in bacteria emitting green fluorescence when stimulated by blue light.[45,46] Since GFP could be cloned to precede or follow any gene of interest—and therefore it would be translated along with it and attached to the protein product in question—the authors suggested that the use of GFP to monitor the gene expression activity might be superior to other techniques used at the time. Tagging a protein with luciferase, for example, required that luciferin and ATP be added at the time of imaging for fluorescence to be detected. In contrast, the use of GFP provided a clear advantage in that the protein could fluoresce at any time if exposed to blue light and therefore

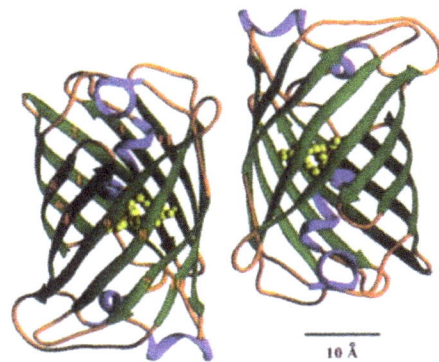

FIGURE 11.14 The structure of green fluorescent protein (GFP). GFP is composed of an 11-stranded β-barrel with an α-helix running through it; also seen are residues 65–67 at the center (light green) forming the chromophore.[44] (Reprinted by permission from Springer Nature Customer Service Centre GmbH: Springer Nature, Nature Biotechnology 14(10), Yang F, Moss LG, Phillips GN. The molecular structure of green fluorescent protein, Copyright © 1996.)

might *be used as a vital marker so that cell growth [...] and movement can be followed in situ, especially in animals that are essentially transparent like C. elegans and zebra fish.*[45]

In the 1990s, Roger Tsien wanted to use FRET to determine if two fluorescent proteins interacted, such that two colors would be visualized if the proteins did not interact, whereas only the second fluorochrome would be visualized if they did; therefore, he *obviously needed a second color.*[47] As such, he noted that the emission spectrum of *Aequorea* differed slightly from that of another species of jellyfish, *Renilla*, in that in the former it was low and broad, and in the latter, high and narrow, conditions that were better suited for his intended FRET.[48] Thus, Heim, Prasher, and Tsien started randomly mutating the GFP DNA hoping to improve its fluorescence spectrum, and about 6000 bacterial colonies were screened for desired properties. Surprisingly, one of these mutants *was excitable by UV light and fluoresced bright blue in contrast to the green of wild-type protein [...] the excitation and emission maxima* [having been] *shifted by 14 and 60nm, respectively.*[48] DNA sequencing revealed that although a number of mutations were present in this clone, only a Tyr to His mutation at position 66 (the middle residue of the chromophore triplet) was essential for the new property. Given the success of these results, the group continued to investigate changes in fluorescence as a result of mutagenesis of GFP using PCR under conditions where the error rates were high, and obtained additional *useful variants*, one of which was enhanced GFP (EGFP), which had brighter green fluorescence.[49,50]

Finally, in 1999, a group from Russia reasoned that *the enormous variety of colors present in coral reefs [...] may be due to GFP-like fluorescent proteins*, so they *attempted to isolate homologous cDNAs*[51]. As such, the authors used five *brightly fluorescent Anthozoa species from the Indo-Pacific area* and analyzed their DNA, looking for sequences similar to that of GFP. In doing so, they identified six proteins with 26–30% homology to GFP, five of which were of different hues of green and one of which was of a red color.[51] To confirm the usefulness of these new proteins, the group transfected human cells and Xenopus embryos and found that, indeed, these proteins were efficiently expressed with no toxicity (Figure 11.15). Using the newly isolated "red" gene, Tsien and his team started mutating it as they had done the GFP gene and in doing so, they engineered a wide array of fluorescent color, many of which are still widely used for a number of different purposes, including FRET imaging, as originally intended by Tsien.

* * * * *

Strangely enough, the next innovation partly has its roots in the development of the inkjet printer in the 1960s. However, it first started with a need for a faster way to count blood cells for patient sample analysis. Until the 1930s, this had been done by serial dilution of a sample until the number of cells to be counted was such that it was manually manageable and then multiplying by the dilution factor to get a final count. However, Andrew Moldavan in Montreal thought this

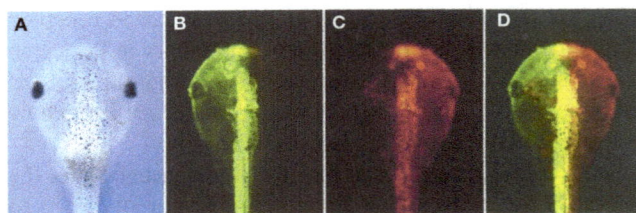

FIGURE 11.15 Visualization of fluorochromes of different colors. mRNA coding for fluorescent proteins were injected into the left or right side of Xenopus embryos and imaged using DIC (A) or fluorescence microscopy (B and C). D is a composite of B and C.[51] (Reprinted by permission from Springer Nature Customer Service Centre GmbH: Springer Nature, Nat Biotechnol., Matz M v, Fradkov AF, Labas YA, et al. Fluorescent proteins from non-bioluminescent anthozoa species, Copyright © 1999.)

could be done more efficiently by adapting *photoelectric methods to the direct counting of microscopical cells in suspension in water.*[52] As such, he suggested that red blood cells could be suspended in water and *forced under pressure to circulate through [a] capillary tube* that was *placed under the high magnification of a field microscope.* The microscope, being fitted with a *photoelectric apparatus*, then counted the number of cells via changes in electric current as each cell passed in front of the lens. Moldavan mentioned a number of issues that needed to be addressed for the realization of such an apparatus, but did not report any results or, in fact, whether or not he actually built such a devise. However, from the first paragraphs of John P. Crosland-Taylor's 1953 paper, it appears that a number of scientists had indeed failed at making Moldavan's vision a reality:

> Attempts to count small particles suspended in fluid flowing through a tube have not hitherto been very successful. With particles such as blood cells the experimenter must choose between a wide tube which allows particles to pass two or more abreast across a particular section, or a narrow tube which makes microscopical observations of the contents of the tube difficult due to the different refractive indices of the tube and the suspending fluid. In addition, narrow tubes tend to block easily.
>
> These difficulties can be overcome by slowly injecting a suspension of the particles into a faster stream of fluid flowing in the same direction. Provided there is no turbulence, the wide column of particles will then be accelerated to form a narrow column surrounded by fluid of the same refractive index[...][53]

Thus, the author built an apparatus in which a cell suspension was discharged through a needle centered between two water streams, which helped to create a single-cell suspension (Figure 11.16).[53] As such, as *the stream of cells emerges from the tapered tip of the needle it is narrowed by the faster peripheral stream (3), created by water entering through tubes (2) and (2a), and as the vortex (5) is approached the stream of the cells narrows further as the velocity is increased.* The author described the cells he saw in the microscope as *a thin luminous stream of cells*

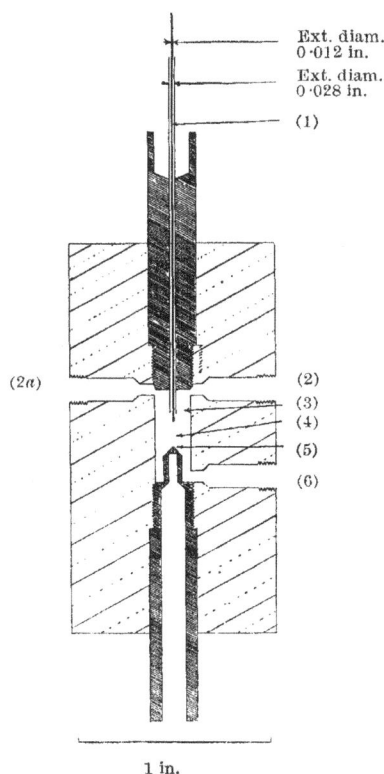

Ext. diam.
0·012 in.

Ext. diam.
0·028 in.

(1)

(2a)

(2)
(3)
(4)
(5)
(6)

1 in.

FIGURE 11.16 Crossland-Taylor's apparatus for counting cells. Cells entered the apparatus through a needle (1) and the stream was kept straight by running water at a higher velocity on either side of the stream (2 and 2a). This also served to create a single-cell suspension which could be observed via a microscope as they passed through (4). Cells then exited the apparatus through (6).[53] (Reprinted by permission from Springer Nature Customer Service Centre GmbH: Springer Nature, Nature, Crosland-Taylor PJ. A device for counting small particles suspended in a fluid through a tube, Copyright © 1953.)

less than 10μ wide, wavering slightly [...] and stressed the importance of using as narrow a needle as possible:

The needle should be wide enough not to block easily, but provided it does not block, should be as narrow as possible to enable the velocity of cells passing through the needle to be substantially greater than the sedimentation effect of gravity. A further reason for the narrow needle is that if a constant flow-rate is used as a measure of the volume of cell suspension to be counted, the fluctuations of flow in a wide-bore needle will be greater for the same minute changes of pressure.[53]

In 1956, Wallace H. Coulter reported the development of an apparatus that could both count cells and provide *a means of obtaining complete cell size distribution data of an accuracy not heretofore possible and in such a short interval of time that the procedure could be used routinely should the need become established.*[54] Coulter's device took advantage of the fact that cells are generally nonconductive entities; therefore, the intensity of an electric current passing through a given solution changes in a manner that is proportional to the size of a cell present in that solution. Thus, diluted samples were drawn into the apparatus from a beaker by means of a siphon and cell counting was activated manually (see link Ref. 54 for images of the apparatus). The threshold—the minimum strength a signal needed to have to be considered a real signal, that is, a cell—could be manually adjusted to correspond to the cell size of interest to exclude debris and electronic noise. This new device allowed blood counts to be obtained much more precisely and in under 5 minutes. Importantly, Coulter found that it could also help determine the presence of ascites (cancer cells) in sample, since these cells were found to be *5 to 8 times the diameter of the red blood cells and produce very high amplitude pulses in relation to the red cell pulses.*[54] Finally, the author noted that the device was already being used in a number of laboratories *and a considerable saving in time as well as an increase in accuracy has been demonstrated in routine use by average medical technicians.*

In 1965, Kamentsky, Melamed, and Derman used the light-absorbing characteristics of cells to determine the heterogeneity of cell populations. As such, the authors developed an apparatus whereby cells were flowed in a pressurized suspension fluid at a rate of 500 cells/sec through a beam of light generated by a mercury lamp. The absorbance of DNA at a wavelength of 2537 Å and the light scattered by each cell at 4100 Å were detected by a microscope and read by an oscilloscope, giving an indication of DNA content and cell size.[55] The authors optimized the proper fixation solutions to be used to standardize the procedure and determined that a number of different cell lines gave quite different scatter patterns. This device reportedly allowed them to detect a *few unusual cells in populations of 100,000 or more cells.*[56]

Two years later, Kamenstky and Melamed extended this work by adapting the device such that cells could be *separated according to functions of multiple simultaneous optical measurements on each cell.*[56] To that end, the authors used four different wavelengths, both in the ultraviolet and the visible range, to characterize stained and unstained cells and displayed these characteristics as a plot of any two measurements. Cells that met the specified criteria as they flowed passed the observation point cells triggered a signal which activated a motor attached to a syringe, *causing the plunger [...] to draw 0.03μl of fluid* which flowed into a channel until the end of the experiment, after which the cells were *flushed out of the side channel and trapped on* a filter, removed, stained, and visualized as appropriate; cells that did not have the selected properties flowed through to a waste container. Although very crude, the device allowed significant purification—from 1:10 000 to 1:5—of selected cells for further analysis.

At the same time, Mack Fulwyler was investigating why homogenous blood cells displayed a bimodal distribution when using a Coulter counter.[57] Thus, he modified the Coulter counter with inkjet technology (then called an ink writing oscillograph) developed by Richard Sweet at Stanford University, which used vibrations to generate ink drops that were subsequently electrically charged such that they were deposited at specific areas on paper. As such, Fulwyler used the Coulter counter to identify cell characteristics and the stream that contained the cells was then vibrated at 72,000 Hertz, thereby creating 72,000 droplets per second.[58] The droplets which contained cells of desired characteristics were then electrically charged such that they were deflected into a collection compartment as they passed through deflection plates; cells not desired for separation were not charged—and therefore, not deflected—and continued in a straight path into a waste container. The charge imparted upon the droplets was electronically determined based on the location of the collecting vessel, that is, the farther the collection vessel for a particular cell, the greater the charge needed to be, allowing cells with various characteristics to be sorted at once.

As such, Fulwyler flowed a mixture of mouse and human red blood cells, which have volumes of about 50 and 100 μm^3, respectively, in his new apparatus and specified cells of 80 μm^3 or greater to be separated. After the separation, the author ran the sorted cells back through the device to determine the degree of purity, and found that cells were, to a large extent, human cells, 63% of which were viable after separation, indicating that they could be cultured if desired.[58] Fulwyler went on to make the prescient suggestion that:

> in principle the system is capable of separating minute particles [...] according to other electronically measurable characteristics such as optical density, reflectivity, or fluorescence. It may be possible also to measure simultaneously two (or more) characteristics of a cell and to make separation dependent on the ratio of such characteristics.[58]

In 1969, the first apparatus with capabilities akin to what we currently use as a cell sorter was developed by Wallace Coulter's brother, Joseph—with whom he started the Coulter Corporation soon after the publication of the Coulter counter paper—at the Los Alamos Scientific Laboratory in California. As such, Van Dilla et al. adapted Crosland-Taylor's apparatus to function with fluorescent dyes, which had become widely used for the labeling of biological samples. The instrument used a 1-watt argon ion light source at 488 nm to excite fluorochromes attached to cells emerging from the Crosland-Taylor flow chamber in a series of droplets.[59] Light emitted from the cells was captured, stored in the memory of a pulse-height analyzer (the amplitude of which was proportional to cell fluorescence), and *a frequency distribution histogram of fluorescent light emission per cell [...] is displayed and read out for further analysis*. As such, the authors stained the DNA of asynchronous Chinese hamster ovary (CHO) cells and ran their samples in their apparatus. Analysis showed a bimodal distribution of DNA, with a high peak at the G1 phase, a valley for S phase, and a smaller peak at twice the fluorescence of the G1 peak for G2/M phase.[59] This was expected since DNA content in this latter phase is twice as much as in G1 phase (since DNA was duplicated in S phase in preparation

for cell division), whereas it progressively increases in S phase as DNA is being duplicated. The authors noted that:

> [t]he noteworthy differences between *[the Coulter counter and our device]* are the improvement in statistical precision evident in our result on about 50,000 cells, as compared with the microspectrophotometer result in 594 cells, and the reduction in measurement time (1 minute for our result compared with many hours for the microspectrophotometer result) [...]
>
> We anticipate that extension of this method is possible and of potential value. Other possibilities include (i) staining with other fluorochromes such as acridine orange for DNA and RNA and brilliant sulfaflavine for protein; (ii) application to fluorescent antibody studies; (iii) application to cancer cell identification; (iv) optical spectra analysis by use of more than one photomultiplier and appropriate filters; (v) multi-parameter analysis by combination with small-angle light scatter or Coulter orifice; (vi) and cell sorting on the basis of these optical sensors.[59]

The same year, Leonard Herzenberg's group at Stanford University developed the fluorescence-activated cell sorter (FACS) closer to what we are familiar with today. Herzenberg combined Fulwyler's cell separation technique—which was in turn based on Sweet's inkjet technology—Crosland-Taylor's coaxial flow system to *confine the cells to the stream axis*, and cell fluorescence technology as used by Kamentsky and Van Dilla et al., and designed a device that could sort cells based on different fluorescently labeled protein markers on the cells.[60] Improvements made to this device were published three years later, in 1972, in which a *cell suspension is injected through the inner nozzle [...] of a coaxial nozzle assembly* under pressure, and an argon laser tuned to a wavelength of 488 nm was directed at the liquid stream and the emitted fluorescence, if present, was directed to a detector through a yellow filter to reduce noise (Figure 11.17a).[60]

The fluid stream was separated into droplets by vibrating the nozzle assembly 40,000 times per second—creating 40,000 droplets per second—and cells that fell within the selected fluorescence parameters were charged and deflected into collection vessels. The duration of the charge, 75 µsec, was long enough to charge three droplets at a time, so proper dilution of the sample was essential for adequate purity since either of the two other droplets could potentially contain unwanted cells. In addition, a precise lag time between fluorescence detection and droplet charging, which was determined to be 150 µsec, was required since these occurred at two separate locations in the device. As such, the authors found that a purity of 85% for sorted samples was achievable in their system. However, they noted that an improved system was under development that also uses a cell's *light scattering properties, and which will inhibit drop charging if an unwanted cell would otherwise be deflected along with a fluorescent cell*.[60]

Bonner et al. tested their device by separating fluorescent from nonfluorescent cells, but reported that the *most*

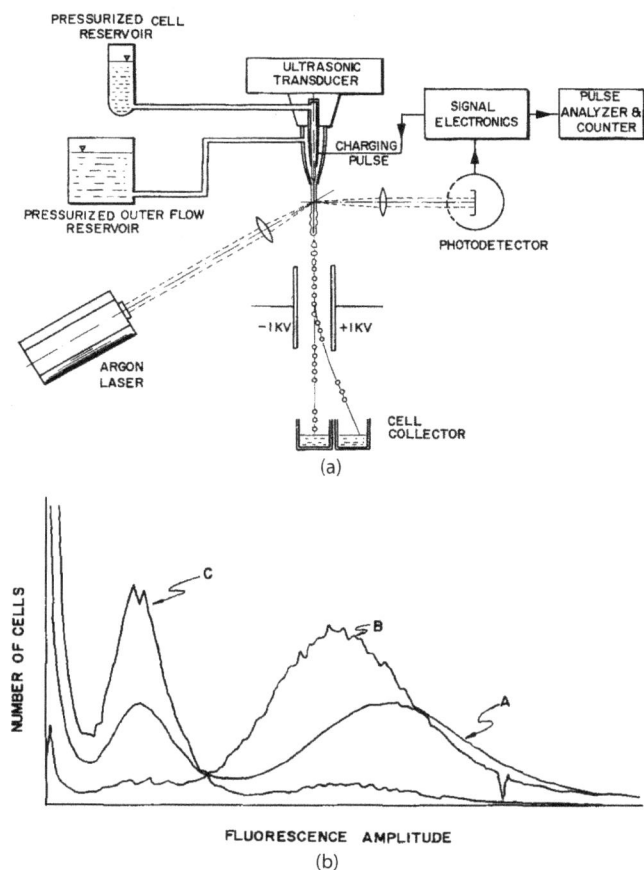

(a)

(b)

FIGURE 11.17 Herzenberg's fluorescence-activated cell sorter (FACS). (a) Diagram of the instrument. Cell under pressure flowed passed an interrogation point as previously described, and cells were sorted based on fluorescence characteristics. (b) Histogram of cells before sorting (line A), of the sorted cells (line B), and of the cells which were discarded (line C).[60] (Reprinted from Bonner WA, Hulett HR, Sweet RG, Herzenberg LA. Fluorescence-activated cell sorting. Review of Scientific Instruments. 1972; 43(3):404–409, with the permission of AIP Publishing.)

interesting [separation] *uses the indirect immunofluores-cent sandwich technique*, in which cell surface proteins were detected with fluorescent-tagged antibodies. In a similar experiment, the authors attached a fluorescent molecule to concanavalin A, a protein that binds carbohydrates on the cell surface, and ran a sample of concanavalin A-labeled human leukocytes in their sorter.[60] The authors found that the sample displayed a bimodal distribution for this marker and set the separation threshold to *the minimum between the two peaks*. As such, they were able to separate human leukocytes based on the degree of concanavalin A labeling, the separation of which was verified by running the sorted samples back into the device (Figure 11.17b). Microscopic assessment determined that *the deflected cells were over 90% granulocytes, the undeflected cells 90% lymphocytes, while the original suspension was about 75% granulocytes and 25% lymphocytes*.[60]

The next year, Coulter's group improved upon the design of their device which was now *capable of multiparameter*

analysis and subsequent sorting of biological cells according to cell volume, single or two-color fluorescence, and light scatter.[61] They described their system as a *multisensor cell analysis and sorting instrument that employs a newly developed flow chamber incorporating Coulter volume and optical sensors* (Figure 11.18).[61] In addition to detecting fluorescence, the authors used the argon laser to assess other cellular characteristics: they could determine cell size by using a laser that was smaller than the diameter of the cell since the length of time that the beam was interrupted was directly proportional to the size of the cell. The instrument used this parameter in combination with the Coulter volume sensor to provide increased resolution and allowed better discrimination of weak fluorescent signals over *background noise created by small fluorescing particles (debris)*. Furthermore, light hitting the cell was deflected by its internal structures and scattered at all angles around the cell; therefore, a detector was placed 90° to the incident light to capture the scattered light and used it as a parameter to characterize cells, since certain types of cells are more granular than others. Finally, their apparatus was designed such that droplets could be given either a positive or negative charge—deflecting cells one way or the other about the stream—which increased the number of cell characteristics that could be sorted at the same time.

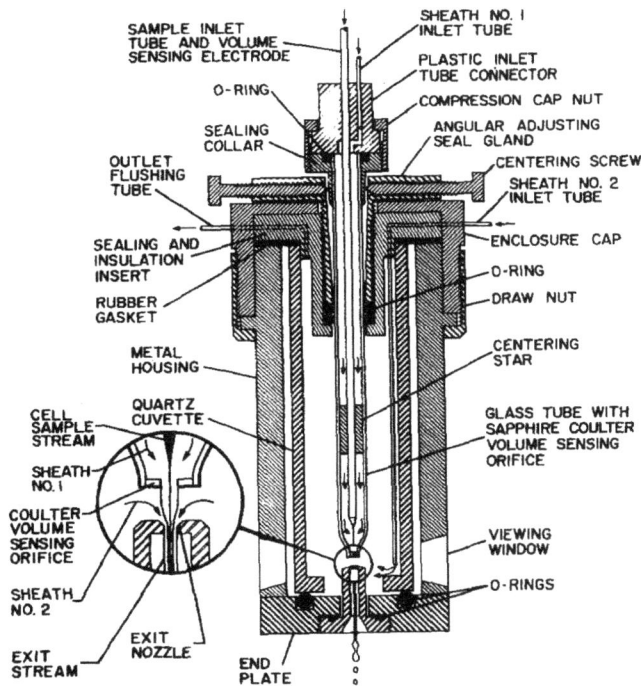

FIGURE 11.18 Coulter's improved, dual laser sorter. This apparatus combined their own cell-counter technology with Crossland-Taylors' coaxial flow, Fulwyler's droplet formation, and fluorescence techniques.[61] (Reprinted from Steinkamp JA, Fulwyler MJ, Coulter JR, Hiebert RD, Horney JL, Mullaney PF. A new multiparameter separator for microscopic particles and biological cells. Review of Scientific Instruments. 1973;44(9): 1301–1310. With the permission of AIP Publishing.)

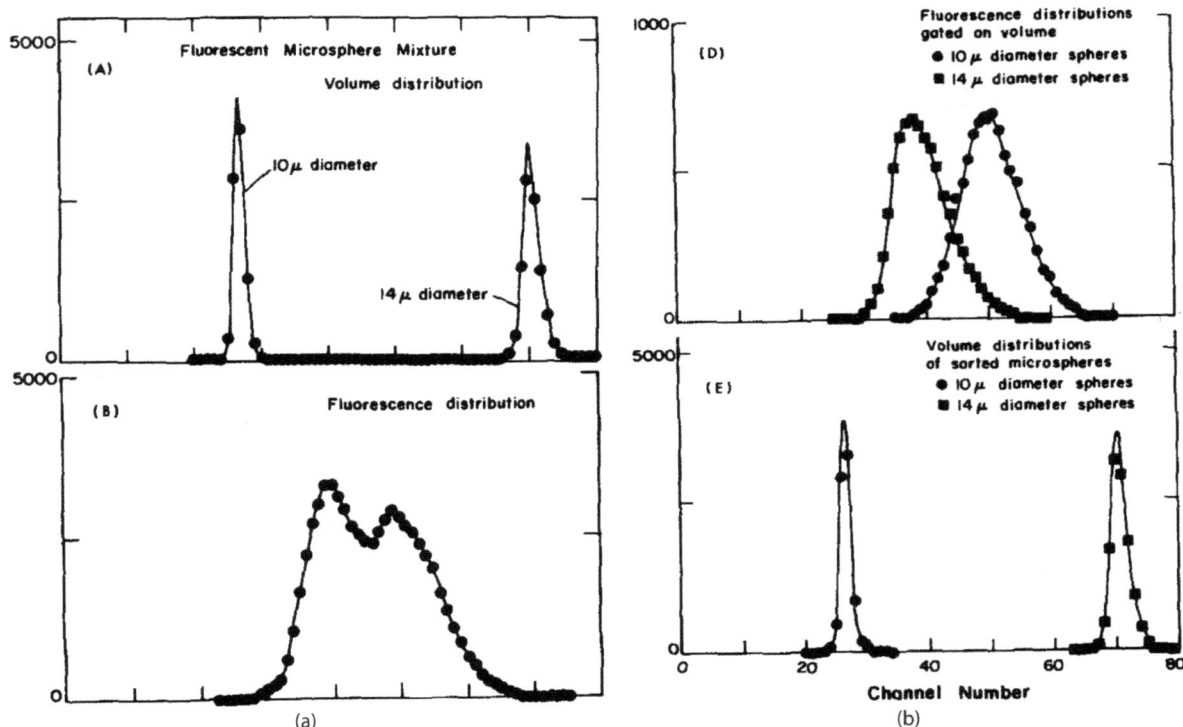

FIGURE 11.19 Sorting results using Coulter's new instrument. (a) Mixed synthetic beads of two volumes displayed distinct volume peaks (A) but overlapping fluorescence peaks (B). (b) Following sorting using volume as a parameter, the sorted populations were run through the cytometer once again to evaluate the distribution of their fluorescence (D) and volume (E).[61] (Adapted from Steinkamp JA, Fulwyler MJ, Coulter JR, Hiebert RD, Horney JL, Mullaney PF. A new multiparameter separator for microscopic particles and biological cells. Review of Scientific Instruments. 1973;44(9):1301–1310, with the permission of AIP Publishing.)

Steinkamp et al. tested the sorting efficiency of their device by first determining the distribution of 10 μm and 14 μm spheres using a Coulter volume sensor and emitted fluorescence, the former of which displayed two distinct peaks and the latter of which displayed overlapping peaks (Figure 11.19a).[61] They then mixed the samples and sorted the mixture into two populations based on volume, and then ran the sorted samples back into the instrument to confirm that the samples were now two completely separate populations (Figure 11.19b). Similar to Bonner et al.'s FACS instrument, the authors determined that a sorting efficiency of around 85% could be reached. Although these primitive instruments only had one laser and two fluorescence detectors, flow cytometers, as these devices would come to be known, are now fitted with up to 10 laser sources and can detect up to 30 separate fluorescent signals.

* * * * *

REFERENCES

1. Tiselius A. Arne Tiselius: Nobel Lecture: Electrophoresis and adsorption analysis as aids in investigations of large molecular weight substances and their breakdown products. NobelPrize.org. Nobel Media AB   Copyright © The Nobel Foundation 1948. https://www.nobelprize.org/prizes/chemistry/1948/tiselius/lecture/

2. Faraday M VI. Experimental researches in electricity. Sixth series. *Philosophical Transactions of the Royal Society of London.* 1834;124:77–122. doi:10.1098/rstl.1834.0008

3. Dunbar BS. Basic theories and principles of electrophoresis. In: *Two-Dimensional Electrophoresis and Immunological Techniques.* Springer US; 1987:1–23. doi:10.1007/978-1-4613-1957-3_1

4. Tiselius A. A new apparatus for electrophoretic analysis of colloidal mixtures. *Transactions of the Faraday Society.* 1937;33(0):524. doi:10.1039/tf9373300524

5. Smithies O. Zone electrophoresis in starch gels: Group variations in the serum proteins of normal human adults. *Biochemical Journal.* 1955;61(4):629–641. doi:10.1042/bj0610629

6. Davis BJ. Disc electrophoresis. II. Method and application to human serum proteins. *Annals of New York Academy of Sciences.* 1964;121:404–427. doi:10.1111/j.1749-6632.1964.tb14213.x

7. Ornstein L. Disc electrophoresis. I. Background and theory. *Annals of New York Academy of Sciences.* 1964;121:321–349. doi:10.1111/j.1749-6632.1964.tb14207.x

8. Shapiro AL, Viñuela E, Maizel JV. Molecular weight estimation of polypeptide chains by electrophoresis in SDS-polyacrylamide gels. *Biochemical and Biophysical Research and Communications.* 1967;28(5):815–820. doi:10.1016/0006-291X(67)90391-9

9. Weber K, Osborn M. The reliability of molecular weight determinations by dodecyl sulfate-polyacrylamide gel electrophoresis. *Journal of Biological Chemistry.* 1969;244(16):4406–4412.

10. Pitt-Rivers R, Impiombato FSA. The binding of sodium dodecyl sulphate to various proteins. *Biochemical Journal.* 1968;109(5):825. doi:10.1042/bj1090825

11. Tanford C, Reynolds J. *Nature's Robots: A History of Proteins.* Oxford University Press; 2001.

12. File: Protein-SDS interaction.png—Wikimedia Commons. Accessed November 1, 2022. https://commons.wikimedia. org/wiki/File:Protein-SDS_interaction.png

13. Laemmli UK. Cleavage of structural proteins during the assembly of the head of bacteriophage T4. *Nature.* 1970; 227(5259):680–685. doi:10.1038/227680a0

14. Maizel JV. SDS polyacrylamide gel electrophoresis. *Trends in Biochemical Sciences.* 2000;25(12):590–592. doi:10.1016/ S0968-0004(00)01693-5

15. Southern EM. Detection of specific sequences among DNA fragments separated by gel electrophoresis. *Journal of Molecular Biology.* 1975;98(3):503–517. doi:10.1016/ S0022-2836(75)80083-0

16. Towbin H, Staehelin T, Gordon J. Electrophoretic transfer of proteins from polyacrylamide gels to nitrocellulose sheets: Procedure and some applications. *Proceedings of National Academy of Sciences U S A.* 1979;76(9):4350–4354. doi:10.1073/pnas.76.9.4350

17. Renart J, Reiser J, Stark GR. Transfer of proteins from gels to diazobenzyloxymethyl-paper and detection with antisera: A method for studying antibody specificity and antigen structure. *Proceedings of National Academy of Sciences U S A.* 1979;76(7):3116–3120. doi:10.1073/pnas.76.7.3116

18. Burnette WN. "Western blotting": Electrophoretic transfer of proteins from sodium dodecyl sulfate-polyacrylamide gels to unmodified nitrocellulose and radiographic detection with antibody and radioiodinated protein A. *Analytical Biochemistry.* 1981;112(2):195–203. doi:10.1016/0003-2697(81)90281-5

19. Meselson M, Weigle JJ. Chromosome breakage accompanying genetic recombination in bacteriophage. *Proceedings of National Academy of Sciences U S A.* 1961;47(6): 857–868. doi:10.1073/pnas.47.6.857

20. Gellert M. Formation of covalent circles of lambda DNA by *E. coli* extracts. *Proceedings of National Academy of Sciences U S A.* 1967;57(1):148. doi:10.1073/pnas.57.1.148

21. Weiss B, Richardson CC. Enzymatic breakage and joining of deoxyribonucleic acid, i. Repair of single-strand breaks in DNA by an enzyme system from *Escherichia coli* infected with T4 bacteriophage. *Proceedings of National Academy of Sciences U S A.* 1967;57(4):1021–1028. doi:10.1073/ pnas.57.4.1021

22. Olivera BM, Lehman IR. Linkage of polynucleotides through phosphodiester bonds by an enzyme from *Escherichia coli. Proceedings of National Academy of Sciences U S A.* 1967;57(5):1426–1433. doi:10.1073/pnas.57.5.1426

23. Gefter ML, Becker A, Hurwitz J. The enzymatic repair of DNA. I. Formation of circular lambda-DNA. *Proceedings of National Academy of Sciences U S A.* 1967;58(1): 240–247. doi:10.1073/pnas.58.1.240

24. Kaiser A, Hogness DS. The transformation of *Escherichia coli* with deoxyribonucleic acid isolated from bacteriophage λdg. *Journal of Molecular Biology.* 1960;2(6):392–IN6. doi:10.1016/S0022-2836(60)80050-2

25. Mandel M, Higa A. Calcium-dependent bacteriophage DNA infection. *Journal of Molecular Biology.* 1970;53(1): 159–162. doi:10.1016/0022-2836(70)90051-3

26. Cohen SN, Chang ACY, Hsu L. Nonchromosomal antibiotic resistance in bacteria: Genetic transformation of *Escherichia coli* by R-factor DNA. *Proceedings of National Academy of Sciences U S A.* 1972;69(8):2110. doi:10.1073/ pnas.69.8.2110

27. Arber W. Host-controlled modification of bacteriophage. *Annual Reviews of Microbiology.* 1965;19(1):365–378. doi:10. 1146/annurev.mi.19.100165.002053

28. Arber W, Dussoix D. Host specificity of DNA produced by *Escherichia coli.* I. Host controlled modification of bacteriophage lambda. *Journal of Molecular Biology.* 1962;5: 18–36. doi: 10.1016/S0022-2836(62)80058-8

29. Meselson M, Yuan R. DNA restriction enzyme from *E. coli. Nature.* 1968;217(5134):1110–1114. doi:10.1038/2171110a0

30. Smith HO, Wilcox KW. A restriction enzyme from *Hemophilus influenzae.* I. Purification and general properties. *Journal of Molecular Biology.* 1970;51(2):379–391. doi:10.1016/0022-2836(70)90149-X

31. Kelly TJ, Smith HO. A restriction enzyme from *Hemophilus influenzae:* II. Base sequence of the recognition site. *Journal of Molecular Biology.* 1970;51(2):393–409. doi:10.1016/0022-2836(70)90150-6

32. Smith HO, Nathans D. Letter: A suggested nomenclature for bacterial host modification and restriction systems and their enzymes. *Journal of Molecular Biology.* 1973;81(3): 419–423. doi:10.1016/0022-2836(73)90152-6

33. Jackson DA, Symons RH, Berg P. Biochemical method for inserting new genetic information into DNA of simian virus 40: Circular SV40 DNA molecules containing lambda phage genes and the galactose operon of *Escherichia coli. Proceedings of National Academy of Sciences U S A.* 1972;69(10):2904–2909. doi:10.1073/pnas.69.10.2904

34. Cohen SN, Chang AC, Boyer HW, Helling RB. Construction of biologically functional bacterial plasmids in vitro. *Proceedings of National Academy of Sciences U S A.* 1973; 70(11):3240–3244. doi:10.1073/pnas.70.11.3240

35. Mullis KB, Faloona FA. Specific synthesis of DNA in vitro via a polymerase-catalyzed chain reaction. *Methods of Enzymology.* 1987;155:335–350. doi: 10.1016/0076-6879 (87)55023-6

36. File: Polymerase chain reaction-en.svg—Wikimedia Commons. Accessed November 1, 2022. https://commons. wikimedia.org/wiki/File:Polymerase_chain_reaction-en.svg

37. Shimomura O. The discovery of aequorin and green fluorescent protein. *Journal of Microscopy.* 2005;217(1):3–15. doi:10.1111/j.0022-2720.2005.01441.x

38. Shimomura O, Johnson FH, Saiga Y. Extraction, purification and properties of aequorin, a bioluminescent protein from the luminous *Hydromedusan aequorea. Journal of Cellular and Comparative Physiology.* 1962;59(3):223–239. doi:10.1002/jcp.1030590302

39. Morise H, Shimomura O, Johnson FH, Winant J. Intermolecular energy transfer in the bioluminescent system of Aequorea. *Biochemistry.* 1974;13(12):2656–2662. doi:10. 1021/bi00709a028

40. Prasher DC, Eckenrode VK, Ward WW, Prendergast FG, Cormier MJ. Primary structure of the *Aequorea victoria* green-fluorescent protein. *Gene.* 1992;111(2):229–233. doi: 10.1016/0378-1119(92)90691-H

41. Cody CW, Prasher DC, Westler WM, Prendergast FG, Ward WW. Chemical structure of the hexapeptide chromophore of the Aequorea green-fluorescent protein. *Biochemistry.* 1993;32(5):1212–1218. doi:10.1021/bi00056a003

42. Shimomura O. Structure of the chromophore of Aequorea green fluorescent protein. *FEBS Letters.* 1979;104(2): 220–222. doi:10.1016/0014-5793(79)80818-2

43. Ormö M, Cubitt AB, Kallio K, Gross LA, Tsien RY, Remington SJ. Crystal structure of the *Aequorea victoria*

green fluorescent protein. *Science (1979)*. 1996;273(5280): 1392–1395. doi:10.1126/science.273.5280.1392

44. Yang F, Moss LG, Phillips GN. The molecular structure of green fluorescent protein. *Nature Biotechnology 1996 14:10*. 1996;14(10):1246–1251. doi:10.1038/nbt1096-1246

45. Chalfie M, Tu Y, Euskirchen G, Ward WW, Prasher DC. Green fluorescent protein as a marker for gene expression. *Science (1979)*. 1994;263(5148):802. doi:10.1126/science.8303295

46. Inouye S, Tsuji FI. Aequorea green fluorescent protein. Expression of the gene and fluorescence characteristics of the recombinant protein. *FEBS Letters*. 1994;341(2-3): 277–280. doi:10.1016/0014-5793(94)80472-9

47. Tsien RY, Roger Y. Tsien - Nobel Lecture: Constructing and Exploiting the Fluorescent Protein Paintbox. NobelPrize. Org. Nobel Media AB © The Nobel Foundation 2008. https://www.nobelprize.org/prizes/chemistry/2008/tsien/lecture/

48. Heim R, Prasher DC, Tsien RY. Wavelength mutations and posttranslational autoxidation of green fluorescent protein. *Proceedings of National Academy of Sciences U S A*. 1994;91(26):12501–12504. doi:10.1073/pnas.91.26.12501

49. Heim R, Cubitt AB, Tsien RY. Improved green fluorescence. *Nature*. 1995;373(6516):663–664. doi:10.1038/373663b0

50. Heim R, Tsien RY. Engineering green fluorescent protein for improved brightness, longer wavelengths and fluorescence resonance energy transfer. *Current Biology*. 1996;6(2): 178–182. doi:10.1016/s0960-9822(02)00450-5

51. Matz M, Fradkov AF, Labas YA, et al. Fluorescent proteins from nonbioluminescent Anthozoa species. *Nature Biotechnology*. 1999;17(10):969–973. doi:10.1038/13657

52. Moldavan A. Photo-electric technique for the counting of microscopical cells. *Science (1979)*. 1934;80(2069):188–189. doi:10.1126/science.80.2069.188

53. Crosland-Taylor PJ. A device for counting small particles suspended in a fluid through a tube. *Nature*. 1953; 171(4340):37–38. doi:10.1038/171037b0

54. Coulter W. High speed automatic blood cell counter and cell size analyzer. In: *Proc. Nat. Electronics Conf. 1956*. Vol 12.; 1957:1034–1042. Accessed October 5, 2022. https://whcf.org/wp-content/uploads/2015/05/1956-WHC-NEC-Paper.pdf

55. Kamentsky LA, Melamed MR, Derman H. Spectrophotometer: New instrument for ultrarapid cell analysis. *Science (1979)*. 1965;150(3696):630–631. doi:10.1126/science.150.3696.630

56. Kamentsky LA, Melamed MR. Spectrophotometric cell sorter. *Science (1979)*. 1967;156(3780):1364–1365. doi:10.1126/science.156.3780.1364

57. Givan AL. The past as prologue. In: *Flow Cytometry: First Principles*. 2nd ed. John Wiley & Sons, Inc.; 2001:1–9. doi:10.1002/0471223948.ch1

58. Fulwyler MJ. Electronic separation of biological cells by volume. *Science (1979)*. 1965;150(3698):910–911. doi:10.1126/science.150.3698.910

59. van Dilla MA, Trujillo TT, Mullaney PF, Coulter JR. Cell microfluorometry: A method for rapid fluorescence measurement. *Science (1979)*. 1969;163(3872):1213–1214. doi:10.1126/science.163.3872.1213

60. Bonner WA, Hulett HR, Sweet RG, Herzenberg LA. Fluorescence activated cell sorting. *Review of Scientific Instruments*. 1972;43(3):404–409. doi:10.1063/1.1685647

61. Steinkamp JA, Fulwyler MJ, Coulter JR, Hiebert RD, Horney JL, Mullaney PF. A new multiparameter separator for microscopic particles and biological cells. *Review of Scientific Instruments*. 1973;44(9):1301–1310. doi:10.1063/1.1686375

12 Cell Signaling Part I
The Role of Phosphorylation

In previous chapters, we saw how the high-energy bonds in ATP serve as the main energy source of cells by transferring its terminal phosphate to molecules. In this and the next chapter, we will discuss how ATP also has another vital role to play: the regulation of protein activity in signaling pathways by the transfer of a phosphate group. In the early 20th century, Harden and Young discovered that phosphate plays a significant role in fermentation, leading to the elucidation of glycolysis and the Kreb cycle (in which ATP is produced). In 1936, Arda Green and Gerti Cori isolated a new molecule, glucose-1-phosphate, from frog muscle extracts and showed that this molecule was the result of glycogen phosphorylation, naming the enzyme that catalyzed the transfer of the phosphate group, phosphorylase (it was later renamed glycogen phosphorylase).[1] Cori's group subsequently found that phosphorylase was widely distributed in mammalian tissue and in yeast, and that the reaction was one which was reversible. However, although the Coris showed that vertebrate phosphorylase required the addition of adenylic acid (i.e., AMP) to proceed, other groups found that this was not required in yeast or potatoes. Cori and Green soon determined that phosphorylase obtained from rabbit muscle extracts readily lost activity at room temperature unless adenylic acid was added, which provided an explanation for the above discrepancy and made them suspect the presence of an inactivating enzyme that converted one form of phosphorylase to the other.

Upon reviewing their phosphorylase preparation protocols, the authors determined that if one of their precipitation steps was not carried out to completion, an enzyme would be present in their crude extracts which would act on phosphorylase to inactivate it, such that *no crystal will be obtained*. Therefore, it was clear *that the manipulations preceding the isoelectric precipitation must be carried out as speedily as possible and that the temperature must be kept low throughout*.[1] Keeping these new findings in mind, they succeeded in purifying active phosphorylase and subsequent analysis revealed that the substance was present in two forms: one *which crystallizes readily and which has activity without addition of adenylic acid*, which they named phosphorylase *a*; and *a much more soluble protein, phosphorylase b, which is inactive without addition of adenylic acid*.[2] This change in form was determined to occur due to the removal of a prosthetic group by an enzyme, the latter of which the authors purified from phosphorylase preparations and which the authors aptly named, "prosthetic group-removing" (PR) enzyme.

The authors determined that PR enzyme was required for the removal of the prosthetic group since it could not be removed by dilution, prolonged dialysis, washing of the crystals, or by exposure to extreme pH; heat treatment of the protein did result in inactivation of the protein, but not in the removal of the prosthetic group. These results indicated to the authors that *the bond between the enzyme protein and the prosthetic group is not easily dissociated*.[2] To test the conditions in which the prosthetic group was removed by PR enzyme, Cori and Green incubated phosphorylase *a* with PR enzyme in the presence or absence of adenylic acid and found that the activity of the former was significantly decreased in the absence of the nucleotide (Figure 12.1a). In addition, they added that *PR enzyme obviously does not destroy the phosphorylase protein [...] but changes it in such a manner that thereafter it requires the addition of adenylic acid for its activity*.[2] According to the authors, these results pointed to the fact that *phosphorylase a contains firmly bound adenylic acid and that the PR enzyme removes it from the protein*, and adopted this as part of their working hypothesis. However, investigations into the composition of trypsin-digested phosphorylase *a* could not identify free adenylic acid, indicating that *[w]hatever is split off [...] cannot be free or unchanged adenylic acid*.[2] In addition, they reported that they were not able to convert phosphorylase *b* into phosphorylase *a* by adding excess adenylic acid under a number of different conditions, which indicated that *the addition of adenylic acid to phosphorylase b does not result in the formation of an undissociable linkage such as is present in phosphorylase a*.

The authors also reasoned that *[i]f adenylic acid is present in phosphorylase a, the enzyme should contain pentose and phosphorus*. Indeed, a pentose sugar was found in purified phosphorylase *a*, whereas it was not detected in phosphorylase *b*.[2] Results also showed that phosphorylase *a* contained much more phosphate than phosphorylase *b*, from which Cori and Cori reasoned that *if this difference were significant, one would expect that PR would liberate phosphate while acting on phosphorylase a*.[3] Consistent with this, further investigation revealed that a significant amount of phosphorus was released when they incubated phosphorylase *a* in the presence of PR enzyme, the authors noting a *parallelism between the per cent conversion of phosphorylase a to b and the liberation of phosphate* (Figure 12.1b). They also added that *the principle involved in the action of the PR enzyme, namely, the splitting off of a prosthetic group from another enzyme, may be of physiological importance* and that *it seems probable that other PR enzymes exist*.

In the following years, the Coris and others continued to elucidate the mechanism involved in the removal of the

DOI: 10.1201/9781003379058-12

Effect of Muscle and Spleen Protein Fraction (PR Enzyme) on Activity of Crystalline Phosphorylase

PR preparations of different concentrations were added to phosphorylase dissolved in glycerophosphate-cysteine buffer at pH 6.8 and incubated at 25° for varying lengths of time before phosphorylase activity was tested in the presence and absence of adenylic acid.

Experiment No.	Origin of PR enzyme	Time of incubation	Phosphorylase activity		
			With adenylic acid; incubated with PR	Without adenylic acid	
				Incubated without PR	Incubated with PR
		min.	*per cent**	*per cent†*	*per cent†*
1	Muscle	75	96	64	37
2	"	97	105	68	21
3	"	300	94	65	12
4	"	10		67	35
		33			10
		72			5
5	Spleen	60	98	65	22
		125	99	68	13
6	"	36	81	63	31
		100	91	64	8
7a	"	20	92	62	6
7b‡	"	20	96	62	26
7b‡	"	60	97	63	12

* The activity of a control sample incubated without PR is taken as 100.

† The activity in the presence of adenylic acid is taken as 100.

‡ One-fourth the PR enzyme concentration of the sample in Experiment 7a.

(a)

Total P in Filtrates and in Protein Precipitate after Incubation of Phosphorylase a with PR Enzyme or with Trypsin

Filtrates were prepared by precipitation of protein with CCl_3COOH in the cold.

Phosphorylase *a* preparation No.	Time of incubation with PR or trypsin (Tr)	Phosphorylase *a* converted to *b*	P in CCl_3COOH filtrate	P in CCl_3COOH precipitate
	min.	*per cent*	γ *per mg. protein*	γ *per mg. protein*
84	45 (PR)	70	0.57	0.31
85	30 (Tr)	90	0.61	
	60 "	100	0.68	0.25
86	60 (PR)	90	0.45*	0.30*
87	60 "	85	0.56	
	120 "	100	0.69	0.20
96	20 "	75	0.51	
	45 "	95	0.65	
	120 "	100	0.67	0.21
Average (for 100 % conversion)			0.68	0.22

*No protein precipitant was used. P was determined in the dialysate and in the fluid remaining in the dialyzing bag, respectively.

(b)

FIGURE 12.1 (a) Adenylic acid increases phosphorylase activity. A reaction mixture was incubated with or without PR enzyme and/or adenylic acid for different lengths of time and phosphorylase activity was assessed as reported by the presence of glucose-1-phosphate. Notice how incubation in the absence of adenylic acid leads to a progressive loss of phosphorylase activity over time in the presence of PR enzyme, whereas it remains constant in its absence. (b) Phosphorus is released upon incubation of phosphorylase a with PR enzyme. A reaction mixture was incubated with or without PR enzyme for the indicated times and the amount of phosphate released was measured by precipitation with CCl_2COOH. Note how a significant amount of phosphate appears in the filtrate (i.e., soluble P_i; second to last column) in the presence of PR enzyme. (Adapted from refs. [2,3].)

prosthetic group from phosphorylase; however, very little was learned about the mechanism of the reverse reaction, that is, the addition of the prosthetic group to phosphorylase. Ten years after the Coris' publications—in 1955—Edmond Fisher and Edwin Krebs (no relation to the Krebs and Fischer we learned about in previous chapters) were the first to report such investigations. By this point, it had been determined that phosphorylase isolated from muscles at rest were predominantly in the *b* form, contrary to reports by the Coris that the *a* form was predominant; once again, the conditions during the extraction were found to be to blame.[4] In fact, the authors, who used cell-free extracts in their studies, found that the conversion of form *b* to form *a* was actually the faster reaction provided that certain divalent cations, such as Ca^{2+} or Mn^{2+}, were included in addition to ATP or ADP (Figure 12.2). Furthermore, the authors suggested that the reaction was catalyzed by an enzyme, which they temporarily named "converting enzyme", since the conversion required the presence of a protein fraction.

The same year, Earl Sutherland's group was studying the inactivation of phosphorylase purified from dog liver extracts and were specifically looking in the incubation media for small molecules that might be released from phosphorylase during inactivation. Their efforts, using new techniques developed by Fiske and Subbarow (the discoverers of ATP we discussed in Chapter 5), confirmed the Cori's findings that inorganic phosphate was indeed released upon incubation with PR enzyme (which they had renamed "inactivating enzyme"), concomitant with the inactivation of phosphorylase *a*.[5] In addition, it was reasoned that since the removal of phosphate from phosphorylase inactivated the enzyme, conversion to the active form should result from the addition of phosphate. Using their finding that epinephrine (also known as adrenaline) increased the activation of phosphorylase purified from dog liver extracts, Rall, Sutherland, and Wosilait incubated extracts with radioactive phosphate in the absence or presence of epinephrine and found that indeed, radioactive phosphate was incorporated into inactive phosphorylase and activated the enzyme; the

authors were also able to purify the enzyme responsible for the phosphorylation of inactive phosphorylase, and named it phosphokinase (it was later renamed to phosphorylase *b* kinase) (Figure 12.3a).[6]

In light of these results, Sutherland and colleagues suggested that the *the observed enzymatic release of phosphate from purified liver phosphorylase is related to a physiological mechanism concerned with the activation and inactivation of liver phosphorylase in intact cells*,[5] and that the

> concentration of active phosphorylase in liver tissue is the result of a balance between two opposing reactions: the inactivation of the enzyme by phosphorylase phosphatase *[inactivating enzyme; also Cori's PR enzyme]* and the reactivation of the resulting dephosphophosphorylase by the phosphokinase system.[6]

Sutherland's group also showed that epinephrine could increase the formation of active phosphorylase in cell-free liver and muscle extracts and that this phenomenon was dependent on the presence of a new molecule—identical to another molecule first described by Cook, Lipkin, and Markham the same year—named cyclic 3′,5′-AMP (cAMP) (Figure 12.3b).[7–9] As such, Rall, and Sutherland used ^{14}C-labeled ATP to show that the formation of cAMP came as a result of cyclization of ATP since cAMP *was formed with apparently the same amount of radioactivity per adenine residue as the original ATP*.[10] However, the requirements for this phenomenon was not investigated, noting that *Further information on the mechanism of both synthesis of cyclic 3,5-AMP and hormone action will depend on purification, or at least simplification of the tissue preparations used.* Finally, the authors identified and partially purified an enzyme, which they called phosphodiesterase, that could inactivate cAMP by its hydrolysis to 5′AMP.[10]

In 1959, Krebs, Graves, and Fischer, working off of Sutherland's group's findings, showed that addition of Ca^{2+} or cAMP increased the activity of phosphorylase *b* kinase in a manner which required an unidentified labile factor.[11] That

Extent of Conversion of Phosphorylase b to a in Rabbits under Various Conditions

The mean values are obtained by dividing the sum of the activity units per ml. of crude extract by the number of rabbits. Reaction system for conversion as described under "Methods;" the activities are determined after 30 minutes incubation at 25°.

Extract	No. of rabbits	Mean activities (units per ml.) after incubation with							
		No additions		10^{-3} M Mn^{++}		10^{-3} M ATP		10^{-3} M Mn^{++} + 10^{-3} M ATP	
		+ AMP	− AMP	+ AMP	− AMP	+ AMP	− AMP	+ AMP	− AMP
5°; Method 1.....	3	1680	345	1950	1500			2260	2190
24–28°; Method 1.	8	1570	130	1840	1470	1730	150	2060	1830

FIGURE 12.2 Requirement of ATP and Mn^{2+} in the conversion of phosphorylase b to phosphorylase a. Muscle extracts were incubated in a reaction mixture with phosphorylase b and the conversion of the latter to phosphorylase a was assessed (shown as activity of the extract in units/ml). The "Extract" column indicates the conditions under which the rabbit muscles were prepared. (Adapted from ref. [4].)

Phosphate Transfer to Dephospho-LP Catalyzed by Phosphokinase

Samples of dephospho-LP, ATP, and Mg⁺⁺ were incubated with and without phosphokinase. Following ammonium sulfate and alcohol precipitation, the samples were incubated with IE. At zero time and at 45 minutes, aliquots were placed in TCA; after centrifugation, the TCA-supernatant solutions were analyzed for inorganic phosphate. Other aliquots were assayed for LP activity.

History of sample	Time of incubation with IE	Composition of sample		P released by IE into TCA-supernatant fluid
		Dephospho-LP	Active LP	
	min.	*mg.*	*mg.*	*γ*
Incubated with phosphokinase..	0	0.25	1.00	
" " " ..	45	1.25	0	0.27
" without phosphokinase......................	0	1.25	0	
Incubated without phosphokinase......................	45	1.25	0	0

(a)

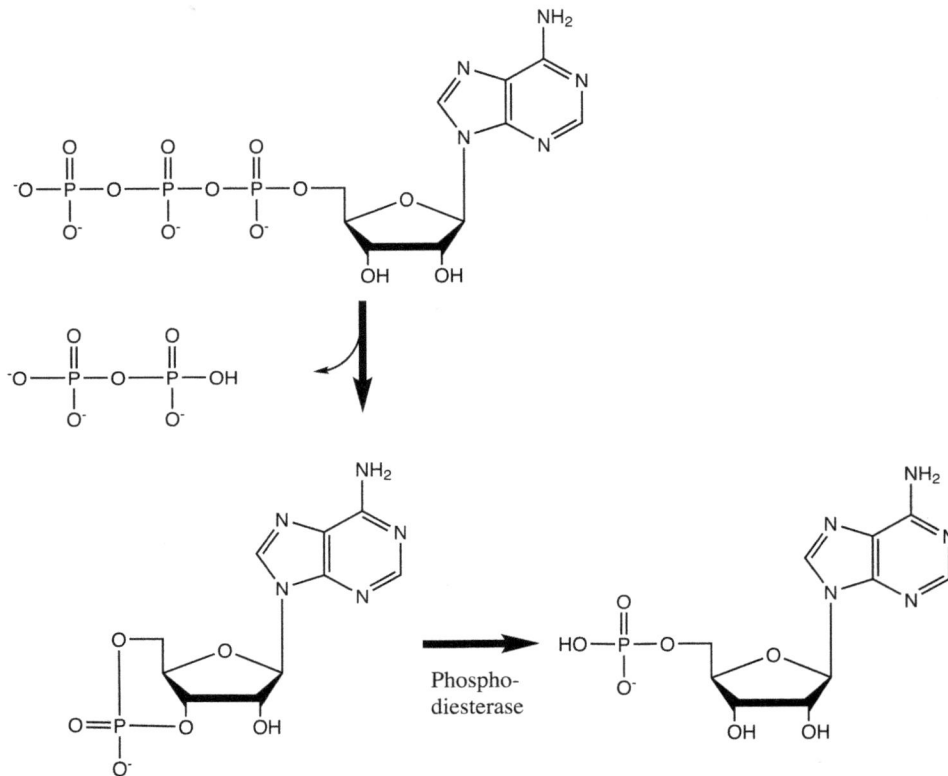

(b)

FIGURE 12.3 Transfer of ^{32}P by phosphokinase inactivates phosphorylase. (a) A reaction mixture was incubated with phosphorylase (dephospho-LP) in the presence or absence of phosphokinase, after which the samples were incubated with phosphorylase phosphatase (inactivating enzyme, IE). Released phosphate and amount of phosphorylase a and b were assessed before and after incubation with IE. Note how phosphate is only released when the sample was incubated with phosphokinase and IE. (b) cAMP synthesis from ATP. ATP loses two phosphates (pyrophosphate, PP_i) to produce cAMP. Phosphodiesterase (PD) hydrolyzes cAMP to AMP.[6]

ATP was this labile factor was shown by the observation that addition of more cAMP after 60 minute's incubation did not reactivate the enzyme unless ATP was also added (Figure 12.4). In addition, it was noted that cAMP itself was consumed in the reaction since addition of ATP alone failed to reactivate the enzyme and that no other nucleotide, cyclic or otherwise, could replace it. In their conclusion, the

authors advanced a *highly speculative hypothesis* to explain their results:

[...] phosphorylase b kinase itself exists in phosphorylated and dephosphorylated forms, which are active and inactive in a manner analogous to phosphorylases a and b. According to this hypothesis an enzyme system consisting

Requirements for activation of phosphorylase b kinase in muscle extracts by adenosine 3′,5′-phosphoric acid

Phosphorylase *b* kinase activities were determined at a final dilution of 1 to 900 of muscle extract. Incubation of the extract was carried out at 30° at pH 7.0. Cyclic 3′,5′-AMP was added to a final concentration of 1×10^{-4} M and ATP to a final concentration of 2×10^{-3} M where shown. The mixture was divided into 3 portions at 60 minutes.

Treatment of extract	Phosphorylase *b* kinase activity at	
	pH 7.0	pH 8.6
	units/ml	*units/ml*
No additions	200	12,400
Incubation for 10 minutes after addition of cyclic 3′,5′-AMP	3,100	12,400
Incubation continued for 30 minutes	0	13,100
Incubation continued for 60 minutes	0	11,200
Incubation continued for 71 minutes (cyclic 3′,5′-AMP readded at 61 minutes)	400	
Incubation continued for 71 minutes (ATP readded at 61 minutes)	300	
Incubation continued for 71 minutes (cyclic 3′,5′-AMP and ATP added at 61 minutes)	3,400	
No additions. Extract incubated for 71 minutes	100	12,900

FIGURE 12.4   Effects of cAMP on phosphorylase kinase activity. Muscle extracts were incubated in a reaction mixture with the indicated treatments and phosphorylase b kinase activity was assessed. Note how addition of cAMP alone for 60 minutes was not enough to regain activity. However, addition of cAMP and ATP completely reactivated the reaction, indicating that ATP was the labile factor. (Adapted from ref. [11].)

of another kinase and a phosphatase might be involved in the activation and inactivation of phosphorylase b kinase. The effects of Ca++, pH, ATP, and cyclic 3′,5′-AMP on the state of kinase activity could involve either the activating or inactivating enzymic reactions.[11]

A few years later, Friedman and Larner determined that the activity of another enzyme, glycogen synthase, was also regulated by its phosphorylation state and stimulated by cAMP.[12] Oddly, unlike phosphorylase and phosphorylase kinase which were activated by phosphorylation, glycogen synthase—which, as the name implies, is involved in glycogen synthesis—was inactivated by phosphorylation. In time, it would be determined that regulation of protein activity by their state of phosphorylation/dephosphorylation would apply to a vast number of different proteins *involved in biosynthesis and degradation and it would appear to be one of the 'principles' that has been established for the regulation of enzymes by phosphorylation-dephosphorylation.*[13] Surprisingly, phosphorylation of both phosphorylase kinase and glycogen synthase was subsequently found to be regulated by the same cAMP-dependent protein kinase and many more targets of this enzyme would eventually be found; as such, the enzyme was renamed cAMP-dependent protein kinase, or PKA (Figure 12.5).[13–15] Until the late 1960s, phosphorylation reactions were thought to be specific for,

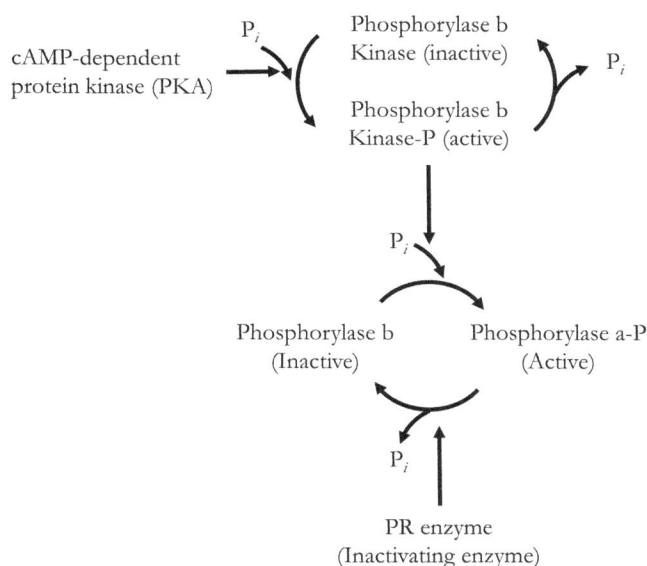

FIGURE 12.5   The phosphorylase pathway as it was known in the late 1950s.

and unique to, the metabolism of carbohydrates and thus, were generally of minimal interest. However, the discovery that PKA had a number of different targets in a variety of different processes brought widespread attention to the phenomenon of phosphorylation. Interest in protein phosphorylation also increased significantly in 1969 when Linn et al. discovered that PKA was involved in regulation of pyruvate dehydrogenase, a *clearcut example of a phospho-dephospho enzyme system well beyond the confines of glycogen metabolism.*[13]

* * * * *

In 1897, Paul Ehrlich, after studying the *mysterious nature* of antitoxins, proposed his "Theory of Immunity", which suggested a mechanism by which toxins caused their injurious effects in animals. He proposed that a group of poisons called toxins—metabolic products of animals, plants, and bacteria, such as snake venoms and tetanus toxin—were composed of two distinct properties: that of being poisonous and that of inducing the production of specific antitoxins in the animal body (these would later be called antibodies).[16] Antitoxins, as the name implies, were substances that were known to counter the effects of toxins, and Ehrlich made the observation that the *relations between toxin and its antitoxin are strictly specific - tetanus antitoxin neutralizes exclusively tetanus toxin, diphtheria serum only diphtherial toxin, snake serum only snake venom*; therefore, he suggested that this specificity was of a chemical nature which he likened to Emil Fischer's "lock and key" analogy for the action of enzymes.[16]

Borrowing from the terminology of chemistry, he proposed that the interaction was as a result of a chemical group on the toxin, which he called haptophore, which bound to side chain atoms of the antitoxin, in effect anchoring the former to the latter on the surface of cells (Figure 12.6).[17] As

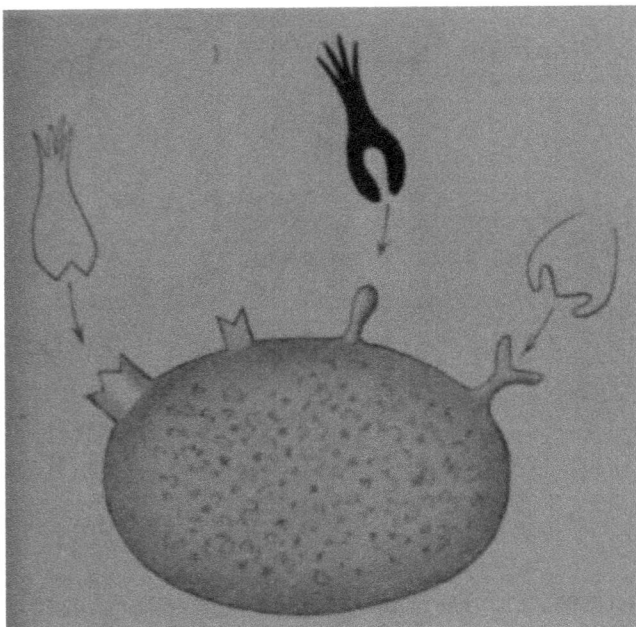

FIGURE 12.6 Antitoxins as proposed by Ehrlich. Each particular toxin binds to a specific cell surface antitoxin like a lock and key to trigger a response. (Adapted from ref. [17].)

such, the effects of toxins were proposed to *be caused by the adhesion of the toxic substance to quite definite cell complexes* which he called *poison receptors, or just receptors*.[16] Furthermore, natural immunity against certain substances was thought to be the result of the cells of an individual not having the appropriate receptors to which the particular toxin could bind. Ehrlich pictured these receptors to be components of the plasma membrane which were up-regulated upon infection by toxins. He also proposed that when the concentration of receptors became too great to be retained on the plasma membrane, they were shed into the blood stream and were then able to neutralize the toxins before they reached the cells and produce their toxic effects:

the antitoxines [sic] represent nothing more than side-chains reproduced in excess during regeneration, and therefore pushed off from the protoplasm; and so coming to exist in a free state.[17]

Ironically, Ehrlich also concluded:

That these *[antitoxins]* are in function specially designed to seize on toxines cannot be for one moment entertained. It would not be reasonable to suppose that there were present in the organism many hundreds of atomic groups destined to unite with toxines, when the latter appeared [...] It would indeed be highly superfluous, for example, for all our native animals to possess in their tissues atomic groups deliberately adapted to unite with abrin, ricin, and crotin, substances coming from the far distant tropics.[17]

Around the same time, the idea began to emerge that substances which cause muscle contractions—such as

adrenaline—did not directly cause the contraction but instead *depends upon the presence in the muscle protoplasm of some substance which is not contractile substance* but which in turn causes the contractions.[18] As such, inspired by Ehrlich's receptor theory in immunity, John N. Langley suggested that:

Since this accessory substance is the recipient of stimuli which it transfers to the contractile material, we may speak of it as the *receptive substance* of the muscle [...]

I conclude then that in all cells two constituents at least must be distinguished, (1) substance concerned with carrying out the chief functions of the cells, such as contraction, secretion, the formation of special metabolic products, and (2) receptive substances especially liable to change and capable of setting the chief substance in action.[18] *[emphasis in original]*

Despite the proposal of cell surface receptors by Ehrlich and Langley at the turn of the 20th century, it would be another 60 years—when Sutherland's group began studying *the enzyme system that catalyzes the formation of cyclic 3',5'-AMP*—before the concept was conclusively shown to have broad applications.[19] By this time, it had been discovered that various hormones, such as adrenaline, could influence the activity of this system, and cAMP activity was found to be involved in a number of enzyme activities well beyond that of phosphorylase and glycogen synthase. In 1962, Sutherland and his team published a series of papers, the first of which reported that an enzyme system, which they named adenyl cyclase, was present in all animals and in most tissues examined.[19] They also found that addition of sodium fluoride, known to inhibit the effects of an ATP phosphatase, significantly stabilized the activity of adenyl cyclase and allowed more accurate measurements of cAMP activity to be obtained. They also found that a high degree of purification of the system was significantly hampered by the fact that most of the adenyl cyclase was found to reside in the nuclear or cell membrane fraction.[19] However, their yields were high enough to show that this enzyme system was indeed responsible for the conversion of ATP into cAMP (Figure 12.7). The same year, the phosphodiesterase

Precursor	Average specific activity		
	ATP	Cyclic 3',5'-AMP	ATP/cyclic 3',5'-AMP ratio
	c.p.m./μmole		
Experiment 1:			
AP³²PP	1,065	711	0.67
APP³²P³²	11,950	250	0.02
Experiment 2:			
AP³²PP	315,000	265,000	0.84

FIGURE 12.7 Results from experiments with radiolabeled ATP showing that adenyl cyclase converts ATP into cAMP. A reaction mixture containing ATP with the indicated radioactive phosphates was incubated with 0.5 mg (exp 1) or 2.6 mg (exp 2) of adenyl cyclase for 30 minutes. The ATP and cAMP were then recovered and radioactivity was assessed.[21]

involved in converting cAMP into 5'AMP, the only *physiological mechanism [...] by which the action of* [cAMP] *could be terminated*, was purified and characterized by Sutherland's group. This was a particularly difficult enzyme to isolate and early attempts at fractionation by usual methods were not successful; the authors found that they had to start with *extremely high protein* concentrations in order for the purification to be successful.[20] Like adenyl cyclase, they found that the enzyme was widely distributed in terms of organs and of species studied.

By the mid-60s, the action of numerous hormones was found to result in alterations in cAMP activity, via activation of either adenyl cyclase or phosphodiesterase. As such, Sutherland's group proposed the concept of second messengers to explain the relationship between hormones, adenyl cyclase, and cAMP:

> As a result of studies with catecholamines *[e.g., adrenaline]* and other hormones, the concept has arisen that a number of hormones act by a two-messenger system. The first messenger in this concept is the hormone or neurohormone, which is released by stimuli which may be varied and complex [...] this first messenger travels to effector cells and causes the release therein of a second messenger. The only second messenger identified to date is cyclic 3', 5'-AMP, but possibly other second messengers exist, even for the same hormones that stimulate adenyl cyclase.[22]

There was also growing evidence that adenyl cyclase was a membrane bound protein and that it may be composed of two components with separate functions (Figure 12.8). As

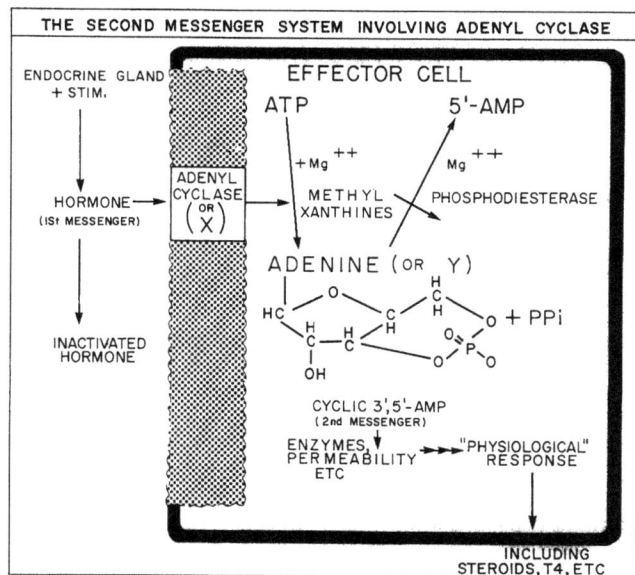

FIGURE 12.8 The second messenger model proposed by Sutherland's group.[24] (Reprinted with permission from Wolters Kluwer Health, Inc., from Sutherland EW, Robison GA, Butcher RW. Some aspects of the biological role of adenosine 3',5'-monophosphate (cyclic AMP). Circulation. 1968;37(2):279–306. Copyright © 1968.)

such, Robison, Butcher, and Sutherland proposed that adenyl cyclase is

> composed of at least two distinct subunits, a regulatory subunit (R), facing the extracellular fluid, and a catalytic subunit (C), the active center of which is in contact with the interior of the cell. According to this model, the hormone or neurohumor would be expected to interact with the regulatory subunit, which in turn might influence the configuration of the catalytic subunit.[23]

This mechanism was novel since the idea of a protein spanning the plasma membrane—that is, a protein that extends completely through the membrane—was a relatively new and unproven concept for which further evidence was only brought to light in the mid-1970s. To that end, Henderson and Unwin obtained electron microscopy images of the purple membrane of *Halobacterium halobium*—the same used by Oesterhelt and Stoeckenius which helped elucidate the electron transfer chain—but using a new technique for *determining the projected structures of unstained crystalline specimens* which allowed them to obtain *a three-dimensional map of the membrane at 7 Å* [0.7 nm] *resolution* revealing *the location of the protein and lipid components, the arrangement of the polypeptide chains within each protein molecule, and the relationship of the protein molecules in the lattice.*[25]

Forgoing the complicated theoretical considerations employed by the authors, it was determined that these membrane proteins were arranged in clusters of seven *rod-shaped features aligned perpendicular to the plane of the membrane*, likely with portions exposed on both sides of the membrane (Figure 12.9).[25] These results gave support to a recently advanced model for cell membranes, called the "Fluid Mosaic Model", which, at the time, was not well supported by direct experimental evidence.

Even in the early 1970s, most researchers did not think there was any generalizations to be made about the composition of cell membranes given the observed diversity between different cells. Still, in 1972, Seymour Singer and Garth Nicolson proposed that an analogy could be made between cell membranes and proteins in that although the latter are very diverse in terms of composition, function, and structure, certain generalizations, such as being globular or fibrous, could be made to better understand them.[26] As such, the authors suggested that:

> a mosaic structure of alternating globular proteins and phospholipid bilayer was the only membrane model among those analyzed that was simultaneously consistent with thermodynamic restrictions and with all the experimental data available.[26]

Here, the authors stressed the importance of hydrophobic and hydrophilic interactions in the maintenance of the integrity and composition of the membrane, as exemplified by the phospholipid bilayer structure: the hydrophobic tails of lipids are sequestered away from the water while

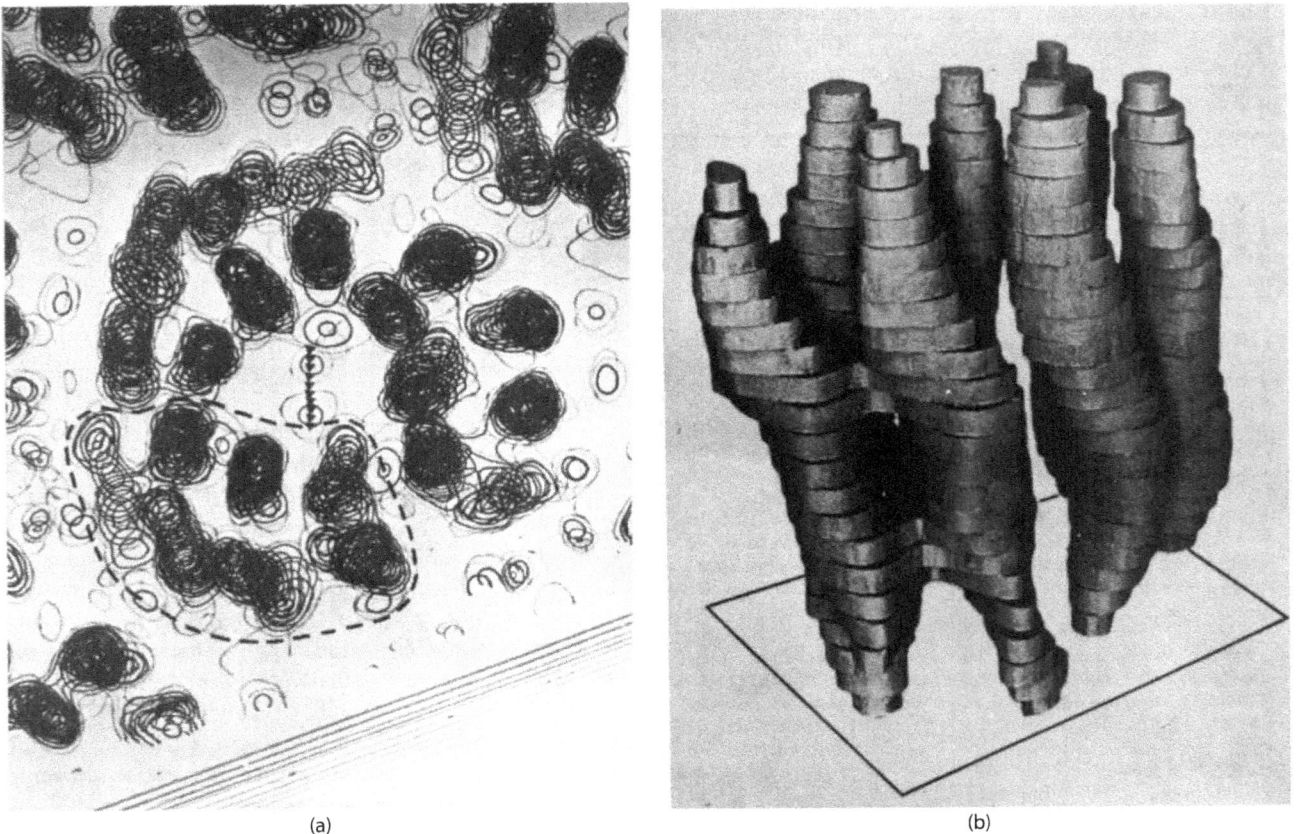

(a)   (b)

FIGURE 12.9 *H. halobium* contains protein clusters that span the plasma membrane. (a) Overhead view of a section of the membrane. Note that there appears to be clusters of proteins, one of which is indicated by the broken line. (b) A model of a single cluster depicted in (a), which corresponds to a vertical section where the middle section is in contact with membrane lipids.[25] (Used by permission from Springer Nature Customer Service Centre GmbH: Springer Nature, Nature 257(5521), Henderson R, Unwin PNT. Three-dimensional model of purple membrane obtained by electron microscopy, Copyright © 1975; Permission conveyed through Copyright Clearance Center, Inc.)

hydrophilic components of lipids interact with the surrounding water. Similarly, hydrophobic parts of proteins would be buried within the bilayer, while hydrophilic parts would protrude from the membrane and interact with the water.[26] This type of protein was called an amphipathic protein, that is, *they are structurally asymmetric, with one highly polar end and one nonpolar*; specifically, this class of proteins was referred to as "integral" proteins, to which globular proteins were postulated to be members.[26]

They also proposed that "peripheral" proteins may only contact the hydrophilic heads of the lipids either on the inside or the outside of the cell, *held to the membrane only by rather weak noncovalent [...] interactions and [...] not strongly associated with the membrane lipid*.[26] In addition, they concluded that evidence supported the fact that both sides of the membrane were not symmetrical; that is, different peripheral proteins were present on the inner compared to the outer side of the membrane. The authors also addressed the question of whether the phospholipids or the proteins were the matrix of the plasma membrane in which the other component was inserted, that is, *which component is the mortar, which the bricks*. Given the available experimental evidence, the authors concluded that *the fluid*

mosaic model of membrane structure [results from] *the free diffusion and intermixing of the lipids and the proteins (or lipoproteins) within the fluid lipid matrix*.[26] Singer and Nicolson summarized their model as such:

In this model, the proteins that are integral to the membrane are a heterogeneous set of globular molecules, each arranged in an amphipathic structure, that is, with the ionic and highly polar groups protruding from the membrane into the aqueous phase, and the nonpolar groups largely buried in the hydrophobic interior of the membrane. These globular molecules are partially embedded in a matrix of phospholipid. The bulk of the phospholipid is organized as a discontinuous, fluid bilayer, although a small fraction of the lipid may interact specifically with the membrane proteins. The fluid mosaic structure is therefore formally analogous to a two-dimensional oriented solution of integral proteins (or lipoproteins) in the viscous phospholipid bilayer solvent.[26]

They also suggested that the fluid mosaic model had functional consequences:

The physical or chemical perturbation of a membrane may affect or alter a particular membrane component or set of components; a redistribution of membrane components can then occur by translational diffusion through the viscous two-dimensional solution, thereby allowing new thermodynamic interactions among the altered components to take effect. This general mechanism may play an important role in various membrane-mediated cellular phenomena that occur on a time scale of minutes or longer.[26]

Thus, the results of Henderson and Unwin concerning the three-dimensional structure of the purple membrane suggested that *the protein is globular, is almost certainly exposed at both sides of the membrane, and is surrounded by lipids which are arranged in separate areas with a bilayer configuration. The purple membrane thus seems to provide a simple example of an "intrinsic" membrane protein.*[25]

* * * * *

In 1969, shortly after Sutherland's group proposed the idea of a second-messenger system, Gill and Garren demonstrated that cAMP could bind specifically to a cytosolic protein, although the nucleotide was not a substrate for the protein (i.e., the protein did not modify cAMP).[27] This prompted the authors to suggest that *perhaps the action of cyclic AMP on the binding protein is similar to that of the cyclic nucleotide on various enzymes, i.e., to alter function, presumably by modifying the macromolecular structure.*[27] Spurred on by this hypothesis, the authors discovered that cAMP in the adrenal cortex stimulated the activation of a protein kinase and investigated its relationship to PKA. They found that this kinase seemed to phosphorylate the same substrates as PKA, though endogenous substrates were not identified, and confirmed that the activity was dependent on cAMP concentration.[28] They also purified the protein by fractionation, which they achieved by taking advantage of its ability to bind to and be activated by cAMP and by the fact that it had kinase activity. Surprisingly, the fraction which mediated binding to cAMP did not have kinase activity (Figure 12.10, 1st line). However, when this fraction was incubated with one which had kinase activity, kinase activity was significantly decreased in absence of cAMP, though it could be restored by addition of cAMP.[28] Finally, their studies showed that the properties of the

kinase protein were similar to those of PKA. From these data, the authors proposed a model in which the

cyclic AMP binding moiety in association with the protein kinase suppresses its activity; the binding of cyclic AMP to its receptor relieves the suppression. The cyclic AMP activation of the protein kinase may result by favoring the dissociation of the binding moiety from the protein kinase.[28]

The same year, Fritz Lipmann's group published a paper which corroborated the findings that cAMP could bind a protein in reticulocyte lysates.[29] In addition, sucrose density fractionation of this enzyme revealed the presence of two distinct peaks of kinase activity, one of which was dependent on the presence of cAMP, while the other, a minor peak, was independent (Figure 12.11a). When purification of the protein was performed in the presence of cAMP, the major peak disappeared and the minor peak became the only peak of kinase activity (Figure 12.11b). Furthermore, sedimentation experiments determined that the cAMP-binding protein was slightly bigger than the protein with kinase activity. As such, these results suggested that:

the heavy protein kinase I fraction is an inactive form and that cAMP binds to and dissociates an inhibitor of a little more than half the molecular weight leaving the light, catalytically active, form behind. We realize that there are limitations to such an interpretation since the experiments were done with an impure system. However, the proposition does constitute an attractive working hypothesis and would present a novel mode of enzyme regulation.[29]

The next year, Kreb's group further defined the relationship between cAMP and these two proteins and showed that the interaction existed in a dynamic equilibrium, which they proposed proceeded as follows:[30]

$$RC + cAMP \rightleftharpoons R \cdot cAMP + C,$$

where R is the regulatory subunit and C is the catalytic subunit. Thus, in the absence of cAMP, R and C form a complex (RC, called a holoenzyme—an enzyme formed from the interaction of two or more proteins) in which C

*Enzyme used		mμm ATP^{32} incorporated	
Binding protein (4s peak)	Kinase protein (7s peak)	−cyclic AMP	+cyclic AMP
9.0 μg +	0	0	.05
0 +	8 μg	.158	.455
4.5 μg +	8 μg	.125	---
9.0 μg +	8 μg	.109	.555
13.5 μg +	8 μg	.093	---

FIGURE 12.10 The "binding protein" fraction of the isolated protein complex inhibits kinase activity. Kinase protein was added alone or in combination with binding protein, along with ATP^{32}, and incorporation of ^{32}P in histones (shown to be a substrate for the kinase) was assessed in the presence or absence of cAMP. Note the gradual decrease in activity with an increase in binding protein.[28] (Reprinted from Biochem Biophys Res Commun. 39(3), Gill GN, Garren LD. A cyclic-3′,5′-adenosine monophosphate dependent protein kinase from the adrenal cortex: Comparison with a cyclic AMP binding protein, 335–343, Copyright © 1970, with permission from Elsevier.)

FIGURE 12.12 Activation of PKA by cAMP. Two cAMP molecules bind to each regulatory subunit of PKA (R), resulting in a conformational change in the protein and the dissociation of the catalytic subunits (C). C then phosphorylates substrates (S) containing the correct amino acid recognition sequence.[31]

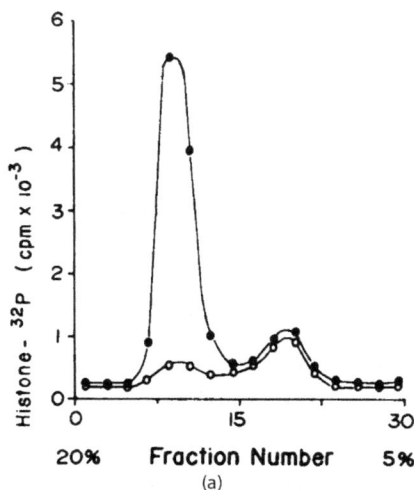

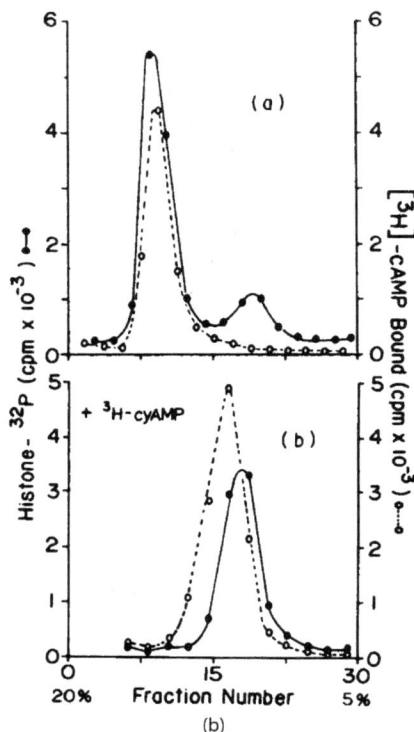

FIGURE 12.11 Kinase activity of a sucrose gradient fractionation experiment has two peaks of kinase activity. (a) Kinase activity of kinase I in the presence (closed circles) or absence (open circles) of cAMP. (b) Kinase I (closed circles) and cAMP-binding (open circles) activity of the protein fractions were assessed in the presence of cAMP following purification of the kinase in the absence (a) or presence (b) of cAMP. In the absence of cAMP during purification (a), cAMP binding overlaps with high kinase activity, whereas it overlaps with the position of the low peak in its presence (b). In addition, kinase activity has only one peak when the kinase is purified in the presence of cAMP (b), and this fraction binds cAMP with high affinity.[29]

is inhibited. Binding of hormones to cell-surface receptors leads to an increase in cAMP, which binds to R and results in the release of C from the complex which goes on to phosphorylate its substrates. Termination of cAMP activity can be accomplished via its hydrolysis by phosphodiesterases,

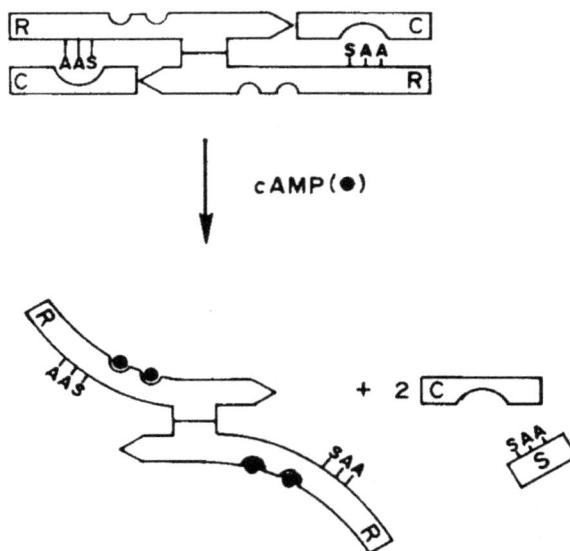

resulting in the reconstitution of RC, and regulation of activity is dictated by the balance between hormone concentration and phosphodiesterase activity.

In the late 1970s, Corbin et al. determined a simple method for the purification of the cAMP-dependent protein complex, PKA, which enabled its properties to be further studied. They confirmed that the purified regulatory subunit was able to inhibit the kinase activity of the catalytic subunit by incubating the two components together and assessing kinase activity.[31] In addition, since the regulatory subunit was purified in the presence of ^{3}H-labeled cAMP, the authors were able to assess the ratio of cAMP bound to each regulatory protein, which they determined to be 2:1. They also found that, although the binding of cAMP to the regulatory subunit was not covalent, it was one with high affinity. As to the involvement of phosphodiesterases, the authors added that it

follows that in the cell the regulatory subunit and phosphodiesterase might compete for cAMP. Although other factors including the catalytic subunit are involved, assuming that cAMP is not saturating for either protein, the physiological action or rate of hydrolysis would depend both on the concentration of each protein and on their affinity for cAMP.[31]

In addition, based on their findings and those of others, Corbin *et al.* suggested that two regulatory subunits bind to two catalytic subunits, forming a holoenzyme which inhibits the activity of the catalytic subunits by blocking their protein interaction sites (Figure 12.12). Binding of 2x cAMP molecules to each regulatory subunit causes a conformational change in the protein which prevents interaction with the catalytic subunits, thereby resulting in their release.

* * * * *

REFERENCES

1. Green AA, Cori GT, With a note by Oncley JL. Crystalline muscle phosphorylase: I. Preparation, properties, and molecular weight. *Journal of Biological Chemistry*. 1943;151(1): 21–29. doi:10.1016/S0021-9258(18)72110-8

2. Cori GT, Green AA. Crystalline muscle phosphorylase: II. Prosthetic group. *Journal of Biological Chemistry*. 1943;151(1):31–38. doi:10.1016/S0021-9258(18)72111-X

3. Cori GT, Cori CF. The enzymatic conversion of phosphorylase a to b. *Journal of Biological Chemistry*. 1945; 158(2):321–332. doi:10.1016/S0021-9258(18)43139-0

4. Fischer EH, Krebs EG. Conversion of phosphorylase b to phosphorylase a in muscle extracts. *Journal of Biological Chemistry*. 1955;216(1):121–132. doi:10.1016/S0021-9258(19)52289-X

5. Sutherland EW, Wosilait WD. Inactivation and activation of liver phosphorylase. *Nature*. 1955;175(4447):169–170. doi:10.1038/175169a0

6. Rall TW, Sutherland EW, Wosilait WD. The relationship of epinephrine and glucagon to liver phosphorylase. III. Reactivation of liver phosphorylase in slices and in extracts. *Journal of Biological Chemistry*. 1956;218(1):483–495. doi:10.1016/S0021-9258(18)65911-3

7. Cook WH, Lipkin D, Markham R. The formation of a cyclic dianhydrodiadenylic acid (i) by the alkaline degradation of adenosine-5′-triphosphoric acid (II). *Journal of American Chemical Society*. 1957;79(13):3607–3608. doi:10.1021/ja01570a086

8. Berthet J, Rall TW, Sutherland EW. The relationship of epinephrine and glucagon to liver phosphorylase. IV. Effect of epinephrine and glucagon on the reactivation of phosphorylase in liver homogenates. *Journal of Biological Chemistry*. 1957;224(1):463–475. doi:10.1016/S0021-9258(18)65045-8

9. Sutherland EW, Rall TW. Fractionation and characterization of a cyclic adenine ribonucleotide formed by tissue particles. *Journal of Biological Chemistry*. 1958;232(2): 1077–1091. doi:10.1016/S0021-9258(19)77423-7

10. Rall TW, Sutherland EW. Formation of a cyclic adenine ribonucleotide by tissue particles. *Journal of Biological Chemistry*. 1958;232(2):1065–1076. doi:10.1016/S0021-9258(19)77422-5

11. Krebs EG, Graves DJ, Fischer EH. Factors affecting the activity of muscle phosphorylase b kinase. *Journal of Biological Chemistry*. 1959;234:2867–2873.

12. Friedman DL, Larner J. Studies on UDPG-α-glucan transglucosylase. III. Interconversion of two forms of muscle UDPG-α-glucan transglucosylase by a phosphorylation-dephosphorylation reaction sequence. *Biochemistry*. 1963;2: 669–675. doi:10.1021/bi00904a009

13. Krebs E. Historical perspectives on protein phosphorylation and a classification system for protein kinases. *Philosophical Transactions of the Royal Society B: Biological Sciences*. 1983;302(1108):3–11. doi:10.1098/rstb.1983.0033

14. Walsh DA, Perkins JP, Krebs EG. An adenosine 3′,5′-monophosphate-dependant protein kinase from rabbit skeletal muscle. *Journal of Biological Chemistry*. 1968;243(13): 3763–3765. doi:10.1016/S0021-9258(19)34204-8

15. Schlender KK, Wei SH, Villar-Palasi C. UDP-glucose: Glycogen α-4-glucosyltransferase I kinase activity of purified muscle protein kinase. Cyclic nucleotide specificity. *Biochimica Biophysica Acta*. 1969;191(2):272–278. doi:10.1016/0005-2744(69)90246-0

16. Ehrlich P Paul Ehrlich - Nobel lecture: Partial cell functions. NobelPrize.org. Nobel Media AB © The Nobel Foundation 1908. https://www.nobelprize.org/prizes/medicine/1908/ehrlich/lecture

17. Ehrlich P. Croonian lecture.—On immunity with special reference to cell life. *Proceedings of the Royal Society of London*. 1900;66(424-433):424–448. doi:10.1098/rspl.1899.0121

18. Langley J. On the reaction of cells and of nerve-endings to certain poisons, chiefly as regards the reaction of striated muscle to nicotine and to curari. *Journal of Physiology*. 1905;33(4-5):374. doi:10.1113/jphysiol.1905.sp001128

19. Sutherland EW, Rall TW, Menon T. Adenyl cyclase: I. Distribution, preparation, and properties. *Journal of Biological Chemistry*. 1962;237(4):1220–1227. doi:10.1016/S0021-9258(18)60312-6

20. Butcher RW, Sutherland EW. Adenosine 3′,5′-phosphate in biological materials. I. Purification and properties of cyclic 3′,5′-nucleotide phosphodiesterase and use of this enzyme to characterize adenosine 3′,5′-phosphate in human urine. *Journal of Biological Chemistry*. 1962;237:1244–1250.

21. Rall TW, Sutherland EW. Adenyl cyclase. II. The enzymatically catalyzed formation of adenosine 3′,5′-phosphate and inorganic pyrophosphate from adenosine triphosphate. *Journal of Biological Chemistry*. 1962;237:1228–1232.

22. Sutherland EW, Robison GA. The role of cyclic-3′,5′-AMP in responses to catecholamines and other hormones. *Pharmacology Review*. 1966;18(1):145–161.

23. Robison GA, Butcher RW, Sutherland EW. Adenyl cyclase as an adrenergic receptor. *Annals of the New York Academy of Sciences*. 1967;139(3):703–723. doi:10.1111/j.1749-6632.1967.tb41239.x

24. Sutherland EW, Robison GA, Butcher RW. Some aspects of the biological role of adenosine 3′,5′-monophosphate (cyclic AMP). *Circulation*. 1968;37(2):279–306. doi:10.1161/01.CIR.37.2.279

25. Henderson R, Unwin PNT. Three-dimensional model of purple membrane obtained by electron microscopy. *Nature*. 1975;257(5521):28–32. doi:10.1038/257028a0

26. Singer SJ, Nicolson GL. The fluid mosaic model of the structure of cell membranes. *Science (1979)*. 1972; 175(4023):720–731. doi:10.1126/science.175.4023.720

27. Gill GN, Garren LD. On the mechanism of action of adrenocorticotropic hormone: The binding of cyclic-3′,5′-adenosine monophosphate to an adrenal cortical protein. *Proceedings of National Academy of Sciences U S A*. 1969;63(2):512–519. doi:10.1073/pnas.63.2.512

28. Gill GN, Garren LD. A cyclic-3′,5′-adenosine monophosphate dependent protein kinase from the adrenal cortex: Comparison with a cyclic AMP binding protein. *Biochemical and Biophysical Research Communications*. 1970;39(3):335–343. doi:10.1016/0006-291X(70)90581-4

29. Tao M, Salas ML, Lipmann F. Mechanism of activation by adenosine 3′:5′-cyclic monophosphate of a protein phosphokinase from rabbit reticulocytes. *Proceedings of National Academy of Sciences U S A*. 1970;67(1):408–414. doi:10.1073/pnas.67.1.408

30. Brostrom CO, Corbin JD, King CA, Krebs EG. Interaction of the subunits of adenosine 3′:5′-cyclic monophosphate-dependent protein kinase of muscle. *Proceedings of National Academy of Sciences U S A*. 1971;68(10):2444–2447. doi:10.1073/pnas.68.10.2444

31. Corbin JD, Sugden PH, West L, Flockhart DA, Lincoln TM, McCarthy D. Studies on the properties and mode of action of the purified regulatory subunit of bovine heart adenosine 3′:5′-monophosphate-dependent protein kinase. *Journal of Biological Chemistry*. 1978;253(11):3997–4003.

13 Cell Signaling Part II
G-protein-coupled Receptors

We saw in the previous chapter how a protein kinase—3′,5′-cAMP-dependent protein kinase (PKA)—was responsible for a variety of phosphorylation events in the cell. In addition, cAMP was shown to bind to regulatory components of PKA, resulting in the dissociation of these subunits from the catalytic subunits, the latter being responsible for the observed phosphorylation events. What remains of this story—and the topic of this chapter—is the mechanism by which ligand-binding to cell surface receptors results in the activation of adenylate cyclase. In the late 1960s, Martin Rodbell's group investigated one aspect of these questions: whether the various hormones which stimulate adenyl cyclase activity bound to the same or separate receptors. To this end, Birnbaumer and Rodbell used plasma membrane-rich fractions, called "ghosts", isolated from liver fat cells; essentially, cells whose contents were removed, leaving behind only their plasma membranes and the proteins associated with them. As such, the authors found that there was no increase in adenyl cyclase activity when a number of hormones (glucagon, ACTH, or epinephrine)—all of which activated adenyl cyclase by themselves—were incubated together (Figure 13.1a).[1] This indicated that each hormone competed for activation of a single adenyl cyclase, since in the alternate scenario each hormone would activate their respective adenyl cyclase and would presumably result in an additive increase in cAMP activity. However, they also determined that each hormone bound to a specific receptor on the cell surface by showing that known inhibitors of a given hormone had no effect on the ability of the other hormones to stimulate adenyl cyclase activity (Figure 13.1b).

A few years later, Rodbell et al. determined that guanyl nucleotides (GTP or GDP) were required components of the adenyl cyclase system. Although preliminary studies had indicated that these nucleotides decreased the binding of glucagon to its receptor, it was also suspected that they might activate the adenyl cyclase system at lower concentrations. However, ATP, at the concentration required for these studies (since the read-out for these experiments was the conversion of ATP into cAMP), was found to have the same effect as guanine nucleotides on the binding of glucagon to its receptor and reduction of ATP concentrations resulted in its premature degradation by ATPases. Therefore, the authors made use of a radiolabeled ATP analogue, AMP-PNP, which was resistant to hydrolysis by ATPases. As such, the authors found that indeed, addition of low concentrations of GTP in combination with glucagon significantly increased the activity of cAMP (Figure 13.2).[2]

This requirement was found to be specific for guanine nucleotides since other nucleotides only had minimal

effects. In addition, results indicated that glucagon and GTP *interact with the adenyl cyclase system in a non-competitive fashion.*[2] At this time, it was still unclear how the binding of hormones to their respective receptors induced activation of adenyl cyclase and increased cAMP formation, but these new findings indicated that there was a *relationship between the effects of guanyl nucleotides on glucagon binding and hormone activation.*[2] In addition, since both glucagon and GTP were required for maximal activation under some conditions, the authors suggested that there were *two regulatory sites, requiring the binding [...] of relationship between the effects of guanyl nucleotides on glucagon binding activation of the enzyme.* Finally, the authors commented on the consequences of these last discoveries on the action of hormones:

> These studies were initiated on the premise, widely held, that hormone receptors receive and transmit information imparted by the hormone to its target cell depending only upon the circulating levels of the hormone. This premise seems untenable for glucagon in view of the finding that guanyl nucleotides play an obligatory role in regulating the response of liver adenyl cyclase to glucagon. Further studies of the actions of the guanyl nucleotides on the liver and other adenyl cyclase systems may provide new insights into not only how hormones regulate target cell metabolism at the receptor level but also how a target cell metabolite regulates the initial response to a hormone.[2]

Although Rodbell's laboratory had shown that GTP was required for the activation of adenyl cyclase by hormone receptors, these results generated significant skepticism since they could not be readily reproduced. However, as Alfred Gilman later put it, *most were not working with the very nice membrane preparations that characterized the Rodbell laboratory.*[3] Nonetheless, the idea that GTP was indeed involved in the adenyl cyclase system was supported a few years later when a guanine nucleotide-binding protein was isolated from avian erythrocyte extracts and experiments suggested it may be involved in the activation of adenyl cyclase.[4] The next year, Cassel and Selinger detected hormone-stimulated GTPase activity (a protein which hydrolyzes GTP to GDP) that seemed to regulate the adenyl cyclase system.[5] These authors began from the knowledge that a non-hydrolysable GTP analog, guanosine-5′-[β,γ-imino] triphosphate (Gpp(NH)p), in combination with epinephrine, could practically irreversibly activate adenyl cyclase, neither of which were required to sustain the activation after a short lag period. This was contrary to the action of GTP and hormones, including epinephrine, for

DOI: 10.1201/9781003379058-13

Additions	Change in adenyl cyclase activity due to hormones
ACTH (6×10^{-8} M)	0.27 ± 0.03
Epinephrine (50×10^{-8} M)	0.28 ± 0.03
Glucagon (10×10^{-8} M)	0.25 ± 0.05
ACTH (3×10^{-8} M) + epinephrine (25×10^{-8} M)	0.40 ± 0.04
Epinephrine (25×10^{-8} M) + glucagon (5×10^{-8} M)	0.36 ± 0.02
ACTH (3×10^{-8} M) + glucagon (5×10^{-8} M)	0.46 ± 0.04

(a)

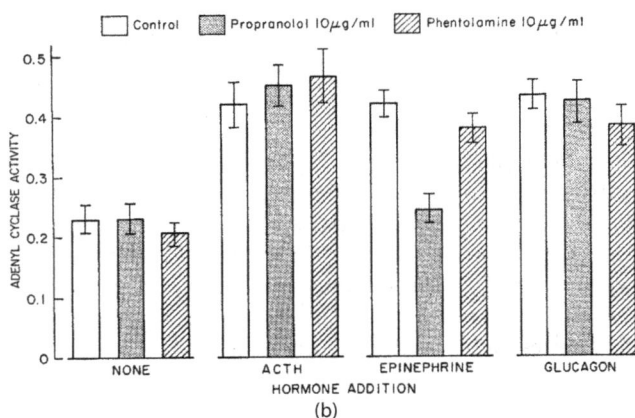

(b)

FIGURE 13.1 Hormones activate only one adenyl cyclase but bind to hormone-specific receptors. (a) A reaction mixture was incubated with the indicated components, and adenyl cyclase activity was assessed. The activity of adenyl cyclase under combination treatment was always lower than the calculated additive activity (obtained by adding the cyclase activity of the single treatments). (b) Cell ghosts were incubated with adenyl cyclase activators (ACTH, epinephrine, or glucagon) in the presence of either propranolol (β-adrenergic antagonist, the epinephrine receptor) or phentolamine (unrelated α-adrenergic receptor antagonist, used as a control). Notice how propranolol only inhibits the effects of epinephrine, while phentolamine has no effect.[1]

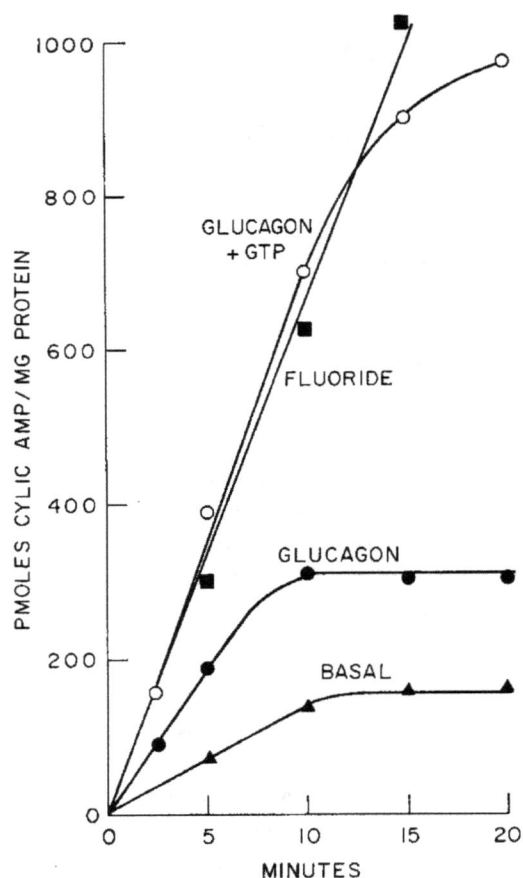

FIGURE 13.2 GTP is required for the efficient formation of cAMP. Liver plasma membranes were incubated with the indicated treatments and the amount of radioactive cAMP was assessed. GTP further increased the effects of glucagon. Fluoride was used as a positive control for cAMP activity.[2]

which activation of the enzyme was *completely reversible and enzyme activity required the continuous presence of both GTP and the hormone.*[5] As such, several investigators suggested that the adenyl cyclase system included a GTPase component which inactivated the enzyme.

To address this, Cassel and Salinger developed an assay to study the hydrolysis of [γ-^{32}P]-GTP by GTPases in turkey erythrocyte membranes in which the activity of other nonspecific nucleoside triphosphatases was minimized. As such, the results of their experiments indicated that a *high affinity, specific GTPase* [which they called basal GTPase] *[...] caused about 60% of the total GTP hydrolysis* upon activation of their system.[5] However, they also found that GTPase activity did not necessarily increase following activation of the adenylate cyclase system, which suggested that it had its own regulatory mechanism. In addition, results indicated that GTPase activity was localized to the *inner surface of the* [...plasma] *membrane* and therefore, that the *membrane contains a GTPase which is coupled to*

the β-adrenergic receptor [...]. The authors proposed the following mechanism to explain their data:

> Both the "basal" and the catecholamine [i.e., adrenaline] stimulated GTPases have a common catalytic component which exists in the membrane in two different states. In one state this component is coupled to the β-adrenergic receptor and has an "inhibited" conformation (catecholamine-stimulated GTPase) whilst in the other state the same component is not coupled to the receptor and is permanently active (basal GTPase). Upon activation by catecholamines, the "inhibited" GTPase gains the active conformation of the basal GTPase [...]
>
> We suggest that the catecholamines activated state of the adenylate cyclase has an increased ability to hydrolyze GTP. Nevertheless, the hydrolysis is slow enough to allow for an almost continuous presence of GTP at the regulatory site. Hydrolysis of GTP is required ultimately in order to allow the system to return to the inactive state.[5]

By the late 1970s, the mechanism of action of hormones was thought to be vectorial in nature; that is, hormone binding to its receptor on the outer surface of the cells was coupled to activation of adenyl cyclase on the inside of the cell. However, the precise mechanism by which this phenomenon occurred was still unclear; specifically, although the idea of a second messenger system—as proposed by Sutherland's group in the late 1960s—was well received, it was still debated whether ligand binding and the catalytic activity resulting in cAMP formation was mediated by a single or separate proteins. In 1977, Limbird and Lefkowitz conclusively answered this question when they identified two cell fractions, one of which mediated adrenaline binding to the β-adrenergic receptor, while the other was responsible for adenylate cyclase activity. As such, the authors concluded that:

> The direct documentation in the present studies that [...] receptor binding and adenylate cyclase activities reside in separable macromolecules implies that each hormone and drug receptor may be a distinct molecule capable of independently interacting with the adenylate cyclase moiety.[6]

The same year, Alfred Gilman's group provided further proof of this separate activity by purifying the two proteins and showing that they had completely different characteristics. The authors also suggested that both *the receptor and the enzyme may be integral membrane proteins that do not interact directly with each other but that communicate by mechanisms that are less subject to stringent stoichiometric restraints.*[7] Therefore, the mechanism by which these two receptors cooperated in the transduction of hormonal signals into cAMP activation became of major interest. Since purification of the receptor and the enzyme were still problematic, Ross and Gilman attempted to *study the mechanism of regulation of catecholamine-sensitive adenylate cyclase by the stepwise resolution of its components from intact membranes*, in the hope to *assay them by the reconstitution of hormone-sensitive adenylate cyclase activity in membranes that have been depleted of one or more factors by genetic or chemical manipulation.*[8] As such, the authors used three cell lines derived from the wild-type S49 cell (a murine lymphoma cell line, Figure 13.3): the AC⁻ cell line was deficient in adenylate cyclase activity but expressed β-adrenergic receptors; the B82 cell line had β-adrenergic receptors without a binding site for agonists and did it respond to these signals but has functional adenylate cyclase; and the UNC cell line had all known components

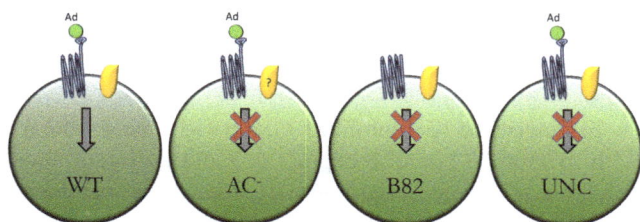

FIGURE 13.3 Cell lines derived from S49. See text for details. β-adrenergic receptor agonists (i.e., adrenaline) are indicated by a green circle; the β-adrenergic receptor is represented by the blue transmembrane protein; adenyl cyclase is in yellow; and transduction of signals to cAMP activation—that is, adenyl cyclase activity—is represented by the arrow.

but β-adrenergic receptor agonists still failed to stimulate adenylate cyclase activity.[8]

As such, the authors found that, although AC⁻ and B82 cells were unresponsive to hormones by themselves, significant stimulation resulted from the incubation of wild-type or B82 protein extracts with intact AC⁻ membranes, which indicated that the β-adrenergic receptors of one preparation could activate the adenyl cyclase in the other. However, this was not the case when UNC extracts were mixed with AC⁻ membranes, indicating that these membranes were not complimentary. These results also suggested that the B82 and wild-type extracts contained additional components necessary for the activation of adenyl cyclase. As such, the AC⁻ cell line was proposed to be deficient, not in adenylate cyclase itself, but rather, in unidentified components of this pathway. Although many aspects of their study remained unexplained at the conclusion of the paper, it was now clear that the β-adrenergic receptor and adenyl cyclase were separate and complimentary components and that the hormone-mediated adenylate cyclase system was composed of the following components: a hormone receptor to which hormones bound; an intermediate component that coupled the receptor to the enzyme; a catalytic component responsible for conversion of ATP to cAMP; a component regulated by guanine nucleotides; as well as additional components still needed to be identified.[9]

The same year, Ross and Gilman reported that adenylate cyclase activity was quickly lost when wild-type extracts were heated at 37°C but could be fully restored by the addition of AC⁻ membranes, which have no adenylate cyclase activity on their own.[9,10] This suggested that a heat-stable component (or at least, stable up to 37°C—they later determined it was stable up to 50°C) was present in wild-type cells that could rescue activity in AC⁻ cells. In addition, the observed activity in the co-incubation system gradually decreased if the AC⁻ membranes were heated at 30°C, which suggested that another component—which was labile at 30°C—was present in AC⁻ cells which could rescue the activity of heat-inactivated wild-type cells (Figure 13.4, basal). Interestingly, the component present in AC⁻ extracts was also found to be required for Gpp(NH)p- (which almost permanently activates adenylate cyclase) and sodium fluoride (NaF)-mediated activation (Figure 13.4). These results could also be reproduced when wild-type cells were treated with N-ethylmaleimide (NEM)—a reagent that reacts with cysteine residues (which suggested that the components were proteins)—strengthening the argument that adenylate cyclase activity was dependent on multiple proteins, and led the authors to conclude that a

> [...] thermostable, N-ethylmaleimide-resistant component (or components) is [...] not only sufficient to restore adenylate cyclase activity to detergent extracts of AC⁻ membranes but will also restore hormonal responsiveness when reconstituted with receptor-replete membranes from AC⁻ cells.[10]

In addition, the authors speculated that the component present in AC⁻ cells was the catalytic component (which they named C) of the adenylate cyclase system since it

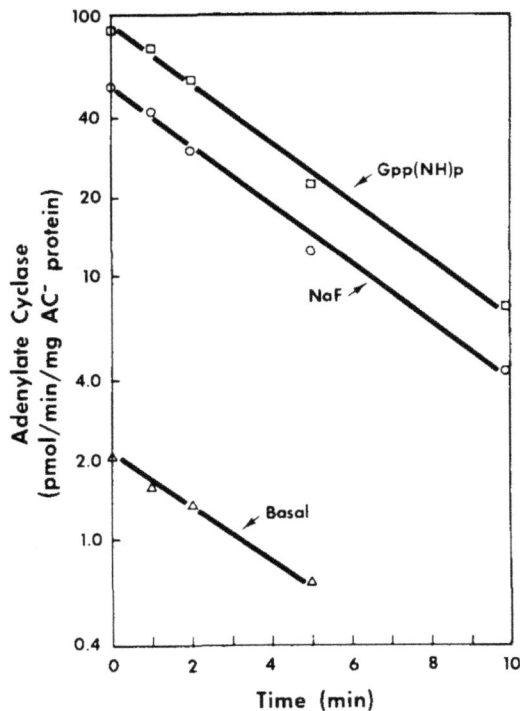

FIGURE 13.4 Differential heat sensitivities of adenylate cyclase components. AC⁻ membrane extracts were heated at 30°C for the indicated times, incubated with heat-inactivated (37°C) wild type extracts, and assayed for adenylate cyclase activity.[9]

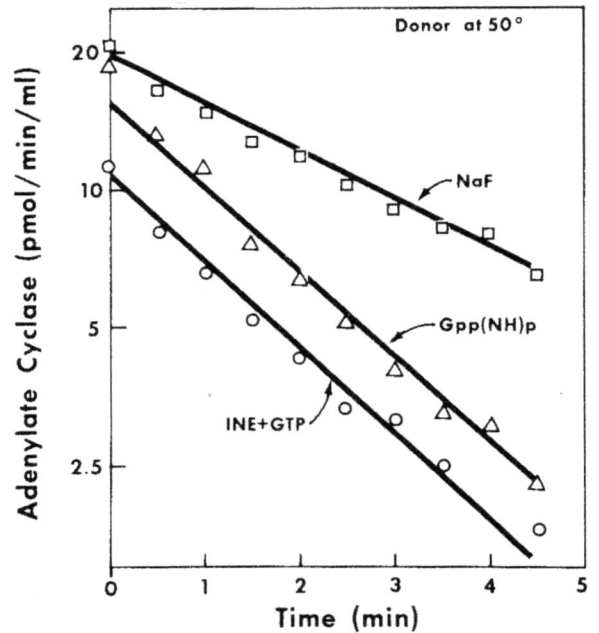

FIGURE 13.5 F/G is a coupling factor between the receptor and C. Wild type extracts were heated at 37°C to inactivate C and further heated at 50°C for the indicated times before being added to AC⁻ membranes (source of C) in the presence of the indicated reagents. The activity in the presence of INE + GTP represents basal activity of the system.[10]

was required for restoration of all activity, whether basal or otherwise.[10] In support of this claim, they found that AC⁻ membranes and detergent extracts could catalyze the formation of cAMP cells in the presence of Mn^{2+} without being subject to regulation by hormones, NaF, or Gpp(NH)p. In addition, the kinetics of inactivation of C and of Mn^{2+}-dependent activity in AC⁻ cells were very similar, which suggested that these two components were the same. Therefore, these results were consistent with their previous suspicion that the AC⁻ cell line did, in fact, express adenylate cyclase but was instead deficient in other component(s), namely, the regulatory component(s) of the system; as such, the AC⁻ cell line was renamed cyc⁻ and the name adenylate cyclase was used to refer to C.

Initial experiments on the kinetics of inactivation of the heat-stable and NEM-resistant component(s) present in wild-type cells—experiments in which wild-type extracts were heated to 37°C to inactivate C and heated further to 50°C for various amounts of time before being added to AC⁻ membranes—suggested the presence of two separate proteins: the decline in activity in the presence of NaF was much slower than that under basal or Gpp(NH) p-mediated conditions (Figure 13.5). As such, the authors named the component responsible for the regulation of NaF-mediated activity, "F", while the one that regulated basal and Gpp(NH)p activity—and which seemed to have a GTP binding site—was named "G". In addition, the authors suggested that *G is required as an essential coupling factor*

between the receptor and C since it was responsible for the basal activity.[10] However, the authors stressed that the current data could not fully address if these components were truly separate proteins or the same protein performing both functions, therefore they often referred to the component(s) as F/G. They suggested that this/these protein(s) were the regulatory component(s) of the system.[10] As such, the authors proposed that:

[…] a reasonable mechanism for regulation of activity is regulation of the interaction of these components *[C and G/F]*. Gpp(NH)p, NaF, or the hormone receptor complex might be involved in modulating an association-dissociation reaction among these proteins in the membrane. At a more practical level, these studies raise the possibility that the apparent lability of solubilized adenylate cyclase, observed by many investigators, might be explained by dissociation of two otherwise active proteins or by the denaturation of only one of several components.[9]

To determine if solubilized C and G/F could interact with membrane-bound receptors to yield a system amenable for adenylate cyclase activity, Ross *et al.* incubated solubilized wild-type cell membrane extracts (which contained solubilized C) with heat-treated AC⁻ cell membranes (to inactivate C, used as a source of membrane-bound receptor), and then assessed the reconstituted mixture for adenylate cyclase activity. Indeed, not only was NaF- and Gpp(NH)p-mediated adenylate cyclase activity in the reconstitution system retained, but it was greater than that

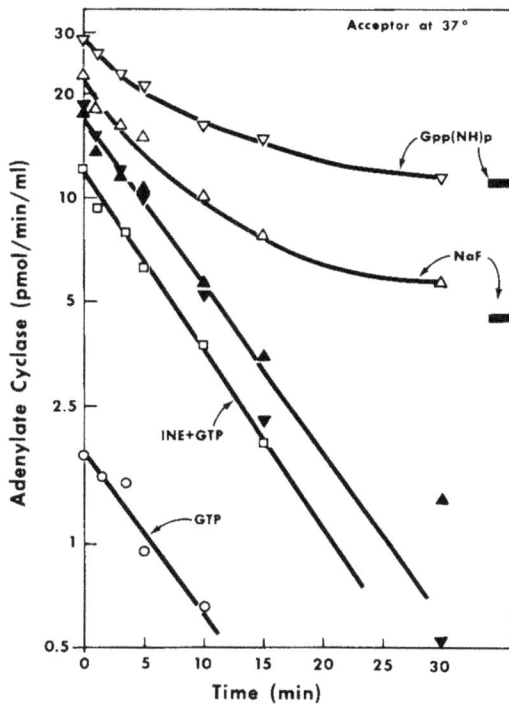

FIGURE 13.6 G/F can only interact with membrane-bound C. AC⁻ membranes were incubated at 37°C for the indicated times (to inactivate C) and then added to wild-type membrane extracts in the presence of the indicated reagents and adenylate cyclase activity was assessed (open symbols). As a control, untreated wild type membranes extracts were incubated with NaF or Gpp(NH)p (black bars to the right) and these values were subtracted from the reconstitution values (black triangles) to account for the activity contributed by the donor membranes.[10]

observed in the wild-type sample alone, a phenomenon that had been previously observed but remained unexplained (Figure 13.6, compare T_0 of open triangles vs solid bars on the right).[10] However, the authors found that this increase was lost if the AC⁻ membranes were heated for an extended period of time (Figure 13.6, compare T_{30} of open triangles vs solid bars on the right); as such, the decrease in activity following subtraction of the wild type activity had a slope reminiscent of the inactivation of membrane-bound C (compare solid triangles in Figures 13.5 and 13.6). This indicated that the observed NaF- and Gpp(NH)p-stimulated activity was mediated solely by the soluble C contributed by the wild-type extracts. As such, the authors reasoned that only the membrane-bound C in AC⁻ membranes—and not the soluble form added from the wild-type extracts—could interact with G/F during hormone-stimulated activation.[10] Again, similar results were obtained when AC⁻ membranes were inactivated with NEM. Thus, the authors concluded that:

> solubilized G/F [...] interacts freely with membrane-bound C and receptor to promote a response of the enzyme to hormone. It is tempting to speculate that C and hormone receptors are integral membrane proteins situated

primarily in the core of the bilayer or bound to it via a hydrophobic extension, and that G/F is a peripheral protein on the cytoplasmic surface of the plasma membrane. Such speculations can be tested in increasingly defined reconstituted systems.[10]

* * * * *

In 1980, Howlett and Gilman determined that G/F had a molecular weight of 130 kDa and sedimented at 4.8 S in untreated cells, but lost about 40 kDa and decreased in sedimentation rate to 3.6–3.8 S when treated with NaF or the non-hydrolyzable GTP analog, Gpp(NH)p.[11] The loss in molecular weight correlated with previous findings that GTP bound to a protein with a molecular weight of 42 kDa and that treatment with Gpp(NH)p resulted in the appearance of an additional protein which sedimented at 4.6 S. As such, the authors proposed that *there is no reason to believe that this* [Gilman's 40 kDa protein] *is not G/F or a component thereof.* The authors also noted that:

> The physical nature of the change responsible for the altered sedimentation behavior is mysterious. If a subcomponent of G/F of 40,000 daltons dissociated from a larger species of 130,000 daltons, one would expect to be able to separate the hypothetical particles either by gel filtration or sucrose density gradient centrifugation.[11]

However, a complicating factor in the purification of these proteins was that it required conditions which caused the protein to revert back to its holoenzyme state. In addition, the molecular weight of the adenyl cyclase complex (which included both C and G/F) had previously been determined to be 250 kDa and that of C to be 200 kDa. Therefore, the fact that G/F had now been determined to be 130 kDa was puzzling and the authors conceded that:

> Unfortunately, no simple, unambiguous explanation can be offered since it would require a relatively generous margin for error for 200,000 + 130,000 to equal 250,000. Some of the obvious explanations include the possibilities that the molecular weights of C and G/F as determined in detergent solution do not correspond to the species that exist in the membrane or that the poorly reversible Gpp(NH)p-activated form of the enzyme is not formed simply by association of C and G/F. Since it has not yet been possible to resolve the unstable C from G/F by other than genetic techniques (the S49 cyc⁻ cell *[G/F deficient]*), the relationship of this protein to the actual catalyst is perhaps suspect.[11]

Later the same year, the same group reported the successful purification of G/F, noting that:

> Purification of G/F was the logical outgrowth of the ability to assay its activity by reconstitution of a functional adenylate cyclase complex. Because the reconstitution of G/F depends on its prior extraction from membranes with detergents, such extracts represent the first point at which the specific activity of the protein can be measured.[12]

As such, Northup et al. used rabbit liver extracts *because large quantities of partially purified plasma membranes can be obtained with moderate ease and the specific activity of G/F is comparable to or better than that observed from several other sources.*[12] The authors described a six-step purification procedure resulting in a 2000-fold increase in G/F activity compared to the crude detergent extracts, and SDS-PAGE analysis revealed the presence of two distinct bands at 45 kDa and 35 kDa, and a minor band at 52 kDa, which agreed with the 130 kDa molecular weight determined for native G/F protein (Figure 13.7a; the 41 kDa protein was thought to be a contaminating protein). These proteins were confirmed to be G/F since their addition to cyc⁻ membranes reconstituted adenyl cyclase activity and thus, provided additional support for the idea that *G/F has a multisubunit structure of one or more of the polypeptides* (Figure 13.7b).[12] Interestingly, the 35 kDa protein was found to be present in excess in these purifications and much of it was released during one of the purification steps. The ratio of this subunit to the other two proteins was determined to be 1:1 in the peak fraction of G/F activity, but it did not contain any G/F activity on its own.[12] In addition, in contrast to their hypothesis that G/F activity came as a result of the association of the subunits, Northup et al. reported that incubation with both the 35 kDa and 45 kDa proteins did not restore activity. As such, the authors noted that proof of *contribution of all three species* to G/F activity would require successful isolation of each of these proteins.

The next year, Gliman's group reported improvements to their purification protocol that proved to be quicker, more efficient, and provided a 15-fold increase in yield.[13]

In their study, Sternweis et al. confirmed the presence of the three previously observed bands using SDS-PAGE but determined the total molecular weight to be 70 kDa for G/F, almost half of what they had previously reported. However, consistent with previous results, activation of G/F resulted in a decrease in its size, this time to about 50 kDa. The authors also reported that the activity of purified G/F varied widely depending on the detergent used for purification and the activation treatment, resulting in increased difficulties in accurately assessing the success of reconstitution experiments; these difficulties might explain much of the confusion that arose when different levels of activity were reported. Nonetheless, they assessed the requirements of these proteins on the activity of adenyl cyclase and the nature of their interaction using GTPγS-activated G/F, which is *a very stable form of G/F*, and by varying the amount of C and of cyc⁻ membrane added. Results showed that activity was *clearly dependent on the concentration of both G/F and C [...] consistent with a simple bimolecular equilibrium in which the complex G/F · C is observed as adenyl cyclase activity: G/F + C ⇌ G/F · C.*[13] In addition, the authors succeeded in the partial resolution of the 45 kDa and 52 kDa subunits and noticed that the latter eluted slightly ahead of the former in one of their purification steps (western blot in Figure 13.8). Assessment of these proteins indicated that although hormone-mediated activity was more efficient in the presence of the 52 kDa protein, this protein was not strictly required for adenyl cyclase activity since NaF-mediated activity was only correlated with expression of the 45 kDa protein (Figure 13.8).[13]

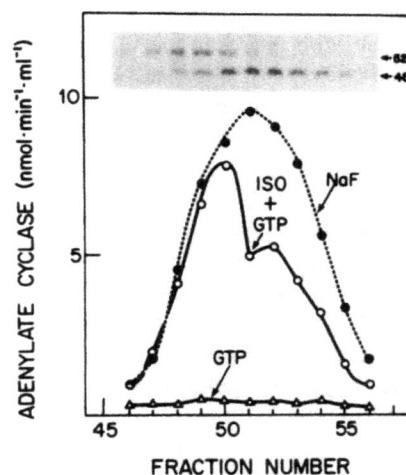

(a) (b)

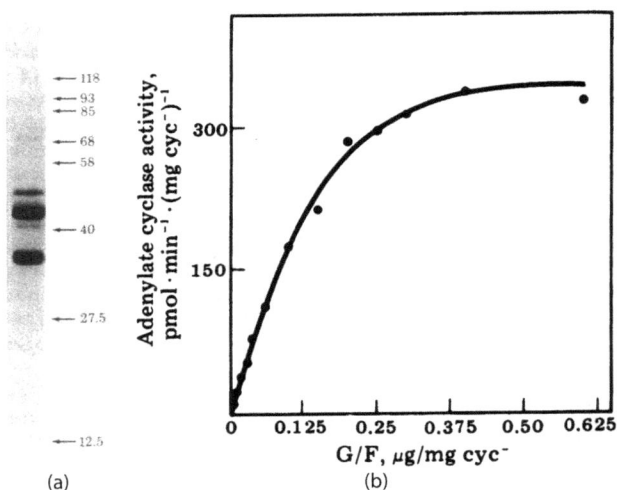

FRACTION NUMBER

FIGURE 13.7 (a) Western blot of purified G/F showing the predominant 35, 45, and 52 kDa proteins. (b) Addition of G/F to a reaction mixture increases adenylate cyclase activity. Purified G/F protein was added to cyc⁻ membranes in the presence of NaF and assessed for adenylate cyclase activity.[12] (Reprinted from Northup JK, Sternweis PC, Smigel MD, Schleifer LS, Ross EM, Gilman AG. Purification of the regulatory component of adenylate cyclase. Proc Natl Acad Sci U S A. 1980;77(11):6516–6520. Used by permission.)

FIGURE 13.8 Differential requirement of the 45 kDa and 52 kDa on NaF- and hormone-stimulated activation of adenylate cyclase. Components of G/F protein were purified and each fraction was assayed for adenylate cyclase activity in the presence of GTP, GTP + isoproterenol (adrenaline analog that stimulates adenylate cyclase activity), or NaF. Note how activity in the presence of NaF changes directly with the presence of the 45 kDa protein, whereas the peak of hormone-mediated activity occurs in the presence of the 52 kDa protein.[13]

Although the 35 kDa protein was not assayed in these experiments and did not seem to be required for activity, the authors stressed their previous finding that the 35 and 45 kDa proteins were consistently present in a 1:1 ratio and that the best evidence for the former being an integral member of G/F was the fact that these proteins *had not been separable under conditions that permit recovery of G/F activity* and that they were the only two proteins purified in turkey erythrocytes.[13] They also justified the smaller molecular weight of 70 kDa reported in this publication by noting that their previous experiments in which they reported a molecular weight of 130 kDa for the complex used an amount of detergent that may have prevented the isolation of a monodispersed protein, resulting in protein aggregation that would have skewed the results. They also discussed—contrary to their initial hypothesis—the possibility that activation of G/F leads to dissociation of the 45 kDa and the 35 kDa species since the subunit remaining after activation (with a total size of 50 kDa) was similar in size to that of the single subunits. However, they could not explain the fact that purification of this complex was performed under conditions that included activating reagents, *yet both 35-kilodalton and 45-kilodalton proteins are retained.*[13]

Finally, the authors admitted that their discussion ignored the 52 kDa protein, justifying that it was present only as a minor constituent of the protein complex and questioned its functional role. Nonetheless, they conceded that although the 45 kDa protein alone was sufficient to reconstitute activation by hormones, the *52-kilodalton subunit appear to reconstitute a more efficient coupling between receptor and the catalytic portion of adenylate cyclase than do preparations that are essentially devoid of this subunit.*[13] In addition, they found that samples in which the 52 kDa protein was enriched had a faster response to the non-hydrolysable GTP analogue, GTPγS, which they thought might reflect the same functional differences observed in response to β-adrenergic agonists and suggested that *the two subunits possess the same functional activities but with altered capacities to carry them out.*[13] This also led to the speculation that *the two subunits are related in structure.*[13]

This study was followed up, once again, by Gilman's group the following year in a paper in which they used radiolabeled GTPγS or Gpp(NH)p to *assess the interactions between pure G/F and guanine nucleotides and kinetic analyses of the binding and activation reactions.*[14] They first observed, as expected, that the interaction of these purine analogues with G/F perfectly correlated with the activity of G/F. Strikingly, they found that *although GTPγS does not undergo any covalent modification during the binding reaction, binding appears to be essentially irreversible in the presence of divalent cation.* In an attempt to clarify this point, the authors further studied the kinetics of G/F activation using different concentrations of GTPγS, which revealed that the steady state level of G/F activation was dependent, whereas its rate of activation was completely independent,

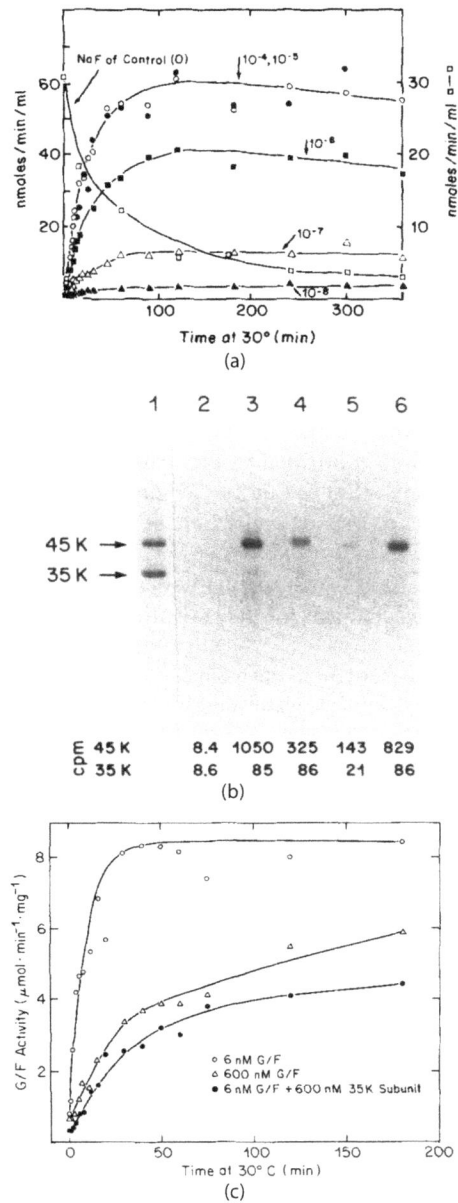

FIGURE 13.9 (a) The rate of G/F activation is independent of the concentration of GTPγS. G/F was added to a reaction mixture along with the indicated concentrations of GTPγS. Samples were removed at the indicated times, added to cyc⁻ membranes, and assayed for adenylate cyclase activity. Note that the slope of the curve at each GTPγS concentration is similar, which indicates that GTPγS activates G/F equally well regardless of its concentration. (b) GTP binds to the 45 kDa subunit. G/F was incubated with [^{32}P]-8-N$_3$GTP (a photolyzable GTP analog, i.e., emits light when incorporated into a protein) for 10 minutes at 4°C with the indicated reagents. The membrane was then irradiated to activate the photolyzable GTP analog. Lane 1 was a control for purified 35 and 45 kDa proteins, whereas lane 2 was not photolyzed. The subunits were then purified from the gel and assessed for radioactivity (shown at the bottom of the gel). Note how GTP was present at the same position as the 45 kDa subunit and that GTPγS, but not ATP, blocked incorporation of [^{32}P]-8-N$_3$GTP (lane 3 vs. 5 vs. 6). (c) G/F activity is inhibited by the 35 kDa subunit. G/F was incubated in a reaction mixture containing GTPγS, with or without excess 35 kDa subunit, and adenylate cyclase activity was assessed.[14]

on the concentration of GTPγS (Figure 13.9a). The authors explained these *somewhat anomalous kinetics* by proposing

a model in which rapid equilibrium binding of GTPγS is followed by a first order, apparently irreversible, activation reaction. In addition, unliganded G/F is subject to a first order, irreversible inactivation. The steady state activation seen is then dependent on the ratio of the two competing first order reaction rates and the fractional saturation by GTPγS.[14]

Kinetic and inhibition experiments using ^{32}P-labeled GTP analogs also revealed the presence of only one GTP-binding site per molecule of G/F, which bound specifically to the 45 kDa subunit (Figure 13.9b). Their hypothesis that activation of the complex resulted from the dissociation of its subunits was confirmed in this paper by adding excess 35 kDa protein to a mixture of G/F and GTPγS and found that indeed, *the rate of activation of G/F was decreased [...] by the addition of the purified 35,000-dalton subunit* (Figure 13.9c).[14] As such, the authors proposed the following revised mechanism for the activation of G/F:

We propose that G/F in the basal state exists as a dimer of the two predominant subunits (45,000 and 35,000 daltons). This species participates in a rapid equilibrium binding reaction with nucleotide [...] Activating nucleotides such as GTPγS and Gpp(NH)p presumably possess a greater affinity for the dissociated form of the 45,000-dalton subunit [...]

A model for regulation of adenylate cyclase by guanine nucleotides that has directed research concepts in the field for some time ascribes a GTPase activity to the nucleotide regulatory site. This model also requires that the hydrolytic product, GDP, remains tightly associated with the site [...] We do not rule out the expression of a GTPase activity of G/F after reinsertion of the protein into a membrane or the requirement for other protein components of the hormone-sensitive adenylate cyclase system for such activity [...]

The ability of GTPγS and Gpp(NH)p to activate adenylate cyclase essentially irreversibly has been ascribed to the resistance of these nucleotides to the action of the putative GTPase. However, if GTP is not hydrolyzed by pure G/F, an explanation for the lack of activation of G/F by GTP must be sought.[14]

* * * * *

Between 1978 and 1981, Michio Ui's group in Japan published a number of papers in which they identified a new protein, called islet-activating protein (IAP), which was isolated from culture medium of *Bordetella pertussis* and which was determined to be one of the pertussis toxins.[15,16] Following investigations on the action of IAP on the adenylate cyclase system, the authors suggested that IAP might *exert its unique influences on the mechanism by which stimulation of membrane receptors leads to inhibition or activation of adenylate cyclase in islet cells.*[17] In support of this hypothesis, Ui's group showed that this process was dependent on GTP and, similar to what had already been shown with G/F, GTPγS and Gpp(NH)p further increased these effects. However, despite the fact that

the adenyl cyclase system was likely involved in this process, their data indicated that the site of action of IAP was neither the catalytic subunit nor the GTP-binding protein of adenyl cyclase, nor was it the α-adrenergic receptor itself. Katada and Ui also reported that the effects of IAP were associated with the incorporation of an ADP-ribose moiety into an unknown 41 kDa membrane protein.[18,19]

In 1983, Gilman's group, which now included Katada (who had made the leap from Japan to the US), partially purified the substrate for IAP, which was found to be a protein complex containing 41 and 35 kDa subunits akin to the proteins in the G/F system.[20] In fact, IAP and G/F were so similar that both substrates readily co-purified; ironically, the 41 kDa protein was long thought to be a pesky contaminant in many of Northup's purifications of G/F—this can actually be seen in the western blot of Figure 13.7 (a) (light band under the major 45 kDa band) when the purification of G/F was initially published.[3] In addition, like the 45 kDa protein in G/F, IAP's 41 kDa protein also contained a GTP-binding site and the molecular weight of its substrate decreased from 80 to 50 kDa in the presence of GTPγS or NaF, consistent with the proposed mechanism of dissociation of G/F upon activation.[20] Given the similar properties of the G/F and IAP proteins, the larger of these was renamed the α-subunit and the smaller the β-subunit.[21] Concluding their paper, the authors suggested that:

It is plausible that the catalytic activity of adenylate cyclase is modulated in a reciprocal fashion by a homologous pair of GTP-binding regulatory proteins, G/F (G$_s$, stimulatory) and the IAP substrate (G$_i$, inhibitory). Also, intriguing is the fact that transducin, a guanine nucleotide-binding regulatory protein of the rod outer segment that interacts with a light-activated cyclic GMP phosphodiesterase *[in the eyes]*, exhibits striking structural and functional similarities. An interesting family of proteins appears to be emerging.[20]

At the same time, Northup, Sternweis, and Gilman described a new,

highly sensitive method for detecting the activity of the β subunit *[the 35 kDa protein]*, both that which has been resolved from G/F during the final step of purification and the polypeptide that is released from the G/F heterodimer by thermal denaturation [...] based on the ability of the 35,000-Da protein to stimulate the rate of deactivation of G/F[...][21]

To this end, the authors incubated a mixture of NaF-activated G/F with or without the β-subunit and found significant deactivation in its presence (Figure 13.10).[21] In addition, the reversibility of this reaction was demonstrated by the fact that the addition of excess NaF to a mixture that was inhibited by the 35 kDa protein resulted in the recovery of about 80% of the initial activity. Given the particular sensitivity of this new assay, the authors tested their genetic models to determine if β-subunit activity could be detected in them and surprisingly, found that both the cyc⁻ (which was thought to

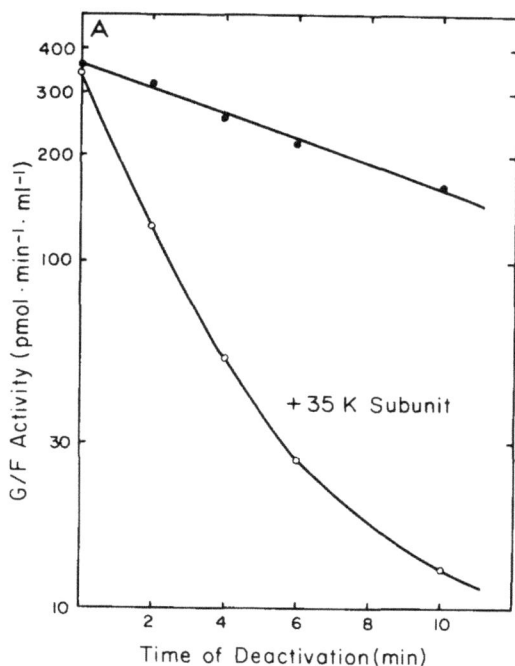

FIGURE 13.10 Purified 35 kDa subunit of G/F inhibits G/F activation. G/F was activated by NaF (closed circles) and purified 35 kDa subunit (i.e., the β-subunit) was added (open circles) for the indicated times.[21]

be deficient in G/F) and the UNC (which was deficient in G/F activity but contained the 45 kDa protein) were positive for such activity, that is, adenylate cyclase activity was inhibited by addition of detergent extracts from these cell lines.[21] In addition, β-subunit activity was found to be severely compromised by addition of the 41 kDa protein of IAP, indicating that association between the two proteins may be possible. Thus, the authors concluded that:

> the G/F deactivating activity in detergent extracts of plasma membranes is indeed the 35,000-Da protein and that the putative GTP-binding protein with which the majority of the β subunit associates appears to be the 41,000-Da of the IAP substrate. This dimer dissociates and its subunits are resolved during chromatography on heptylamine-Sepharose in the presence of *[activating compounds]*. In *[their]* absence, the *[protein]* fractionates as a dimer of 41,000- and 35,000-Da polypeptides [...]
>
> This finding lends further support to the hypothesis that G/F and the AIP substrate share common β-subunits [...]
>
> The fact that two separate GTP-binding regulatory proteins, G/F and the IAP substrate, share a common β subunit raises many interesting questions about the function of the β subunit in the regulation of adenylate cyclase activity. We believe that we have developed convincing evidence that the β subunit exerts an inhibitory influence on the activation of G/F by fluoride and guanine nucleotides [...][21]

The experiments above were performed using partially resolved G/F subunits, that is, each preparation was contaminated with small amounts of the other subunit. In an accompanying paper, the authors reported the successful

resolution of the two G/F subunits.[22] As such, they confirmed that purified α-subunit alone was sufficient for the reconstitution of the G/F activity in the presence of adenylate cyclase and likewise, that the β-subunit was sufficient for its deactivation. In addition, they used a sedimentation procedure to confirm that deactivation of α-subunit results in the reformation of the α·β subunit, that is, the original G/F complex.

The next year, Hildebrandt et al. reported the purification of yet another protein contained in the G/F complex, which previously escaped detection due to its small size. This discovery was spurred on by the transducin protein, a protein complex analogous to G/F (now referred to as N_s, for stimulatory) and IAP (now referred to as N_i, for inhibitory), which was found to be composed of three proteins: a 39 kDa α-subunit, a 35 kDa β-subunit, and a 5 kDa γ-subunit.[23] To that end, the authors modified their approach to gel electrophoresis since the standard method developed by Laemmli did not allow visualization of very small proteins. As such, they *set-up a urea-polyacrylamide gel electrophoresis system* [SDS-DUPAGGE] *[...] which allowed for visualization of well separated* α *and* β *subunits [...] and of well focused polypeptides of M_r between 2,500 and 10,000.* In doing so, they were able to consistently resolve a band of approximately 5 kDa in preparations of N_s, N_i, and of those of a new complex containing a 40 kDa protein and a β-subunit but neither of the known α-subunits (Figure 13.11a). In addition, the finding that the 5 kDa protein seemed to be present whenever the β-subunit was present suggested to the authors that it remained bound to it and was analogous to the γ-subunit of transducin. The authors concluded their paper as such (Fig. 3.11b):

> The findings reported here, together with data currently published in the literature, suggest adenylyl cyclase systems as "3-component systems" to which stimulatory and inhibitory receptors couple by interacting with distinct N coupling proteins, each of which is a heterotrimer of αβγ composition [...] Presently, we do not know if all of the γ subunits associated with N_s, N_i, and 40K protein are identical or not. All have about the same molecular weight on SDS-DUPAGGE, suggesting that they may be identical; but, this parameter does not detect subtle differences. Likewise, we do not yet have any idea as to its function. The γ subunit of transducin was discovered quite early in the purification of that regulatory protein, but its function is also still unknown.[23]

* * * * *

In previous chapters, we discussed the first experiments that elucidated an important aspect of a cell's life called signal transduction—a signaling cascade usually mediated by phosphorylation—which is transmitted from one protein to another. This story had its beginnings in sugar metabolism whereby the sequential phosphorylation and cleavage of molecules was required for the synthesis of ATP and led to the recognition that phosphorylation was

(a)   (b)

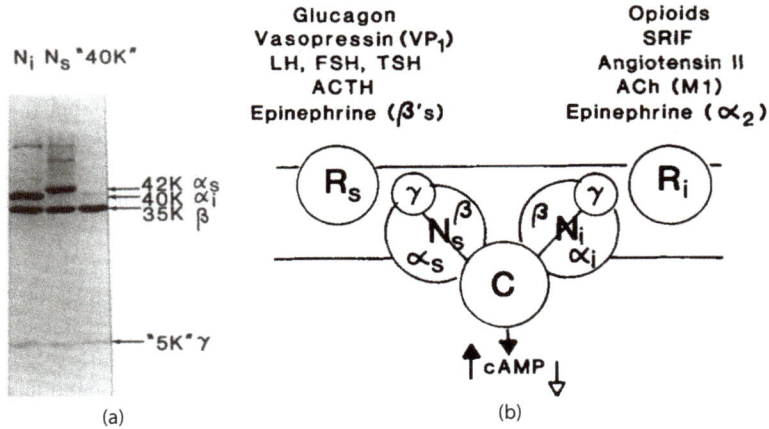

FIGURE 13.11 (a) Resolution of the α-, β-, and γ-subunits of G/F (N_s), IAP (N_i), and 40K fractions; (b) Model of the N_s and N_i systems as suggested by Hildebrandt et al. in 1984. Ligand binding to a stimulatory receptor (R_s) activates Ns (the G/F complex), which, in turn, activates adenylate cyclase (C), and stimulates cyclization of ATP (i.e., cAMP production). In contrast, ligand binding to an inhibitory receptor (R_i) leads to the activation of N_i (IAP) and the inhibition of adenylate cyclase activity, resulting in a decrease in cAMP formation. (Adapted from ref. [23].)

an important aspect of cell signaling in general. This modification by phosphorylation was shown to occur in the following manner:

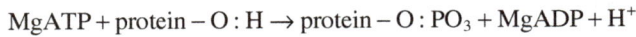

$$MgATP + protein - O:H \rightarrow protein - O:PO_3 + MgADP + H^+$$

whereas phosphatases performed the opposite role, that is, dephosphorylation of proteins:

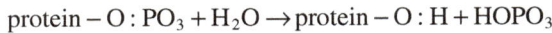

$$protein - O:PO_3 + H_2O \rightarrow protein - O:H + HOPO_3$$

The first kinase involved in cell signaling to be discovered, cyclic AMP-dependent protein kinase (later renamed protein kinase A, or PKA), came as a result of these initial studies. However, the cellular events leading to PKA activation came following studies focused on the effects of hormones on cells (Figure 13.12). These studies revealed that in the uninduced cell, trimeric G-protein complexes (i.e., protein complexes composed of α-, β-, γ-subunits)—anchored to the plasma membrane by their lipid tails but otherwise existing in the cytosol—move around in the plasma membrane by Brownian motion as dictated by the fluid mosaic model and sometimes collide with various transmembrane receptors (this mode of coupling was termed "collision coupling").[24] Binding of adrenaline to a β-adrenergic receptor (a type of seven-transmembrane protein also known as G-protein coupled receptor (GPCR)) induces a conformational change in the receptor, which facilitates the exchange of GDP for GTP on the α-subunit of a G_s type of trimeric G-proteins (in the case of adrenaline) associated with the receptor. This releases the βγ-subunits from the complex and allows the GTP-bound, activated, α-subunit to diffuse along in the plasma membrane until it bumps into and activates adenylyl cyclase (the C component), which catalyzes the formation of cAMP from ATP. Two cAMP molecules then go on to bind to each of the regulatory subunits of PKA, resulting in a change in conformation in these

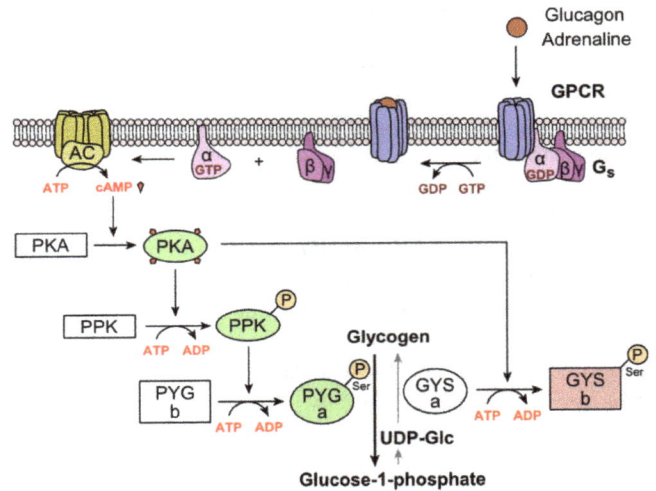

FIGURE 13.12 Pathway leading to the activation of PKA by adrenaline. Adrenaline binds to its receptor, a G-protein coupled receptor, which catalyses the exchange of GDP for GTP in a trimeric G-protein and results in the separation of the α-subunit from the complex. This protein goes on to activate adenylate cyclase (AC), which catalyses the cyclization of ATP to cAMP, four of which go on to bind the regulatory subunits of PKA, thereby releasing the catalytic subunits. These subunits, which are kinases, phophorylate a host of proteins which contain the appropriate amino acid recognition site. One of these substrates, phosphorylase kinase (PPK), phosphorylates the inactive form of glycogen phosphorylase (glycogen phosphorylase b (PYG b)), which results in its activation (glycogen phosphorylase a (PYG a)) and the production of glucose-1-phosphate from glycogen, which goes on to enter glycolysis for the production of ATP. PKA also phosphorylates and inactivates glycogen synthase, thereby inhibiting synthesis of glycogen from glucose-1-phosphate. As one can imaging these events are triggered when the body requires energy to do work. This pathway is also regulated by enzymes whose activation leads to the dephosphorylation of glycogen synthase and glycogen phosphorylase a, thereby increasing the synthesis of glycogen, as would occur after eating to ensure that the sugars consumed are stored for later use.[26]

subunits and the release of the two catalytic subunits. The latter go on to phosphorylate any protein with a proper recognition sequence, the canonical sequence being Arg-Arg-X-Ser/Thr, where X represents any amino acid.

Regulation of this signaling cascade can be achieved in a number of ways, the first of which is an intrinsic GTPase activity in the α-subunit leading to hydrolysis of GTP to GDP; this implies that activation of this pathway has its own built-in "timer". Alternatively, ligand-binding of other GPCRs which activate G_i α-subunits (such as found in IAP) leads to the inactivation of adenylate cyclase, a decrease in cAMP, and consequently, the inactivation of PKA; similarly, phosphodiesterases can also modulate PKA activity downstream of hormone binding by hydrolyzing cAMP. A number of other mechanisms are also involved but will not be discussed here; as expected, these events are very tightly regulated by a variety of pathways which make sure that the precise level of phosphorylation ensues given the sum of all signals detected by the cell. Although PKA was the first kinase to be discovered, it was certainly not the last. Among other important kinases are PKB (or Akt), PKC (activated by calcium), and PKG (activated by cGMP), each with their own specific agonists and substrates. In 2002, Manning et al. published a paper describing all protein kinases

known at the time: 518 genes divided into nine groups, a protein family constituting almost 2% of all human genes (one of the largest in the human genome).[25]

While PKA phosphorylates specifically on serine or threonine residues (called protein-serine/threonine kinases), other types also exist (protein-tyrosine kinases and protein-tyrosine-like kinases), each with their own specific recognition motif.[25] In addition, although PKA and the other kinases mentioned above are soluble proteins found in the cell's cytoplasm, many others are transmembrane receptors which activate pathways by directly phosphorylating downstream proteins. Either way, phosphorylation represents the main event in the mechanism of action of signal transduction pathways, which can result in conformational changes in a protein which allows it to associate with other proteins, or the phosphate may simply add an element (such as a charge) required for association with other proteins (remember that protein–protein interactions are analogous to a lock and key). In contrast, phosphorylation can also interfere with protein association by creating steric hinderance or adding a negative charge which causes a repulsion between proteins which are normally associated.

In previous sections, we discussed how binding of GTP to the α-subunit of trimeric G-proteins leads to their

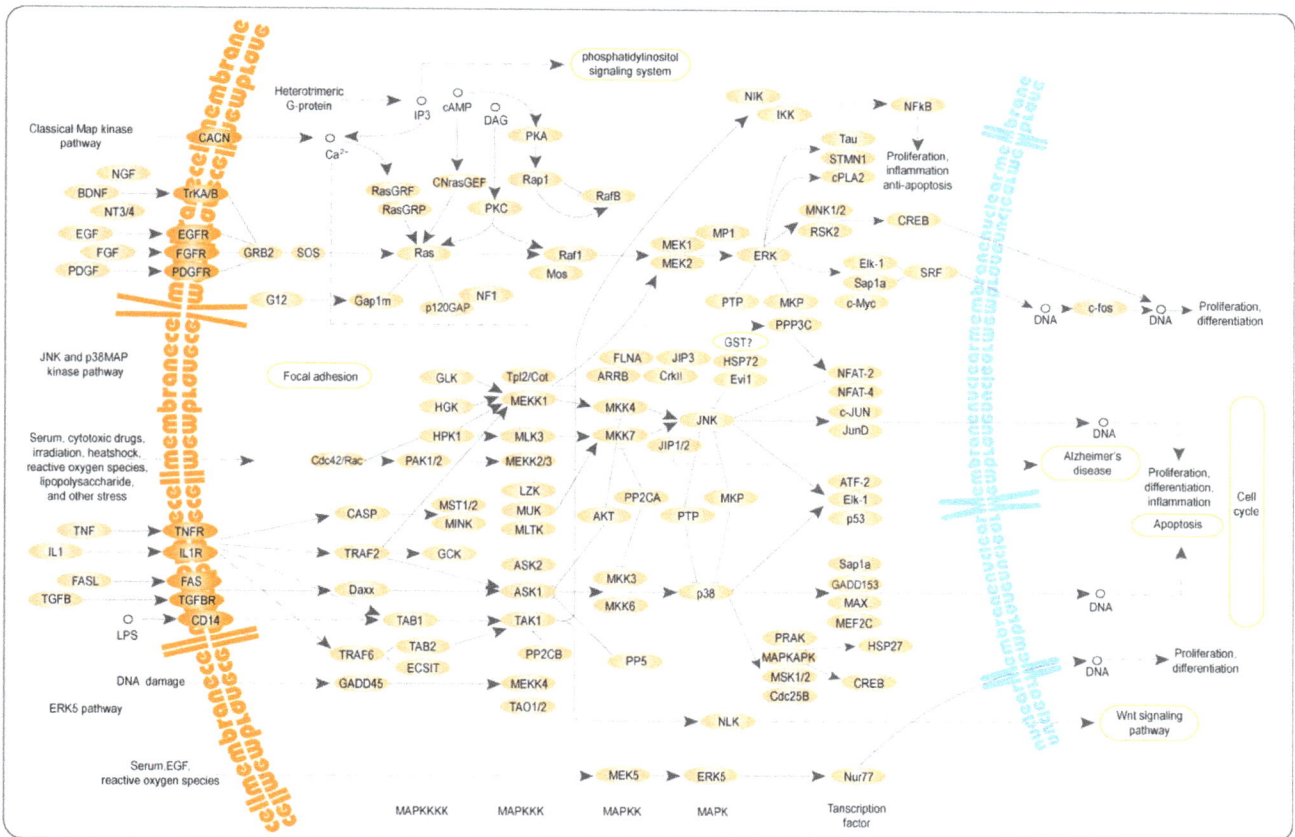

FIGURE 13.13 Signaling pathways. Shown here is the MAP kinase signaling pathway, which comprises different levels phosphorylation. Proteins, many of which are transcription factors, are phosphorylated by MAP kinases (MAPK), which are phopshorylated by MAP kinase kinase (MAPKK), which are in turn phosphorylated by MAP kinase kinase kinsae (MAPKKK), themselves being activated by other kinases, which are recruited or activated upstream by binding of extracellular singaling molecules to cell membrane receptors.[27]

activation, which is terminated by the hydrolysis of GTP to GDP. It was later discovered that other proteins—called G-proteins (to be contrasted with the trimeric G-proteins)—also act in a GTP-dependent manner, though these proteins are not part of a trimeric complex. One such example is the G-protein, Ras, which is activated by SOS. The latter is a type of protein called GTP exchange factor (GEF) which catalyzes the exchange of GDP for GTP in Ras, leading to activation of this protein (since all pathways are tightly regulated in cells, other proteins act as GTPase-activating proteins (GAPs), which catalyze the hydrolysis of protein-bound GTP and leads to inactivation of the enzyme). Activation of Ras by GTP enables Raf-1 to bind to and be activated by Ras by inducing a conformational change that reveals a catalytic site in Raf-1. Once activated, Raf-1 phosphorylates and activates another protein, MEK, which phosphorylates and activates, ERK. ERK then phosphorylates and activates specific transcription factors—proteins which initiates transcription of specific genes. The MEK/ERK pathway described above represents one of many signal transduction pathways used by cells to regulate protein activity, the example of which is shown in Figure 13.13.

This ends our discussion of signaling pathways, though we will come back to many of these concepts in subsequent chapters. The complex task of elucidating the many pathways in cells has been a great focus of biological research for the past several decades. Only by learning the specific role of each protein in a cell and the pathways in which they are involved can we begin to understand the consequences of modifying these signals to better our health or to treat diseases. However, the more we learn about these pathways, the more we realize the complexity by which these systems operate, all of which work together to ensure the survival of the cell within its environment and consequently, the survival of the tissue and the organism of which they are a part.

* * * * *

REFERENCES

1. Birnbaumer L, Rodbell M. Adenyl cyclase in fat cells. II. Hormone receptors. *Journal of Biological Chemistry.* 1969;244(13):3477–3482. doi:10.1016/S0021-9258(18)83396-8

2. Rodbell M, Birnbaumer L, Pohl SL, Krans HM. The glucagon-sensitive adenyl cyclase system in plasma membranes of rat liver. V. An obligatory role of guanylnucleotides in glucagon action. *Journal of Biological Chemistry.* 1971;246(6):1877–1882. doi:10.1016/S0021-9258(18)62390-7

3. Gilman AG. Alfred Gilman: Nobel lecture: G proteins and regulation of adenylyl cyclase. NobelPrize.org. Nobel Media AB © The Nobel Foundation 1994. https://www.nobelprize.org/prizes/medicine/1994/gilman/lecture/

4. Pfeuffer T, Helmreich EJ. Activation of pigeon erythrocyte membrane adenylate cyclase by guanylnucleotide analogues and separation of a nucleotide binding protein. *Journal of Biological Chemistry.* 1975;250(3):867–876. doi:10.1016/S0021-9258(19)41866-8

5. Cassel D, Selinger Z. Catecholamine-stimulated GTPase activity in turkey erythrocyte membranes. *Biochimica Biophysica Acta.* 1976;452(2):538–551. doi:10.1016/0005-2744(76)90206-0

6. Limbird LE, Lefkowitz RJ. Resolution of beta-adrenergic receptor binding and adenylate cyclase activity by gel exclusion chromatography. *Journal of Biological Chemistry.* 1977;252(2):799–802. doi:10.1016/S0021-9258(17)32788-6

7. Haga T, Haga K, Gilman AG. Hydrodynamic properties of the beta-adrenergic receptor and adenylate cyclase from wild type and variant S49 lymphoma cells. *Journal of Biological Chemistry.* 1977;252(16):5776–5782. doi:10.1016/S0021-9258(17)40090-1

8. Ross EM, Gilman AG. Reconstitution of catecholamine-sensitive adenylate cyclase activity: Interactions of solubilized components with receptor-replete membranes. *Proceedings of National Academy of Sciences U S A.* 1977;74(9):3715–3719. doi:10.1073/pnas.74.9.3715

9. Ross EM, Gilman AG. Resolution of some components of adenylate cyclase necessary for catalytic activity. *Journal of Biological Chemistry.* 1977;252(20):6966–6969. doi:10.1016/S0021-9258(19)66920-6

10. Ross EM, Howlett AC, Ferguson KM, Gilman AG. Reconstitution of hormone-sensitive adenylate cyclase activity with resolved components of the enzyme. *Journal of Biological Chemistry.* 1978;253(18):6401–6412. doi:10.1016/S0021-9258(19)46947-0

11. Howlett AC, Gilman AG. Hydrodynamic properties of the regulatory component of adenylate cyclase. *Journal of Biological Chemistry.* 1980;255(7):2861–2866. doi:10.1016/S0021-9258(19)85819-2

12. Northup JK, Sternweis PC, Smigel MD, Schleifer LS, Ross EM, Gilman AG. Purification of the regulatory component of adenylate cyclase. *Proceedings of National Academy of Sciences U S A.* 1980;77(11):6516–6520. doi:10.1073/pnas.77.11.6516

13. Sternweis PC, Northup JK, Smigel MD, Gilman AG. The regulatory component of adenylate cyclase. Purification and properties. *Journal of Biological Chemistry.* 1981;256(22):11517–11526. doi:10.1016/S0021-9258(19)68431-0

14. Northup JK, Smigel MD, Gilman AG. The guanine nucleotide activating site of the regulatory component of adenylate cyclase. Identification by ligand binding. *Journal of Biological Chemistry.* 1982;257(19):11416–11423. doi:10.1016/S0021-9258(18)33775-X

15. Katada T, Ui M. Slow interaction of islet-activating protein with pancreatic islets during primary culture to cause reversal of alpha-adrenergic inhibition of insulin secretion. *Journal of Biological Chemistry.* 1980;255(20):9580–9588. doi:10.1016/S0021-9258(18)43431-X

16. Hazeki O, Ui M. Modification by islet-activating protein of receptor-mediated regulation of cyclic AMP accumulation in isolated rat heart cells. *Journal of Biological Chemistry.* 1981;256(6):2856–2862. doi:10.1016/S0021-9258(19)69693-6

17. Katada T, Ui M. Islet-activating protein. A modifier of receptor-mediated regulation of rat islet adenylate cyclase. *Journal of Biological Chemistry.* 1981;256(16):8310–8317. doi:10.1016/S0021-9258(19)68845-9

18. Katada T, Ui M. Direct modification of the membrane adenylate cyclase system by islet-activating protein due to ADP-ribosylation of a membrane protein. *Proceedings of National Academy of Sciences U S A.* 1982;79(10):3129–3133. doi:10.1073/pnas.79.10.312

19. Katada T, Ui M. ADP ribosylation of the specific membrane protein of C6 cells by islet-activating protein associated with modification of adenylate cyclase activity. *Journal of Biological Chemistry.* 1982;257(12):7210–7216. doi:10.1016/S0021-9258(18)34558-7

20. Bokoch GM, Katada T, Northup JK, Hewlett EL, Gilman AG. Identification of the predominant substrate for ADP-ribosylation by islet activating protein. *Journal of Biological Chemistry.* 1983;258(4):2072–2075. doi:10.1016/S0021-9258(18)32881-3

21. Northup JK, Sternweis PC, Gilman AG. The subunits of the stimulatory regulatory component of adenylate cyclase. Resolution, activity, and properties of the 35,000-Dalton (beta) subunit. *Journal of Biological Chemistry.* 1983;258(18):11361–11368. doi:10.1016/S0021-9258(17)44426-7

22. Northup JK, Smigel MD, Sternweis PC, Gilman AG. The subunits of the stimulatory regulatory component of adenylate cyclase. Resolution of the activated 45,000-Dalton (alpha) subunit. *Journal of Biological Chemistry.* 1983;258(18):11369–11376. doi:10.1016/S0021-9258(17)44427-9

23. Hildebrandt JD, Codina J, Risinger R, Birnbaumer L. Identification of a gamma subunit associated with the adenylyl cyclase regulatory proteins Ns and Ni. *Journal of Biological Chemistry.* 1984;259(4):2039–2042. doi:10.1016/S0021-9258(17)43308-4

24. Tolkovsky AM, Levitzki A. Mode of coupling between the β-adrenergic receptor and adenylate cyclase in turkey erythrocytes. *Biochemistry.* 1978;17(18):3795–3810. doi: 10.1021/bi00611a020

25. Manning G, Whyte DB, Martinez R, Hunter T, Sudarsanam S. The protein kinase complement of the human genome. *Science (1979).* 2002;298(5600):1912–1934. doi:10.1126/science.1075762

26. File: Glucagon Activation.png—Wikimedia Commons. Accessed October 31, 2022. https://commons.wikimedia.org/wiki/File:Glucagon_Activation.png

27. File: MAPKpathway.jpg—Wikimedia Commons. Accessed November 1, 2022. https://commons.wikimedia.org/wiki/File:MAPKpathway.jpg

14 The Secretory Pathway

We saw in previous chapters that Palade and Claude had observed that parts of the endoplasmic reticulum (ER) had a "studded" appearance—due to the presence of ribosomes (these sections are called the rough ER, RER)—and this was subsequently found to be the site of protein translation. We also discussed that the first step in the translation of all proteins is the binding of ribosomes to mRNA, which initiates translation in the cytosol. However, what happens next depends on the type of protein being translated: transmembrane proteins need to be embedded into a membrane; secreted proteins need to be discharged from the cell; and cytosolic proteins—which swim freely in the cytosol (or in some cases associate with the cytosolic domain of membranes)—simply get translated in the cytosol with little fanfare. Thus, translation of the former two needs to be redirected to the ER and from there, they need to be transported to their final destinations. As such, this chapter will focus on the mechanism by which these events occur.

By the end of the 1960s, Palade and others had demonstrated, using electron microscopy, that the synthesis of proteins originating from the RER led to the formation of large secretory vesicles called zymogen granules (Figure 14.1). In addition, it was established that the Golgi apparatus concentrated proteins into densely packed *condensing vacuoles*—Golgi structures in which proteins are highly concentrated—before being converted to zymogen granules. However, the lack of resolution in these studies could not conclusively determine if the proteins exiting the RER passed through the Golgi apparatus, as Palade suggested, or simply diffused throughout the cytosol before being incorporated into zymogen granules.[1] A limitation factor in these studies was that:

> [...] cell fractionation was imperfect—there was extensive intercontamination of fractions; and incomplete—a number of subcellular components were unaccounted for. This applied especially to the Golgi complex which appears to be extensively involved in a number of operations in the secretory cycle.[1]

Therefore, Palade and his team improved the experimental techniques and advancements in electron microscopy imaging enabled them to further study the fate of synthesized proteins upon their exit from the RER.

As such, Jamieson and Palade switched from an *in vivo* labeling approach, the conditions of which were found to be less than optimal, to an *in vitro* system, in which nascent proteins from guinea pig pancreas slices were labeled with radioactive leucine for 10 minutes, followed by incubation without radioactive tracer for a period of time at 37°C, a condition which allowed synthesis to resume. This technique enabled them to follow nascent, radiolabeled, proteins

as they moved throughout the cell since only proteins that were synthesized while the tracer was present were labeled.[1,2] Thus, the authors first investigated the kinetics of intracellular transport of secretory proteins by assessing protein labeling in cellular fractions at different time intervals, thereby [checking] *the postulated transfer of secretory proteins from the rough ER to small vesicles at the periphery of the Golgi complex.*[1] Results indicated that rough microsomes—which represented the ER fraction—were labeled first, followed by the smooth microsomal fraction—which contained the Golgi apparatus (Figure 14.2).[1] In addition, the authors confirmed that proteins were confined to the inner spaces of the secretory vesicles by extracting the contents of the vesicles and assessing radioactivity. As such, the authors concluded that *following segregation in the cisternal spaces of the rough ER, secretory proteins remain within and are transported through membrane-enclosed spaces of the exocrine cell.*[1]

Upon investigation of the kinetics of zymogen labeling, the authors found that, whereas radioactivity declined rapidly in the Golgi apparatus after the end of the pulse, labeling of the zymogen fraction gradually increased, which suggested that translocation proceeds from the former to the latter.[2] In addition, the fact that protein content in the postmicrosomal supernatant remained relatively constant indicated that the decrease in proteins in the microsomal fraction was not a result of ineffective fractionation. Electron microscopy images supported the idea that protein-filled vesicles proceeded from the RER to the Golgi apparatus, to condensing vacuole, then to zymogen granules, before finally fusing with the cell membrane and expelling their contents in a process now known as exocytosis (Figure 14.3).[2,3] Thus, from these data, Palade concluded that it was:

> clear that the small vesicles of the Golgi complex mediate the transport of secretory proteins from the rough ER to condensing vacuoles. This implies transport in bulk or in mass since each vesicle must contain and carry a large number of protein molecules. The transport must also be discontinuous; otherwise, concentration of the solution of secretory proteins at the next step (condensing vacuoles) would not be possible. With these restrictions in mind, we can inquire into cellular mechanisms possibly involved in this operation [...]
>
> Radioautographic observations on intact cells of slices after 1 hr and especially after 2 hr postpulse incubation conclusively demonstrate the transformation of condensing vacuoles into zymogen granules and, finally, the discharge of protein into the acinar lumen. Hence, we now have a reasonably complete elucidation of the entire intracellular pathway of secretory proteins from their site of synthesis on attached ribosomes to their ultimate discharge into the duct system of the gland [...]

DOI: 10.1201/9781003379058-14

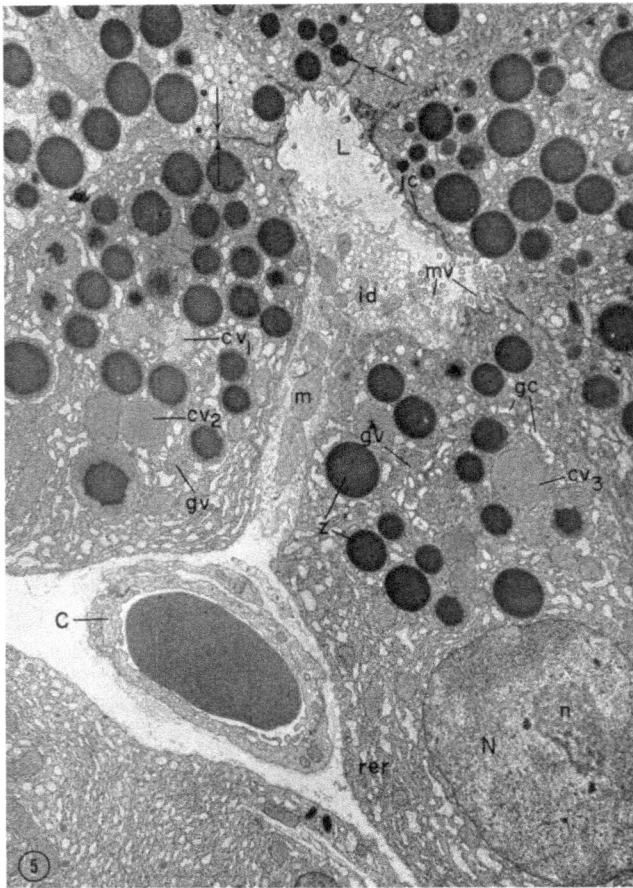

FIGURE 14.1 Electron microscopy images of zymogen granules. Normal pancreatic acinar cells showing large zymogen granules (dark circles).[1] (Used with permission of the Rockefeller University Press, from Jamieson JD, Palade GE. Intracellular transport of secretory proteins in the pancreatic exocrine cell. I. Role of the peripheral elements of the Golgi complex. J Cell Biol. 1967;34(2):577–596, Copyright © 1967; Permission conveyed through Copyright Clearance Center, Inc.)

In addition, these studies provide direct or indirect evidence that several types of intracellular transport, some of them new, participate in the cycle. These include: vectorial transport of newly synthesized proteins across the membranes of the rough ER; transport in bulk of materials between cell compartments; concentration of cell products within membrane-bounded structures possibly by intracellular ion pumps; and finally, transport of secretory proteins from zymogen granules to the acinar lumina. The forces, molecular events, and control mechanisms operating at each step remain to be elucidated by future work.[2]

* * * * *

The next question posed was, if translation of all proteins begins in the cytoplasm, what mechanisms are in place for the transfer of proteins to the RER given that membranes are impermeable to such large molecules? A successful attempt at answering this question came in 1971

when Günter Blobel, one of Palade's protégés, suggested the "signal peptide hypothesis", which postulated that:

> all mRNA's to be translated on bound ribosomes contain a unique sequence of codons to the right of the initiation codon (henceforth referred to as the signal codons); translation of the signal codons results in a unique sequence of amino acid residues on the amino terminal end of the nascent chain (henceforth referred to as the signal sequence); the latter triggers attachment of the ribosome to the membrane.[4]

To test this hypothesis, Blobel started from his mentor's and Claude's discovery in the 1950s and 1960s that the ER was studded with ribosomes and that these junctions were the site of protein synthesis, and therefore reasoned that:

> the ribosome membrane junction may function in the transfer of proteins across the membrane: by topologically linking the site of synthesis with the site of transfer, the protein would transverse the membrane only in status nascendi in an extended form before assuming its native structure, thus maintaining the membrane's role as a diffusion barrier to proteins.[4]

Early investigations using *in vitro* translation of isolated mRNA had shown that the molecular weight of protein products was larger than expected due to the presence of about 20 amino acids—which were found to subsequently be removed—at the N-terminus of proteins. Thus, Blobel and Dobberstein used two kinds cell-free systems in their experiments: an *initiation system* in which polypeptide are synthesized *de novo* and which consisted of purified ribosomal subunits, purified antibody IgG light chain mRNA (mRNA which codes for a subunit of an antibody complex), as well as all the required elements for translation to proceed; and a *readout system*, in which previously started polypeptides are completed and which consisted of the same system as the initiation system, but instead used purified free ribosomes, rough microsome fractions (ribosomes derived from the ER along with associated lipids), or purified detached ribosomes, that is, ribosomes which were attached to the ER membrane but whose lipids were removed in the purification process—all of which contained polypeptides of various lengths since translation had been initiated to various degrees. As such, the authors used SDS-PAGE analysis and determined that mRNA translated in the initiation system synthesized a product that was 4 KDa heavier than the expected product for IgG light chain, which they named "precursor".[4] When they compared protein products in the initiation system to that of the readout system, they found that the latter synthesized proteins at both the expected and at the heavier size, which suggested that *ribosomes are heterogeneous with respect to their content of processed and unprocessed nascent light chains*.[4] In addition, neither product was synthesized when the readout system was used with free ribosomes, which suggested a requirement for attachment to the ER for synthesis of this protein.

Specific Radioactivities of Proteins in Microsomal and Submicrosomal Fractions and in the Post-Microsomal Supernate

Exp. no.	Incubation time		Total microsomes cpm/mg protein	Smooth microsomes cpm/mg protein	Rough microsomes cpm/mg protein	$\dfrac{\text{SA smooth}}{\text{SA rough}}$	Post-microsomal supernate cpm/mg protein
	Pulse	Chase					
	min	*min*					
1	3		2750	2220	5080	0.44	—
	3	+ 7	2090	5100	2480	2.06	—
	3	+17	1450	2580	1850	1.39	- —
2	3		2180	1680	2740	0.61	360
	3	+ 7	2020	3500	1670	2.10	470
	3	+17	1365	1850	1350	1.37	440
3	3		3380	1720	3950	0.43	390
	3	+17	3510	3770	2050	1.85	650
4	3		3670	2390	5370	0.44	213
	3	+17	3200	5780	2590	2.23	450
5	3		3050	980	4540	0.22	300
	3	+17	1950	3480	2030	1.71	480
6	3		1860	470	2310	0.20	210
	3	+17	2140	1612	2420	0.65	280
7	3		1370	920	2300	0.40	370
	3	+17	1070	2000	1120	1.78	460
8	3		3400	1460	3800	0.38	990
	3	+17	2530	3770	2300	1.64	970
	3	+57	1750	2190	1470	1.49	930

FIGURE 14.2 Kinetic studies of the movement of proteins through microsomal fractions. Rough and smooth microsomes were isolated after a 3-minute pulse with radioactive leucine and an additional chase of seven or seventeen minutes (i.e., after the removal of radioactive leucine) and radioactivity was assessed in each fraction, in total microsomes, and in the supernatant.[1] (Used with permission of the Rockefeller University Press, from Jamieson JD, Palade GE. Intracellular transport of secretory proteins in the pancreatic exocrine cell. I. Role of the peripheral elements of the Golgi complex. J Cell Biol. 1967;34(2):577–596, Copyright © 1967; Permission conveyed through Copyright Clearance Center, Inc.)

A time-course experiment using the read-out system showed that, initially, only the low molecular weight protein was synthesized and this species came to a steady state after about 9 minutes' incubation. In contrast, the heavier protein band only started appearing around about 9 minutes and continued to increase until the end of the experiment. Keeping the signal hypothesis in mind, the authors interpreted this as indicating that proteins whose translation was well underway at the beginning of the experiment translated proteins with a cleaved N-terminus—that is, the signal sequence had already been cleaved since the ribosome was already attached to the ER—whereas proteins whose translation was just getting started resulted in the synthesis of full-length proteins prior to cleavage of the signal sequence. That both processed and unprocessed proteins were isolated in these studies was explained by the fact that the *in vitro* system using detached ribosomes did not contain the necessary enzymes for protein processing. Consistent with this idea, the authors found that cleavage activity was retained in experiments performed using rough microsomes, which suggested to them that this type of protein modification occurred relatively early in the synthesis process but following a substantial translation of

the mRNA.[4] As such, they suggested that *the processing activity is part of the membrane.* In addition, the universality of their signal hypothesis was demonstrated when they synthesized IgG light chain protein from mRNA and ribosomes isolated from a variety of sources. Conclusions from this series of experiments were summarized by Blobel and Dobberstein as follows:

> [...] the essential feature of the signal hypothesis [...] is the occurrence of a unique sequence of codons, located immediately to the right of the initiation codon, which is present only in those mRNA's whose translation products are to be transferred across a membrane. No other mRNA's contain this unique sequence. Translation of the signal codons results in a unique sequence of amino acid residues on the amino terminal of the nascent chain. Emergence of this signal sequence of the nascent chain from within a space in the large ribosomal subunit triggers attachment of the ribosome to the *[ER]* membrane, thus providing the topological conditions for the transfer of the nascent chain across the membrane [...]
>
> Ribosome attachment to, as well as detachment from, the membrane are likely to involve a complex sequence of events [...] The signal sequence of the nascent chain emerging from within a tunnel in the large ribosomal

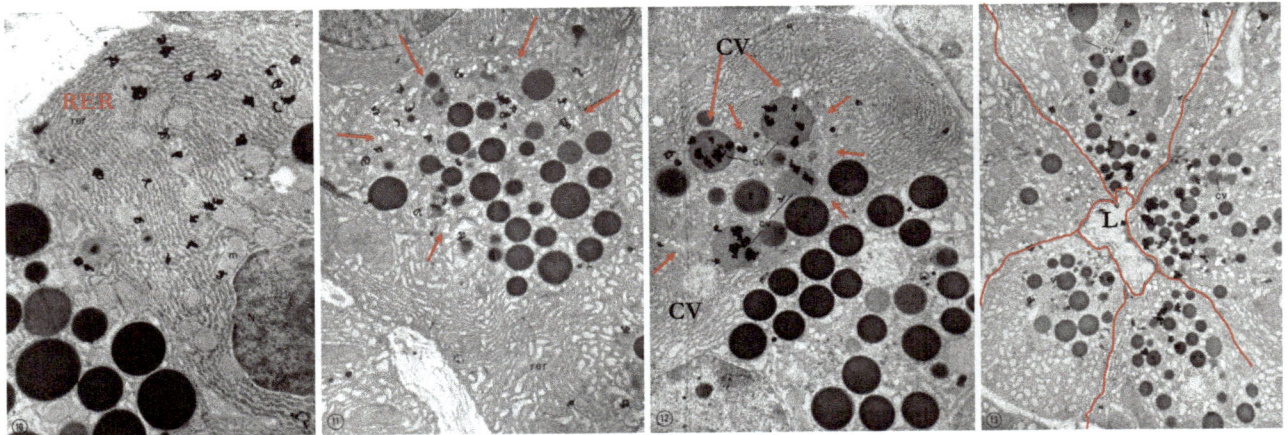

(a)

Distribution of Radioautographic Grains over Cell Components

	% of radioautographic grains					
	3-min (pulse)	Chase incubation				
		+7 min	+17 min	+37 min	+57 min	+117 min
Rough endoplasmic reticulum	**86.3**	43.7	37.6	24.3	16.0	20.0
Golgi complex*						
Peripheral vesicles	2.7	**43.0**	37.5	14.9	11.0	3.6
Condensing vacuoles	1.0	3.8	19.5	**48.5**	35.8	7.5
Zymogen granules	3.0	4.6	3.1	11.3	32.9	**58.6**
Acinar lumen	0	0	0	0	2.9	7.1
Mitochondria	4.0	3.1	1.0	0.9	1.2	1.8
Nuclei	3.0	1.7	1.2	0.2	0	1.4
No. of grains counted	300	1146	587	577	960	1140

The boldfaced numbers indicate maximum accumulation of grains over the corresponding cell component.
* At no time were significant numbers of grains found in association with the flattened, piled cisternae of the complex.

(b)

FIGURE 14.3 (a) Movement of nascent proteins through cell components. Cells were labeled with radioactive leucine and prepared for electron microscopy analysis to assess the location of proteins at the indicated times. Cells were imaged after a 3-minute pulse (a), after a 7-minute chase (b), after a 37-minute chase (c), and after a 117-minute chase (d). Notice that the grains are mostly contained in the rough ER after pulsing, then move to the periphery of the Golgi apparatus after 7 minutes (indicated by arrows), to the vacuoles after 37 minutes (indicated by arrows), and near the lumen (where proteins are excreted, L and lined in red) after 117 minutes, at which time the ER is nearly devoid of grains. The periphery of the cells is denoted in red. (b) The table shows the quantification of the grains in the various cell components in the above experiment.[2] (Used with permission of The Rockefeller University Press, from Jamieson JD, Palade GE. Intracellular transport of secretory proteins in the pancreatic exocrine cell. II. Transport to condensing vacuoles and zymogen granules. J Cell Biol. 1967;34(2):597–615, Copyright © 1967; Permission conveyed through Copyright Clearance Center, Inc.)

subunit may dissociate one or several proteins which have been found to be associated with the large ribosomal subunit of free ribosomes. Dissociation of these proteins may in turn uncover binding sites on the large ribosomal subunit. At the same time the emerging signal sequence also recruits two or more membrane receptor proteins and causes their loose association so as to form a tunnel in the membrane *[Figure 14.4]*. This association is stabilized by each of these membrane receptor proteins interacting with the exposed sites on the large ribosomal subunit, with the latter playing the role of a cross-linking agent. Binding of the ribosome would link the tunnel in the large ribosomal subunit with the newly formed tunnel in the membrane in continuity with the transmembrane space. After release of the nascent chain into the transmembrane space, ribosome

detachment from the membrane would eliminate the crosslinking effect of the ribosome on the membrane receptor proteins. The latter would be free again to diffuse as individual proteins in the plane of the membrane. As a result of their disaggregation, the tunnel would be eliminated. The tunnel, therefore, would not constitute a permanent structure in the membrane.[4]

* * * * *

Endocytosis is a process whereby the cell membrane invaginates and pinches off from the membrane to form intracellular vesicles (Figure 14.5).[5] Although this process was first reported by Roth and Porter, who described the uptake of yolk protein in mosquito oocytes, in 1964,[6] it

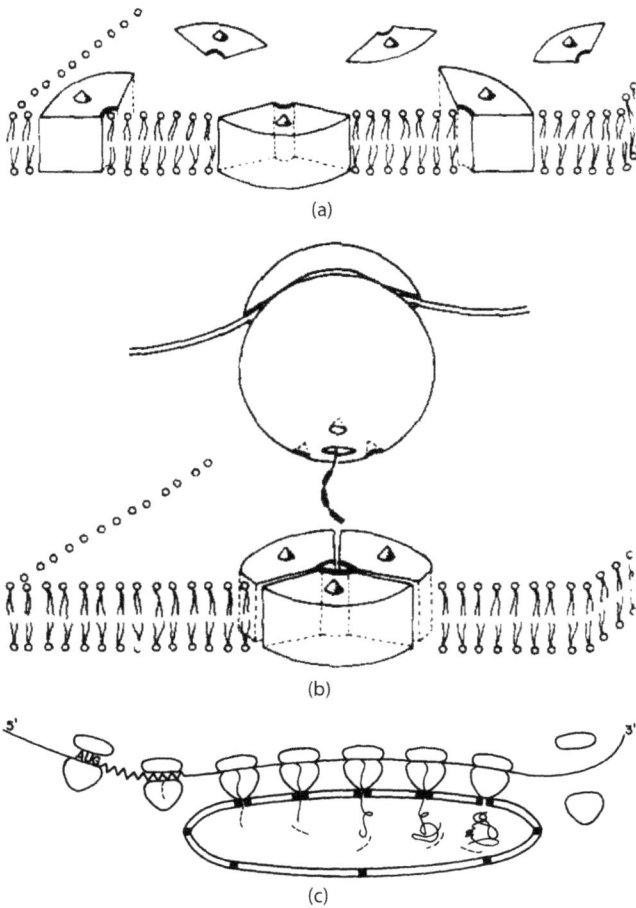

(a)

(b)

(c)

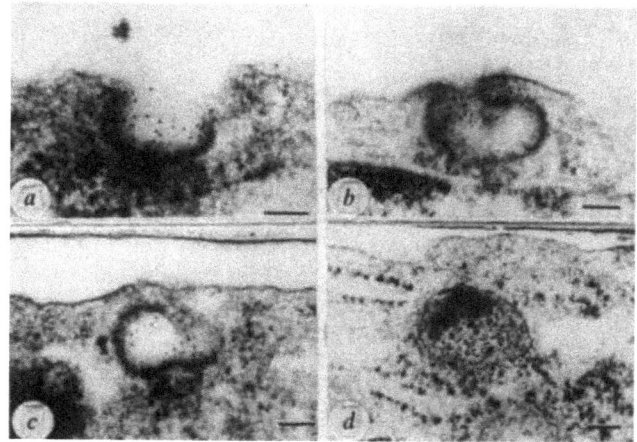

FIGURE 14.5 Electron microscopy images of endocytosis. Ferritin binding to coated pits (a), its internalization (b and c), and its delivery to lysosomes (d).[5] (Reprinted by permission from Springer Nature Customer Service Centre GmbH: Springer Nature, Nature, Goldstein JL, Anderson RGW, Brown MS. Coated pits, coated vesicles, and receptor-mediated endocytosis, Copyright © 1979.)

FIGURE 14.4 Model for protein synthesis proposed by Blobel and Dobberstein in 1975. Synthesis of transmembrane and secreted proteins begins in the cytosol, and if a signal peptide is present, a signal recognition protein complex binds to it and relocates the mRNA, along with the nascent protein, to the ER membrane, where multiple other proteins associate to form a pore through which the nascent protein is directed (a and b). A new ribosome can also bind to the mRNA—and initiate translation of a new polypeptide—as soon as the 5′ end of the mRNA extends out of the first ribosome far enough for another to bind to the recognition sequence, producing what is called "a polysome" or polyribosome" (c).[4] (Used with permission of the Rockefeller University Press, from Blobel G, Dobberstein B. Transfer of proteins across membranes. I. Presence of proteolytically processed and unprocessed nascent immunoglobulin light chains on membrane-bound ribosomes of murine myeloma. J Cell Biol. 1975;67(3):835–851 Copyright © 1975; Permission conveyed through Copyright Clearance Center, Inc.)

was later found that endocytosis could occur for a number of other reasons: as a mechanism of nutrient uptake in protozoa; to take-in proteins that have bound to cell surface receptors; and as a reuptake mechanism of secreted factors by presynaptic nerve terminal. Similar vesicles were also found to be present in a number of other processes such as budding from presynaptic nerve terminals and in the Golgi region of rat vas deferens; however, in this case, the vesicles were formed on the outside of the

membrane.[7] Microscopy images showed that these vesicles usually [appeared] *to be bounded by a smooth membrane but some are observed with 'coats' on their cytoplasmic surfaces*, while in others *there appeared to be a lattice-like network around each vesicle, the whole having an external diameter of about 600 Å.*[7]

In 1975, Barbara Pearse investigated the composition of these coated vesicles and developed a simple centrifugation protocol for their isolation.[7] As such, she found that the structures were of 550 Å to 850 Å in diameter, and 75% of its composition came from a single 180-KDa protein by SDS-PAGE electrophoresis. Pearse proposed the name *clathrin* for this protein and, using electron microscopy, determined that these vesicles were formed from *a closed network of hexagons and 12 pentagons [...] built from 108 identical subunits* (Figure 14.6).[8] She also suggested that:

clathrin itself may be able to pinch off a vesicle through interactions between its subunits and the membrane phospholipids. The coat may also be able to interact with other molecules, to ensure formation of vesicles at particular sites on the membrane and for their transport to specific cellular locations.[7]

* * * * *

In 1970, Randy Schekman joined the laboratory of Arthur Kornberg, who won the Nobel Prize for his discovery of DNA polymerase—the first enzyme whose function was shown to be to replicate DNA.[10] Shortly before Schekman's arrival, Kornberg had visited a number of membrane biologist's laboratories, including that of Palade, and became interested in membrane proteins, hoping they might provide an insight into DNA replication. Schekman credits a new postdoctoral fellow in Kornberg's team,

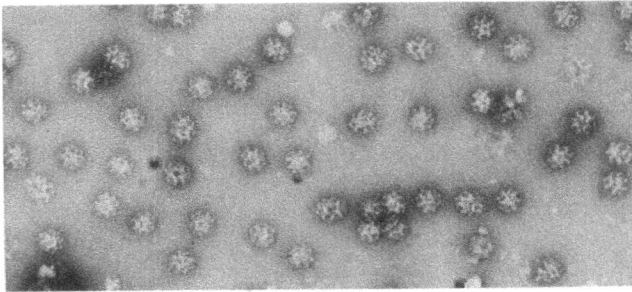

(a)

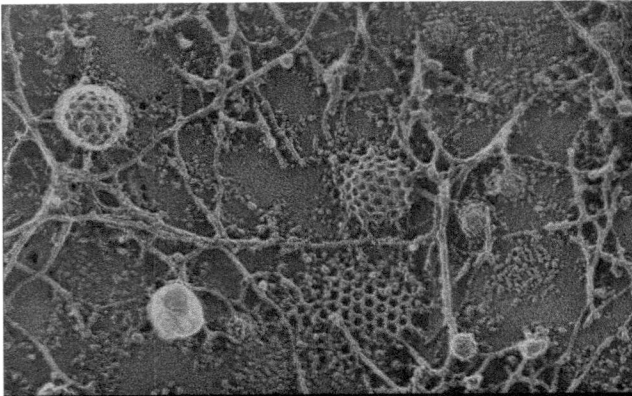

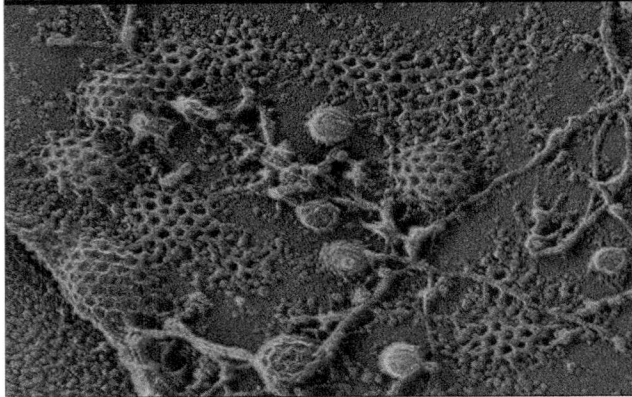

(b)

FIGURE 14.6 Clathrin-coated pits. Pearse's electron micros-copy image of clathrin-coated pits (a) and a more recent image of clathrin-coated pits clearly showing the pentagon structure of the coat (b).[8,9] (Figure part (a) adapted from J Mol Biol. 97(1) Pearse BM. Coated vesicles from pig brain: Purification and biochemical characterization, 93–98, Copyright © 1975 with permission from Elsevier. Figure part (b) adapted with permission from John Wiley and Sons, from Miya Fujimoto L, Roth R, Heuser JE, Schmid SL. Actin assembly plays a variable, but not obliga-tory role in receptor-mediated endocytosis. Traffic. Copyright © Munksgaard 2000.)

Bill Wicker, as an invaluable source of information about membrane biology, stating that:

> [he] and I shared endless hours in conversation about our work but importantly, I learned a great deal from him about what was or was not known about how membranes are put together. In this context, I read the work of Palade and his associates David Sabatini and Phillip Siekevitz who were

then exploring the mechanism of vectorial membrane translocation of secretory proteins as they are made on ribosomes associated with the ER.[10]

As Schekman approached the end of his graduate studies in 1974, he began to ponder the area of research he would embark on as a new, independent researcher. As such, he noticed that:

> the beginnings of a revolution in genetic engineering and recombinant DNA were just emerging [...] The tools of molecular cloning were in prospect, thus it was appealing to consider how they may be applied to uncover essential genes in any number of cellular processes.[10]

In 1975, shortly after Schekman joined Jonathan Singer's laboratory (of "Fluid Mosaic Model of Membrane Structure" fame) as a postdoctoral fellow, Blobel and Dobberstein published their breakthrough papers on the signal hypothesis and ushered *Palade's pathway into the molecular era*.[10] Thus, Schekman *believed a unique oppor-tunity lay in the evaluation of plasma membrane assembly in S. cerevisiae and my reading of the literature focused on what was known before 1975*, which he remembers was not very much.[10] As such, he learned that *[v]esicles implicated in secretion were seen by thin section electron micros-copy to localize to the cytoplasm of an early cell bud [...]* and he thought that *it seemed reasonable to suppose that these vesicles were responsible for secretion and localized plasma membrane assembly*, which excited him greatly.[10] Amazingly, just as Schekman was starting his postgradu-ate work in Singer's laboratory, he was offered a post as Assistant Professor with the faculty of Berkley, and he was so excited that he *foolishly accepted* [the position] *over the phone with no further negotiation! And so within the first few months of my postdoctoral training, I had the luxury of planning my future career without the responsibilities of the job*.[10]

In his new position, Schekman set out to study the syn-thesis of membrane proteins, though at the time, *there were no tools available to study the localization of a newly synthesized plasma membrane protein*.[10] Starting with meager funding of $35 000 for two years (!) and a small University grant, Schekman got started and within a few months, he had hired a graduate student named Peter Novick, whose abilities, by Schekman's own admission, were *technically superior* to his own.[10] Schekman had the idea of isolating yeast cells that were defective in their secretory pathway using a genetic approach and developed a system that allowed him to find very rare mutations that would likely be lethal (lethal mutations are particularly hard to isolate because their identification is reliant upon the cell dying). It should be noted at this point that yeast mutations can occur in such a way that a particular yeast strain might behave normally at a *permissive* temperature, but have a particular defect when grown at the restrictive temperature; these are known as *temperature sensitive* mutations (*ts*) and are usually a result of the inability of a particular protein to fold correctly at the restrictive temperature due to a genetic mutation, but folds properly at the permissive temperature. As such, after several attempts at finding the right method for the isolation of *ts* mutants of the secretory pathway, Novick and Schekman finally succeeded in identifying yeast strains which did not secrete invertase and acid phosphatase.

To achieve this, Novick and Scheckman started from baker's yeast (*Saccharomyces cerevisiae*) mutated with 3% ethyl methane-sulfanate for 60 minutes at 37°C.[11] Cells were then allowed to expand and were spread on agar plates, grown at 25°C (the permissive temperature), and replica-plated (i.e., transferred to a new plate by pressing a fresh plate against the plate on which yeast were grown in order to produce an exact replica of yeast colonies on the second plate), before being grown at 37°C (the restrictive temperature). In doing so, they found that 87 colonies out of 16,000 were *ts* mutants, that is, they died when incubated overnight at 37°C. They then replica-plated the original plates again to test them for acid phosphatase and invertase secretion and found two clones that were defective in secretion, which they named sec 1-1 and sec 2.[10] These clones were determined to have a more than five-fold decrease in secretion upon temperature shift to 37°C, but surprisingly, there was a concomitant increase in intracellular protein content (Figure 14.7a). Furthermore, the vesicles were subsequently secreted when the temperature was decreased to 25°C, even in the presence of cyclohexamide (which inhibits protein synthesis). These data indicated that the secreted pool of proteins was not newly synthesized proteins but rather came from the release of accumulated vesicles, and suggested that proteins were indeed synthesized at the restrictive temperature but accumulated inside the cells rather than being released.[11]

To further confirm these findings, Novick and Schekman used the fact that sulfate permease expression on the cell surface of yeast was such that it was repressed when cells were grown in media containing 1.5 mM methionine but derepressed when cells were transferred to sulfate-free media.[11] As such, although permease activity in wild-type cells was detected about 2 hours after derepression at the restrictive temperature, activity was not detected in *sec* mutants until incubation was returned to 25°C—even in the presence of cycloheximide—again suggesting that permease accumulated inside the cell due to a defect in the secretion pathway and was subsequently released at the permissive temperature. The hypothesis that protein-filled vesicles accumulated inside cells upon inhibition of the secretory pathway was supported by electron microscopy studies which showed a significant increase in small, dense vesicles inside *sec* mutants at the restrictive temperature, and these were shown to be filled with acid phosphatase using antibody labeling (Figure 14.7 b - d).[11]

Following the publication of these results, Novick and Schekman continued to look for other mutants using their system but did not find any. Therefore, they opted to adopt a

new screening procedure. Following what Schekman called *a brilliant string of observations*,[10] Novick reasoned that secretion mutants might be more dense than their wild-type counterparts because of the accumulation of intracellular proteins, so he tested the feasibility of a *beautiful experiment* by mixing 5 million wild-type strains with 500 million sec 1 strains for several hours to ensure sufficient mixing and then centrifuged them in a density gradient.[12] What he found was that the two strains nearly perfectly separated, with the mutants accumulating at the bottom of the gradient. When they repeated the experiment with a new set of mutated cells, the authors found that about 15% of the mutants isolated in this manner had a temperature-sensitive secretion defect, totaling 485 strains, which were further classified into three classes, the first of which (188 strains in total) *showed an accumulation of invertase at the nonpermissive temperature.*[12]

These strains were found to have a similar phenotype to the sec 1 and sec 2 mutants previously isolated, in terms of their permease activity and in their accumulation of acid phosphatase and invertase in intracellular vesicles, though the type of vesicles varied (Figure 14.8). Thus, the authors separated these mutants into three classes: one class accumulated cytosolic membrane-enclosed vesicles; the second class *developed a more extensive network of ER than was seen in wild-type cells*; the third class produced new structures not previously observed, which the authors named *Berkley Bodies* (Bb). Genetic testing revealed that 23 different genes were mutated in the various *sec* mutants, though the authors acknowledged that this was likely a minimum since the *density selection has the disadvantage that it eliminates mutants that die rapidly at 37.*[12] Therefore, the authors reasoned that:

> at least 23 gene products are required for the transport of secretory proteins from the site of synthesis to the cell surface. Thermosensitive defects in these gene products also block incorporation of a plasma membrane permease and stop bud growth. Taken together, these observations suggest that membrane growth and secretion are accomplished by parallel if not identical pathways. Furthermore, membrane-enclosed organelles accumulate in 22/23 of the mutants at 37° but not at 25°C. We propose that these structures are intermediates in the secretory pathway; their soluble contents are destined for secretion by exocytosis and their membranes will be incorporated into the plasma membrane by fusion.[12]

Now, there still remained the question as to where in the pathway each gene acted. Thus, Schekman and his group employed a screening technique that had been recently developed by another group and which involved phenotype analysis of double mutants, reasoning that if organelles that accumulated secretory enzymes in the single mutants

> represent stages in the passage of secretory proteins along a linear pathway, a double mutant should accumulate the organelle corresponding to the earliest block [...] If the

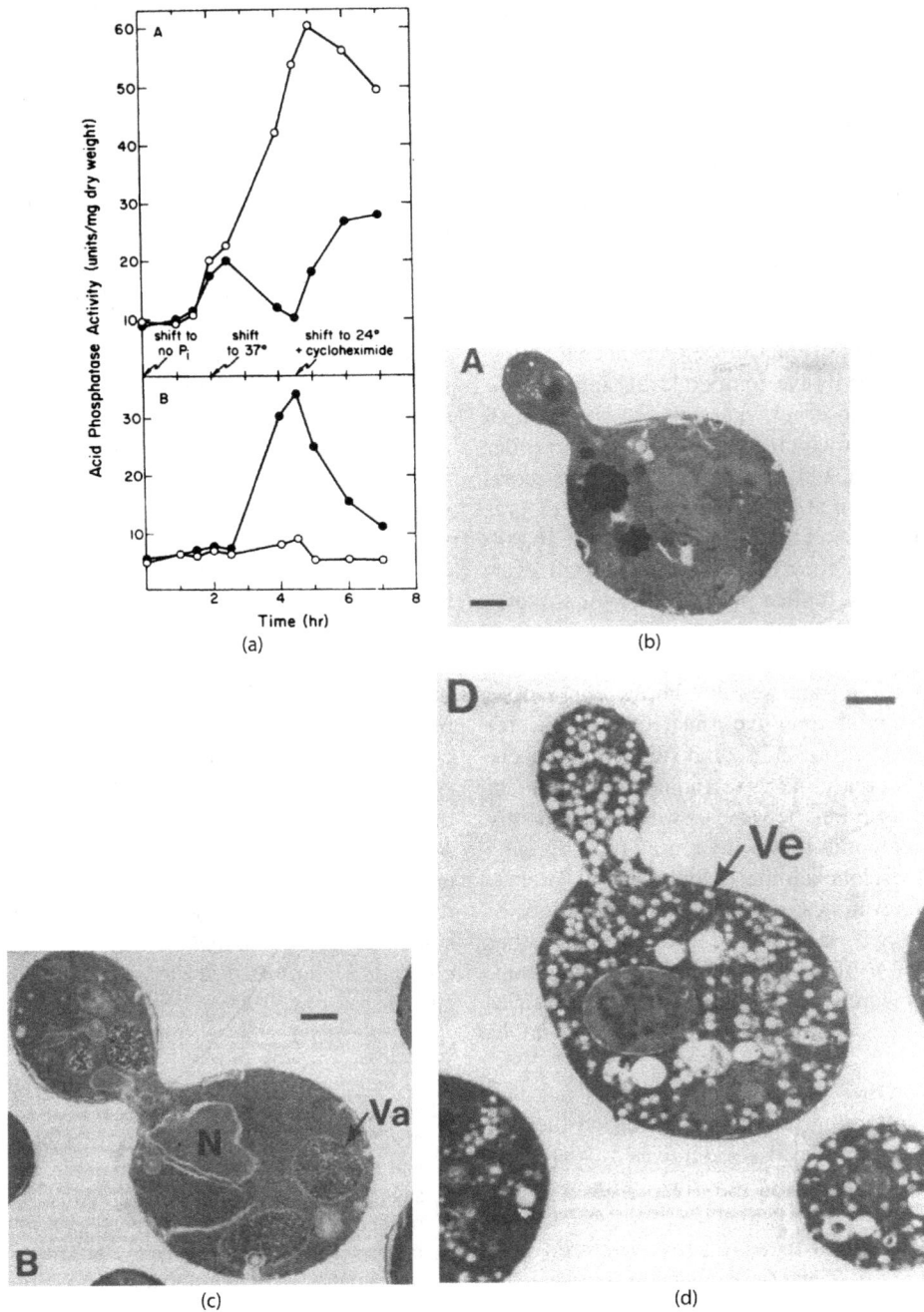

FIGURE 14.7 Cell growth and acid phosphatase secretion are defective in yeast mutants. (a) Wild type (open circles) and temperature-sensitive mutant (closed circles) yeast were grown at 25°C and transferred to phosphate-deficient media at time 0, before shifting to 37°C. Cycloheximide was added to inhibit protein synthesis. Samples were harvested and assessed for secreted (A) and intracellular (B) acid phosphatase. Electron microscopy images show wild type cells grown at 37°C (b) and a sec mutant grown at 25°C (c) or at 37°C (d).[11] (Adapted from Novick P, Schekman R. Secretion and cell-surface growth are blocked in a temperature-sensitive mutant of *Saccharomyces cerevisiae*. Proc Natl Acad Sci U S A. 1979;76(4):1858–1862. Used with permission.)

gene products operate in independent pathways that contribute to the same process, the phenotypes of both single mutants would appear in the double mutant.[13]

Thus, with an aim to *establish the order of genes product function*, they constructed double mutants by mating yeast cells with single mutations that produced one of the three

distinct phenotypes. They found that the gene products operated in a linear fashion, but that some genes products were involved in the same stage of the process; a summary of their findings is shown in Figure 14.9.

* * * * *

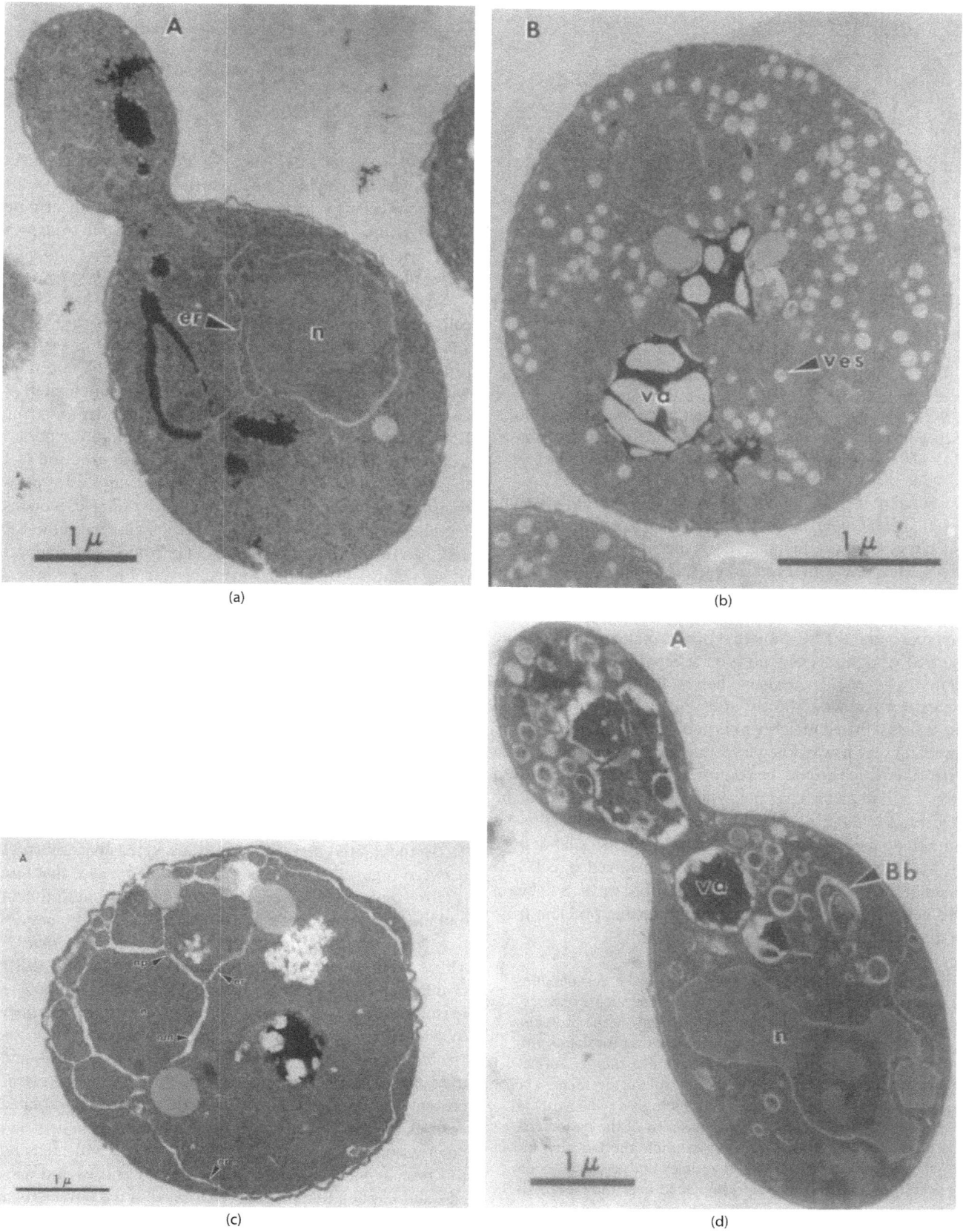

FIGURE 14.8 Phenotypes of protein accumulation defects in the yeast mutants isolated by Schekman's group. *Sec* mutants grown at 25°C (a, no defects) or at 37°C (b, accumulation of white vesicles); accumulation of proteins in the ER (thick white lines, c); Berkley bodies (Bb, d).[12] (Adapted from Novick P, Field C, Schekman R. Identification of 23 complementation groups required for post-translational events in the yeast secretory pathway. Cell. 1980;21(1):205–215, Copyright © 1980, with permission from Elsevier.)

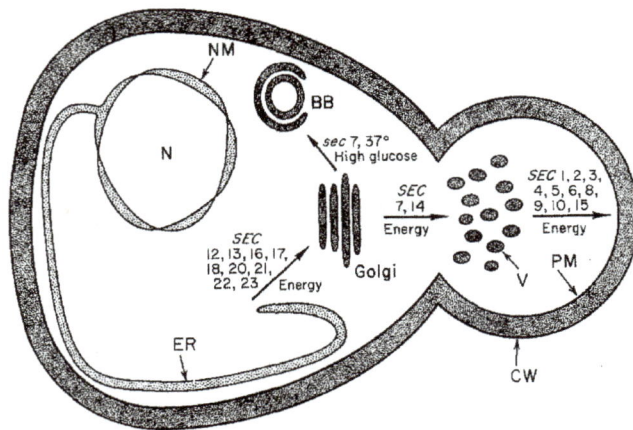

FIGURE 14.9 Diagram of the yeast secretory pathway as proposed by Novick, Ferro, and Schekman. N: nucleus; NM: nuclear membrane; BB: Berkley body; PM: plasma membrane; V: vesicle; CW: cell wall.[13] (Reprinted from Cell 25(2), Novick P, Ferro S, Schekman R. Order of events in the yeast secretory pathway, 461–469, Copyright © 1981, with permission from Elsevier.)

Schekman had planned to develop a cell-free system from the moment he and Novick had isolated their first secretory pathway mutants; however, his students and fellows were not convinced about the prospects of such an effort so he was not able to get this project off the ground.[10] James Rothman was also convinced of the merits of this approach and this was the first thing he attempted when he started work at his new position at Stanford. Rothman was a first-year medical student when he first learned about the secretory pathway, shortly after Palade's discoveries, and he was fascinated by it.[14] He started working on the "sorting problem", as it was known in the late 1970s, when, much like Randy Schekman, he received an offer to join Stanford while still in medical school, which, also like Schekman, he accepted without pause. However, the University allowed him to do postdoctoral work in Harvey Lodish's laboratory at MIT, where he learned techniques involved in cell-free membrane systems. He arrived at Stanford in 1978 to begin his new position, and approached the sorting problem from a different perspective from Schekman:

Applying the mindset of a physicist to the complexities and mysteries of cell biology afforded me such a perspective and the approach to productively tackle the problem. Physicists seek universal laws to explain all related processes on a common basis, and achieve this by formulating the simplest hypothesis to explain the facts. The prevailing opinion among cell biologists was without doubt that the anatomical arrangement of the endomembrane systems in the cell—for example the fact that the transitional ER (from which vesicles bud to carry secretory products to the Golgi) is placed near the Golgi—is vital to ensure the delivery of cargo [...] But the simplest idea is rather the opposite— that intrinsic chemical specificity enables specific cargo delivery, and that the observed anatomy arises as the consequence of chemical specificity in operation.

[...] The great virtue of the simpler idea is the remarkable prediction it makes: that accurate vesicle traffic can in principle take place accurately in cell-free extracts [...] Once reconstituted, cell-free transport could be used as an assay to permit the underlying enzyme proteins to be discovered and purified according to their functional requirements.[14]

Thus, Rothman and his first graduate student, Erik Fries—who had helped uncover the mechanism of viral entry into cells for his Ph.D. thesis—set out to determine how *to achieve conditions under which the same transport of nascent proteins can proceed in cell-free extracts, for only then can the tools of biochemistry be fruitfully applied.*[15]

To that end, they used Chinese hamster ovary (CHO) cells and infected them with vesicular stomatis virus (VSV), since this virus was known to synthesize only one glycoprotein, called VSV G (or protein G, not to be confused with the G-proteins we discussed in the previous chapter)— a transmembrane protein which is synthesized in the ER, transported to the Golgi apparatus, and finally exported to the plasma membrane.[15] The authors also used the fact that the ER and Golgi apparatus were known sites of post-translational modifications of proteins; for example, proteins are glycosylated (i.e., addition of carbohydrates) in the ER and then further modified in the Golgi complex by removal of six mannose sugars and the subsequent addition of terminal sugars, such as N-acetylglucosamine (GlucNac).[16] In addition, the enzyme endoglycosidase H (endo H) had previously been shown to cleave proteins exiting the ER which contain unmodified carbohydrates but not proteins whose carbohydrates had been modified by Golgi-resident enzymes by addition of GlucNac. As such, Rothman used this enzyme as a tool by which to assess the progress of proteins through the ER and the Golgi apparatus (i.e., proteins which had just exited the ER would be endo H sensitive while those which had been modified in the Golgi apparatus would be endo H resistant).

By the 1980s, it was generally accepted that transport between organelles was enabled by small vesicles that *bud off from one membrane and fuse with another* and Rothman and Fines reasoned that clathrin-coated vesicles *are attractive candidates for these presumed transport vehicles.*[16] Thus, they isolated clathrin-coated vesicles from VSV G-infected CHO cells according to Pearse's protocol— reasoning that if *G protein is transported intracellularly by coated vesicles, then this polypeptide should be demonstrable in highly purified coated vesicle preparations*—and used SDS-PAGE electrophoresis to visualize the proteins present in them. As expected, the authors found that isolated vesicles of VSV-infected CHO cells contained clathrin and other virion proteins, including VSV G, which indicated that this protein was a major constituent of clathrin-coated vesicles and implied that the latter was used in the intracellular transport of the former (Figure 14.10a).[16] The authors also used a pulse-chase experiment to determine if these vesicles were indeed transported from one organelle to another. As such, VSV G-infected CHO cells were incubated with

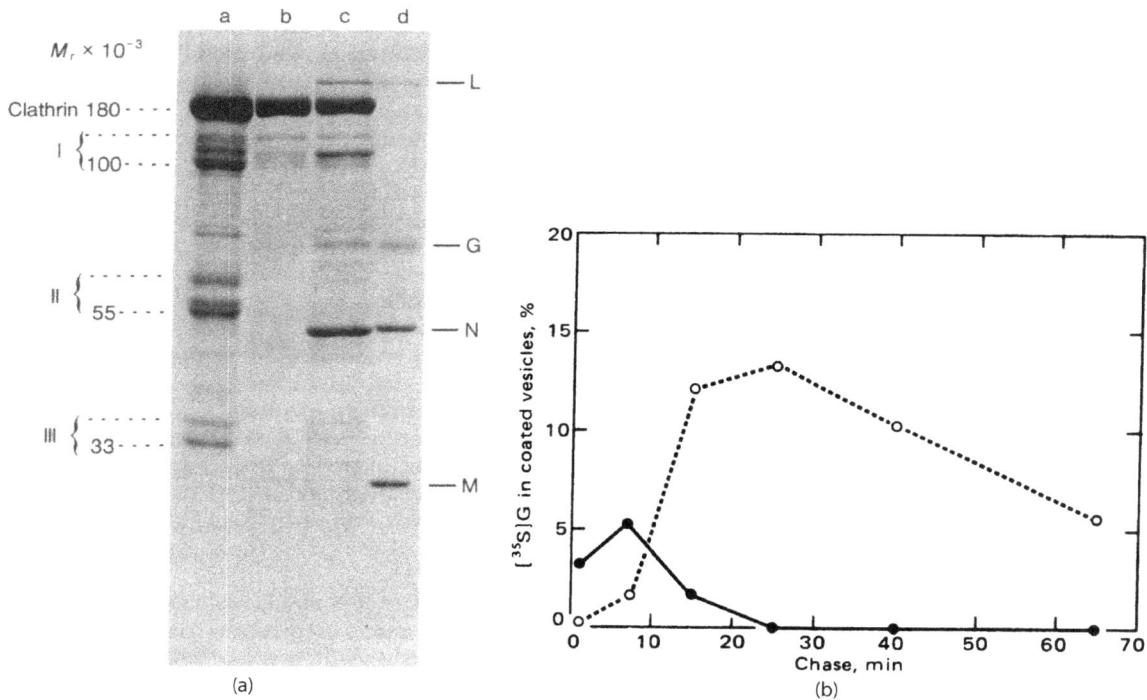

FIGURE 14.10 (a) SDS-PAGE analysis of coated vesicle proteins. Coated vesicles were isolated from calf brain (lane a), CHO cells (lane b), or VSV-infected CHO cells (lane c) and subjected to SDS-PAGE. Purified VSV G proteins (lane d) were used as control. (b) Kinetics of passage of VSV G protein through cellular organelles. Coated vesicles were isolated at various times following ^{35}S-labeling of VSV G proteins, and VSV G proteins were extracted and subjected to endo H treatment. Endo H-sensitive (ER proteins, closed circles) and endo H-resistant. (Golgi apparatus proteins, open circles) are plotted against time.[16] (Adapted from Rothman JE, Fine RE. Coated vesicles transport newly synthesized membrane glycoproteins from endoplasmic reticulum to plasma membrane in two successive stages. Proc Natl Acad Sci U S A. 1980;77(2):780–784. Used by permission.)

^{35}S-methionine for a 5-min pulse (to label newly translated proteins), and vesicles were isolated at various times following a chase with unlabeled methionine, digested with endo H, and the amount of ^{35}S-methionine was quantified in the endo H-sensitive and -resistant species.[16] They found that the movement of VSV G through coated vesicles occurred in two distinct waves: one immediately following the chase which was endo H-sensitive, followed by another which was endo H-resistant (Figure 14.10b), noting that:

> The time course of these successive waves fits well with the interpretation that they represent the successive transit steps between reticulum and Golgi apparatus and Golgi and plasma membrane, respectively.
>
> [...] [T]he early stage corresponds to coated vesicles that bud off from the ER and fuse with the Golgi apparatus, thereby delivering *[VSV-G]* to this organelle, within which Endo H resistance is conferred. The late stage would then correspond to coated vesicles that bud from the Golgi apparatus and fuse with the plasma membrane.[16]

Next, Fries and Rothman tested a novel cell-free system using a mutant strain of CHO cells isolated by another group, CHO 15B, which was found to lack an enzyme—called UDP-GlcNAc glycosyltransferase—required to modify oligosaccharides to the endo H-resistant type. Thus, they incubated ER and Golgi apparatus extracts isolated from

VSV G-infected, ^{35}S-methionine-pulse-labeled CHO 15B cells (the "donor") with similar extracts from unlabeled, uninfected, wild-type, CHO cells (the "acceptor"), expecting to detect endo H-resistant VSV G (Figure 14.11).[15] Keep in mind that in this system, only the non-labeled wild-type cells could modify glycoproteins in the Golgi apparatus. Therefore, the appearance of labeled, endo H-resistant, VSV G in the system could only occur as a result of a transfer of vesicles from the ER of CHO 15B extracts to the Golgi apparatus of wild-type CHO cell extracts. To their dismay, they could not detect any labeled endo H-sensitive G protein and concluded that *G could not be transferred all of the way from its site of synthesis in the rough ER to the Golgi complex in vitro*, which they explained as follows:

> Transport vesicles appear to bud off specifically from the smooth "transitional elements" of ER. After a brief pulse label followed by homogenization, most of G would not be in vesicles derived from these transitional elements, but rather would be in the rough microsomes. These vesicles would be "dead ends" *in vitro*, unable to package G into the transport vesicles needed to travel to Golgi. It might therefore be essential to prepare extracts from cells in which pulse-labeled G had been trapped in the "transitional elements" of ER or at an even later stage of maturation in order to obtain efficient transport (involving fewer steps) *in vitro*.[15]

"DONOR" GOLGI-CONTAINING FRACTION FROM VSV-INFECTED 15B MUTANT **"ACCEPTOR" GOLGI-CONTAINING FRACTION FROM UNINFECTED WILD-TYPE CELLS**

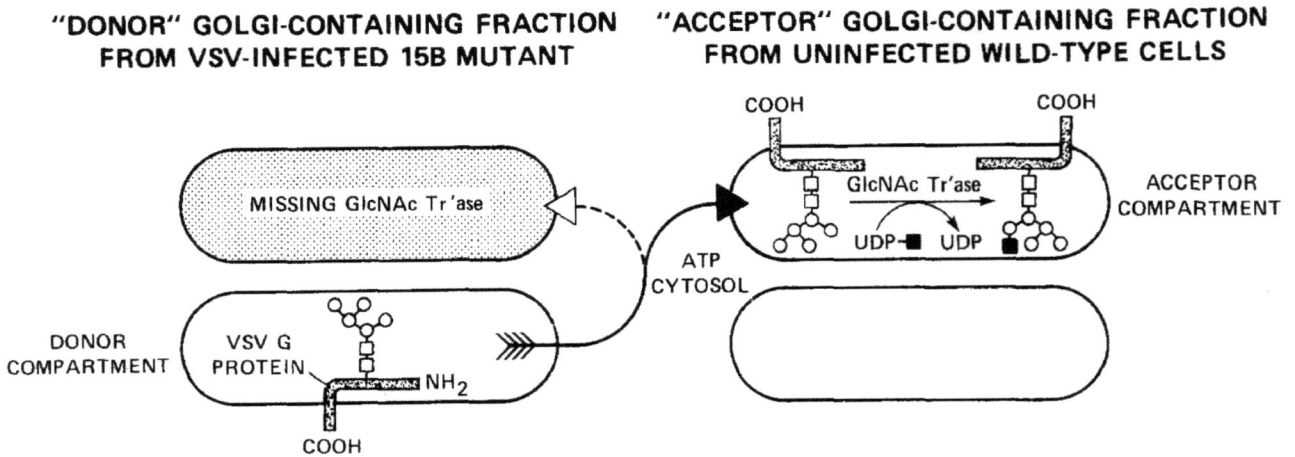

FIGURE 14.11   Diagram of Rothman's novel assay to follow the transport of proteins through the Golgi apparatus. A donor membrane—deficient in GlucNac transferase—was infected with VSV G protein and transferred to a reaction mixture containing wild type, uninfected acceptor membrane. Thus, the only way that GlucNac could be added to VSV G protein to become endo H-resistant was by transfer of the protein from one compartment to another *in a dissociative fashion*.[17] (Reprinted from Cell 39(2 pt 1), Balch WE, Dunphy WG, Braell WA, Rothman JE. Reconstitution of the transport of protein between successive compartments of the Golgi measured by the coupled incorporation of N-acetylglucosamine, 405–416, Copyright © 1984, with permission from Elsevier.)

To accomplish this, they treated cells with a reversible inhibitor of oxidative phosphorylation, CCCP, which had been shown to block *the formation of the vesicles that carry secretory proteins from the transitional elements of the ER to the Golgi complex*. The authors found that a 20-minute treatment with CCCP immediately after pulse labeling resulted in the accumulation of *radioactively labeled secretory proteins in the ER and its transitional elements [...] prior to passage through the Golgi complex*.[15] Then, they incubated these extracts with untreated, uninfected, CHO cells, along with ATP, an ATP-regenerating system, and the enzymes and substrates required for glycoprotein modification in the Golgi apparatus.

Under these conditions, they found that about 60% of VSV G was converted to an endo-H resistant type (i.e., residing in the Golgi), its conversion being dependent on membrane-enclosed vesicles since incubation with cytosolic extracts devoid of vesicles could not replicate the phenomenon.[15] In addition, kinetic studies revealed that the endo H-sensitive VSV G slowly decreased with time, with a concomitant increase in resistant VSV G, suggesting a movement of VSV G from the ER to the late stages of the Golgi apparatus (Figure 14.12). Finally, the movement of proteins from the ER through the Golgi apparatus was determined to require energy in the form of ATP, which they suggested was a requirement for the transport process rather than the processing of oligosaccharides. These results led to the hypothesis that VSV G *is transported in the cell-free extract by coated vesicles that bud from ER-derived vesicles (from 15B cells) and then fuse with vesicles of the Golgi complex (from wild type)*.[15] As Rothman suspected earlier:

A bona fide signal in this revised cell-free reaction explicitly requires that proximity relationships in the cell are not essential for transport, since transport would take place between organelles derived from separate cells. So any signal would suggest that inherent chemical specificity is the key to sorting, not intracellular anatomy.[14]

As already mentioned, the initial thought was that the ER was the vesicle donor since incubation step with CCCP was used to block vesicle transfer from this compartment. However, the next year, Fries and Rothman showed that the donor was, in fact, the Golgi apparatus when they increased the chase time to 5–10 minutes, thereby allowing ^{35}S-labeled VSV G to leave the ER and reach the early stages of the Golgi apparatus.[18] Support for this claim came from the fact that:

the donor membranes have greatest activity when G protein is independently known to be present in the Golgi stack; the G protein that is donated appears to have already

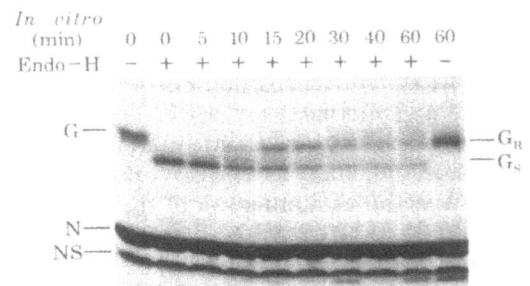

FIGURE 14.12   VSV G protein is made endo H-resistant as it passes through the Golgi apparatus. Pulse-labeled 15B cells were infected with VSV G and cell extracts were incubated with wild-type cell extracts. Samples were then either left untreated or treated with endo H and subjected to SDS-PAGE.[15] (Reprinted from Fries E, Rothman JE. Transport of vesicular stomatitis virus glycoprotein in a cell-free extract. Proc Natl Acad Sci U S A. 1980;77(7):3870–3874. Used with permission.)

entered Golgi membranes as judged by the trimming of its oligosaccharides and the coincidence of its distribution on a sucrose gradient with that of markers of Golgi membranes. Almost all of the newly synthesized G protein in the cell can be donated *in vitro* at one time of chase or another and nearly half can be processed at 10 min of chase. This makes it seem likely that the donor is a major reservoir of G protien, such as the Golgi stacks, rather than a transit form, such as transport vesicles.[18]

The authors also found that the pool of proteins that could be transferred from the early Golgi complex compartment, now called the *cis* face of the Golgi (*i.e.*, from the CHO 15B extracts), to the later compartment, now called the *trans* face (*i.e.*, to the wild type CHO extracts), was extremely transient, such that *[o]nly about half of a pulse of G protein reaches the CHO cell surface in 1 h; yet almost all of the pulse has entered the nontransferable pool by 20 min of chase.*[18] Thus, the authors hypothesized that:

each of the two pools might reside in a distinct compartment of the Golgi complex. G protein would then pass successively from the first compartment (housing the transferable pool) to the second (housing the nontransferable pool). This would perhaps correlate with the cis to trans direction of protein transport through the Golgi stack. In a cell, G protein would generally be transported from the first to the second compartment within the same Golgi complex. The inter-Golgi transport observed *in vitro* would represent transfer from the first compartment of one Golgi complex to the second compartment of another. Because transport from the first to the second compartment would be vectorial, G protein residing in the second compartment *in vivo* might not be transferable to another Golgi *in vitro*. Given that the appearance of G protein in the transferable pool coincides with the trimming of its oligosaccharides, and that processing *in vitro* requires that G protein in the transferable pool arrive at the site of terminal glycosylation in a wild-type Golgi complex, it would then be expected that the α-1,2-mannosidase responsible for trimming would be concentrated in the first Golgi compartment, and that terminal glycosyltransferases would be concentrated in the second Golgi compartment.[18]

* * * * *

By the mid-1980s, the Golgi apparatus had been determined to consist of several compartments, called stacks, which were defined as *cis*, *medial*, and *trans*, respectively, based on the order in which proteins encountered them as they exited the ER, the latter two being sites of carbohydrate addition.[17] Work in Rothman's laboratory had used this aspect of protein transport as a readout for the progress of proteins through the Golgi apparatus, since entry into the later compartments conferred endo H-resistance to proteins. However, this required the used of SDS-PAGE gels and autoradiography, which required almost a week to complete.[17] In 1984, Rothman and his group published three papers back-to-back in which they significantly improved their cell-free assay and extended their findings.[17,19,20] Their

new protocol used the incorporation of ^{3}H-labeled GlcNAc into proteins in the acceptor Golgi membrane as a read-out for protein entry into the *trans* Golgi stacks. Following incubation of the system, VSV was isolated using antiserum, collected on filter paper, and assessed for radioactivity. This method provided a more direct, a much quicker (within a day), and a more sensitive way to assess protein transport through the Golgi apparatus.[17]

As such, the group confirmed that transport between Golgi stacks occurs via a series of vesicle budding and fusion events, the specificity of which was imparted by biochemical processes rather than merely by proximity between the stacks.[19] Additional support for this proposition came from electron microscopy studies, which revealed that the Golgi stacks from donor and acceptor fractions remained *as discreet populations* throughout the process, indicating that the compartments themselves were not altered and did not fuse. Furthermore, Glick and Rothman found that transport could be blocked when both the donor and the acceptor systems were pre-treated with the reagent, N-methylmaleimide (NEM), which suggested that a NEM-sensitive factor (NSF), which could be supplied by either the donor or the acceptor membrane, was required for vesicular transport.[21] In summary, the authors found that vesicle transfer within the Golgi apparatus occurred in three distinct steps (Figure 14.13):

Priming (stage 1), needed to make G protein available for transfer, would correspond to the formation and/or completion of vesicles containing G protein. Transfer of G protein to the acceptor cisternae (step 2) would be accomplished by the diffusion of the detached transport vesicles and their subsequent attachment to target cisternae. The NEM-resistant acceptor intermediate, which results from this transfer (step 2) would be a prefusion complex in which the G-protein-containing transport vesicles have attached but not yet fused. Finally, fusion (within step 3) of the attached vesicles would make G protein available to the GlcNAc transferase in the acceptor cisternae, permitting glycosylation.[20]

Shortly after their publications in 1984, *the great electron microscopist* Lelio Orci reached out to Rothman after reading his papers and suggested they collaborate.[14] Contrary to Rothman's initial hypothesis—which postulated that clathrin-coated vesicles were intermediates in protein transport—one of the first findings of this collaboration was that this process involved a protein coat that was distinct from clathrin coats (Figure 14.14).[22] Rothman's group later purified the coat proteins and found that, in contrast to clathrin coats which were composed of one single protein, the non-clathrin coats of Golgi-derived vesicles were composed of proteins with molecular weights of 160 KDa, 37 KDa, 34 KDa, and a family of 94–102 KDa proteins.[23] In a subsequent paper, they increased the purity of the vesicles by modifying their purification protocol and identified four new proteins, similar in size to that of clathrin coat-associated proteins, which they simply named coat

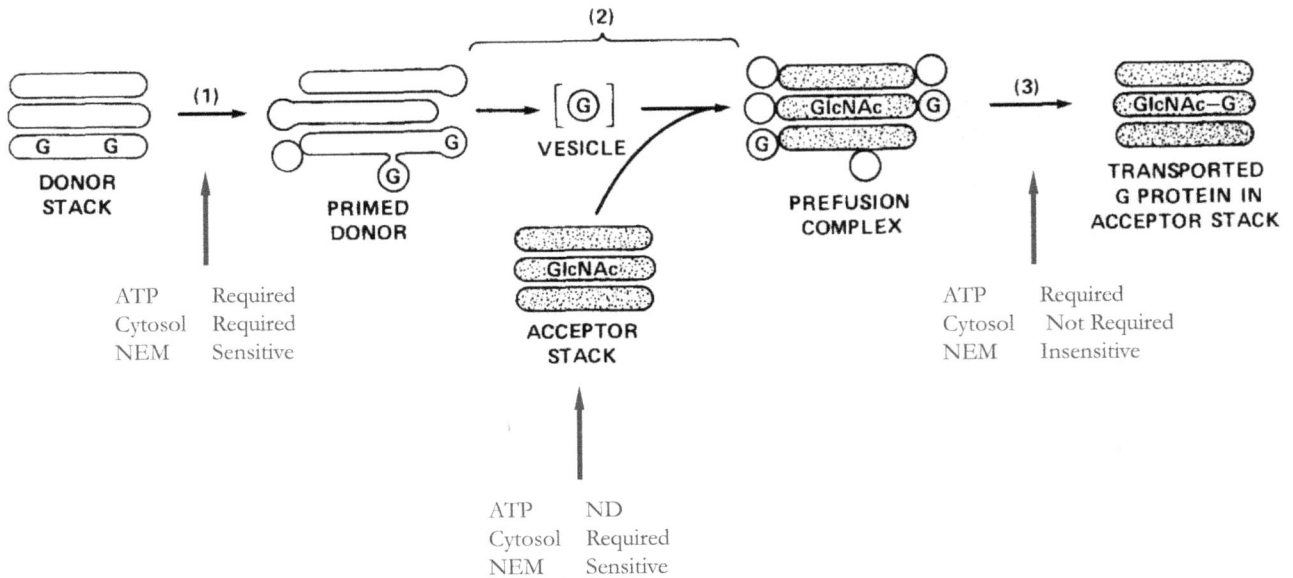

FIGURE 14.13 Diagram of protein transport through the Golgi apparatus. Priming of the Golgi apparatus was proposed to enable the formation of vesicles containing VSV G protein, which bud off of the stacks. These vesicles then fused with acceptor stacks, where GlcNAc was added to G protein. Also shown is the ATP and cytosol requirements of the various stages, as well as their sensitivity to NEM.[20] (Adapted from Cell 39(3 pt 2), Balch WE, Glick BS, Rothman JE. Sequential intermediates in the pathway of intercompartmental transport in a cell-free system, 525–536, Copyright © 1984, with permission from Elsevier.)

proteins (COP).[24] One of these proteins, α-COP, had a similar molecular weight to clathrin, whereas another, β-COP, shared striking similarities to the clathrin-associated protein, β-adaptin, whose role it is to bind clathrin. Thus, the authors proposed that β-COP might function as an α-COP-binding protein during vesicle fusion. Rothman's group also

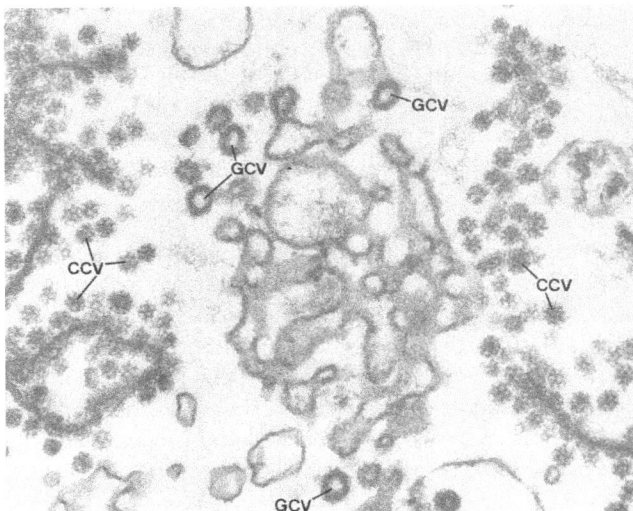

FIGURE 14.14 Vesicular transport is not mediated via clathrin-coated vesicles. Electron microscopy image of clathrin-coated vesicles (CCV) and the new types of vesicles (GCV) discovered by Orci, Glick, and Rothman.[22] (Reprinted from Cell 46(2), Orci L, Glick BS, Rothman JE. A new type of coated vesicular carrier that appears not to contain clathrin: Its possible role in protein transport within the Golgi stack, 171–184, Copyright © 1986, with permission from Elsevier.)

identified a cytosolic protein complex composed of a number of much smaller proteins, which, together with the COP proteins, formed what they called the Golgi coat promoter, or "coatomer", and which they proposed was a precursor to Golgi coated vesicles.[25]

As for the mechanism by which vesicle budding occurred, Rothman remembers that the *essential step in establishing the budding mechanism stemmed from my finding that transport was inhibited by a non-hydrolyzable analog of GTP, GTPγS*.[14] Peter Novick (formerly part of Schekman's laboratory and now leading his own laboratory) had recently found that the protein product of the sec 4 yeast mutant, which had a *massive accumulation of post-Golgi transport vesicles that fail to fuse with the plasma membrane*,[26] had significant homology with mammalian GTP-binding proteins. As we saw in the previous chapter, guanine-nucleotide binding proteins, such as G proteins and small GTPases, were known to be involved in regulatory functions of proteins and to be regulated by the binding of GTP and GDP. As such, these proteins are in an inactive state when bound by GDP and are activated by an exchange of GDP for GTP, which is mediated by proteins called "guanine nucleotide exchange factor" (GEF). In contrast, inactivation of the protein is initiated by GTPase-activating proteins, GAPs, which accelerate the hydrolysis of GTP to GDP. Therefore, Rothman proposed *the exciting possibility that guanine nucleotide-binding proteins may participate in constitutively operating pathways of intracellular protein transport*.[26] To test this possibility, Malançon et al. incubated a cell-free system with two activators of G proteins, GTPγS and AlF$_4$, both of which were found to inhibit transport across the Golgi apparatus; in contrast, ATPγS had no effect, demonstrating the specific requirement for GTP in this process.[26] Electron microscopy studies showed that incubation with GTPγS resulted in the accumulation of vesicles near the Golgi apparatus, indicating that vesicles could still bud off of the donor membrane. In addition, the authors found that although GTPγS prevented the attachment of vesicles to membranes, it did not prevent fusion from proceeding.[26]

The same year, Glick and Rothman continued to explore the effects of NEM they had started a few years before and, while attempting to purify NEM-sensitive factor (NSF), they determined that it was efficiently released from Golgi membranes after about 5 minutes incubation with ATP and Mg^{2+}; this enabled the large-scale purification of NSF from the supernatant of CHO 15B cells, which was found to be composed of a single protein with a molecular weight of 75 KDa.[27] They elucidated the precise stage of action of NSF in vesicular transport by incubating donor or acceptor fractions with NSF (stage I) and quenching its activity with NEM before mixing the fraction with its membrane partner in the presence or absence of NSF (Stage II) (Figure 14.15a). As such, the authors found that transport was only observed when NSF was present in stage II, regardless of whether the donor or acceptor membrane was treated with NEM in stage I.[28] In contrast, if the fractions were incubated together in stage I, transport

was readily observed in stage II without addition of NSF (Figure 14.15b). Furthermore, incubation in the absence of NSF resulted in the accumulation of vesicles in the cytoplasm. Together, these data indicated that NSF played a role specifically at the vesicle fusion stage. Finally, a previous study had defined a new transport intermediate, the low cytosol-requiring intermediate (LCRI), *by the fact that its formation, but not its consumption, was largely blocked when the concentration of CHO cytosol was lowered below the standard 0.5 mg/ml*, which indicated that a factor present in the cytosol was required for its formation.[28] Thus, Malhotra et al., noting that transport became NSF-independent at about the same time-frame as it progressed through the LCRI, tested the requirement of NSF for the formation of LCRI and determined that, indeed, NSF was required in this process. From these data, the authors concluded that *NSF acts after the budding and targeting of transport vesicles have already formed an acceptor-dependent intermediate*.[28]

The next year, Rothman's group investigated the mechanism by which NSF associated with membranes by incubating it with purified Golgi membranes lacking NSF activity (by treatment with NEM) under various conditions, isolating the membranes, and measuring the amount of bound NSF using a transport assay in which the extent of vesicular

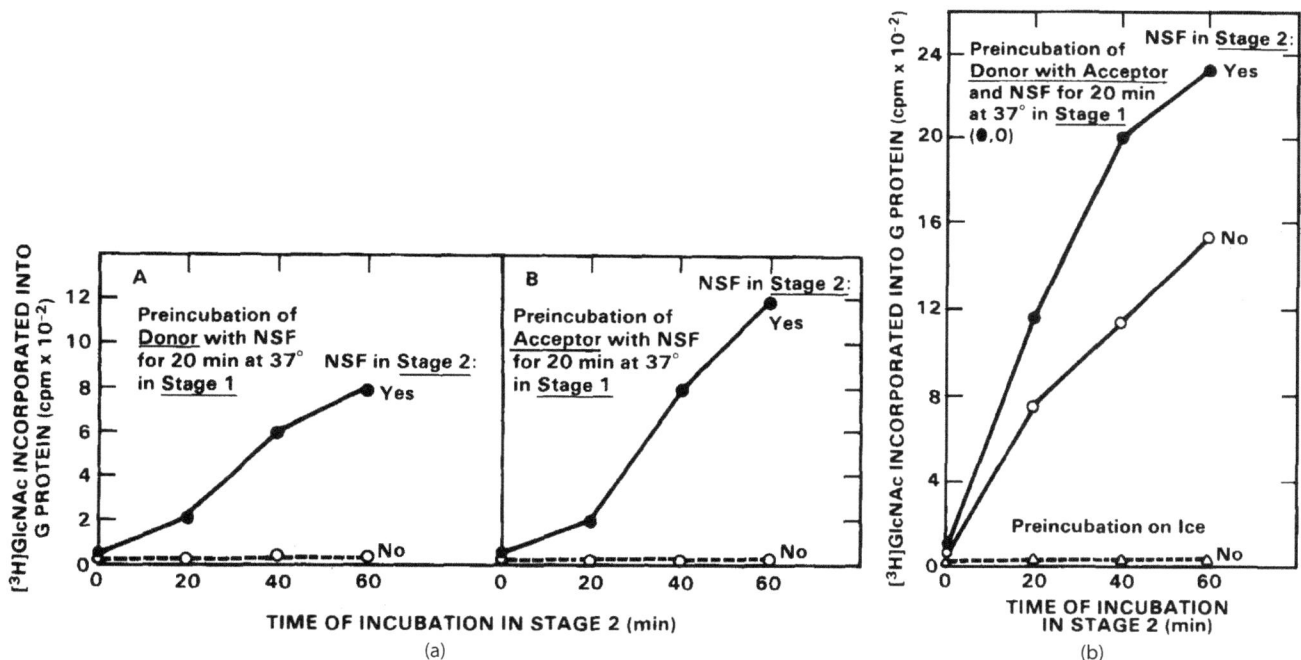

FIGURE 14.15 Transport beyond the NSF-dependent stage requires coincubation of donor and acceptor membranes. (a) (Panel A) CHO cell donor membranes were incubated with NSF for 20 minutes and inactivated with NEM (Stage 1). Samples were then transferred to untreated (i.e., presence of NSF, closed circles) or NEM-treated (i.e., absence of NSF, open circles) acceptor membranes and incubated in the presence of [³H]GlucNAc (Stage 2). Aliquots were removed at the indicated times and assayed for radioactive VSV G protein; (Panel B) same as in panel A, but acceptor membranes were treated with NSF in stage 1 and added to untreated or NEM-treated donors. (b) Same as in (a) but donor and acceptor membranes were incubated together with NSF in stage 1, and then a portion was either untreated (closed circles) or treated with NEM (open circles).[28] (Reprinted from Cell 54(2), Malhotra V, Orci L, Glick BS, Block MR, Rothman JE. Role of an N-ethylmaleimide-sensitive transport component in promoting fusion of transport vesicles with cisternae of the Golgi stack, 221–227, Copyright © 1988, with permission from Elsevier.)

transport reported the amount of NSF present. As such, the authors determined that while only minimal transport occurred in the absence of cytosol, addition of cytosol led to an increase in the amount of membrane-bound NSF in a manner that was proportional to the amount of added membrane (though this was saturable) (Figure 14.16A, B).[29] This suggested that a soluble, cytosolic factor, which they named "soluble NSF attachment protein" (SNAP), was also required for association of NSF to membranes. Thus, the authors reasoned that:

[if] the mechanism of stimulation by cytosol is an enzymatic one, then the extent of binding should depend on the length and temperature of incubation when cytosol is limiting. On the other hand, if cytosol stimulation reflects the formation of a stoichiometric complex of the cytosolic factor(s) with NSF, the extent of binding will be limited by the amount of cytosol present rather than the length of the incubation. The amount of NSF bound to membranes (when NSF is in excess) should then be strictly proportional to the concentration of cytosol in the binding incubation until a plateau is reached at which all available binding sites are titrated (e.g., saturation). In addition, if binding involves specific receptor sites on the Golgi membranes, the maximum extent of binding with excess cytosol should be proportional to the membrane concentration.[29]

Experiments testing these possibilities determined that the association was not enzymatic but rather involved *a high-affinity, stoichiometric interaction between the cytosolic factor* [i.e., SNAPs], *a membrane receptor, and NSF* (Figure 14.16C). In addition, they determined that the membrane receptor for NSF was an integral membrane protein and that *binding of NSF to Golgi membranes [...] reflects the formation of a high-affinity NSF/SNAP/membrane receptor complex*, which was required for vesicle fusion to Golgi membranes.[29] As such, the authors suggested that (Figure 14.17):

[...] docking of the vesicle at the appropriate target membrane initiates the assembly of this fusion "protein machine". The scaffold upon which the machine assembles would be the initial complex formed between a targeting protein(s) on the vesicle and docking protein(s) on the target membrane (in this case a Golgi cisterna). The cytosol- and ATP-dependent maturation of the vesicle-cisterna complex through the series of prefusion complexes would then correspond to the sequential and ATP-dependent recruitment of free subunits to build the fusion machine [...] Once the machine is fully formed, a cytosol-independent and most likely ATP-driven bilayer fusion could then be catalyzed. After fusion is over, the machine would disassemble (perhaps as a result of the major structural rearrangements that would necessarily accompany the fusion process), releasing the subunits for additional rounds.[28]

Following its discovery, a number of studies determined that, in addition to inter-cisternal vesicular transport, NSF was also required for the fusion of ER-derived vesicles to the *cis.* face of the Golgi apparatus as well as for at least one fusion step of receptor-mediated endocytosis.[30] However, progress to find other proteins involved in vesicular transport using their approach was hindered by the complexity of the system and the crudeness of the extracts:

Golgi membranes are isolated in a very gentle fashion in order to maintain them in a transport competent state; this leads to high concentrations of cytosol-derived transport activities associated with the membranes. Many such components can be expected to cycle between cytosol and membranes. This has confounded attempts to develop by fractionation an assay devoid of a particular

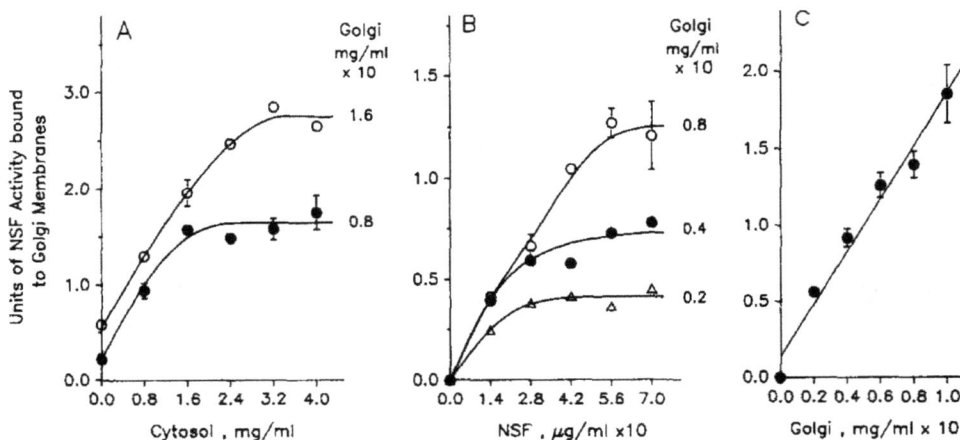

FIGURE 14.16 Binding of NSF to membranes requires a cytosolic component. (A) NSF activity was assessed following incubation of Golgi membrane proteins and titrating concentrations of cytosolic proteins. (B) Same as in (A) but using titrating concentrations of NSF. (C) NSF and cytosolic proteins were incubated with titrating concentrations of Golgi membrane proteins and assayed for NSF activity. Note how, in (A) and (B), NSF activity has the same slope regardless of the concentration of membrane added. In contrast, activity in (C) is directly proportional to the added membrane concentration.[29] (Used with permission of the Rockefeller University Press, from Weidman PJ, Melançon P, Block MR, Rothman JE. Binding of an N-ethylmaleimide-sensitive fusion protein to Golgi membranes requires both a soluble protein(s) and an integral membrane receptor. J Cell Biol. 1989;108(5):1589–1596, Copyright © 1989.)

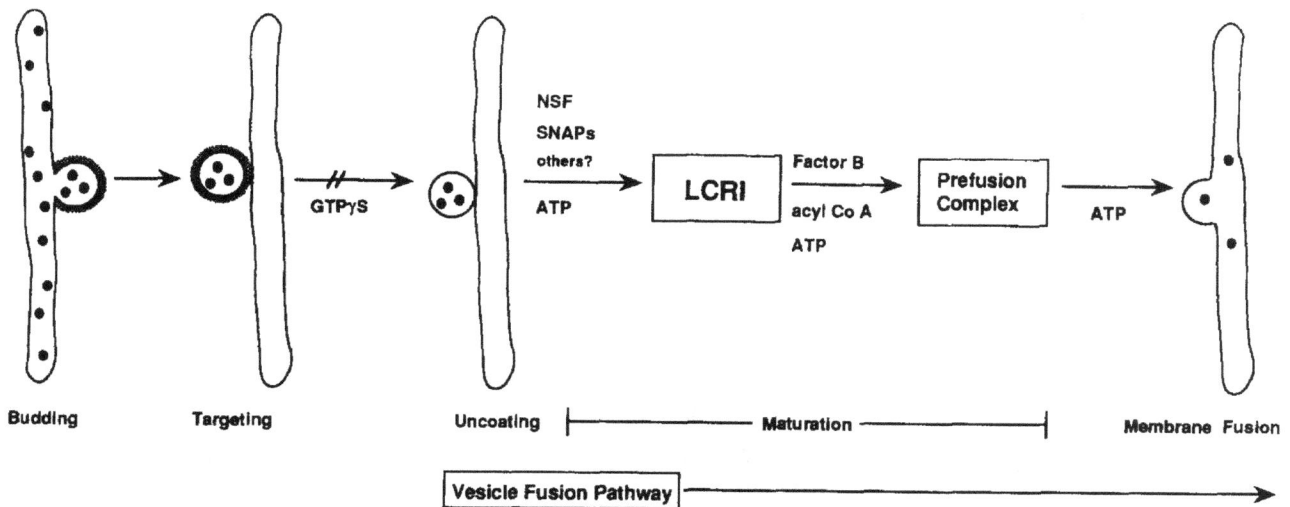

FIGURE 14.17 Proposed mechanism for transport vesicle fusion in the Golgi apparatus. Vesicle contact with the target membrane, which results in the uncoating of the vesicle, was proposed to initiate the fusion machinery required for fusion to take place. The various components required for this process are shown.[30] (Reprinted from Cell 61(4), Clary DO, Griff IC, Rothman JE. SNAPs, a family of NSF attachment proteins involved in intracellular membrane fusion in animals and yeast, 709–721, Copyright © 1990, with permission from Elsevier.)

transport activity; although a factor may be removed from the cytosol by fractionation, it is readded by addition of the membranes.[30]

Thus, in their attempts to produce Golgi membrane extracts that were *stripped of their peripherally associated proteins* and devoid of activity, Clary and Rothman found that treatment of extracts with 1M KCl (which they called K Golgi extracts) led to a significant decrease in activity which could be restored by addition of purified, inactivated NSF along with CHO cell cytosol.[31] Surprisingly, the authors found that, although yeast extracts could normally be substituted for CHO extracts to enable transport, this was not the case when K Golgi was used (Figure 14.18a). Thus, they concluded that:

> Yeast cytosol can provide all the soluble but apparently not all of the peripheral protein components needed to reconstitute Golgi transport with animal cell membranes. This provides an opportunity to use yeast cytosol as a "biochemical mutant" to develop a complementation assay for the activities that are lacking. Specifically, incubation of K Golgi membranes with yeast cytosol and purified NSF provides an assay that will reveal only those components present in animal cell cytosol not provided in a functional form by yeast cytosol (other than NSF).[31]

Therefore, Clary and Rothman tested fractionated bovine brain cytosol and found two fractions, named Fr1 and Fr2, which, together, restored activity in this system.[31] Thus, their new cell-free system was now composed of K Golgi membranes (i.e., membranes devoid of peripheral proteins), NSF, yeast cytosol, and the two bovine fractions, Fr1 and Fr2 (Figure 14.18b). Fr2 was further purified and the authors determined that its activity was provided by three separate

but related proteins, named Fr2-α, -β, and -γ. They found that each one of these proteins could reconstitute transport activity on their own when added to the system instead of Fr2, which indicated that they performed similar functions.

Next, Clary, Griffith, and Rothman sought to elucidate exactly how *NSF* [interacts] *with membranes to promote fusion*. Thus, the authors first modified the Fr2 fraction in an attempt to replace the yeast extracts so that their system was entirely composed of mammalian components. As such, they added different Fr2 fractions to bovine brain cytosol fractions that lacked Fr2 activity and determined which could efficiently restore transport without addition of yeast extracts. As such, they found two fractions, which they called Fr3 and Fr4, that together could efficiently substitute for yeast extracts in their system.[30] They next investigated at which stage of vesicle transport these Fr2 proteins acted. They reasoned that a logical place to begin was with the known intermediate step made visible by the use of GTPγS, which was shown to inhibit the fusion of coated vesicles following budding:

> if the requirement for Fr2 activity can be localized after the GTPγS block, it would imply that Fr2 is part of the fusion pathway; if Fr2 is required before the GTPγS block, it would be implicated in the budding or targeting reactions.[30]

Thus, they incubated their cell-free system in two stages: the first was a system composed either of a complete set of components—which reported the maximum transport possible under the conditions—or a system deficient in one each of Fr1, Fr2, or NSF; in the second stage, bovine cytosol and NSF were supplemented to drive transport and GTPγS was added to block it before the vesicle fusion step (Figure 14.19a). As such, they found that transport proceeded close-to-normal in

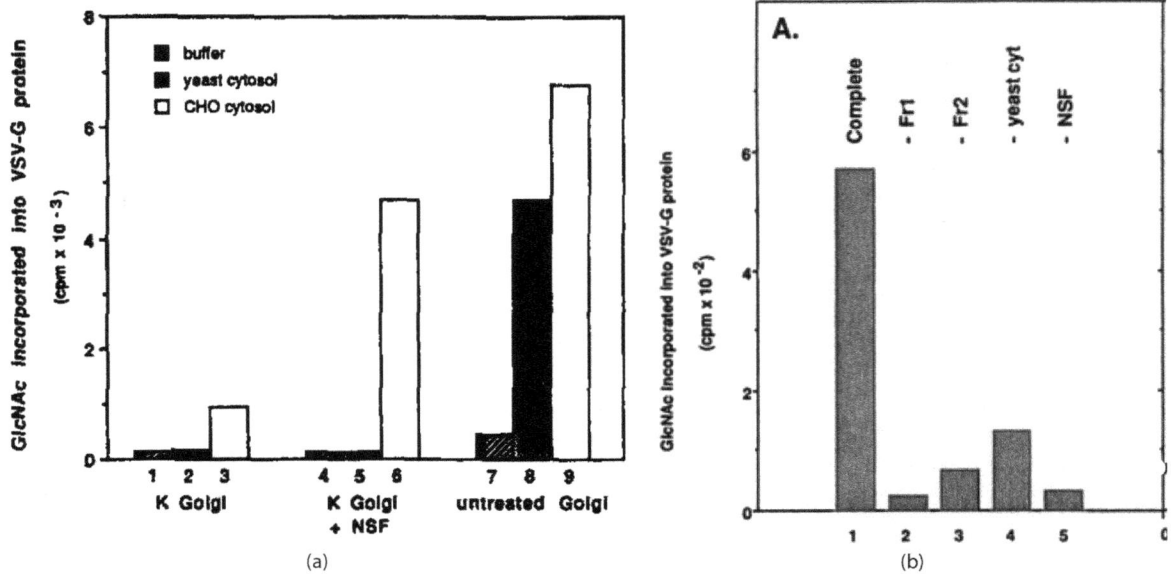

(a)

(b)

FIGURE 14.18 Identification of animal transport components not present in yeast extracts. (a) Yeast cytosol was incubated with K Golgi, K Golgi + NSF, or untreated Golgi membranes, and samples were assayed for [³H]GlcNAc incorporation into VSV G protein. Buffer alone or CHO cytosol were used as controls; (b) K Golgi, yeast cytosol, NSF, Fr1, and Fr2 are all required for reconstitution of Golgi transport. (Adapted from ref. [31].)

systems deficient in NSF or Fr2—at approximately 70% of maximum—whereas little-to-no transport was observed in the absence of Fr1. This indicated that Fr2, like NSF, acted after the inhibition point of GTPγS and thus, likely in the fusion step, whereas Fr1 acted prior to it, in an early event of vesicular transport.[30] Specifically, the authors found that Fr2 was required in parallel to NSF for the formation of the LCRI.

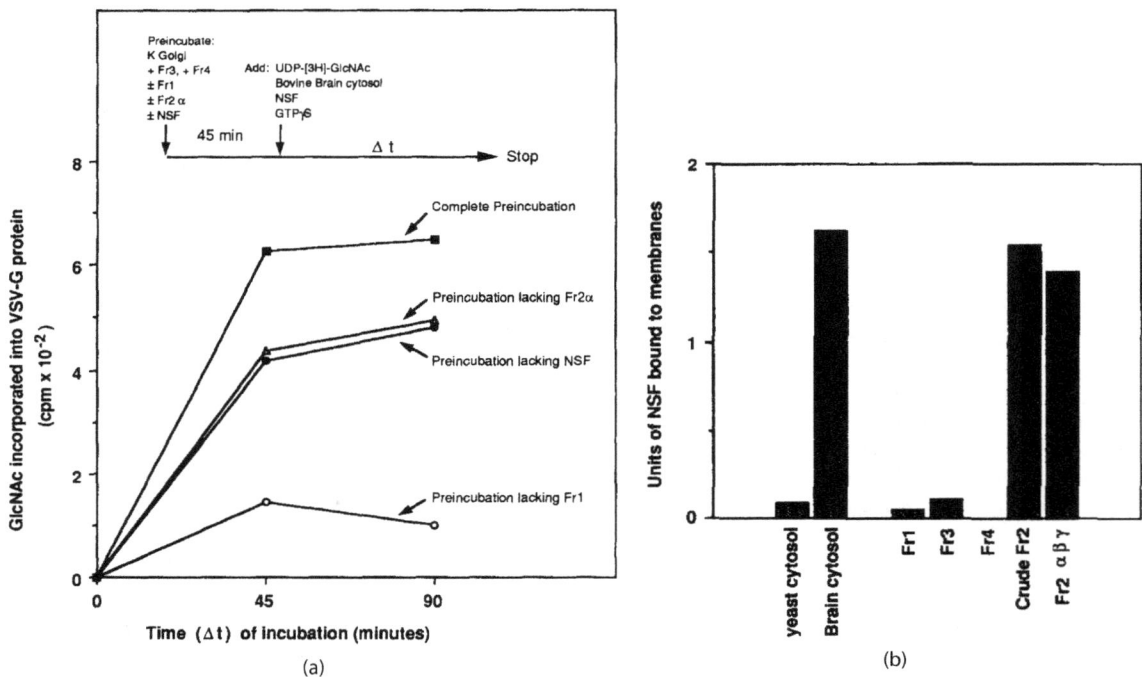

(a)

(b)

FIGURE 14.19 Elucidation of the stage of action of NSF and Fr2-α. (a) A reaction mixture was pre-incubated with all but one component, as indicated (Stage 1), followed by addition of [³H]GlcNAc, NSF, cytosol, and GTPγS (to block transport, stage 2), and samples were assayed for the incorporation of GlcNAc into VSV G proteins. (b) The Fr2 family has SNAP activity. NSF was added to a reaction mixture with the indicated components and assayed for NSF binding to purified Golgi membranes.[30] (Reprinted from Cell 61(4), Clary DO, Griff IC, Rothman JE. SNAPs, a family of NSF attachment proteins involved in intracellular membrane fusion in animals and yeast, 709–721, Copyright © 1990, with permission from Elsevier.)

In addition, the Fr2 proteins were found to be of similar molecular weights to SNAPs; therefore, he wondered whether the proteins in the Fr2 fraction might, in fact, be the SNAP proteins. Since the combination of Fr1, Fr2, Fr3, and Fr4 contained all the necessary ingredients (aside from NSF) to enable transport in the presence of K Golgi, the authors reasoned that the SNAP proteins must be contained in one of these Fr pools. As such, Clary, Griffith, and Rothman found that the only pool that efficiently enabled the binding of NSF to membranes was the Fr2 pool (Figure 14.19b).[30] Surprisingly, they also found that all three Fr2 proteins, -α, -β, and -γ, had SNAP activity in concentrations similar to that required for transport, showing for the first time that SNAPs were required for the fusion events of intra-Gogi vesicular transport. Thus, they renamed the Fr2 proteins SNAP-α, SNAP-β, and SNAP-γ.

* * * * *

Rothman next turned his attention to the role of GTP in Golgi vesicular transport, which by this point, had been shown to be important in both yeast and mammals. A first step was the finding that ADP-ribosylation factor (ARF)—a GTP-binding protein that catalyzes the activation of adenylate cyclase in *Vibrio cholerae*—was concentrated in the Golgi apparatus.[32] As such, Serafini et al. used a new technique that took advantage of the fact that many GTP-binding proteins could refold following separation by SDS-PAGE and transfer to nitrocellulose, and therefore retained their ability to bind GTP. Thus, the authors found several GTP-binding proteins that co-purified with the COP proteins of the coatomer, two of which were shown to be of the ARF family of proteins by western blotting. They also determined that approximately three ARF proteins were present in each coatomer, which was consistent with other coat subunits and which implied that ARF was a coat protein.[32]

To test this hypothesis, Serafini et al. determined the amount of ARF present in vesicles under conditions that led to the accumulation of coated or uncoated vesicles, that is, in the presence of GTPγS or NEM, respectively. As such, experiments revealed that ARF was present in abundance in vesicles accumulated in the presence of GTPγS, but not in the presence of NEM, which further supported the idea that ARF was a component of the coat protein complex (Figure 14.20a).[32] Electron microscopy studies also confirmed the presence of ARF proteins in GTPγS-enriched coated vesicles. In addition, the authors found that ARF binding to Golgi membranes could be abolished by pretreatment of the membrane with heat or trypsin (a protease), which indicated that a membrane receptor might be involved in the process. These data indicated that ARF was *a structural component of coated vesicles*, and suggested *a precise role for ARF in cell biology- in the formation and uncoating of coated transport vesicles.*[32] Thus, the authors proposed a model (Figure 14.20b)

in which the Golgi membrane contains an ARF-specific nucleotide exchange protein. Nucleotide exchange would thus occur only at the Golgi membrane, and the ARF[GTP]

thus generated would bind [...] to the only nearby bilayer, that of a Golgi cisterna. This would give rise to the "receptor"- and GTP-dependent binding of ARF to Golgi cisternae that we observe, and this pool of ARF would presumably serve as the precursor for coat assembly [...] Thus, in this hypothesis, a localized nucleotide exchange protein would catalyze membrane insertion of ARF, triggering coat assembly for budding.

The coat structure of the resulting vesicle would be stable until such time as the vesicle encountered an ARF-specific GAP, presumably at the target membrane. Hydrolysis of the bound GTP could destabilize the coat, as ARF[GDP] has no affinity for lipid bilayers. With ARFs present throughout the coat structure, the conformational change in ARF during GTP hydrolysis could have a profound effect. When ARF is loaded with GTPγS during budding, coats would be permanently stable and coated vesicles would accumulate, as observed. Upon GTP hydrolysis, ARF[GDP] would be released into the cytosol and thus recycled.[32]

Shortly thereafter, Orci et al. determined that the coatomer and ARF were the only cytosolic proteins required for the assembly of COP-coated vesicles.[33] The extraction of Golgi membrane was also refined, resulting in greater quantities of purified COP-coated vesicles without relying on the use of GTPγS to accumulate them. Electron microscopy performed under these conditions revealed the formation of coated buds which were continuous with the Golgi membrane, and Ostermann et al. showed that these structures were indeed functional, that is, that they matured and detached from the Golgi to form COP-coated vesicles.[34] The authors also found that ARF was a stable component of the coat complex, whether in the presence of GTP or GTPγS, which implied that:

the hydrolysis of GTP bound to ARF, which is a requirement for uncoating, does not occur spontaneously; in other words, hydrolysis of ARFs bound to GTP must require other cytosolic factors, docking of the vesicle to its target membrane, or both.[34]

They also found that acyl-CoA was required for the pinching off of the bud during formation of the vesicle. As such the authors proposed the following scheme for the formation of COP-coated vesicles (Figure 14.21):

The first step [...] in the assembly of a COP-coated vesicle from a Golgi cisterna is the formation of a site committed to budding, as marked by activated ARF protein bound to its membrane receptor. ARF protein is recruited from the cytosol, where it resides in its GDP-bound form. Nucleotide exchange (catalyzed by a protein at the Golgi surface) produces ARF (GTP), which inserts into the local lipid bilayer via its myristic acid chain, whose exposure is triggered by GTP binding. ARF then becomes more tightly bound when it associates with its receptor, creating high affinity binding sites for coatomer. At this stage, the bound ARFs are not clustered, and the cisternal membrane is still flat.

When coatomer binds, coated buds form that need only be pinched off at their base (a process we term fission) to

A. Possible mechanism of ARF binding to Golgi

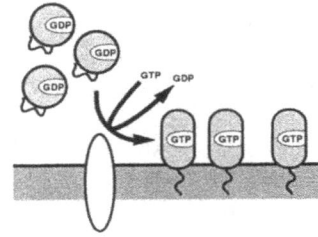

B. Possible GTP-driven cycle of budding and uncoating

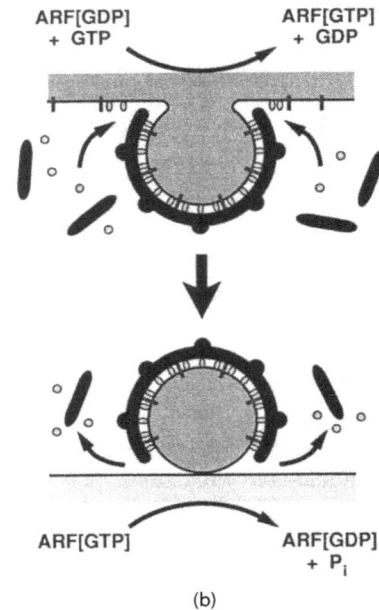

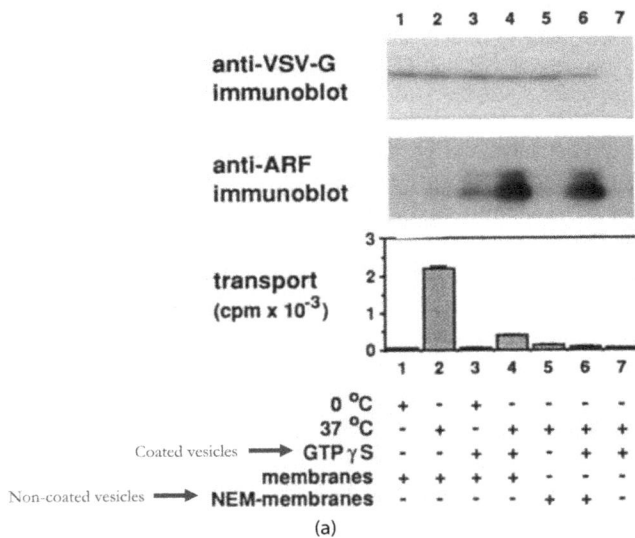

(a) (b)

FIGURE 14.20 (a) ARF binding to Golgi membranes is coincident with the appearance of coated vesicles. Donor and acceptor membranes were mixed in a 1:1 ratio and left untreated (+ membranes) or treated with NEM (+ NEM membranes), either at 0°C or 37°C and in the presence or absence of GTPγS as indicated. Samples were subjected to SDS-PAGE and blotted for VSV G (top membrane) or ARF (bottom membrane). Note how a significant amount of ARF binds to membranes under conditions in which coated membranes accumulated compared to those in which uncoated membranes accumulated (compare lane 4 to 5). Although NEM treatment eliminates transport (as expected, bar graph), it does not eliminate ARF binding to membranes (compare lane 4 to 6). Lane 7 acted as a control as it did not contain membranes. (b) Proposed mechanism of ARF and GTP in the budding and uncoating of transport vesicles. GEFs are proposed to be required to catalyze the exchange of GDP for GTP in ARF, allowing it to bind to the membrane (A), and aid in vesicle coating (B). GAPs are also proposed to be resident Golgi membrane proteins which aid in the disassembly of coats at target membranes by hydrolyzing ARF-GTP and causing its release from the membrane.[32] (Modified from Cell 67(2), Serafini T, Orci L, Amherdt M, Brunner M, Kahn R A, Rothman JE. ADP-ribosylation factor is a subunit of the coat of Golgi-derived COP-coated vesicles: A novel role for a GTP-binding protein, 239–253, Copyright © 1991, with permission from Elsevier.)

yield complete and fusion-competent coated vesicles. The only nucleotide required to form coated buds is GTP, which remains bound to ARF protein and unhydrolyzed; GTPγS can substitute for GTP and an ARF mutant that can bind but not hydrolyze GTP can substitute for wild-type ARF. Coatomer binding simultaneously results in the formation of buds and the clustering of ARF (presumably still bound to its receptor) with coatomer into buds; little bound coatomer remains outside of assembled coats [...]

The simplest interpretation of the requirement for fatty acyl-CoA in fission is that acylation of a fusion protein facing the lumen triggers fission when the coat brings the lumenal surfaces of the lipid bilayers into proximity at the base of the bud [...]

The simple view that vesicles having different coats bud from different organelles using similar underlying mechanisms seems increasingly tenable. This is suggested by certain similarities in the sequences of clathrin and COP coated vesicle proteins and by the requirement for ARF in binding reactions involved in the assembly of both types of coats from Golgi membranes.[34]

In a subsequent paper, Tanigawa et al. tested the hypothesis that vesicle uncoating was mediated by the hydrolysis of ARF-bound GTP *in the appropriate biological context*, that is, following budding and prior to fusion.[35] The authors were inspired by another GTP-binding protein, Ras, whose Q61L

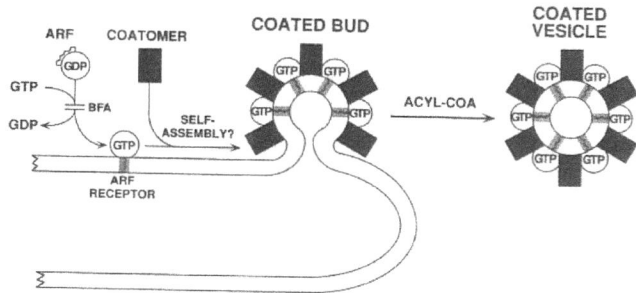

FIGURE 14.21   Summary of steps involved in the formation of COP-coated vesicles.[34] (Reprinted from Cell 75(5), Ostermann J, Orci L, Tani K, et al. Stepwise assembly of functionally active transport vesicles, 1015–1025, Copyright © 1993, with permission from Elsevier.)

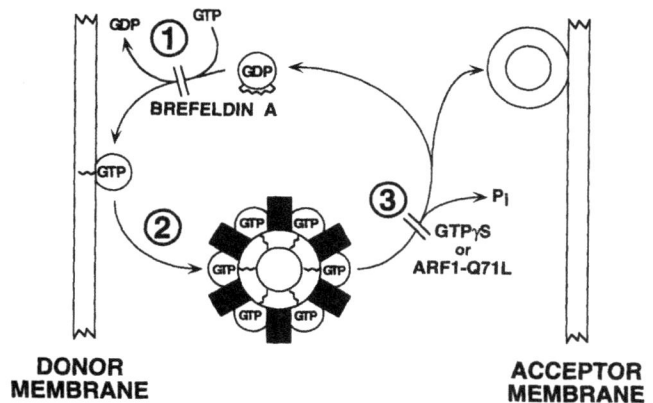

FIGURE 14.22   Model for the recycling of ARF during COP-coated vesicle trafficking. ARF (circles with GTP or GDP) binds to donor membrane after GDP is exchanged to GTP by an unknown GEF, enabling coatomer (black rectangles) binding to the membrane and budding off of a vesicle. Hydrolysis of ARF-bound GTP causes uncoating of the vesicle and attachment/fusion to acceptor membrane, a process which is blocked by the non-hydrolyzable GTPγS.[35] (Used with permission of the Rockefeller University Press, from Tanigawa G, Orci L, Amherdt M, Ravazzola M, Helms JB, Rothman JE. Hydrolysis of bound GTP by ARF protein triggers uncoating of Golgi-derived COP-coated vesicles. J Cell Biol. 1993;123(6 pt 1):1365–1371, Copyright © 1993.)

mutation was shown to inhibit GTP hydrolysis but not GTP binding; thus, they mutated the corresponding amino acid in ARF (mARF1-Q71L) and added either wild-type or mutant ARF to their cell-free system. They found that in the presence of wild-type ARF, the binding of coatomer to Golgi membranes increased significantly in the presence of GTPγS compared to that of GTP; in contrast, binding increased in the presence of both GTP and GTPγS when membranes were incubated with mARF1-Q71L. Thus, the authors concluded that *mARF1-Q71L binds to membranes but does not hydrolyze GTP, in contrast to wildtype mARF1.* Consistent with these results, protein transport was also inhibited in the presence of mARF1-Q71L.[35] Finally, the authors presented an updated model for their results (Figure 14.22):

Switching into the bilayer-binding GTP form is required for mARF to bind to specific sites in Golgi membranes ("ARF receptors") and to form complexes with coatomer before coated vesicle budding […]

The Golgi cisternal membranes remain flat when only ARF has bound, but when coatomer binds, coated regions of membrane are deformed into the shape of a vesicle. ARF can only bind Golgi membranes in its […] GTP-bound form, and this activated ARF is generated when cytosolic mARF[GDP] encounters a Golgi-bound nucleotide exchange factor in the presence of free GTP […]

Given these properties, it is predictable that when ARF hydrolyzes its GTP, the protein will switch back to the GDP conformation […] and dissociate from the membrane of the coated vesicle. Because coatomer binding to membranes is ARF-dependent, coatomer release into the cytosol would be expected to follow dissociation of ARF.[35]

At the same time, Rothman's laboratory continued to elucidate the mechanism of interaction between NSF and SNAP proteins. As such, they found that these proteins sedimented into a 20S particle complex along with a SNAP receptor (SNARE) upon solubilization of membranes with detergents.[36] In addition, NSF was shown to be an ATPase with two binding sites for ATP, and either ATP alone or Mg-ATPγS was required for stabilization of the NSF-SNAP-SNARE complex; incubation with Mg-ATP

resulted in the rapid dissociation of the complex, which was proposed to be *an intrinsic step in the fusion machinery.*[36] Thus, Söllner et al. used myc-tagged NSF to purify SNAREs from the 20S particle in the presence of ATPγS and subsequently used Mg-ATP to release NSF and SNAPS (Figure 14.23a). The eluted proteins were run on SDS-PAGE and sequenced, and they were identified as syntaxin A (band C in the Figure 14.23b) and B (band A), SNAP-25 (this protein had been previously named as such by coincidence, band E), and VAMP (band F).

Of particular interest was SNAP-25, since it was known to be part of a protein family involved in vesicle binding in nerve synapses, some of which bound to SNAREs on transport vesicles (v-SNAREs), while others bound to SNAREs on target membranes (t-SNAREs). Thus, the authors suggested the SNARE hypothesis for the specificity with which vesicles bind to a target membrane (Figure 14.24):

[…] each transport vesicle contains one or more members of the v-SNARE superfamily obtained on budding from a corresponding donor compartment, and every target compartment in the cell contains one or more members of the t-SNARE superfamily. Specificity in membrane transactions would be assured by the unique and non-overlapping distribution of v-SNAREs and tSNAREs [sic] among the different vesicles and target compartments. In the simplest view, that is, if there were no other source of specificity, only when the complementary v-SNARE and t-SNARE pairs engage would a productive fusion event be initiated.[36]

Subsequent studies found that the SNARE complex, that is, binding of v-SNARE to t-SNARE, was sufficient to

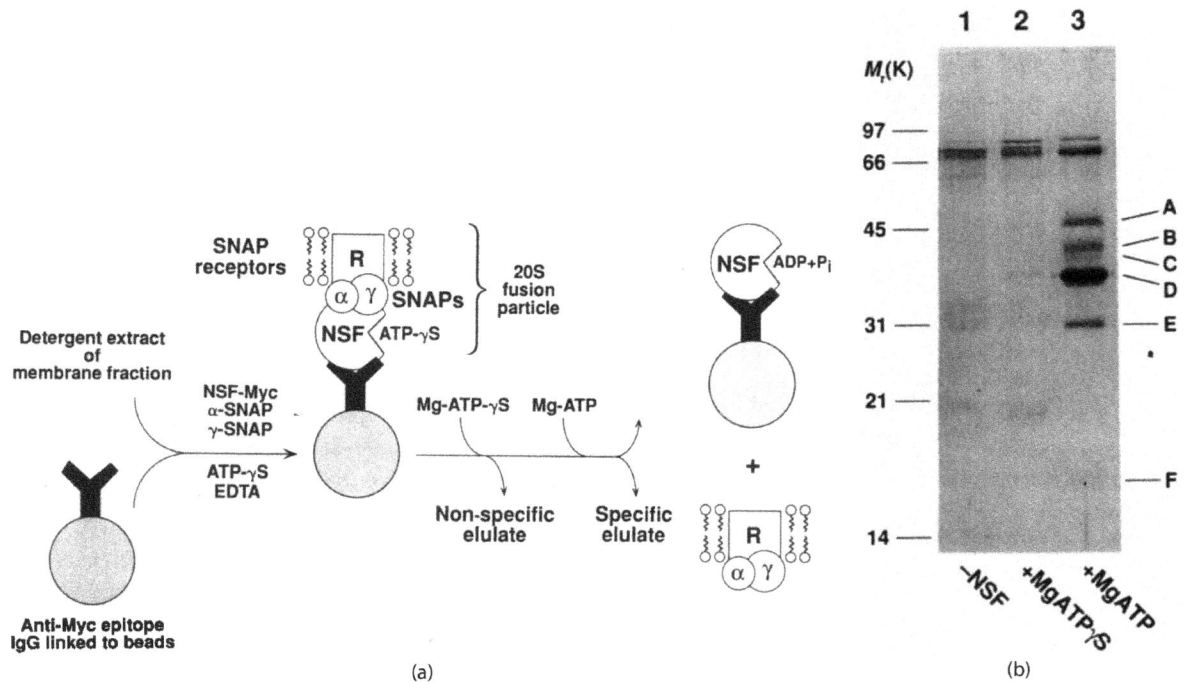

(a)

(b)

FIGURE 14.23 Söllner et al.'s procedure to purify SNAP proteins. (a) Myc-tagged NSF and SNAP proteins were incubated in a mixture that induced the formation of a complex composed of these proteins and SNAP receptors, and an anti-Myc antibody (bound to a sepharose bead) was used to precipitate the complex out of solution. The receptor and SNAP proteins were released from the bead-bound NSF by addition of Mg-ATP, and the eluate was subjected to SDS-PAGE. (b) SDS-PAGE gel of the experiment: lane 3 has the elution of SNAP proteins from the IP, identified as bands A–F. Lanes 1 and 2 were used as controls, the former being the eluate from a reaction in the absence of NSF, whereas the latter was a reaction with NSF but in the presence of GTPγS. (which prevents the release of the SNAP complex from NSF).[36] (Adapted by permission from Springer Nature Customer Service Centre GmbH: Springer Nature, Nature, Söllner T, Whiteheart SW, Brunner M, et al. SNAP receptors implicated in vesicle targeting and fusion, Copyright © 1993.)

enable vesicle fusion, whereas NSF and SNAP play a vital role in sustaining ongoing fusion by separating the SNARE complex after fusion is complete.[14]

* * * * *

FIGURE 14.24 The SNARE hypothesis as proposed by Rothman. v-SNARES on vesicles bind to their counter receptors, t-SNARES on target membranes, allowing the fusion of former to the latter.[37] (Adapted from Cell 92(6), Weber T, Zemelman B v., McNew JA, et al. SNAREpins: Minimal machinery for membrane fusion, 759–772, Copyright © 1998, with permission from Elsevier.)

I will end this chapter by briefly reviewing one of the first visualizations of protein trafficking in real time. In 1997, Jennifer Lippincott-Schwartz used the new technique of tagging proteins with GFP—enabled by the work of researchers who worked out the techniques for cloning and by those who studied and isolated the protein responsible for the fluorescence emitted by jellyfish—to visualize and track the movement of proteins within cells. In her experiments, Lippincott-Schwartz attached GFP to a temperature-sensitive mutant protein of Rothman's well-characterized VSV G protein (ts045), which was retained in the ER at 40°C as a consequence of misfolding, but proceeded normally through the Golgi apparatus when the temperature was returned to 32°C. Thus, Presley et al. transfected a plasmid containing this mutant in CHO cells, and by switching from a restrictive to a permissive temperature, the authors were able to follow the path of VSV G from the ER to the Golgi apparatus, and finally to the cell surface.[38] As expected, the authors found that VSV G accumulated in the ER when incubated at 40°C for 12 hours but flowed to the Golgi apparatus following a decrease in temperature to 32°C (Figure 14.25a). In addition, they found that small vesicles—smaller than the pre-Golgi structures—began to *arise randomly within the ER* within 1-5 minutes of the temperature shift *at widely dispersed sites within the reticular ER [...] before translocating into the Golgi region*.[38] In another experiment, cells were first

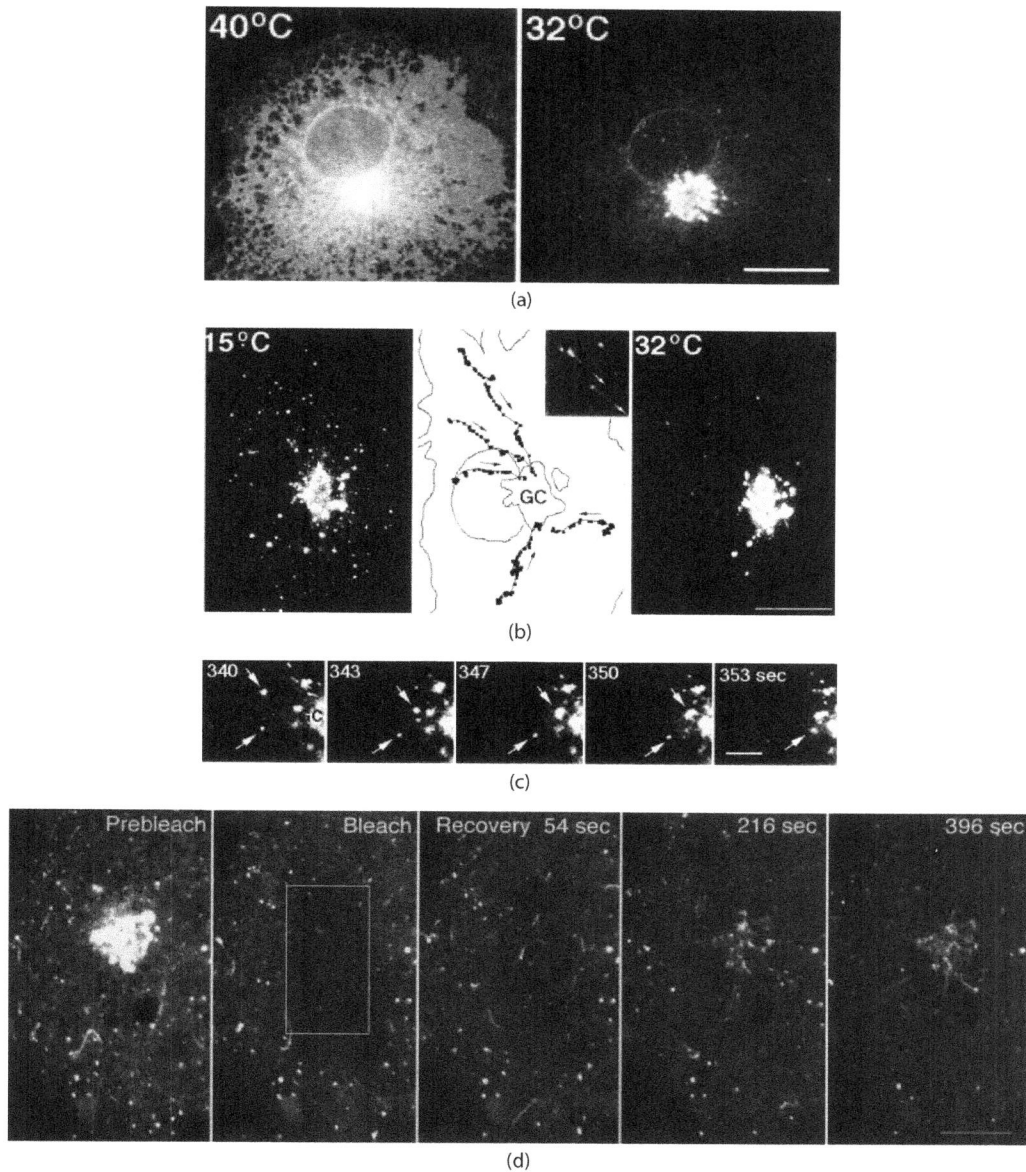

FIGURE 14.25 GFP tagging of VSV G protein allows tracking of its transport in real time. GFP was cloned into a VSV G-containing plasmid and then transfected into CHO cells. Fluorescence microscopy was used for visualization within the cell. (a) CHO cells were incubated at 40°C to concentrate proteins in the ER (left panel) and subsequently released the block by incubation at 32°C (right panel), at which time proteins moved "en mass" to the Golgi apparatus. (b) Cells were incubated at 15°C to block protein transport, which resulted in the accumulation of proteins in pre-Golgi intermediates (left panel, scattered dots). Upon switching to a permissive temperature of 32°C, the dots disappeared and accumulated in the Golgi (right panel). Also shown is the movement of several pre-Golgi structures to the Golgi (middle panel and c). (d) Tracking of VSV G protein transport by photobleaching of the Golgi apparatus. Fluorescent proteins in the Golgi apparatus were photobleached with a high-intensity laser and the movement of pre-Golgi clusters were followed over about 6 minutes.[38] (Adapted by permission from Springer Nature Customer Service Centre GmbH: Springer Nature, Nature, Presley JF, Cole NB, Schroer TA, Hirschberg K, Zaal KJM, Lippincott-Schwartz J. ER-to-Golgi transport visualized in living cells, Copyright © 1997.)

incubated at 15°C to accumulate proteins in the pre-Golgi structures—*presumably owing to a rate-limiting step in membrane transport through these intermediates*—and subsequently raised to 32°C to alleviate the block. Under these conditions, vesicles were found to travel towards, and cluster in, the Golgi complex in a *stop-and-go fashion* at a velocity of up to 1.4 µm/s (Figure 14.25b, c).[38]

The authors also used a technique called photobleaching, which involved shining light on a cell to extinguish the fluorescence emitted from GFP—fluorescent proteins lose their ability to fluoresce the longer they are exposed to light—creating an area of darkness where the light had been shone. This technique allowed the authors to determine the degree to which fluorescent particles repopulated the bleached area (Figure 14.25d). As such, Presley et al. found that pre-Golgi structures could be seen translocating into the Golgi complex almost immediately after photobleaching, and within 3 to 5 minutes of photobleaching, *the profile of VSVG-GFP-containing Golgi membranes was similar to the pre-bleach image.*

* * * * *

REFERENCES

1. Jamieson JD, Palade GE. Intracellular transport of secretory proteins in the pancreatic exocrine cell. I. Role of the peripheral elements of the Golgi complex. *Journal Cell Biology.* 1967;34(2):577–596. doi:10.1083/jcb.34.2.577
2. Jamieson JD, Palade GE. Intracellular transport of secretory proteins in the pancreatic exocrine cell. II. Transport to condensing vacuoles and zymogen granules. *Journal of Cell Biology.* 1967;34(2):597–615. doi:10.1083/jcb.34.2.597
3. Palade G. Intracellular aspects of the process of protein synthesis. *Science (1979).* 1975;189(4200):347–358. doi:10.1126/science.1096303
4. Blobel G, Dobberstein B. Transfer of proteins across membranes. I. Presence of proteolytically processed and unprocessed nascent immunoglobulin light chains on membrane-bound ribosomes of murine myeloma. *Journal of Cell Biology.* 1975;67(3):835–851. doi:10.1083/jcb.67.3.835
5. Goldstein JL, Anderson RGW, Brown MS. Coated pits, coated vesicles, and receptor-mediated endocytosis. *Nature.* 1979;279(5715):679–685. doi:10.1038/279679a0
6. Roth TF, Porter KR. Yolk protein uptake in the oocyte of the mosquito *Aedes aegypti.* L. *Journal of Cell Biology.* 1964;20:313–332. doi:https://doi.org/10.1083/jcb.20.2.313
7. Pearse BM. Coated vesicles from pig brain: Purification and biochemical characterization. *Journal of Molecular Biology.* 1975;97(1):93–98. doi:10.1016/S0022-2836(75)80024-6
8. Pearse BM. Clathrin: A unique protein associated with intracellular transfer of membrane by coated vesicles. *Proceedings of National Academy of Sciences U S A.* 1976;73(4):1255–1259. doi:10.1073/pnas.73.4.1255
9. Miya Fujimoto L, Roth R, Heuser JE, Schmid SL. Actin assembly plays a variable, but not obligatory role in receptor-mediated endocytosis. *Traffic.* 2000;1(2):161–171. doi:10.1034/J.1600-0854.2000.010208.X
10. Schekman RW, Randy W. Schekman: Nobel lecture: Genes and proteins that control the secretory pathway. NobelPrize.org. Nobel Media AB © The Nobel Foundation 2013. https://www.nobelprize.org/prizes/medicine/2013/schekman/lecture/
11. Novick P, Schekman R. Secretion and cell-surface growth are blocked in a temperature-sensitive mutant of *Saccharomyces cerevisiae. Proceedings of National Academy of Sciences U S A.* 1979;76(4):1858–1862. doi:10.1073/pnas.76.4.1858
12. Novick P, Field C, Schekman R. Identification of 23 complementation groups required for post-translational events in the yeast secretory pathway. *Cell.* 1980;21(1):205–215. Accessed March 30, 2019. http://www.ncbi.nlm.nih.gov/pubmed/6996832
13. Novick P, Ferro S, Schekman R. Order of events in the yeast secretory pathway. *Cell.* 1981;25(2):461–469. doi:10.1016/0092-8674(81)90064-7
14. Rothman, E. Rothman: Nobel Lecture: The Principle of Membrane Fusion in the Cell. NobelPrize.org. Nobel Media

AB © The Nobel Foundation 2013. https://www.nobelprize.org/prizes/medicine/2013/rothman/lecture/
15. Fries E, Rothman JE. Transport of vesicular stomatitis virus glycoprotein in a cell-free extract. *Proceedings of National Academy of Sciences U S A.* 1980;77(7):3870–3874. doi:10.1073/pnas.77.7.3870
16. Rothman JE, Fine RE. Coated vesicles transport newly synthesized membrane glycoproteins from endoplasmic reticulum to plasma membrane in two successive stages. *Proceedings of National Academy of Sciences U S A.* 1980;77(2):780–784. doi:10.1073/pnas.77.2.780
17. Balch WE, Dunphy WG, Braell WA, Rothman JE. Reconstitution of the transport of protein between successive compartments of the Golgi measured by the coupled incorporation of N-acetylglucosamine. *Cell.* 1984;39(2 Pt 1):405–416. Accessed March 29, 2019. http://www.ncbi.nlm.nih.gov/pubmed/6498939
18. Fries E, Rothman JE. Transient activity of Golgi-like membranes as donors of vesicular stomatitis viral glycoprotein in vitro. *Journal of Cell Biology.* 1981;90(3):697–704. doi:10.1083/JCB.90.3.697
19. Braell WA, Balch WE, Dobbertin DC, Rothman JE. The glycoprotein that is transported between successive compartments of the Golgi in a cell-free system resides in stacks of cisternae. *Cell.* 1984;39(3 Pt 2):511–524. doi:10.1016/0092-8674(84)90458-6
20. Balch WE, Glick BS, Rothman JE. Sequential intermediates in the pathway of intercompartmental transport in a cell-free system. *Cell.* 1984;39(3 Pt 2):525–536. doi:10.1016/0092-8674(84)90459-8
21. Glick BS, Rothman JE. Possible role for fatty acyl-coenzyme A in intracellular protein transport. *Nature.* 1987;326(6110):309–312. doi:10.1038/326309a0
22. Orci L, Glick BS, Rothman JE. A new type of coated vesicular carrier that appears not to contain clathrin: Its possible role in protein transport within the Golgi stack. *Cell.* 1986;46(2):171–184. doi:10.1016/0092-8674(86)90734-8
23. Malhotra V, Serafini T, Orci L, Shepherd JC, Rothman JE. Purification of a novel class of coated vesicles mediating biosynthetic protein transport through the Golgi stack. *Cell.* 1989;58(2):329–336. doi:10.1016/0092-8674(89)90847-7
24. Serafini T, Stenbeck G, Brecht A, et al. A coat subunit of Golgi-derived non-clathrin-coated vesicles with homology to the clathrin-coated vesicle coat protein β-adaptin. *Nature.* 1991;349(6306):215–220. doi:10.1038/349215a0
25. Waters MG, Serafini T, Rothman JE. "Coatomer": A cytosolic protein complex containing subunits of non-clathrin-coated Golgi transport vesicles. *Nature.* 1991;349(6306):248–251. doi:10.1038/349248a0
26. Melançon P, Glick BS, Malhotra V, et al. Involvement of GTP-binding "G" proteins in transport through the Golgi stack. *Cell.* 1987;51(6):1053–1062. doi:10.1016/0092-8674(87)90591-5
27. Block MR, Glick BS, Wilcox CA, Wieland FT, Rothman JE. Purification of an N-ethylmaleimide-sensitive protein catalyzing vesicular transport. *Proceedings of National Academy of Sciences U S A.* 1988;85(21):7852–7856. doi:10.1073/pnas.85.21.7852
28. Malhotra V, Orci L, Glick BS, Block MR, Rothman JE. Role of an n-ethylmaleimide-sensitive transport component in promoting fusion of transport vesicles with cisternae of the Golgi stack. *Cell.* 1988;54(2):221–227. doi:10.1016/0092-8674(88)90554-5
29. Weidman PJ, Melançon P, Block MR, Rothman JE. Binding of an N-ethylmaleimide-sensitive fusion protein

to Golgi membranes requires both a soluble protein(s) and an integral membrane receptor. *Journal of Cell Biology.* 1989;108(5):1589–1596. doi:10.1083/jcb.108.5.1589

30. Clary DO, Griff IC, Rothman JE. SNAPs, a family of NSF attachment proteins involved in intracellular membrane fusion in animals and yeast. *Cell.* 1990;61(4):709–721. doi:10.1016/0092-8674(90)90482-T

31. Clary DO, Rothman JE. Purification of three related peripheral membrane proteins needed for vesicular transport. *Journal of Biological Chemistry.* 1990;265(17):10109–10117. Accessed March 30, 2019. https://www.jbc.org/article/S0021-9258(19)38786-1/pdf

32. Serafini T, Orci L, Amherdt M, Brunner M, Kahn RA, Rothman JE. ADP-ribosylation factor is a subunit of the coat of Golgi-derived COP-coated vesicles: A novel role for a GTP-binding protein. *Cell.* 1991;67(2):239–253. doi:10.1016/0092-8674(91)90176-Y

33. Orci L, Palmer DJ, Amherdt M, Rothman JE. Coated vesicle assembly in the Golgi requires only coatomer and ARF proteins from the cytosol. *Nature.* 1993;364(6439):732–734. doi:10.1038/364732a0

34. Ostermann J, Orci L, Tani K, et al. Stepwise assembly of functionally active transport vesicles. *Cell.* 1993;75(5):1015–1025. doi:10.1016/0092-8674(93)90545-2

35. Tanigawa G, Orci L, Amherdt M, Ravazzola M, Helms JB, Rothman JE. Hydrolysis of bound GTP by ARF protein triggers uncoating of Golgi-derived COP-coated vesicles. *Journal of Cell Biology.* 1993;123(6 Pt 1):1365–1371. doi:10.1083/jcb.123.6.1365

36. Söllner T, Whiteheart SW, Brunner M, et al. SNAP receptors implicated in vesicle targeting and fusion. *Nature.* 1993;362(6418):318–324. doi:10.1038/362318a0

37. Weber T, Zemelman B, McNew JA, et al. SNAREpins: Minimal machinery for membrane fusion. *Cell.* 1998;92(6):759–772. doi:10.1016/S0092-8674(00)81404-X

38. Presley JF, Cole NB, Schroer TA, Hirschberg K, Zaal KJM, Lippincott-Schwartz J. ER-to-Golgi transport visualized in living cells. *Nature.* 1997;389(6646):81–85. doi:10.1038/38001

15 The Mechanism of Cell Death

So far, we have seen how proteins are central to cellular processes such as the metabolism of food into energy, the secretory pathway, and transduction of extracellular signals to intracellular signals which provide behavioral cues for the receiving cell. However, an equally important aspect for multicellular organisms is the process of cell death. Although cells themselves were not specifically known to be responsible for these processes—cells were not discovered until the 19th century—cell death had been observed since ancient times, such as that which occurs following minor skin injuries or as a result of more serious conditions, such as gangrene—when tissues die due to a lack of blood supply—or as a consequence of a variety of infections, such as those produced by flesh-eating bacteria. As a result of these conditions, cells undergo necrosis, a type of cell death whereby cell membranes are ruptured and the cell's contents are released into the extracellular space. These proteins, which are not normally found outside the cell, attract cells of the of immune system, such as white blood cells, and other cells—called macrophages—responsible for clearing out debris, and this increase in cell density and cellular debris in a given area results in what we know as inflammation: swelling, redness, and the sensation of heat. This process is responsible for the tissue damage that occurs following injuries.

However, cell death can also proceed in a controlled, voluntary manner that is vital for an organism's survival, in a process called "apoptosis". Shortly after the recognition that organisms and tissues were composed of cells, that is, after Schwann and Schleiden's publication of the cell theory in the mid-19th century, Karl Vogt reported observing dead or dying cells during his studies of toads' nervous system.[1] Soon thereafter, Rudolph Virchow—who added the last tenet of the cell theory—broached the subject of cell death in his famous "Cellular Pathology", contrasting necrosis with a newer type of cell death, necrobiosis, in which

> [...] death is brought on by (altered) life—a spontaneous wearing out of living parts—the destruction and annihilation consequent upon life— natural as opposed to violent death (mortification) *[or necrosis]*.
>
> [...] in necrosis we conceive the mortified part to be preserved more or less in its external form. Here *[necrobiosis]* on the contrary the part vanishes, so that we can no longer perceive it in its previous form. We have no necrosed fragment at the end of the process, no mortification of the ordinary kind, but a mass in which absolutely nothing of the previously existing tissues is preserved. The necrobiotic processes, which must be completely separated from necrosis, are in general attended by softening as their ultimate result.[2]

In 1885, Walther Flemming, who gave chromatin and mitosis their names, published an important paper in which he noted that the nucleus of cells in the lining of regressing ovarian follicles broke up and ultimately disappeared, and called this *chromatolysis*.[3] By the early 1900s, Gräper concluded that *[c]hromatolysis must exist in all organs in which cells must be eliminated* and that neighboring cells helped to eliminate the debris by engulfing it.[3] Although this line of research was largely ignored, it did not escape the realm of embryologists who had attributed a number of developmental processes in a variety of organisms to cell death, *such as invaginations and evaginations, separation of parts from each other, migration of rudiments, closure of tubes, vesicles, etc.*[4] However, although the important concept of cell "suicide", that is, apoptosis, per se, first emerged in the 1950s, it was not until John Kerr published his seminal paper on the subject in 1971 that the phenomenon was fully accepted as a legitimate cellular process.

Kerr developed an interest in cell death after graduating as a medical student specializing in pathology at the University of Queensland in Australia when he relocated to London in 1962 to begin a Ph.D.[5] There, his supervisor suggested he repeat some experiments performed by Rous and Larimore in 1920, in which the lobes of rabbit livers were found to shrink following interruption of their blood supply, with a concomitant increase in the size of the rest of the organ.[6] Kerr's experiments in rats had similar results as those of Rous and Larimore, and he concluded that the shrinkage resulted from two separate processes: the development of necrosis around the area farthest from the obstructed vein, the debris of which was removed by phagocytosis (engulfment of debris and dead cells by specialized cells called phagocytes, such as the aforementioned macrophages), and the progressive deletion of cells in other areas of the liver still receiving blood by a process he determined to be quite different from necrosis.[5] The deletion of these cells was most prominent during the rapid shrinkage of the liver and came to equilibrium as the liver stabilized into its new shape. Most importantly—though only inconspicuously noted near the beginning of the paper—this process was determined to also occur in healthy rat livers, albeit at a much-reduced rate, and the term "shrinkage necrosis" was suggested for the phenomenon.[5]

Upon his return to Queensland in 1965, having taken a position in the Pathology Department, his first order of business was to further investigate this new phenomenon. When the University acquired its first electron microscope in 1967, Kerr set out to study the *ultrastructural events involved in the evolution of shrinkage necrosis*, and found that *rounded bodies that still lay free in the extracellular space comprised membrane-bounded fragments of condensed parenchymal cell cytoplasm in which the closely packed organelles were well preserved*. These were

DOI: 10.1201/9781003379058-15

determined to arise *by a process of budding-off of pro-tuberances that developed on the surface of condensing cells.*[5] In a subsequent paper, Kerr suggested that:

> [...] whilst severe damage to a tissue causes classical necrosis, a moderately noxious environment induces scattered cells to undergo shrinkage necrosis [...] the prolific cellular budding that occurs in shrinkage necrosis is likely to be the result of inherent activity of the cells themselves.[5]

His group also determined that shrinkage necrosis often occurred substantially in basal cell carcinomas and a number of other tumors, the fragments of which were taken up and digested by the surrounding tumor cells. From these and other results, Kerr and his group concluded that shrinkage necrosis *is involved in normal cellular turnover* and suggested that *death of both normal and neoplastic cells may be a pre-ordained, genetically determined phenomenon.*[5]

During a study leave from Queensland in 1971, Kerr spent a year at the University of Aberdeen in Scotland, where he was introduced to the long history of cell death during normal embryonic and fetal development—a topic already familiar to embryologists but largely unknown to biologists outside this field.[5] During his time abroad, Kerr found that the morphology of cells during fetal development was strangely similar to those of shrinkage necrosis he observed in his own studies, and this led to a paper, published in 1972, which introduced the concept of apoptosis to the masses.[7] The authors correctly stated early in the paper that *we suspect that further work will confirm it as a general mechanism of controlled cell deletion, which is complementary to mitosis in the regulation of animal cell populations* and suggested *that it be called 'Apoptosis' [...] (αποπτωσισ)* [which] *is used in Greek to describe the 'dropping off' or 'falling off' of petals from flowers, or leaves from trees".*[7] As such, they described apoptosis as affecting *scattered single cells, and [...] manifested histologically by the formation of small, roughly spherical or ovoid cytoplasmic fragments [...]* [which] *we shall call apoptotic bodies.* As noted earlier, the process was described to occur in two separate phases:

> [...] the first comprises the formation of apoptotic bodies, the second their phagocytosis and degradation by other cells [...]
>
> The formation of apoptotic bodies involves marked condensation of both nucleus and cytoplasm, nuclear fragmentation, and separation of protuberances that form on the cell surface to produce many membrane-bounded, compact, but otherwise well-preserved cell remnants of greatly varying size [...] Fully developed apoptotic bodies show closely packed organelles, which may themselves be condensed, but which are apparently intact, both chemically and structurally. Lucent cytoplasmic vacuoles and dense masses of nuclear material are seen in some bodies [...]
>
> In all the tissues so far studied, the majority of the apoptotic bodies have been found within the cytoplasm of intact cells. This suggests that they are rapidly phagocytosed [...]
>
> Apoptosis is well suited to a role in tissue homoeostasis, since it can result in extensive deletion of cells with little

tissue disruption. Following fragmentation of an affected cell, the remains are rapidly disposed of by nearby intact cells [...] Moreover, the process is economical in terms of re-utilization of cell components.[7]

The authors also described that following ingestion by other cells, apoptotic bodies undergo a series of structural changes resulting from the fusion of phagosomes—membrane-enclosed areas of a cell that contain phagocytosed fragments—with lysosomes, organelles that contain enzymes which *play a vital role in the further degradation of phagocytosed bodies.*[7] However, little was known about the triggers and inhibitors of apoptosis, though it was clear that:

> in certain circumstances apoptosis is an inherently programmed event, determined by *intrinsic "clocks"* specific for the cell type involved. Thus in avian embryonic tissue explants apoptosis occurred in susceptible zones "on schedule" [...]
>
> It is tempting to speculate that it might involve stimulation of messenger RNA and protein synthesis. If indeed apoptosis depends on expression of part of the genome, which is normally repressed in viable cells, the initiation of such a stereotyped series of changes by a wide variety of stimuli would be understandable.[7] Emphasis in original.

* * * * *

Despite Kerr's specific description of apoptosis in the early 1970s, it would require almost twenty years before wide-spread interest about apoptosis was aroused from the scientific community, at which time the biochemical and genetic aspects of apoptosis had already started to accumulate. This area of investigation actively began in the early 1980s, when Hedgecock, Holston, and Thomson used the roundworm, *C. elegans*, to study apoptosis, as it was particularly well-suited for these studies: "programmed cell death", as it was called in these studies, had previously been described during the development of *C. elegans*, the organism losing about 10% of its 1000 or so cells during normal development; this organism had patterns of cell division, migration, and cell deaths that were well characterized and invariant amongst individuals; it had a very short life span (about three days) and large brood sizes, which increased the rate at which studies could be performed; and finally, cell death was readily identified morphologically using simple microscopy.[8] As such, it was discovered that certain genetic mutations in *C. elegans*, called *ced-1* and *ced-2*, prevented the engulfment and clearance of cells that had undergone apoptosis and importantly, these genes were found to have no other role than that involved in apoptosis, which suggested that apoptosis was a process specifically designed for the elimination of unwanted cells.[9]

A few years later, Ellis and Horvitz attempted to *isolate mutations that perturbed the normal pattern of programmed cell deaths* by treating *ced-1* mutant *C. elegans* with a mutagen and *obtained in this way [...] two strains in which the characteristic cell deaths of ced-1 were not seen,* which they called *ced-3* and *ced-4*.[8] In contrast to *ced-1* and

-2 mutants, in which cells underwent cell death but were not eliminated, the *ced-3* and *-4* mutants were found to directly impact cell death by inhibiting it (Figure 15.1a). These cells did not divide, but some were able to differentiate into mature functional cells; as such, these adult individuals had about 100 cells that were not present in wild-type worms! Surprisingly, these extra cells did not seem to impact the survival of the mutants in any detectable way. From these results, the authors concluded that the *ced-3* and *-4* mutations occur upstream of *ced-1* and *-2* in a genetic pathway which regulated cell death and clearance of the ensuing cell debris, since the latter mutants yielded dead cells that were not cleared, whereas in the former, no cell death occurred to begin with (Figure 15.1b).[8] Finally, the authors pondered the rationale for programmed cell death in the development of worms given that the presence of these cells did not seem to be detrimental:

Thus, we are left with the question of what selective advantage may be associated with the extensive programmed cell death that occurs during *C. elegans* neurogenesis. Even small effects on behavior might provide a selective advantage. Alternatively, the elimination of unnecessary cells that would otherwise differentiate may decrease energy consumption and hence increase fitness.[8]

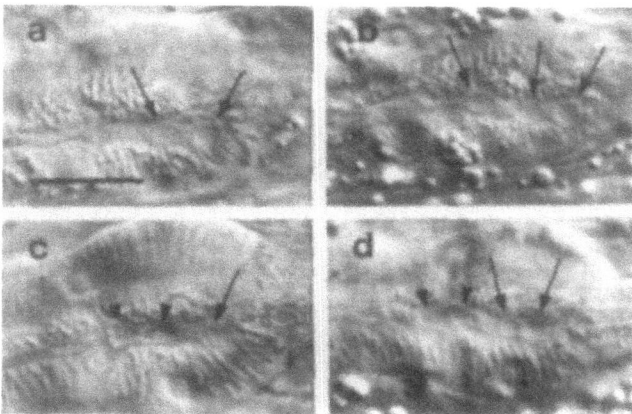

(a)

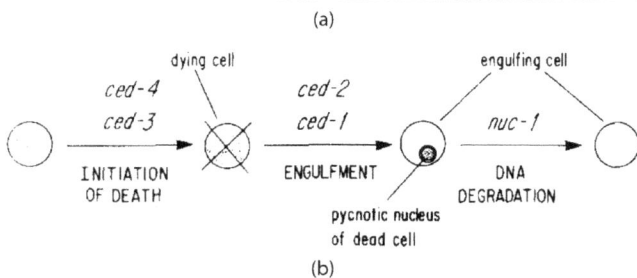

(b)

FIGURE 15.1 (a) Adult *ced-3* mutants have excess cells. Arrows in wild-type *C. elegans* (a and c) and the corresponding arrows in the *ced-3* mutants (b and d) indicate the location cells. The extra arrows in the mutant *C. elegans* represent excess cells that did not die. (b) Proposed pathway for *ced-1/-2* and the new mutants, *ced-3* and *ced-4*.[8] (Reprinted from Cell 44(6), Ellis HM, Horvitz HR. Genetic control of programmed cell death in the nematode *C. elegans*, 817–829, Copyright 1986, with permission from Elsevier.)

In parallel with the studies performed in *C. elegans*, Vaux, Cory, and Adams were investigating the first gene that had a direct impact on apoptosis in humans. This gene first came to light in the early 1980s when Tsujimoto et al. found that a common alteration in many B-cell blood cancers was caused by the amplification of a gene normally found on chromosome 18, which the authors named "B-cell lymphoma/leukemia 2" (Bcl-2) (the mechanism of amplification of Bcl-2 will be addressed in the next chapter).[10,11] After Cleary, Smith, and Sklar successfully cloned the Bcl-2 gene in 1986,[12] Vaux, Cory, and Adams transduced Bcl-2 into normal mouse cells—which resulted in the overexpression of the protein—and found that this conferred a strong survival advantage compared to cells expressing the normal amount of the protein.[13] In addition, Bcl-2 cooperated with a gene which was known to control proliferation, *c-myc*, since transduction of both resulted in unhindered growth—cells essentially became immortal—compared to those transduced with either gene alone, which died after a few weeks. From this, the authors concluded that:

[...] the unique ability of bcl-2 to dissociate cell survival from proliferation suggests that a haematopoietic growth factor provides two factors: one keeps the cell alive, the other stimulates division. Our results suggest that the bcl-2 protein acts within the 'survival' signal pathway, whereas *c-myc* lies on that involved in proliferation.[13]

Shortly after, another protein, p53—already known to be a tumor suppressor protein (a protein whose function prevents the growth of cancer; more on this in the next chapter)—was found to perform its role specifically by inducing apoptosis. Yonish-Rouach et al. transfected p53 in a cell line known to be p53-deficient and found that cell viability quickly decreased following its expression.[14] Morphological assessment and the presence of DNA fragmentation in p53-transfectants indicated that cell death occurred via apoptosis. The authors also noted that:

Expression of p53 is absent in many human myeloid leukaemic cell lines and in primary cells from leukaemic patients. Normally *in vivo* myeloid progenitor cells might continuously die by apoptosis unless the appropriate differentiation or proliferation signal is present. This would keep the number of circulating cells low until they are required more abundantly. If wild-type p53 is involved in this apoptosis, then loss of p53 could relieve this apoptotic programme so that a pretransformed, neoplastic population is generated, where the cells do not die but keep cycling until the accessory oncogenic activation. This mechanism could account for the development of myeloid leukaemia.

[...] Thus, whereas most oncogenes *[such as c-myc]* probably contribute to neoplasia by promoting cell proliferation, cancer may also be induced by the activation of genes whose products allow the survival of a cell that should otherwise die. So if tumour suppresor genes *[like p53]* mediate apoptosis, their inactivation will have the same effect.[14]

* * * * *

By the late-1980s, it was generally accepted that apoptosis required protein synthesis, that NAD levels sharply decreased at the onset of apoptosis, and that DNA was cleaved during the process, a phenomenon which could be observed as "DNA laddering", that is, DNA samples from apoptotic cells look like the rungs of a ladder when they are run on an agarose gel, indicating the presence of a wide variety of DNA sizes in the sample (in contrast to intact DNA which is of fairly uniform size; see Figure 15.4b, lane 1 vs 2 for an example). However, little was known about how phagocytic cells recognized apoptotic cells. Several groups had already noticed that a loss of plasma membrane phospholipid asymmetry was involved in the recognition by macrophages but the specific mechanism of such a phenomenon could not be identified.

In 1988, Allen et al. showed that the presence of the phospholipid, phosphatidylserine (PS), on the outer leaflet of the plasma membrane was responsible for the recognition of red blood cells by phagocytes,[15] and in 1992, Fadok et al. found that this process was required for the engulfment of apoptotic cells by macrophages. As such, the latter irradiated thymocytes to induce apoptosis and separated apoptotic from non-apoptotic cells by centrifugation, which they then added to macrophages to confirm that only apoptotic cells were readily engulfed (Figure 15.2a).[16] They also found that phagocytosis could be inhibited by preincubating the macrophages with PS liposomes—small unilamellar vesicles composed of PS—or derivatives of the polar head group of PS, which suggested that recognition was specific for PS rather than any negatively charged

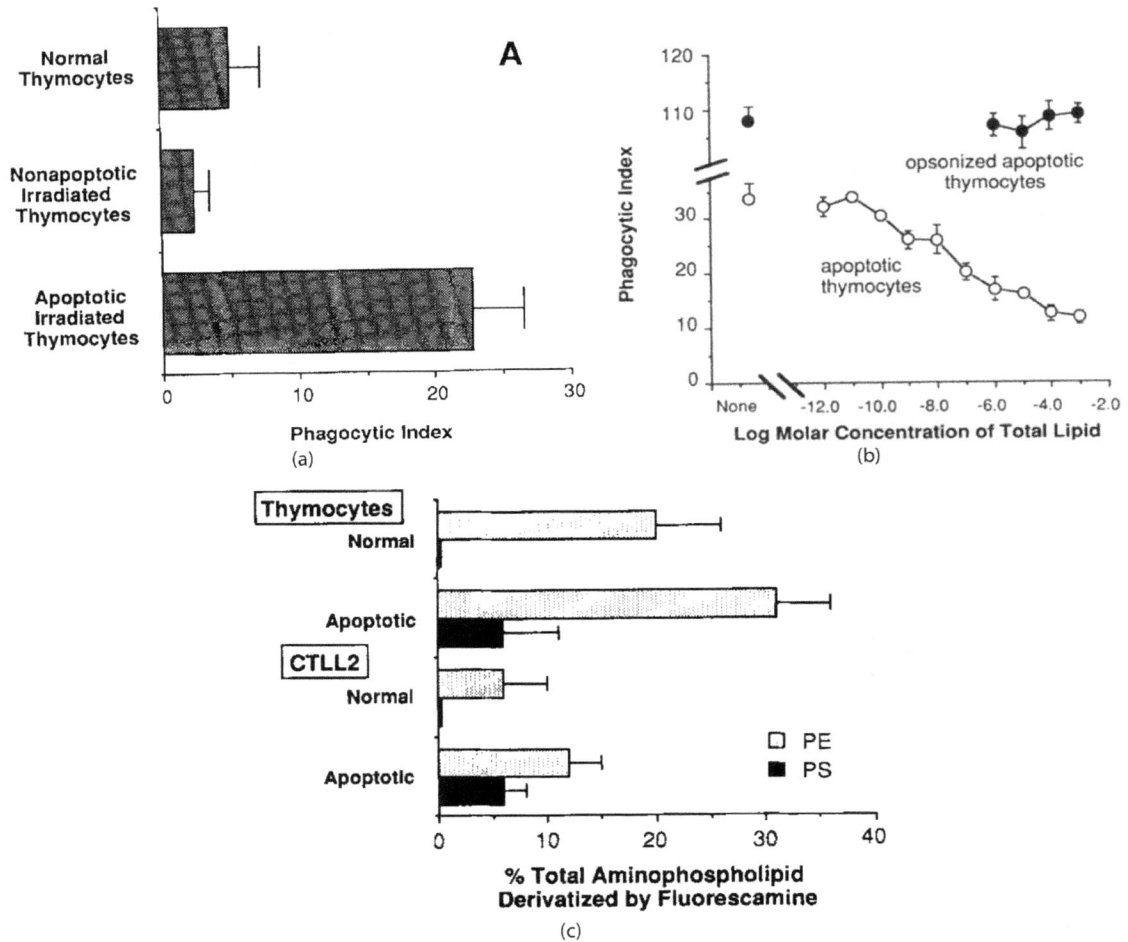

FIGURE 15.2 (a) Apoptotic cells, but not healthy cells, are phagocytosed by macrophages. Cells were left untreated or were irradiated to induce cell death and then added to macrophages. The percentage of phagocytosed cells was then assessed. (b) Apoptotic cells were added to macrophages which were pre-incubated with PS-containing liposomes, and phagocytosis was assessed. Opsonized apoptotic cells were used as a positive control for phagocytosis (black circles). (c) Untreated cells or cells undergoing apoptosis were incubated with an amine derivatizing agent, fluorescamine, and phospholipids were extracted, separated, and the percent derivatized lipids—phosphatidylserine (PS) or phosphatidyletholamine (PE)—were assessed, as compared to total lipids (presence of derivitized lipids indicated that lipids were present on the outer membrane). Note that although PE was found on the cell surface of both live and apoptotic cells, PS was only found on the surface of apoptotic cells.[16] (Adapted from Fadok VA, Voelker DR, Campbell PA, Cohen JJ, Bratton DL, Henson PM. Exposure of phosphatidylserine on the surface of apoptotic lymphocytes triggers specific recognition and removal by macrophages. Journal of Immunology. 1992;148(7):2207–2216. Copyright 1992. The American Association of Immunologists, Inc.)

or hydrophobic component of the plasma membrane (Figure 15.2b). Furthermore, the interaction was found to be highly specific for phosphatidyl-L-serine, to the extent that even phosphatidyl-D-serine was not recognized. The authors also used three different methods to confirm that PS was exposed to the surface of apoptotic cells, one of which used a chemical shown to add phosphorus exclusively to PS on the outer membrane. As such, they found this chemical derivative of PS was present on outer membrane leaflets only when they became apoptotic and that this was specifically responsible for the recognition of apoptotic cells by macrophages (Figure 15.2c).[16]

The next year, Kaufmann and his colleagues identified another phenotype of apoptotic cells: cleavage of the nuclear protein, poly(ADP-ribose) polymerase (pADPRp, later renamed PARP). Before this, PARP was known to catalyze the cleavage of NAD—leading to the polymerization of the resulting ADP-ribose into protein-linked chains (hence, its name) with nicotinamide as a by-product—and to be involved in DNA repair and inflammation.[17] Paradoxically, it had been suggested that the degradation of DNA following exposure to cytotoxic drugs was a result of PARP activation—which occurs following the cleavage of the native 110 kDa PARP into an 85 kDa and a 25 kDa fragment—since activation occurs in a *time course that paralleled the DNA fragmentation* (see Figure 15.3, for an example of PARP cleavage in a western blot).[17] However, their results indicated that although PARP was indeed a *component of the apoptotic process [...] pADPRp activity and consumption of NAD do not play an obligate role in DNA fragmentation or pADPRp cleavage [...] instead, the cleavage of pADPRp appears to represent one of several proteolytic events that commonly occur during apoptosis.*[17] The authors also attempted to determine which enzyme was responsible for PARP cleavage, but none were found

to prevent apoptosis; thus, upstream regulators of PARP remained unknown.

* * * * *

In 1993, Yuan et al. cloned and characterized the *C. elegans* gene, CED-3, and determined that it shared a significant degree of amino acid homology to the mammalian proteins, interleukin-1-converting enzyme (ICE) and NEDD2.[18] The former is a cysteine protease that had previously been found to cleave inactive pro-interleukin-1β (IL-1β) to produce a mature, active IL-1β in response to bacterial infections, but it was not known to be involved in apoptosis.[18] Given the similarities between their amino acid sequences, Yuan et al. tested if a CED-3 mutation—occurring within a highly conserved region of the sequence—which was known to abrogate its function, might have an impact when inserted into ICE. Indeed, the authors found that this mutation inhibited ICE-mediated IL-1β cleavage. Given these similarities, the authors suggested *not only that CED3 might function as a cysteine protease but also that ICE might function in programmed cell death in vertebrates.*[18] The latter was confirmed the same year when they transfected rat cells with ICE and found an increased rate of apoptosis.[19] Importantly, they also showed that transfection of the pro-survival protein, Bcl-2, inhibited ICE-mediated apoptosis, which indicated that these proteins were part of the same apoptosis pathway. As such, the authors proposed that *in living cells, Bcl-2 is active, which may directly or indirectly inhibit the activity of ICE; in cells undergoing programmed cell death, Bcl-2 is inactive, and thus ICE is activated, which in turn causes cells to die.*[19]

Following Kaufmann et al.'s paper on the cleavage of PARP during apoptosis, it was found that *cleavage [of PARP] occurred C-terminal to Asp and that the protease responsible resembled ICE in its susceptibility to chemical*

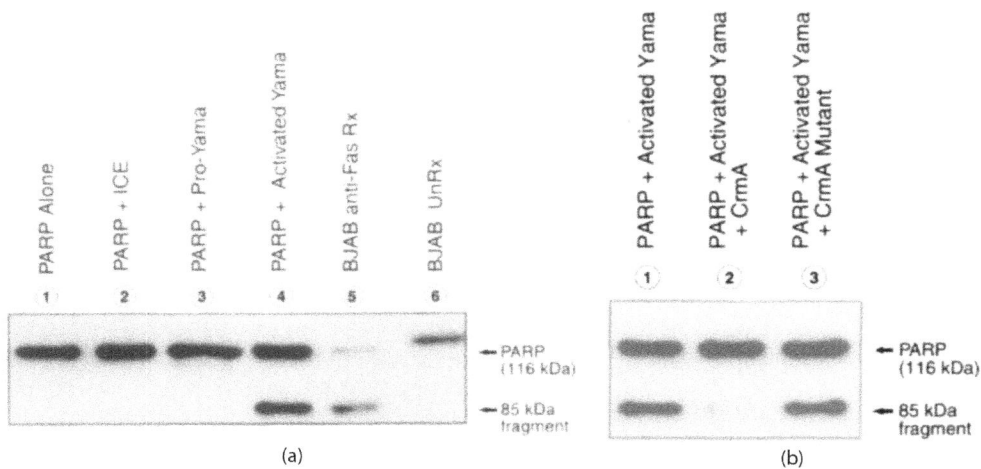

FIGURE 15.3 Cleavage of PARP by ICE-mediated activation of pro-yama is inhibited by CrmA. PARP was incubated in a reaction mixture with or without CrmA, as indicated, and subjected to SDS-PAGE. (a) Lanes 5 and 6 were used as positive and negative controls, respectively, for cleaved PARP. Note how PARP is only cleaved by the activated form of yama (compare lane 3 and lane 4). (b) Mutated CrmA (lane 3) does not inhibit Yama-mediated PARP cleavage.[20] (Adapted from Cell, 81(5), Tewari M, Quan LT, O'Rourke K, et al. Yama/CPP32 beta, a mammalian homolog of CED-3, is a CrmA-inhibitable protease that cleaves the death substrate poly(ADP-ribose) polymerase, 801–809, Copyright © 1995, with permission from Elsevier.)

inhibitors, but was distinct from ICE, since purified ICE did not cleave PARP.[20] Therefore, Tewari et al. set-out to identify the protease in question, taking *advantage of the conservation of the pentapeptide motif QACRG that encompasses the catalytic site Cys of ICE and is conserved among members of the CED-3/ICE protein family* (by this point, a number of proteases with similar function had been discovered, all with this conserved motif). As such, the authors used PCR with primers binding to the conserved motif and also searched databases for proteins containing the conserved sequence, their efforts resulting in the identification of a protein which they designated Yama (after the Hindu God of death), and whose sequence was found to be identical to CPP32β, one of the known members of the ICE/CED-3 protein family. They also determined that although the full-length protein (named pro-Yama) could not cleave PARP, ICE-mediated cleavage of pro-Yama resulted in its activation and endowed it with proteolytic activity that effectively cleaved PARP in a cell-free system (Figure 15.3a).[20] In addition, they found that Yama-mediated PARP cleavage could be inhibited by incubation with CrmA, a known inhibitor of apoptosis (Figure 15.3b). Thus, the authors noted that:

[s]ince PARP cleavage is a biochemical event observed in virtually every form of PCD *[programmed cell death]* examined, the protease responsible might be expected to play a central, universal role in mammalian apoptosis, a role perhaps akin to that of CED-3 in *C. elegans*. At this juncture, Yama is a candidate for such a protease. Additional studies will be required, however, before Yama can be definitively implicated as a death protease [...]

The finding that CrmA inhibits the proteolytic activity of Yama is significant, as it suggests that the well-documented ability of CrmA to inhibit apoptosis might be explained by its inhibition of Yama. Also, the finding that CrmA inhibits PARP cleavage *in vivo* is consistent with its inhibiting Yama [...]

Importantly, our studies indicate that Yama is a zymogen that requires proteolytic activation. This has important ramifications, as it suggests a mode of regulation. Yama is likely synthesized as an inactive proenzyme that is activated by another protease during apoptosis. Hence, a role for a second protease, likely also Asp specific, is suggested. As our studies indicate that ICE itself can fulfill this role *in vitro*, this provides a possible explanation of how overexpression of exogenous ICE induces apoptosis in mammalian cells [...] since Yama is likely activated by cleavages at specific Asp residues, other members of the ICE family, such as ICH-1L/Nedd2, are potential candidates for this regulatory role [...]

Taken together, the data presented are consistent with the cell death pathway being a proteolytic cascade in which apoptotic signals from diverse stimuli converge to activate proteolytically a common protease, perhaps Yama, that in turn cleaves PARP and probably other death substrates [...] The identification of Yama/CPP32β as a CrmA-inhibitable protease that cleaves PARP represents a focus point that may be useful for identifying other upstream and downstream components of the death pathway.[20]

A few weeks later, another group published a paper which confirmed Tewari et al.'s findings. This group, who used the profile of inhibition of PARP cleavage by a variety of protease inhibitors as a starting point, generated a more potent and specific inhibitor of PARP cleavage, ac-DEVD-CHO, which they used to purify the enzyme responsible for PARP cleavage and identified it as CPP-32 (the same protein as YAMA).[21] In 1996, it was suggested that this new protein family be called "caspase", where *the 'c' is intended to reflect a cysteine protease mechanism, and 'aspase' refers to their ability to cleave after aspartic acid, the most distinctive catalytic feature of this protease family.*[22] As such, ICE was renamed caspase 1, NEDD2 (the other protein that Yuan *et al.* found had amino acid homology to CED-3) was renamed caspase 2, and Yama/CPP32β was renamed caspase-3. To date, this family comprises 12 members, each named in the order in which they were discovered and which were found to act in a long pathway of pro-apoptosis proteins activated by the serial cleavage of its members.

* * * * *

A significant step forward in the elucidation of how apoptosis was regulated came from a series of publications—within a two-year span—from Xiaodong Wang's laboratory at Emory University in Atlanta. In 1996, Wang's group established a cell-free system with which to study apoptosis and, as such, reported the separation of the cytosol into two fractions, which they called "Apoptotic protease activating factor" (Apaf)-1 and -2, both of which were required for the cleavage and activation of caspase-3 (Yama/CPP-32) and the laddering of DNA.[23] Upon purification of Apaf-2, the authors noted that the protein had a pinkish color to it so they decided to measure its spectrophotometric absorbance and found that it was similar to that of cytochrome *c* (a component of mitochondria previously identified to be part of the electron transfer chain, see Chapter 9); the identity of Apaf-2 as cytochrome *c* was confirmed by amino acid sequencing. To further support this claim, the authors substituted cytochrome *c* for the Apaf-2 fraction in the system and found that it was sufficient to induce apoptosis (Figure 15.4).

This discovery was somewhat surprising since these proteins were purified from cytosolic fractions and cytochrome *c* was known to be a mitochondrial protein. In addition, at this point in time, other than the known localization of Bcl-2 in the outer mitochondrial membrane, there was little evidence that mitochondria were involved in apoptosis. Therefore, the authors repeated their purification process using a milder protocol in an attempt to *protect the mitochondrial integrity*; however, they found that the cytosol fraction purified in this way *was incapable of initiating the dATP-dependent activation of CPP32* [caspase-3] *unless purified cytochrome c was added.*[23] Surprisingly, cytochrome *c* was readily detected in the cytosol of cells undergoing apoptosis but not in that of healthy cells, which suggested that it was a required step in the mechanism of apoptosis.[23] Given the known role of Bcl-2 on the inhibition

(a)

(b)

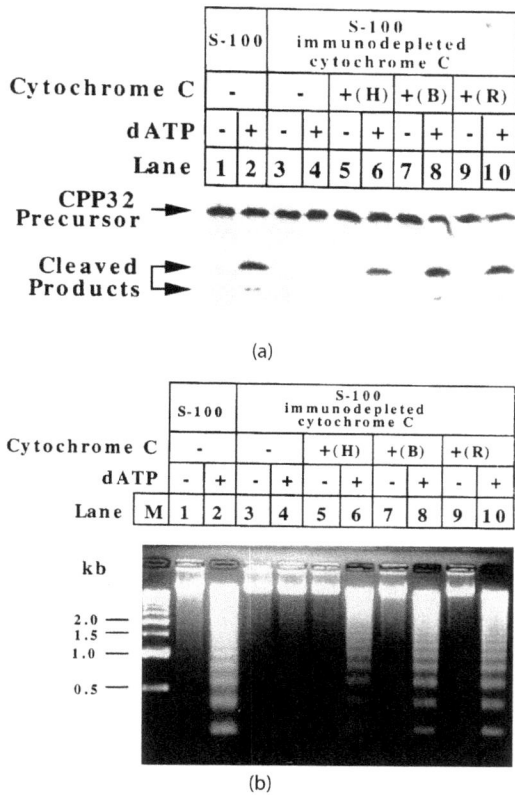

FIGURE 15.4 Confirmation of Apaf-2 as cytochrome c. Cleavage of caspase-3 (CPP32, a) and DNA laddering (b) was triggered by the addition of dATP. Cytochrome c was depleted from cell extracts (S-100) and APAF-2 (H) or cytochrome c purified from bovine heart (B) or rat liver (R) was added as indicated.[23] (Adapted from Cell, 86 (1), Liu X, Kim CN, Yang J, Jemmerson R, Wang X. Induction of apoptotic program in cell-free extracts: Requirement for dATP and cytochrome c, 147–157, Copyright © 1996, with permission from Elsevier.)

of caspase-3 activation and apoptosis, Wang's laboratory tested, in a second paper, its involvement in the release of cytochrome c from mitochondria into the cytosol. Indeed, the authors found that Bcl-2 could block the release of cytochrome c in apoptotic cells, which indicated that *cytochrome c is released from mitochondria early in apoptosis before mitochondrial depolarization, activation of caspases, and DNA fragmentation* and *appears to be independent of any noticeable structural changes in the mitochondria.*[24]

The authors also optimized their purification of APAF-1 and found that it was composed of two proteins, which they named Apaf-1 and Apaf-3.[25] Apaf-1 was found to share significant similarity to *C. elegans ced-3* and *ced-4* proteins, so it was transfected in human embryonic kidney (HEK) cells but caspase-3 cleavage was not detected, which indicated that *Apaf-1 alone was not sufficient to trigger caspase-3 activation.*[25] However, caspase-3 was cleaved when it was incubated with cell extracts from apoptotic cells overexpressing Apaf-1 (Figure 15.5a). They also found that cytochrome c formed a complex with Apaf-1 since the former coimmunoprecipitated with the latter (Figure 15.5b). Thus,

Apaf-1 functioned upstream of caspase-3 but downstream of Bcl-2, since the latter had previously been found to block the release of cytochrome c from mitochondria.

The final paper from Wang's laboratory published in 1997 identified Apaf-3 as caspase-9 while searching for matches in a database using tryptic digestion peptides of Apaf-3.[22] The authors confirmed this by western blotting when a caspase-9 antibody reacted with purified Apaf-3. In addition, caspase-9 was found to be activated by co-incubation with Apaf-1 and cytochrome c in a manner that was independent of caspase-3 activation, although this protein was confirmed as a target of caspase-9. Finally, the authors used co-IP experiments to show that caspase-9 interacted directly with Apaf-1 in a manner that was dependent on cytochrome c (Figure 15.5c).[22] Thus, the experiments performed in 1996/97 in the Wang laboratory elucidated a chain of events that occurs during apoptosis, beginning with the release of cytochrome c from mitochondria into the cytosol upon induction of apoptosis, which then binds to Apaf-1 and forms the "apoptosome" complex that cleaves and activates caspase-9, which in turn leads to the cleavage and activation of caspase-3 (Figure 15.6). Finally, this process was shown to be inhibited by Bcl-2 in a manner that remained to be determined.

* * * * *

Earlier in this chapter, we discussed how Bcl-2 was found to have anti-apoptotic effects on cells, but its mechanism of action still remained to be elucidated. In 1995, Stanley Korsmeyer's group—one of the first to study Bcl-2 and who would go on to contribute a great deal in the elucidation of the apoptosis pathway—reported that a 21 kDa protein co-IPed with Bcl-2 in B cells; thus, they named it "Bcl-2-associated X Protein" (Bax).[27] Following cloning and sequencing of the gene, the authors determined that Bax had as much as 43% homology with Bcl-2, which indicated that it may play a role in apoptosis. However, the fact that Bax RNA was present in a variety of normal tissues and cell lines also indicated that the presence of the protein was not a *de novo* response following apoptosis induction, but rather, that Bax protein was already present at steady state levels. They also found a significant increase in apoptosis when they overexpressed Bax in conjunction with IL-3 deprivation (a growth factor, the deprivation of which induces apoptosis) compared to IL-3 deprivation alone (Figure 15.7a). However, co-transfection of Bcl-2 and Bax only *partially countered the Bax accelerated cell death. In multiple experiments, the rate of cell death roughly paralleled the ratio of* Bcl-2 to Bax.[27]

Given that Bcl-2 and Bax appeared to have reciprocal effects on apoptosis, Oltvai, Milliman and Korsmeyer had a closer look at the interaction between these two proteins. As such, they found that a *substantial amount of endogenous [...] Bax was coprecipitated* when cells were transfected with HA-tagged Bax, while most of the Bcl-2 was not associated with HA-Bax (Figure 15.7b, lane 5).[27] However, the authors found that the amount of Bax that associated with Bcl-2 could

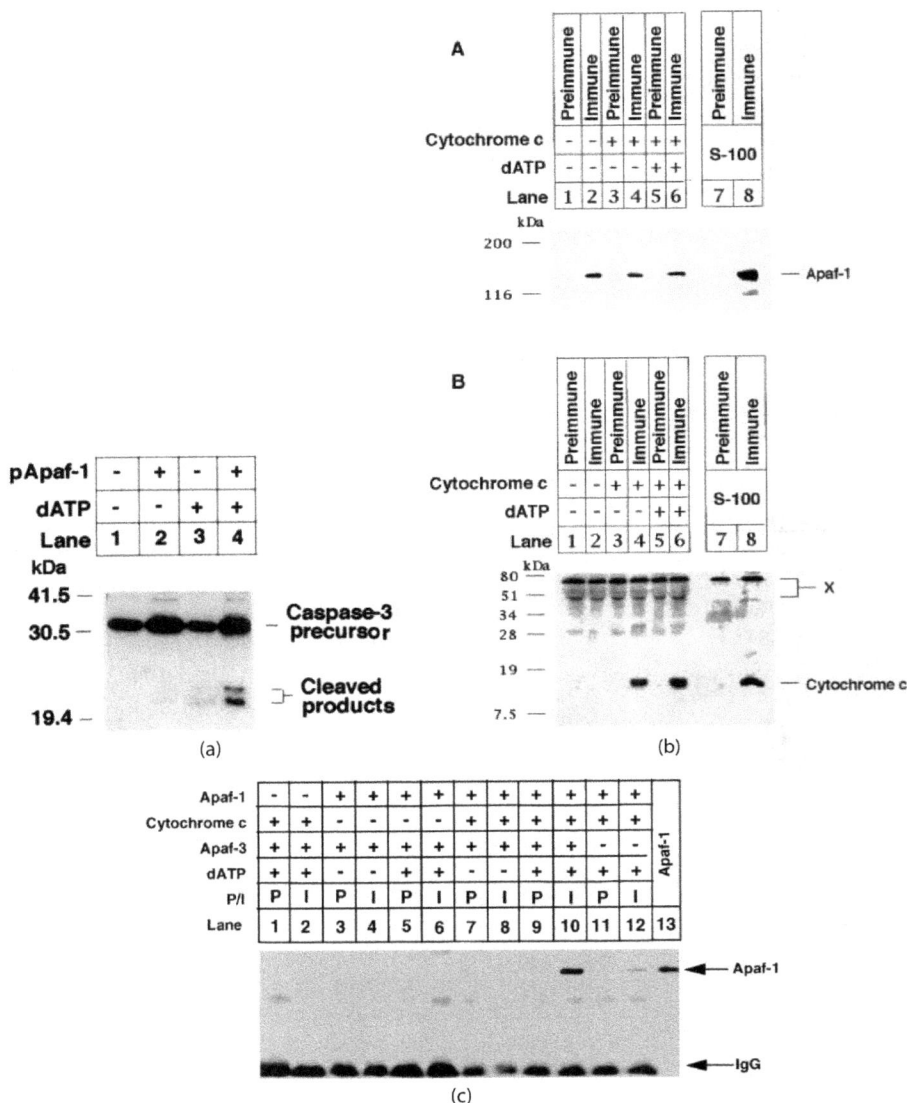

FIGURE 15.5 Cleavage of caspase-3 by Apaf-1 upon induction of apoptosis. (a) A reaction mixture of extracts from cells undergoing apoptosis was incubated with or without excess Apaf-1 and caspase-3 cleavage was assessed SDS-PAGE. (b) Purified cytochrome c, Apaf-1, and dATP were added to a reaction mixture and cytochrome c was immunoprecipitated, subjected to SDS-PAGE, and probed using anti-Apaf-1 (A) or anti-cytochrome c (B) serum. Cell lysates (S-100) were also treated as above as a control. "Preimmune" is serum containing nonspecific antibodies, used here as a control.[25] (c) The indicated components were added to a reaction mixture and caspase-3 was immunoprecipitated using an appropriate antibody. The samples were subjected to SDS-PAGE and probed using pre-immune serum (P) or immune serum (I) against Apaf-1.[22] (Figure parts (a) and (b) adapted from Cell, 90 (3), Zou H, Henzel WJ, Liu X, Lutschg A, Wang X. Apaf-1, a human protein homologous to *C. elegans* CED-4, participates in cytochrome c-dependent activation of caspase-3, 405–413, Copyright © 1996, with permission from Elsevier. Figure part (c) adapted from Cell, 91 (4), Li P, Nijhawan D, Budihardjo I, et al. Cytochrome c and dATP-dependent formation of Apaf-1/caspase-9 complex initiates an apoptotic protease cascade, 479–489, Copyright © 1996, with permission from Elsevier.)

be increased by overexpressing cells with Bcl-2 (Figure 15.7b, lane 2–4), which indicated that *Bax homodimerizes and that the overexpression of Bcl-2 competes for Bax by heterodimerization.*[27] As such, the authors concluded that:

These data favor a model in which the inherent ratio of Bcl-2 to Bax determines the susceptibility to death following an apoptotic stimulus. When Bcl-2 is in excess, cells are protected. However, when Bax is in excess and Bax homodimers dominate, cells are susceptible to apoptosis [...]

Bax might function as a death effector molecule that is neutralized by Bcl-2. In this scenario, Bcl-2 might simply be an inert handcuff that disrupted the formation of Bax homodimers.[27]

Over the next few years, a number of other proteins were identified as having homology to Bcl-2 and Bax, the homologous region of which Korsmeyer's group referred to as "Bcl-2 homology" (BH) domains.[28] Thus, a new family of proteins, the Bcl-2-like protein family, was now emerging

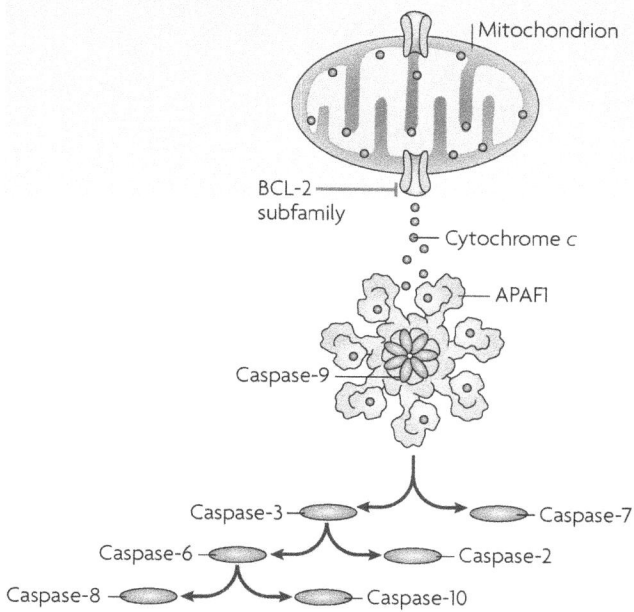

FIGURE 15.6 Current model for the assembly of the apoptosome. Apoptosis-mediated permeabilization of mitochondria results in the release of cytochrome c and ATP into the cytosol, where they bind to Apaf-1 and induce a conformational change in the latter, which enables its oligomerization. Formation of this multi-protein complex activates caspase-9, which then binds to the complex and results in the activation of the apoptosome, which in turn cleaves and activates caspase-3.[26] (Adapted by permission from Springer Nature Customer Service Centre GmbH: Springer Nature, Nature Reviews Molecular Cell Biology, Taylor RC, Cullen SP, Martin SJ. Apoptosis: Controlled demolition at the cellular level, Copyright © 2008.)

and it was concluded that *these genes are likely to be sequential members of a single death pathway or regulators of parallel death pathways*, and whose interaction was mediated at least in part by their BH domains.[27,29] One of these proteins, Bad, was found using a yeast-two-hybrid system, a system that screens for interaction between proteins based on a "bait" made up of an amino acid sequence thought to provide the interaction in question; in this case, a fragment of Bcl-2 containing the BH3 domains was used. This interaction was confirmed by co-IP, which also revealed an interaction with another pro-survival protein of the Bcl-2 protein family, Bcl-x_L.[30] However, it was determined that the binding of Bad to Bcl-x_L was of much greater affinity than to Bcl-2, which led to the finding that Bad, at physiological levels, could only counter the apoptosis-protection effects mediated by Bcl-x_L. Furthermore, Bad had no effect on cells without a death-inducing stimulus, suggesting that it *probably does not function as a singular, downstream death effector molecule*.[30] Thus, Korsmeyer and his group extended their model and proposed that:

> [...] both Bcl-x_L and Bcl-2 must dimerize with Bax to repress death. The data with Bad further support this thesis and adds a third layer of complexity. Overall, the data argue for

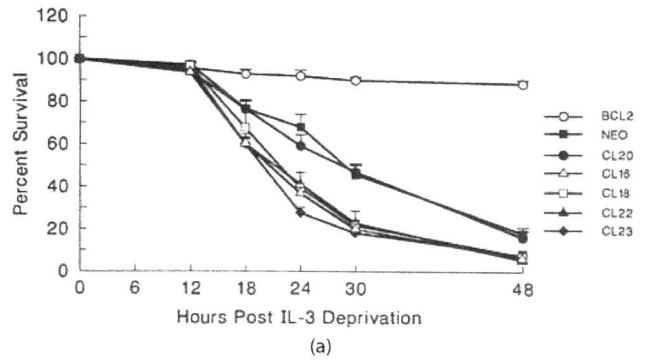

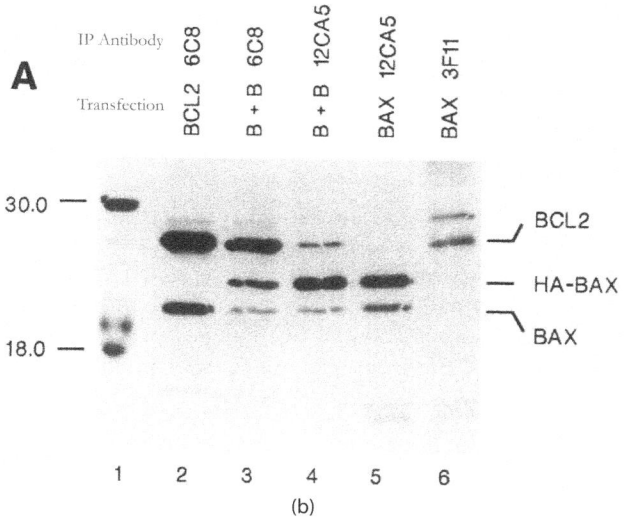

FIGURE 15.7 (a) Overexpression of Bax induces apoptosis. Cells were transfected with Bax and cultured in the absence of IL-3. Survival was assessed at the indicated times in wild-type cells (open circles) and in several clones. (b) Bax homodimerizes and heterodimerizes with Bcl-2. Mouse cells were transfected with human Bcl-2 (Bcl-2), HA-tagged murine Bax (Bax), or both (B + B) as indicated, before labeling proteins with [^{35}S] methionine. Cells were then immunoprecipitated with anti-human Bcl-2 (6C8), HA-specific (12CA5), or anti-mouse Bcl-2 (3F11) antibodies and subjected to SDS-PAGE.[27] (Adapted from Cell 74(4) Oltval ZN, Milliman CL, Korsmeyer SJ. Bcl-2 heterodimerizes *in vivo* with a conserved homolog, Bax, that accelerates programmed cell death, 609–619, Copyright © 1993, with permission from Elsevier.)

a simple competition in which Bad binds Bcl-x_L, displacing Bax into homodimers. Susceptibility to cell death is best correlated with the percent of Bax in heterodimers versus that in homodimers. If roughly half of Bax is complexed in heterodimers [...] cells will be protected from death.[30]

Shortly thereafter, X-ray crystallographic studies of Bcl-x_L revealed that *the α helical structure of BCL-X_L proved* [to be] *reminiscent of membrane insertion domains in the bacterial toxins*.[29] Therefore, a number of groups investigated the potential for Bcl-2 family proteins to form ion pores in phospholipid membranes and it was found that several, including Bax, did indeed possess this ability.[29,31] In one of these studies, Antonsson et al. used Bax-studded liposomes loaded with a fluorescent dye and determined that Bax-oligomerization-induced dye efflux from the liposome in a Bax-concentration-dependent manner.[31] In addition, the authors found that pre-incubation of liposomes with Bcl-2 inhibited dye efflux by Bax, supporting the idea that the formation of Bax channels was involved in the apoptosis pathway and leading the authors to suggest that:

> [m]embrane insertion and pore formation of Bax might promote cell death by allowing the passive flux of ions and small molecules across intracellular membranes in which the protein is localized. Localization of Bax to mitochondrial membranes may trigger a permeability transition and consequent disruption of the transmembrane potential, two critical events during the early stages of apoptosis.[31]

Since the ability of Bax to form ion channels in the above experiments was shown using artificial lipid membranes, it was still unclear whether the monomeric or multimeric form was active in inducing apoptosis *in vivo*. This was addressed by Korsmeyer's group in 1998 when they used the same fractionation technique which led them to the discovery of Bax but now, they assessed the presence of Bax in the mitochondrial fraction. As such, Gross et al. determined that although Bax was found in both the cytosol and mitochondria in healthy cells (Figure 15.8a),

> following a death stimulus [...] most (~70%) of BAX moves from the cytosol to the mitochondrial HM fraction [...] Moreover, following a death signal, most of the mitochondrial BAX converts from being alkali sensitive to alkali resistant, indicative of an integral membrane position.[32]

The authors also determined that cytosolic Bax was in the monomeric form, whereas the mitochondrial species was in multimeric form. This was achieved by cross-linking mitochondrial membrane proteins in isolated mitochondria (to prevent any changes in mono- vs. multimeric forms) with a membrane-impermeable reagent, BS[3], and assessing the molecular weight of Bax by SDS-PAGE (Figure 15.8b). In addition, they also found that overexpression of Bcl-2 prevented the relocalization and dimerization of Bax in mitochondrial membranes. In contrast, forced dimerization of Bax resulted in a significant increase in Bax translocation to mitochondria and apoptosis, indicating that Bax dimerization alone was sufficient to induce apoptosis.[32] From these data, the authors suggested that the *dimerization and mitochondrial membrane insertion of BAX may relate to the ability of BAX to form distinct ion conductive channels* (Figure 15.9).

The next year, Desagher et al. were investigating the mechanism by which Bax-induced cytochrome *c* release from mitochondria and found that Bid, another protein of the Bcl-2 family which had been shown to interact with Bax, induced a change in Bax conformation upon treatment with apoptosis-inducing agents.[33] In their system, the authors isolated mitochondria, incubated them with or without Bid, and stained them with an anti-Bax antibody; using flow cytometry, they found a five-fold increase in Bax fluorescence on the surface of mitochondria compared to samples without Bid. In addition, treatment with Bid resulted in the release of cytochrome *c* from mitochondria, and co-IP experiments using mutated Bax revealed that its BH3 domain was particularly important for these phenomena as it mediated the interaction with Bid. Finally, the authors showed that Bcl-2 and Bcl-x$_L$ blocked the induction of apoptosis by Bid.

In 2000, the same group determined that binding of Bid to Bax was required for its insertion into the membrane since

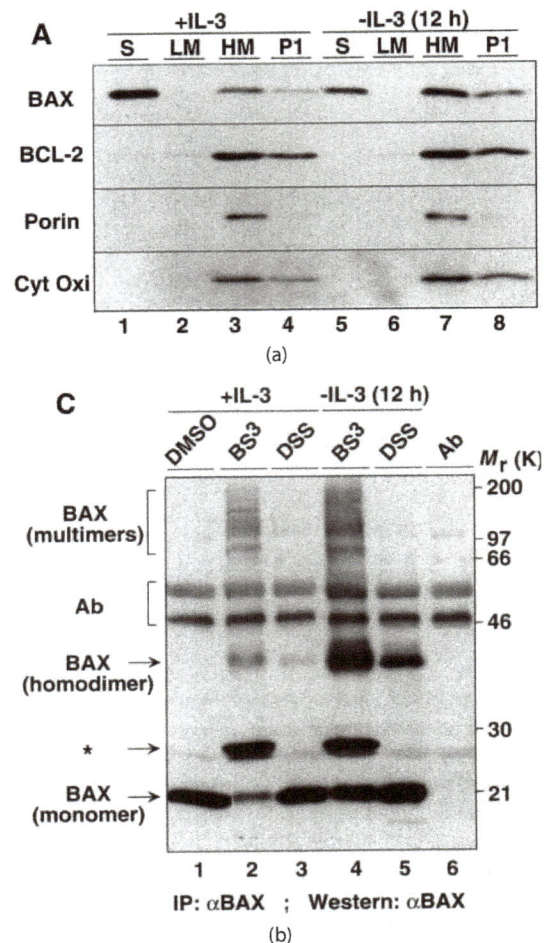

FIGURE 15.8 (a) Bax relocalizes to mitochondria during IL-3 deprivation-induced apoptosis. Apoptosis was induced by IL-3 deprivation and the cytosolic (S), light membrane (ER and plasma membrane, LM), heavy membrane (intact mitochondria, HM), and pellet (residual whole cell, nuclei, and mitochondria, P1) fractions were separated, subjected to SDS-PAGE, and probed for the indicated proteins using western blotting. (b) Mitochondrial Bax form multimers. Apoptosis was induced as above, the HM fraction was collected and treated with DMSO (as a control) or a membrane-impermeable (BS[3]) or -permeable (DSS) cross-linker. Bax was then immunoprecipitated using an appropriate antibody, subjected to SDS-PAGE and probed for Bax following transfer to a membrane.[32] (Adapted with permission from John Wiley and Sons, Gross A, Jockel J, Wei MC, Korsmeyer SJ. Enforced dimerization of BAX results in its translocation, mitochondrial dysfunction and apoptosis. EMBO J. Copyright © 1998 European Molecular Biology Organization.)

Susceptible Cell

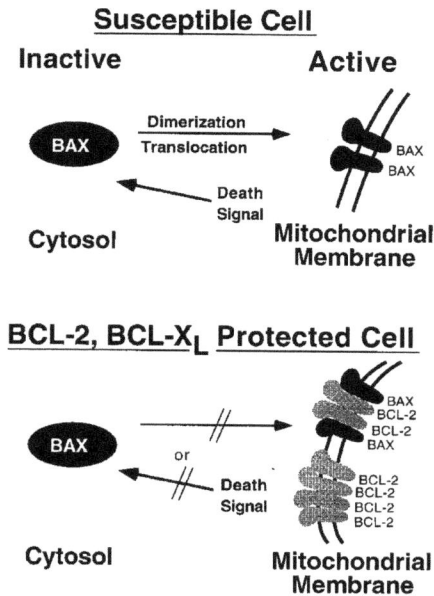

FIGURE 15.9 Model proposed for the role of Bax and Bcl-2 in apoptosis. Apoptosis signals result in the insertion of BAX in the mitochondrial membrane and its dimerization. The latter creates pores that allow mitochondrial proteins, such as cytochrome c, to diffuse into the cytosol. The presence of Bcl-2 in the membrane prevents this from occurring.[32] (Reprinted with permission from John Wiley and Sons, Gross A, Jockel J, Wei MC, Korsmeyer SJ. Enforced dimerization of BAX results in its translocation, mitochondrial dysfunction and apoptosis. EMBO J. Copyright © 1998 European Molecular Biology Organization.)

Bid mutations that inhibited interactions between these proteins prevented insertion of Bax into mitochondria.[34] Since cytosolic Bax had previously been found to be monomeric and mitochondrial Bax to be dimeric, Eskes et al. asked if Bax dimerization was required for its insertion into the mitochondrial membrane. As such, the authors labeled cells with the protein cross-linking reagent, BS[3] as Gross et al. had done, and found that although Bax was detected as a monomer in the absence of apoptosis, higher molecular bands were detected in the presence of Bid, suggesting that *Bax dimerization (or oligomerization) precedes its membrane integration and the efflux of cytochrome c from mitochondria.*[34] As with their previous experiments, they also showed that these effects could be inhibited by Bcl-2 and Bcl-x_L. Therefore, the authors concluded that their

[...] results are consistent with a model of cellular apoptosis in which Bid interacts with Bax to trigger a change in Bax conformation leading to dimerization (or oligomerization) and integration into the outer mitochondrial membrane.[34]

The same year, Korsmeyer and his group characterized the pores formed by Bax by assessing the permeability of a sugar molecule, dextran—the size of which could be precisely varied—through these pores. As such, they incubated dextran with different concentrations of Bax and, using intricate calculations, found that although two Bax molecules

were sufficient to form an ion channel, four molecules were required for larger molecules like dextran to pass through.[35] Since the size of the pores formed by four Bax molecules was theoretically large enough to allow the movement of cytochrome c out of the mitochondria, the authors labeled cytochrome c with a fluorescent tag, FITC, and assessed whether or not it could be released from Bax-containing vesicles. This was accomplished by incubating the vesicles in a solution containing a high concentration of non-labeled cytochrome c, which prevented the release of the intravesicular, FITC-labeled, cytochrome c. Indeed, it was determined that extravesicular fluorescence—that is, FITC-cytochrome c (FCC) released from the vesicles—increased in inverse proportion to the concentration of unlabeled, extravesicular cytochrome c, indicating that *the release is effected by a pore mechanism and indicates the possibility of a specific interaction with the BAX pore.*[35] As such, the authors suggested that Bax *rapidly forms a pathway for release of cytochrome c from liposomes that does not require further proteins* and proposed that:

BAX-mediated release of cytochrome c from mitochondria requires the establishment of a sufficient density of BAX molecules in the mitochondrial membrane to form a large pore. In this way, activation of BAX pores has a critical function in the commitment to cell death.[35]

Korsmeyer's group also confirmed Eskes et al.'s findings that Bid was required for the release of cytochrome c from mitochondria and in the process, found another protein, Bcl-2 homologous antagonist killer (Bak) that seemed to play a similar role to that of Bax. Previous studies had found that Bid was a target of caspase-8, the action of which resulted in a truncated version of the protein (tBid) which could induce the oligomerization of Bax. And by this time, Bid seemed to be *the one molecule absolutely required for the release of cytochrome c in loss-of-function approaches.*[36] However, since mice only expressed minimal levels of Bax—and therefore the effects of Bid could not be ascribed to its interaction with Bax—Wei et al. did a survey of Bcl-2 family members in mice and found that Bak, another protein in the Bcl-2 protein family *structurally similar to Bax*, was expressed abundantly. Thus, the authors found that this protein was a mitochondrial membrane resident protein whose interaction with tBid was required for tBid-mediated cytochrome c release in their system. As such, the authors suggested that—*oligomerized BAK itself provides a pore for cytochrome c release*, which was

[...] reminiscent of the BAX oligomerization noted following growth factor deprivation. One major difference is that these pro-apoptotic molecules are initially in separate subcellular compartments in viable hepatocytes. The BAX present in viable hepatocytes is rather exclusively in the cytosol, whereas BAK is an integral membrane protein even prior to a death signal [...] We expect that tBID's activation of full pro-apoptotic members BAX and/or BAK will prove to be cell type and death signal selective.[36]

The following year, findings that expression of tBid alone could not induce cell death led Wei et al. to hypothesize that *tBid induced the activation of BAX and BAK, resulting in mitochondrial dysfunction including the release of cytochrome c.*[37] Indeed, the authors found that tBid expression induced the independent oligomerization of Bak and Bax and caused oligomerization and insertion of Bax in the mitochondrial membrane (Bak already being a mitochondria-resident protein), as well as the release of cytochrome *c* from mitochondria. Therefore, the authors concluded that *activation of proapoptotic BAX, BAK, or both is required for tBID-induced cytochrome c release in vivo.* They also tested the consequences of the induction of apoptosis in mice deficient in either Bak, Bax, or both, and found that expression of either protein alone decreased both the survival of the mice as well as their mean survival time.[37] In contrast, all mice survived in the absence of both proteins. Cellular fractionation of hepatocytes extracted from mice under these conditions showed that Bax translocated from the cytosol to mitochondria during apoptosis in Bak-deficient, but not in Bid-deficient mice, indicating that Bid functioned upstream of Bax and Bak.[37] In addition, the authors determined that *postmitochondrial events were prevented in DKO* [double knock-out] *cells, including the activation of effector caspase activity*, following induction of apoptosis by a number of different stimuli. However, they found that cells deficient in Bid were still susceptible to these signals, indicating that *they are not dependent on BID and that BID is not the sole activator of BAX or BAK.*[37]

* * * * *

We saw earlier in this chapter that the tumour-suppressor p53 was a protein discovered for its ability to induce apoptosis following DNA damage. p53 was subsequently found to be a regulator of Bax transcription shortly after the latter was co-IPed with Bcl-2 by Yang et al.[38] However, Nobuyuki Tanaka and his group at the University of Tokyo realized that since DNA damage-induced apoptosis still occurred in Bax-deficient mice, there must be other proteins regulated by p53 other than Bax. As such, in 2000, Oda et al. looked for such genes by comparing the RNA profile of p53-expressing and p53-deficient mouse cells following DNA damage and identified a gene which they called Noxa (Latin for damage).[39] Noxa had two BH3 motifs characteristic of the Bcl-2 protein family and had been shown to localize to mitochondrial membranes. However, although its function was determined to be dependent on the BH3 motifs—mutations in these domains rendered Noxa completely inactive—Noxa did not associate with Bax. The authors also found that Noxa co-IPed with Bcl-2 and Bcl-xL when apoptosis was induced, *collectively suggesting the selective interaction of Noxa with the anti-apoptotic Bcl-2 subfamily of proteins.*[39] Furthermore, they determined that expression of Noxa induced the release of cytochrome *c* from mitochondria and the subsequent cleavage and activation of caspases. A search for a human homolog revealed

that it was identical to a gene previously identified as *APR*, whose function was still unknown. Thus, the authors assessed if the human homolog functioned in the same manner as murine Noxa and found that, indeed, its expression was increased following p53-mediated apoptosis and that *human Noxa also induced apoptosis in various cells [...] in a BH3 motif-dependent manner.*[39]

Although the mechanism of action of Noxa remained elusive for many years, it was known that in certain tissues overexpression of Noxa alone was sufficient to induce apoptosis. In addition, Noxa itself did not bind to Bax or Bak, which indicated that Noxa-induced Bax-mediated apoptosis in an indirect manner. In 1993, Kozopas et al. sought to *identify genes that increase in expression* early in the differentiation of hematopoietic cell lines by inducing cells to differentiate and screening the mRNAs that were upregulated in differentiated cells.[40] This led to the identification of "myeloid cell leukemia 1" (Mcl-1), whose protein was found to have significant similarity with the newly identified Bcl-2 protein family. As such, the authors suggested that *MCL1 might similarly have an influence on cell viability/death in the early stages of induction of differentiation.*[40] Indeed, the same group had reported that Mcl-1 expression alone could prolong the life of Chinese hamster ovary (CHO) cells undergoing apoptosis (Figure 15.10), although not as effectively as Bcl-2, and that it could efficiently prevent Bax-mediated apoptosis in yeast.[41] However, although Mcl-1 associated with Bax in a yeast two-hybrid system, they found that Mcl-1 and Bax did not associate in

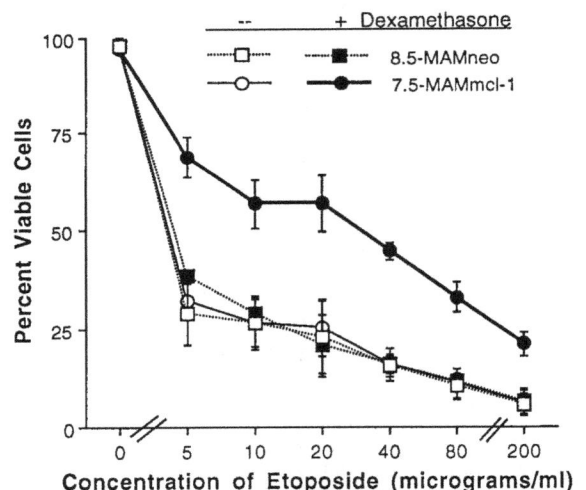

B   Etoposide - Concentration/Response Curve

FIGURE 15.10 Mcl-1 expression protects cells against etoposide-mediated apoptosis. Cells were transfected with dexamethasone-inducible Mcl-1 construct and dexamethasone was added to induce Mcl-1 expression with or without addition of the chemotherapeutic, etoposide.[41] (Adapted from Blood, 89(2), Zhou P, Qian L, Kozopas KM, Craig RW. Mcl-1, a Bcl-2 family member, delays the death of hematopoietic cells under a variety of apoptosis-inducing conditions, 630–643, Copyright © 1997, with permission from Elsevier.)

the murine 32D cell line and that Mcl-1 had *minimal effect on viability* in this same cell line, raising

> [...] the question of whether Mcl-1 could affect cell viability in systems other than CHO and, more specifically, whether its function in hematopoietic cells related to enhancement of viability. The issue was also raised as to whether Mcl-1 could associate with Bax in intact hematopoietic cells as it can in the more artificial systems.[41]

Therefore, the authors investigated the role of Mcl-1 in hematopoietic cells under a number of different apoptotic stimuli. Indeed, IP experiments indicated that Mcl-1 interacted with Bax, which supported *the significance of the interaction observed in more artificial systems such as the yeast two-hybrid system.*[41] Next, they incubated Mcl-1-transfected cells with the chemotherapeutic drug, etoposide, and results showed that cell viability was significantly increased and cell death delayed in Mcl-1 overexpressing cells compared to control cells. Similar results were obtained using other apoptosis-inducing agents, such as growth factor deprivation and UV irradiation. Thus, these data suggested that:

> Mcl-1 decreases the proportion of cells entering the apoptotic death pathway but that, once this process is initiated, Mcl-1 does not affect the secondary loss of membrane integrity that marks the *[progression of the apoptosis pathway]* [...]
>
> Overall, the above results show similarities between Mcl-1 and Bcl-2 in that Mcl-1 affords protection of viability to *[hematopoietic]* cells exposed to a variety of exogenously applied cytotoxic agents, exerting its effects at an early phase of apoptosis.
>
> [...] the results described above show that Mcl-1 can promote cell viability without promoting cell growth under a variety of conditions that produce cell death by apoptosis.[41]

In the early years of the 21st century, several laboratories determined that a decrease in Mcl-1 expression, which was found to be initiated via proteasomal degradation, was sufficient to promote apoptosis and was, in fact, a requirement for the induction of apoptosis following exposure to cytotoxic agents.[42,43] However, the mechanism by which it was regulated was still unknown. In 2005, Zhong et al. started from the fact that *proteins targeted for proteasome degradation are usually modified by a polyubiquitin chain* and, therefore, sought to determine if this was responsible for Mcl-1's short half-life.[44] Indeed, they found the presence of a species of Mcl-1 with a higher-than-expected molecular weight by western blotting when cells were incubated with a proteasome inhibitor (to prevent degradation of ubiquitinated proteins), which was confirmed to be ubiquitinated Mcl-1 using ubiquitin-specific antibodies. They also used protein fractionation to isolate the protein responsible for the ubiquitination of Mcl-1, which they named "Mcl-1 ubiquitin ligase E3" (MULE), and co-IP experiments confirmed that the two proteins interacted

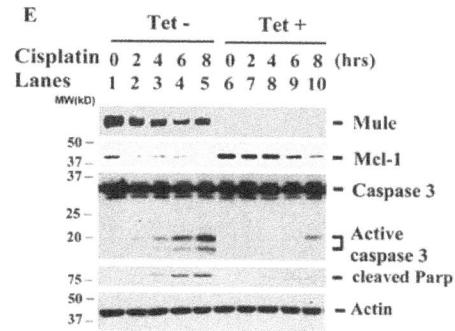

FIGURE 15.11 MULE expression degrades Mcl-1 and induces apoptosis. Cells in which MULE had been knocked down (Tet+) were incubated with the chemotherapeutic cisplastin to induce cell death for the indicated time. Cell extracts were collected and subjected to SDS-PAGE and western blotting with the indicated antibodies.[44] (Adapted from Cell 121(7), Zhong Q, Gao W, Du F, Wang X. Mule/ARF-BP1, a BH3-only E3 ubiquitin ligase, catalyzes the polyubiquitination of Mcl-1 and regulates apoptosis, 1085–1095, Copyright © 2005, with permission from Elsevier.)

in vivo.[44] After cloning MULE, the authors purified recombinant proteins to confirm that MULE could indeed ubiquitinate Mcl-1 *in vitro*. They also found that the BH3 domains were required for interaction between MULE and Mcl-1, which indicated that MULE was a new member of the Bcl-2 protein family. In addition, depletion of MULE using a specific antibody *dramatically compromised Mcl-1 ubiquitination in the extracts*.

The authors also demonstrated that Mcl-1 degradation was inhibited in cells lines in which MULE had been knocked-out, and the effects of treatment by three separate chemotherapeutic drugs were significantly decreased in these cells (Figure 15.11).[44] Finally, caspase-3 activation and PARP cleavage were inversely correlated to Mcl-1 expression. These data established that the short half-life of Mcl-1 was due to it being ubiquitinated by MULE and subsequently degraded by the proteasome. However, the authors also reported that in some systems, Mcl-1 decreased upon induction of apoptosis despite the presence of proteasome inhibitors or the absence of MULE, suggesting that additional regulatory mechanisms existed.

At that time, it was assumed that upon induction of apoptosis, all BH3-only proteins (pro-apoptosis proteins, that is, Bim, Bad, Noxa, Puma, etc.) bound equally well to the anti-apoptosis proteins (Bcl-2, Bcl-x$_L$, Mcl-1, etc.), but this had not been formally tested. Therefore, Chen et al. used immobilized BH3-domain-containing peptides from various BH3-only proteins to determine the extent to which they would bind to pro-survival proteins. As such, they found that although the BH3 domains of Bim and Puma bound effectively to all pro-survival members, those of other BH3-only proteins were highly selective in their association (Figure 15.12a, b).[45] Specifically, Noxa was found to efficiently interact only with Mcl-1 and to a lesser extent, to A1, whereas Bad was more specific for Bcl-2, Bcl-x$_L$, and Bcl-w, while still, binding of Bik and Bid to A1 and Bcl-x$_L$

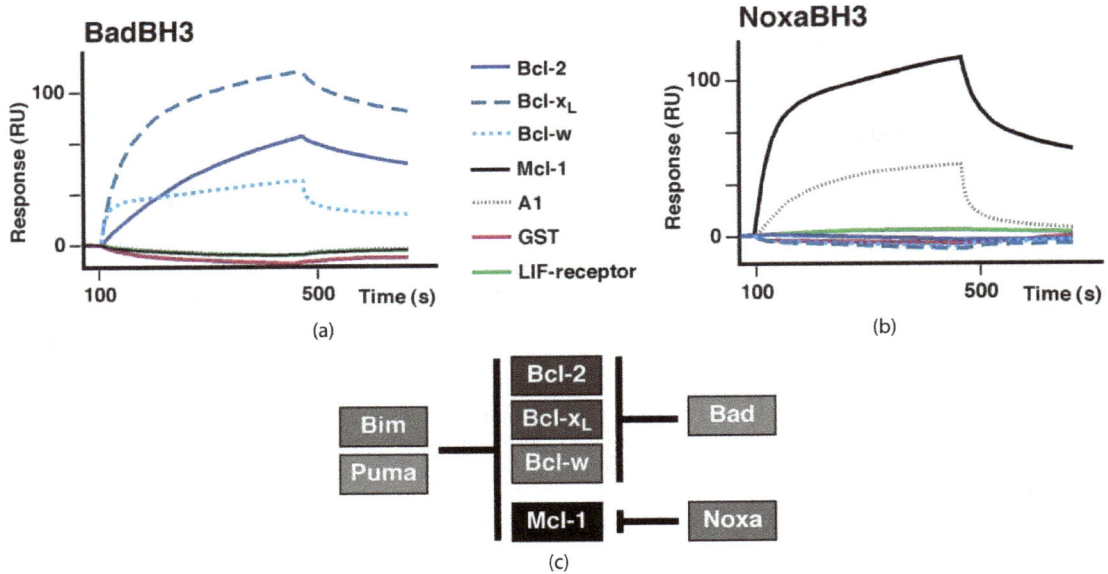

FIGURE 15.12 Differential binding of BH3-only proteins to anti-apoptosis proteins. BH3 peptides from Bad (a) or Noxa (b) were immobilized and the binding affinity of each pro-survival protein was determined; (c) scheme of interaction between members of the Bcl-2-like protein family as proposed by Chen et al.[45] (Adapted from Mol Cell. 17(3), Chen L, Willis SN, Wei A, et al. Differential targeting of prosurvival Bcl-2 proteins by their BH3-only ligands allows complementary apoptotic function, 393–403, Copyright © 2005, with permission from Elsevier.)

was more efficient than to Bcl-2 and Mcl-1 (Figure 15.12c). Importantly, the authors noted the complementary binding of Bad and Noxa, the former binding preferentially to Bcl-2 and Bcl-x_L and the latter exclusively to Mcl-1 and A1. Thus, the authors hypothesized that:

[...] if, as our results suggest, apoptosis requires neutralization of all the relevant prosurvival proteins, BH3-only proteins with complementary binding profiles should cooperate in killing cells [...] Thus, the neutralization of Mcl-1 by Noxa should be complemented by coexpression of a BadBH3 to neutralize Bcl-2 and Bcl-x_L [...] Likewise, the NoxaBH3 should augment killing induced by Bik, which binds to Mcl-1 ~40-fold less well than to Bcl-x_L.[45]

Indeed, Chen et al. found that cell death increased significantly when cells were co-transfected with Noxa and either Bad or Bik compared to the weak killing induced by any one of these proteins alone. From these data, the authors concluded that:

Both Bad and Bik, which predominantly bind Bcl-2 and Bcl-x_L, can potently cooperate with Noxa, which selectively binds Mcl-1, to augment the weak killing observed when any of the three BH3-only proteins is expressed alone [...]

Our evidence that BH3-only proteins have distinct subsets of prosurvival targets [...] strongly suggests that Bim and Puma are particularly potent because they can neutralize all the prosurvival proteins, whereas the less potent BH3-only proteins have a more restricted binding spectrum [...]

Our functional complementation experiments in fibroblasts strongly support the hypothesis that neutralization

of Mcl-1 as well as Bcl-xL and/or Bcl-2 is required for efficient killing [...]

The results presented here argue that particular subsets of BH3-only proteins and their prosurvival targets probably have distinct biological functions. The ability of Bim and Puma to efficiently antagonize all the Bcl-2-like proteins [...] is probably reflected in the marked phenotypes observed on disruption of their genes. Conversely, the much subtler phenotypes in mice lacking Bik [...], Bad [...], or Noxa [...] may reflect the more restricted binding profiles of those proteins.[45]

* * * * *

In this chapter, we discussed how a cell's survival is highly dependent on the delicate balance between a number of proteins of the Bcl-2-like protein family. As such, pro-survival proteins (Bcl-2, Bcl-x_L, and Mcl-1, etc.) are expressed in healthy cells to ensure neutralization of pro-apoptosis activators (BIM and BID), preventing them from activating the effector proteins, BAX and BAK. (Figure 15.13). However, under certain circumstances, such as following exposure to chemotherapeutic drugs, BH3-only sensitizer proteins (Noxa, Bad, Bid, Puma, etc.) are upregulated and bind to pro-survival proteins, which results in the release of activator proteins and their binding to Bax and Bak (and in the case of BAX, induces its translocation to mitochondria). This induces a conformational change in these proteins which results in their oligomerization and the formation of pores in the mitochondrial membrane, leading to the release of cytochrome *c* and other mitochondrial-resident proteins from mitochondria into the cytosol. In addition, these pores allow protons (i.e., H[+]) to move freely in and out of mitochondria and reach an equilibrium, thereby eliminating the membrane potential that is crucial for the synthesis of ATP. Once outside of mitochondria, cytochrome *c* associates with caspase-9 and Apaf-1 to form the apoptosome complex, which goes on to cleave and activate caspase 3 and other downstream caspases. Caspase activation initiates the cleavage of a whole host of proteins, thus preventing the cell from performing cellular functions. Importantly, caspase activation results in the loss of phospholipid asymmetry in the plasma membrane, favoring the exposure of phosphatidylserine (PS) to the outside of the cell. PS is then recognized by macrophages and triggers the phagocytosis of these apoptotic cells, which prevents the uncontrolled inflammation which is characteristic of necrotic cell death.

The pathway just described is called the intrinsic apoptosis pathway, since intracellular deficiencies, such as DNA damage, are what initiates the pathway. Another manner in which apoptosis is triggered is via the extrinsic apoptosis pathway. This pathway is initiated by the binding of signaling molecules, such as "apoptosis stimulating factor" (Fas) or "tumor necrosis factor" (TNF), to cell surface receptors— appropriately called "death receptors"—which induce the binding and activation of downstream signaling molecules, including caspase 8 (Figure 15.13b). Activation of caspase 8 cleaves and activates effector caspases, as well as Bid, with the same results already discussed for the intrinsic pathway. Therefore, whether apoptosis is initiated via the intrinsic or extrinsic pathway, it always leads to the activation of caspases and the permeability of the mitochondrial plasma membrane.

* * * * *

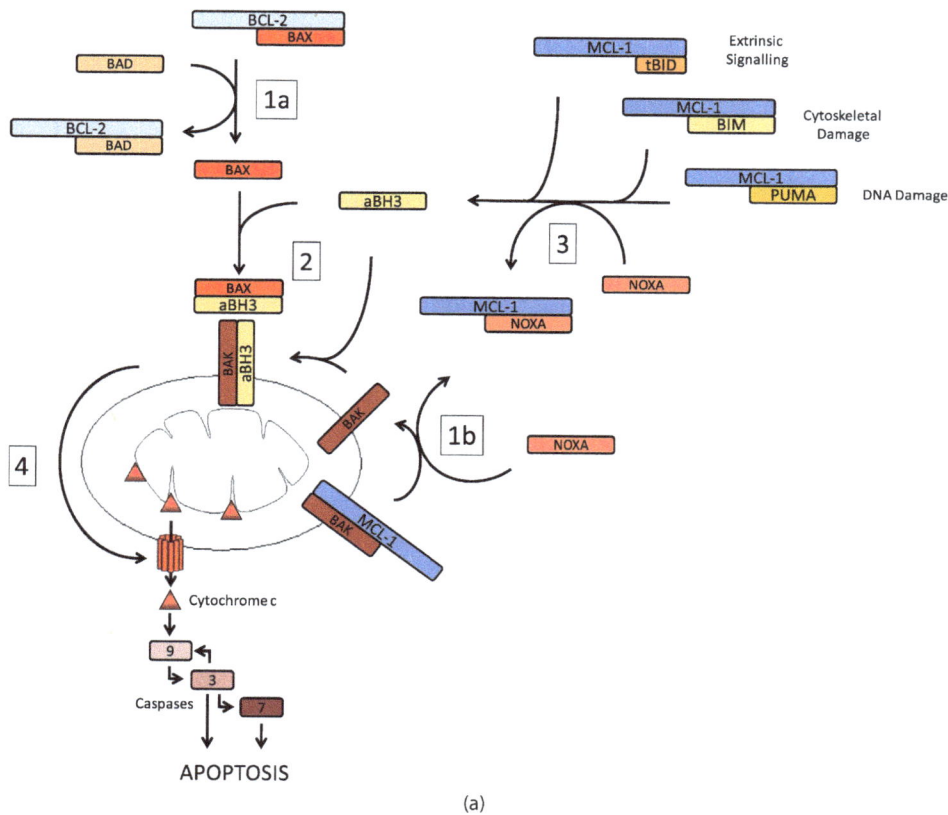

(a)

FIGURE 15.13 Apoptosis pathways. Apoptosis can be induced by intrinsic factors, such as DNA damage or ER stress, or extrinsic factors, such as extracellular ligands which bind to specialized "death receptors". The extrinsic pathway induces apoptosis by first activating caspase-8, which either directly cleaves caspase-3 or cleaves BID produce a truncated BID (tBID), a BH3-only protein which induces the oligomerization of BAX or BAK in mitochondria. In contrast, the intrinsic pathway acts directly on BH3-only proteins by inducing the upregulation of BH3-only sensitizers (BAD and NOXA), which bind to anti-apoptosis proteins (Bcl-2 and Mcl-1, respectively) and compete with binding of pro-apoptosis, BH3-only activators (aBH3 proteins, BIM, tBID). The release of aBH3 proteins from anti-apoptosis proteins allows these proteins to interact with the effector proteins, BAK and BAX, resulting in their activation and oligomerization. This latter step forms pores in the mitochondrial membrane thorough which cytochrome c and ATP escape into the cytosol, initiating the formation, along with APAF1 and procaspase-9, of the apoptosome.[46,47] (Figure (a) reprinted by permission from Springer Nature Customer Service Centre GmbH: Springer Nature, Nat Rev Mol Cell Biol. Tait SWG, Green DR. Mitochondria and cell death: Outer membrane permeabilization and beyond, Copyright © 2010. Figure (b) reprinted with permission from John Wiley and Sons, Thomas LW, Lam C, Edwards SW. Mcl-1: The molecular regulation of protein function. FEBS Lett. Copyright © 2015 Federation of European Biochemical Societies.) (*Continued*)

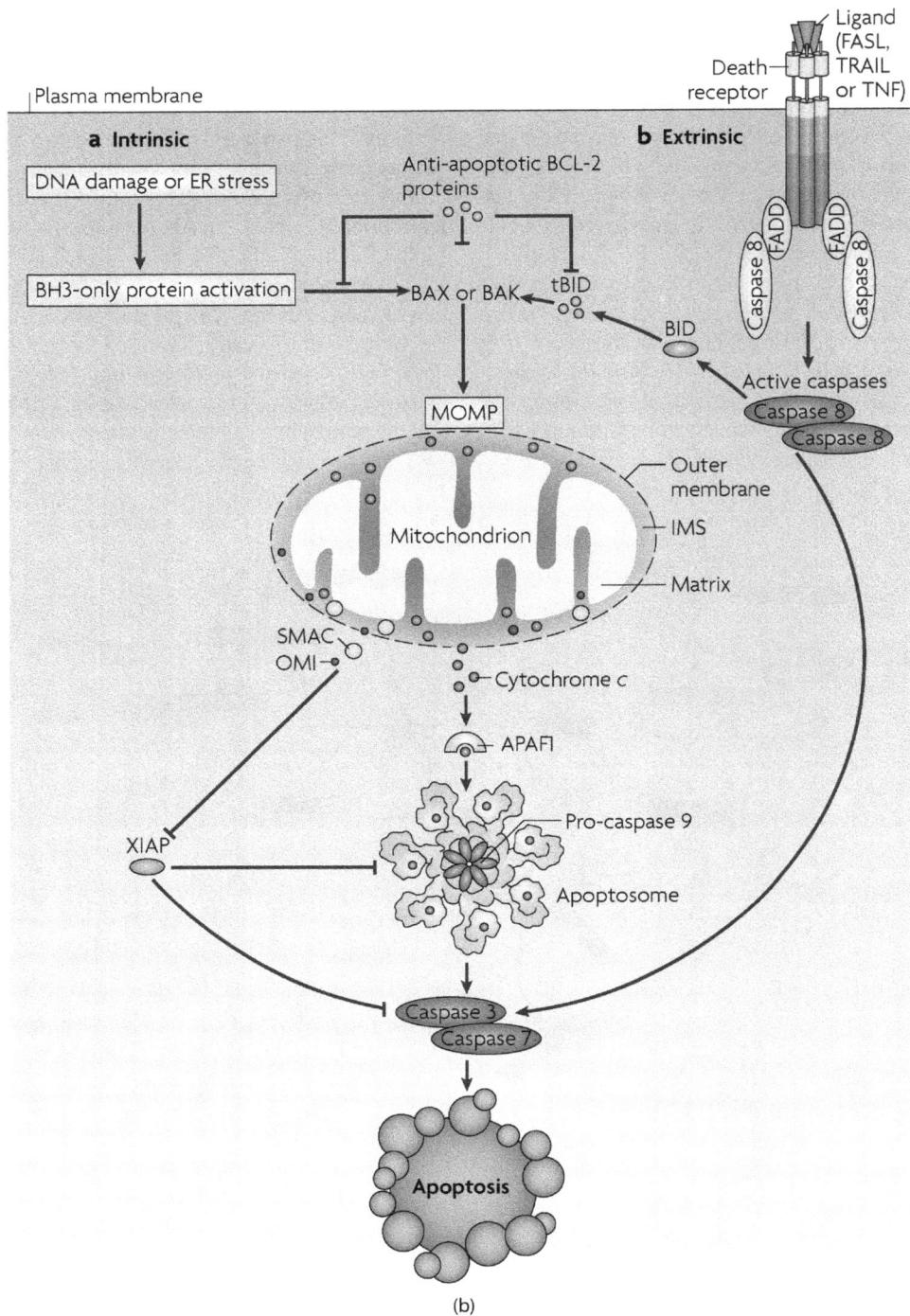

(b)

FIGURE 15.13   (*Continued*)

REFERENCES

1. Diamantis A, Magiorkinis E, Sakorafas GH, Androutsos G. A brief history of apoptosis: From ancient to modern times. *Oncology Research and Treatment*. 2008;31(12):702–706. doi:10.1159/000165071

2. Virchow R. *Cellular Pathology as Based upon Physiological and Pathological Histology*; 1860. Accessed March 29, 2019. https://archive.org/details/dli.ministry.10993

3. Majno G, Joris I. Apoptosis, oncosis, and necrosis. An overview of cell death. *American Journal of Pathology*. 1995;146(1):3–15. Accessed March 29, 2019. http://www.ncbi.nlm.nih.gov/pubmed/7856735

4. Glucksmann A Cell deaths in normal vertebrate ontogeny. *Biological Reviews of Cambridge Philosophical Society*. 1951;26(1):59–86. doi:10.1111/j.1469-185X.1951.tb00774.x

5. Kerr JFR. A personal account of events leading to the definition of the apoptosis concept. In: Kumar S, ed. *Apoptosis: Biology and Mechanisms*. Springer, Berlin Heidelberg; 1999:1–10. doi:10.1007/978-3-540-69184-6_1

6. Rous P, Larimore LD. Relation of the portal blood to liver maintenance : A demonstration of liver atrophy conditional

on compensation. *Journal of Experimental Medicine*. 1920;31(5):609–632. doi:10.1084/jem.31.5.609

7. Kerr JF, Wyllie AH, Currie AR. Apoptosis: A basic biological phenomenon with wide-ranging implications in tissue kinetics. *British Journal of Cancer*. 1972;26(4):239–257. doi:10.1038/bjc.1972.33

8. Ellis HM, Horvitz HR. Genetic control of programmed cell death in the nematode *C. elegans*. *Cell*. 1986;44(6):817–829. doi:10.1016/0092-8674(86)90004-8

9. Vaux DL. Apoptosis timeline. *Cell Death and Differentiation*. 2002;9(4):349–354. doi:10.1038/sj.cdd.4400990

10. Tsujimoto Y, Finger LR, Yunis J, Nowell PC, Croce CM. Cloning of the chromosome breakpoint of neoplastic B cells with the t(14;18) chromosome translocation. *Science (1979)*. 1984;226(4678):1097–1099. doi:10.1126/science.6093263

11. Tsujimoto Y, Cossman J, Jaffe E, Croce CM. Involvement of the Bcl-2 gene in human follicular lymphoma. *Science (1979)*. 1985;228(4706):1440–1443. doi:10.1126/science.3874430

12. Cleary ML, Smith SD, Sklar J. Cloning and structural analysis of cDNAs for bcl-2 and a hybrid Bcl-2/immunoglobulin transcript resulting from the t(14;18) translocation. *Cell*. 1986;47(1):19–28. doi:10.1016/0092-8674(86)90362-4

13. Vaux DL, Cory S, Adams JM. Bcl-2 gene promotes haemopoietic cell survival and cooperates with c-myc to immortalize pre-B cells. *Nature*. 1988;335(6189):440–442. doi:10.1038/335440a0

14. Yonish-Rouach E, Resnftzky D, Lotem J, Sachs L, Kimchi A, Oren M. Wild-type p53 induces apoptosis of myeloid leukaemic cells that is inhibited by interleukin-6. *Nature*. 1991;352(6333):345–347. doi:10.1038/352345a0

15. Allen TM, Williamson P, Schlegel RA. Phosphatidylserine as a determinant of reticuloendothelial recognition of liposome models of the erythrocyte surface. *Proceedings of National Academy of Sciences U S A*. 1988;85(21):8067–8071. doi:10.1073/pnas.85.21.8067

16. Fadok VA, Voelker DR, Campbell PA, Cohen JJ, Bratton DL, Henson PM. Exposure of phosphatidylserine on the surface of apoptotic lymphocytes triggers specific recognition and removal by macrophages. *Journal of Immunology*. 1992;148(7):2207–2216. Accessed March 29, 2019. http://www.ncbi.nlm.nih.gov/pubmed/1545126

17. Kaufmann SH, Desnoyers S, Ottaviano Y, Davidson NE, Poirier GG. Specific proteolytic cleavage of poly(ADP-ribose) polymerase: An early marker of chemotherapy-induced apoptosis. *Cancer Research*. 1993;53(17):3976–3985. Accessed March 29, 2019. http://www.ncbi.nlm.nih.gov/pubmed/8358726

18. Yuan J, Shaham S, Ledoux S, Ellis HM, Horvitz HR. The *C. elegans* cell death gene ced-3 encodes a protein similar to mammalian interleukin-1 beta-converting enzyme. *Cell*. 1993;75(4):641–652. doi:10.1016/0092-8674(93)90485-9

19. Miura M, Zhu H, Rotello R, Hartwieg EA, Yuan J. Induction of apoptosis in fibroblasts by IL-1 beta-converting enzyme, a mammalian homolog of the *C. elegans* cell death gene ced-3. *Cell*. 1993;75(4):653–660. doi:10.1016/0092-8674(93)90486-a

20. Tewari M, Quan LT, O'Rourke K, et al. Yama/CPP32 beta, a mammalian homolog of CED-3, is a CrmA-inhibitable protease that cleaves the death substrate poly(ADP-ribose) polymerase. *Cell*. 1995;81(5):801–809. doi:10.1016/0092-8674(95)90541-3

21. Nicholson DW, Ali A, Thornberry NA, et al. Identification and inhibition of the ICE/CED-3 protease necessary for mammalian apoptosis. *Nature*. 1995;376(6535):37–43. doi:10.1038/376037a0

22. Li P, Nijhawan D, Budihardjo I, et al. Cytochrome c and dATP-dependent formation of Apaf-1/caspase-9 complex initiates an apoptotic protease cascade. *Cell*. 1997;91(4):479–489. doi:10.1016/s0092-8674(00)80434-1

23. Liu X, Kim CN, Yang J, Jemmerson R, Wang X. Induction of apoptotic program in cell-free extracts: Requirement for dATP and cytochrome c. *Cell*. 1996;86(1):147–157. doi:10.1016/s0092-8674(00)80085-9

24. Yang J, Liu X, Bhalla K, et al. Prevention of apoptosis by Bcl-2: Release of cytochrome c from mitochondria blocked. *Science (1979)*. 1997;275(5303):1129–1132. doi:10.1126/science.275.5303.1129

25. Zou H, Henzel WJ, Liu X, Lutschg A, Wang X. Apaf-1, a human protein homologous to *C. elegans* CED-4, participates in cytochrome C-dependent activation of caspase-3. *Cell*. 1997;90(3):405–413. doi:10.1016/s0092-8674(00)80501-2

26. Taylor RC, Cullen SP, Martin SJ. Apoptosis: Controlled demolition at the cellular level. *Nature Reviews. Molecular Cell Biology*. 2008;9(3):231–241. doi:10.1038/nrm2312

27. Oltval ZN, Milliman CL, Korsmeyer SJ. Bcl-2 heterodimerizes in vivo with a conserved homolog, Bax, that accelerates programmed cell death. *Cell*. 1993;74(4):609–619. doi:10.1016/0092-8674(93)90509-O

28. Yin XM, Oltvai ZN, Korsmeyer SJ. BH1 and BH2 domains of Bcl-2 are required for inhibition of apoptosis and heterodimerization with Bax. *Nature*. 1994;369(6478):321–323. doi:10.1038/369321a0

29. Schlesinger PH, Gross A, Yin XM, et al. Comparison of the ion channel characteristics of proapoptotic BAX and antiapoptotic BCL-2. *Proceedings of Nationall Academy of Sciences U S A*. 1997;94(21):11357–11362. doi:10.1073/PNAS.94.21.11357

30. Yang E, Zha J, Jockel J, Boise LH, Thompson CB, Korsmeyer SJ. Bad, a heterodimeric partner for Bcl-XL and Bcl-2, displaces Bax and promotes cell death. *Cell*. 1995;80(2):285–291. doi:10.1016/0092-8674(95)90411-5

31. Antonsson B, Conti F, Ciavatta A, et al. Inhibition of Bax channel-forming activity by Bcl-2. *Science (1979)*. 1997;277(5324):370–372. doi:10.1126/science.277.5324.370

32. Gross A, Jockel J, Wei MC, Korsmeyer SJ. Enforced dimerization of BAX results in its translocation, mitochondrial dysfunction and apoptosis. *EMBO Journal*. 1998;17(14):3878–3885. doi:10.1093/emboj/17.14.3878

33. Desagher S, Osen-Sand A, Nichols A, et al. Bid-induced conformational change of bax is responsible for mitochondrial cytochrome c release during apoptosis. *Journal of Cell Biology*. 1999;144(5):891–901. doi:10.1083/jcb.144.5.891

34. Eskes R, Desagher S, Antonsson B, Martinou JC. Bid induces the oligomerization and insertion of Bax into the outer mitochondrial membrane. *Molecular Cell Biology*. 2000;20(3):929–935. doi:10.1128/MCB.20.3.929-935.2000

35. Saito M, Korsmeyer SJ, Schlesinger PH. BAX-dependent transport of cytochrome c reconstituted in pure liposomes. *Nature Cell Biology*. 2000;2(8):553–555. doi:10.1038/35019596

36. Wei MC, Lindsten T, Mootha VK, et al. tBID, a membrane-targeted death ligand, oligomerizes BAK to release cytochrome c. *Genes Development*. 2000;14(16):2060–2071. doi:10.1101/gad.14.16.2060

37. Wei MC, Zong WX, Cheng EH, et al. Proapoptotic BAX and BAK: A requisite gateway to mitochondrial dysfunction and death. *Science (1979)*. 2001;292(5517):727–730. doi:10.1126/science.1059108

38. Miyashita T, Reed JC. Tumor suppressor p53 is a direct transcriptional activator of the human bax gene. *Cell*. 1995;80(2):293–299. doi:10.1016/0092-8674(95)90412-3

39. Oda E, Ohki R, Murasawa H, et al. Noxa, a BH3-only member of the Bcl-2 family and candidate mediator of p53-induced apoptosis. *Science (1979).* 2000;288(5468): 1053–1058. doi:10.1126/science.288.5468.1053

40. Kozopas KM, Yang T, Buchan HL, Zhou P, Craig RW. MCL1, a gene expressed in programmed myeloid cell differentiation, has sequence similarity to BCL2. *Proceedings of National Academy of Sciences U S A.* 1993;90(8): 3516–3520. doi:10.1073/pnas.90.8.3516

41. Zhou P, Qian L, Kozopas KM, Craig RW. Mcl-1, a Bcl-2 family member, delays the death of hematopoietic cells under a variety of apoptosis-inducing conditions. *Blood.* 1997;89(2):630–643. doi:10.1182/blood.V89.2.630

42. Leuenroth SJ, Grutkoski PS, Ayala A, Simms HH. The loss of Mcl-1 expression in human polymorphonuclear leukocytes promotes apoptosis. *Journal of Leukocyte Biology.* 2000;68(1):158–166. doi:10.1189/jlb.68.1.158

43. Nijhawan D, Fang M, Traer E, et al. Elimination of Mcl-1 is required for the initiation of apoptosis following ultraviolet irradiation. *Genes and Development.* 2003;17(12):1475–1486. doi:10.1101/gad.1093903

44. Zhong Q, Gao W, Du F, Wang X. Mule/ARF-BP1, a BH3-only E3 ubiquitin ligase, catalyzes the polyubiquitination of Mcl-1 and regulates apoptosis. *Cell.* 2005;121(7): 1085–1095. doi:10.1016/j.cell.2005.06.009

45. Chen L, Willis SN, Wei A, et al. Differential targeting of prosurvival bcl-2 proteins by their bh3-only ligands allows complementary apoptotic function. *Molecular Cell.* 2005;17(3):393–403. doi:10.1016/j.molcel.2004.12.030

46. Tait SWG, Green DR. Mitochondria and cell death: Outer membrane permeabilization and beyond. *Nature Reviews. Molecular Cell Biology.* 2010;11(9):621–632. doi:10.1038/nrm2952

47. Thomas LW, Lam C, Edwards SW. Mcl-1: The molecular regulation of protein function. *FEBS Letters.* 2010; 584(14):2981–2989. doi:10.1016/j.febslet.2010.05.061

16 The Biology of Cancer

Although cancer is thought to have existed long before humans walked the earth, the earliest written record of cancer was found in the Edwin Smith Papyrus, circa 3000 B.C.[1] At that time, tumors were treated with the use of cautery, knives, salts, and arsenic paste by the Egyptians, whereas others used herbal medicines, such as teas and boiled cabbage, and the pastes of a number of metals. Surprisingly, many of these early treatments remained in use as recently as the 1800s. In the 2nd century A.D., a renowned Roman surgeon to the gladiators, Claudius Galen, believed that cancer came about as a result of the accumulation of thick black bile and therefore, treated patients with purgatives to rid the body of it. Realizing that these treatments did not cure cancer, physicians in 5th-century Constantinople started to perform amputations to get rid of cancer, the first of which was that of the breast but many other organs soon followed.[1] By this time, cancers were known to involve surrounding nerves, muscles, and blood vessels, so the usual practice was to excise as much of the surrounding tissue as possible to ensure that all of the tumor was removed in an attempt to prevent its recurrence. Though cancer treatment had changed little by the 1300s, French physicians did advance for the first time the idea that cancer might arise by external agents that entered the body, leading to the demise of Galen's theory, which had held for a millennium.[1]

In the 16th century, Paracelsus—the alchemist about whom we learned in Chapter 1 and who changed the role of the discipline to one of medical intervention rather than one of a greedy search for the philosopher's stone—introduced a number of treatments for cancer, but these were hardly an improvement over practices of the time.[2] He pioneered such treatments as the systematic use of mercury, lead, sulfur, and arsenic as chemotherapeutics, though he warned that these chemical treatments were poisons themselves and therefore, administered doses had to be carefully monitored. His followers subsequently published a collection of his papers in which an environmental cause for cancer was implied, in that miners and those who smelted metal ores had increased rates of lung cancer. However, it would be another two hundred years until the harmful effects of metal absorption would become widely known and accepted in the scientific community. In the 1600s, two independent physicians noticed that the rate of some cancers, such as breast cancer, were increased within certain families, and from this, they concluded that cancer was contagious. As such, they advocated that cancer patients be isolated from the population and removed as far from the city or town as possible. In the 1700s, the French surgeon, Deshaies Gendron, proposed what turned out to be a fairly accurate description of the origin of cancer as we understand it today, proposing that cancer arises *from the transformation and continuous growth of glandular, lymphatic, vascular, and solid structures in the body.*[2] However, his suggestions for appropriate treatments remained as was common at the time, that is, the *wide surgical excision* of the tumor.

For the next century or so, the landscape of knowledge of cancer centered around its clinical aspects (i.e., presentation and treatment); however, along with the invention of the microscope came cellular descriptions of cancer, which resulted in a staggering increase in the reporting of different types of tumors and in the improvement in surgical techniques for the removal of tumors, which remained the most effective manner with which to treat the disease. However, the establishment of Schwann's cell theory in 1838 set about monumental changes in the way cancer was perceived. Particularly, Schwann's mentor, Johannes Müller, published a treatise in which he described cancer as an aggregation of newly formed, abnormal cells in diseased organs, with a potential to be destructive by spreading to other parts of the body by way of blood vessels.[3] Importantly, he also described tumor necrosis, that is, the death of tumor cells, and attributed the regression of cancers, which was frequently observed, to this phenomenon. A few years later, Rudolph Virchow, who introduced the third tenet of the cell theory—that all cells are derived from preexisting cells—extended this idea to the origin of cancer and suggested that it arose from normal, preexisting cells that suffered chemical injuries.[4]

In the early 20th century, Peyton Rous, a research pathologist at Rockefeller Institute for Medical Research in New York city, reported a most surprising discovery which would bring about new ideas on the genesis of cancer. It was known at the time that tumors could be transplanted from one animal to another, usually within the same species, but increasingly it was found to be transmittable between species. In 1908, Ellerman and Bang reported the first successful transplantation of a fowl tumor (leukemia) in chickens and passaged it from one animal to another six successive times; the authors also determined that the disease was caused by a filterable virus.[5,6] However, their findings were largely ignored since leukemias were not thought to be neoplastic diseases at the time. In 1910, Rous reported the successful transmission of chicken sarcoma—that is, cancers of the connective tissues such as bone, cartilage, and fat—between fowls.[5] To that end, the authors injected pieces of a *large, irregularly globular mass* of the diseased hen into two other hens of the same flock and one of them developed *a large nodule* 35 days later. The *new growth*, Rous reported, *has proved itself a neoplasm of classical behavior.* Following the first transplantation, the tumor was successfully propagated three more times by the publication of the paper but only in hens that were *intimately related* [to the] *fowls of the pureblood stock in which it* [the tumor] *was first noticed.*[5]

DOI: 10.1201/9781003379058-16

Rous then ground up a sample of the tumors, centrifuged them, and filtered them to clear the debris, and found that out of ten recipients of this filtrate, four individuals developed sarcomas of *characteristic growths [...] and transplantation into other chickens proved successful.*[7,8] Thus, he concluded that *the tumor resulting from injection of a filtrate itself furnished material capable of producing tumors after injection* and that *the first tendency will be to regard the self-perpetuating agent active in this sarcoma of the fowl as a minute parasitic organism.*[7] These results were apparently too good to be true, as it took more than 15 years for them to be deemed valid by the scientific community.[6] Numerous attempts were made to study these special viruses but experiments yielded quantities of virus too small to work with and their purification was problematic since *eliminating non-specific cell structures from viral suspensions was tremendously difficult.*[9] For these and other reasons, many beginners were discouraged from starting this line of studies and thus, progress was slow.

However, it was discovered in the early 1960s that host cell DNA was required for the replication of viral genetic material and to produce transformation of the cell; and by the late 1960s, the viral cause of certain cancers was well established in several different organisms, including chickens, rats, mice, and humans. These and other results led Huebner and Todaro to propose a hypothesis in which viral infection of a cell led to the expression of cancer-causing viral genes, which they called "oncogenes":

> The central hypothesis implies, therefore, that the cells of many if not all vertebrates carry vertically transmitted (inherited) RNA tumor virus information (virogenes) which serves as an indigenous source of oncogenic information (oncogenes) which transforms normal cells into tumor cells [...][10]

Interestingly, it was also found that about 10% of healthy wild-type mice treated with a carcinogen had neoplastic cells which contained a particular viral antigen. In addition, homology was found between this viral RNA-coded gene and uninfected genomic DNA of a number of different species, suggesting that these viral oncogenes were already present in normal, healthy cells, which led to the proposition by Huebner and Todaro that:

> [...] the genes coding for the unique C-type viral functions may not be expressed under normal conditions because of potent repressors for expression. Viewed in this light, the application of radiation, chemical carcinogens, and the natural aging process are believed to "switch on" the viral genome, perhaps by decreasing the level of repressor activity [...]
>
> Its *[viral RNA]* demonstration in nine different species and in three classes of vertebrates, together with the evidence of vertical rather than horizontal transmission as the chief mode of spread, suggests that this virus genome is an essential part of the natural evolutionary inheritance of vertebrate cells.
>
> [...] the "oncogenic" DNA viruses may function, in part, by activating previously repressed oncogenes in the

cells they infect; thus, as part of the over-all hypothesis presented it is possible that the oncogenic DNA viruses may also serve as carcinogens, derepressing C-type RNA virus information indigenous to the cells.

Our hypothesis suggests that the cells of most or all vertebrate species have C-type RNA virus genomes that are vertically transmitted from parent to offspring. Depending on the host genotype and various modifying environmental factors, either virus production or tumor formation or both may develop at some time in these animals and/or in their cells grown in culture. This hypothesis implies that the occurrence of most cancer is a natural biological event determined by spontaneous and/or induced derepression of an endogenous specific viral oncogene(s). Viewed in this way, ultimate control of cancer will therefore very likely depend on delineation of the factors responsible for derepression of virus expression and of the nature of the repressors involved. We believe that the hypothesis provides a rational basis for a unifying theory and is consistent with the phenomena of radiation and chemically induced cancer as well as the stochastic occurrence of spontaneous cancer. The availability of *in vitro* test systems to study the derepressed virus in cells in culture should make it possible to analyze this phenomenon at the cellular and molecular level.[10]

It was subsequently found by Raymond Erickson's group that the transformation of cells by Rous' virus was the result of the expression of a single viral gene, named *src* (short for sarcoma), which coded for a 60KDa phosphoprotein with kinase activity.[11,12] This suggested to Collett et al. that the kinase activity of *src* may be responsible in cellular transformation following viral infection *through abnormal phosphorylation of cellular proteins.*[13] As such, immunoprecipitation experiments using sera from rabbits and mice infected with the newly discovered avian sarcoma virus (ASV) revealed that normal, uninfected rats indeed expressed a protein which was structurally and functionally similar to viral src, in that it was a phosphoprotein with kinase activity.[13] Since homologs of this protein were also found in a number of other species, the authors suggested that:

> [...] these proteins may have a common function in some important facet of cellular metabolism [...] Furthermore, after recombination with transformation-defective ASV containing partial src gene deletions, these normal cell sequences may give rise to transforming viruses with the ability to induce uncontrolled cellular growth. Thus, it can be suggested that the role of the normal cell src-related protein may be related to some basic mechanism of cell growth control. The finding of an associated protein kinase activity with the normal cell homologues of *[src]* implicates phosphorylation-dephosphorylation modification as being critical in this function [...]
>
> [...] the viral gene product is present in substantially greater amounts in transformed cells than is the normal cell protein in uninfected cells, suggesting that cellular transformation may merely be a consequence of a quantitative difference in expression of the two genes. This fact, coupled with the similarity in structure of the cellular and viral gene products and the nature of the transforming viruses recovered by Hanafusa and coworkers,

supports the possibility that the phosphoprotein products of both the normal cellular gene and the viral gene have identical functions. It follows then that the biochemical events in ASV-induced oncogenesis may be qualitatively identical to those in normal cells but perhaps occur to a greater degree *[compared to wild type]* to produce the transformed phenotype. If this is the case, then quantitative changes in the phosphorylation-dephosphorylation of specific cell proteins may provide insights not only about viral transformation but also about normal cellular processes.[13]

* * * * *

By the 1970s, some investigators had suggested that cancers might originate in discreet steps, and as a result of as few as two genetic mutations; but direct evidence to support such a hypothesis was lacking.[14] Therefore, Alfred Knudson studied 48 cases of retinoblastoma (tumors of the eyes) in children, which had been suggested to be caused by either a germline (inherited) or a somatic (acquired) mutation of the retinoblastoma gene (*Rb.*).[14] Published in 1971, his paper began by estimating the percentage of inheritance of retinoblastoma mutations:

The percentage of all cases that are bilateral is approximately 25-30 [...] All bilateral cases should be counted as hereditary because the proportion of affected offspring closely approximates the 50% expected with dominant inheritance. On the other hand, of the 70-75% of all cases that are unilateral, only 15-20% are thought to be hereditary; thus, 10-15% of all cases are unilateral and hereditary. The percentage of all retinoblastoma patients with the hereditary form is, therefore, in the range 35-45; among these, 25-40% are unilateral and 60-75% are bilateral. In contrast, 55-65% of all retinoblastoma cases are of the nonhereditary form and all are unilateral.[14]

Thus, using a number of intricate calculations and insightful deductions, Knudson estimated the rate at which a second mutation might occur in cells that already had an inherited mutation to be 1 in 50,000,000 per year. Surprisingly, this was similar to the rate he had calculated for spontaneous mutations and was consistent with the fact that the age at diagnosis of bilateral cases was about half that of unilateral cases—since it would take twice as long to get two spontaneous mutations, bilateral cases having inherited one mutation. In addition, he found that new cases of bilateral retinoblastoma decreased exponentially with age, which he suggested *reflects the occurrence of a second event at a constant rate in a declining population of embryonal cells.*[14] Finally, Knudson summarized his results:

Those patients that inherit one mutation develop tumors earlier than do those who develop the nonhereditary form of the disease; in a majority of cases those who inherit a mutation develop more than one tumor. On the other hand, the probability that an individual not inheriting a mutation would develop more than one tumor is vanishingly small, so that nonhereditary cases are invariably unifocal [...][14]

Two years later, David Comings published a paper summarizing what was known about carcinogenesis and proposed a hypothesis that unified all the observations.[15] Taking Colette et al.'s conclusions about *scr* being a gene endogenous to vertebrates a step further, he suggested that mutated genes resulting from exposures to carcinogens must already be present since these agents were but simple chemicals and could not possibly introduce a new gene into a genome. In addition, he noted that oncogenes must have *some necessary function during some stage of the cell cycle, or some stage of embryogenesis,* such as for a *burst of cell division during cleavage division, or organogenesis,* otherwise cells would not keep them around given their propensity to cause *so much mischief.*[15] Furthermore, he added that *spontaneous transformation or transformation by chemicals or x-rays is the result of a double mutation of a pair* of genes, an event rare enough for survival. As such, he suggested that:

This mechanism is consistent with the correlation between mutagenesis and carcinogenesis, with the increasing incidence of cancer with age, with the additiveness of chemical and radiation effects on the production of tumors and cell transformation, and with the somatic mutation theories of cancer.[15]

Comings also noted that the above was consistent with what had been observed in inherited cancers as suggested by Knudson for retinoblastoma:

In the present model, his *[Knudson]* dominantly inherited factor would be *[a mutated]* [...] retina-specific [...] gene, and his added somatic mutation would be a mutation of the homologous locus resulting in *[a double mutant]* and [...] development of a retinoblastoma. Since the different [...] genes may be only partially tissue specific, there might be a higher incidence of tumors in nontarget tissues. This is seen in some cases. For example, there is an increase in the frequency of other primary tumors in patients with retinoblastoma. This type of mechanism could also explain the existence of cancer in certain families, including those in which the tumors involve different tissues.[15]

Finally, results of other studies had suggested that the transformation of cells was caused by a *balance of factors carried on specific chromosomes for the expression or suppression of transformation.*[15] Therefore, Comings correctly concluded that this *may represent either the deletion of additional* [tumor suppressor] *genes or duplication of* [oncogenes].

* * * * *

Now that a general mechanism for tumor formation had been proposed, it remained to be determined if it was correct and, if so, exactly how this might be accomplished. In the next sections, we will discuss several manners in which normal genes and their protein products can be modified in a cell and the effects that these changes can have on the function of these protein, as well as how these changes lead to tumor formation. In 1960, Nowell and Hungerford reported that most patients with chronic granulocytic leukemia (CGL) had a small deletion in one of their chromosomes.[16-18] The next year, Tough et al. named this truncated chromosome the Philadelphia chromosome, Ph[1], in reference to the city in which the abnormality was first noticed (Figure 16.1).[19] Nowell and Hungerford suggested that it was:

> [...] possible that the chromosome abnormality observed in chronic granulocytic leukemia may be a primary change rather than a secondary phenomenon appearing during the course of the disease. It was observed in one of our patients [...] at the time of diagnosis, prior to any therapy and before she developed any symptoms referable to the hematopoietic system [...]

The apparent presence of this chromosome change from the outset of the disease, plus the consistent finding of the same abnormal chromosome in association with this one particular form of leukemia, strongly suggests that it is this change in the genetic apparatus which confers on these cells their neoplastic character.[18]

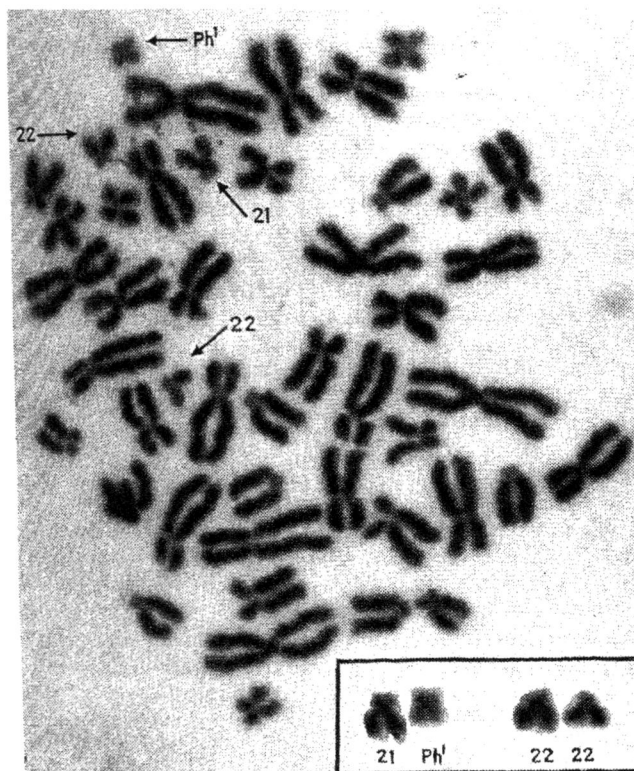

FIGURE 16.1 Image of chromosomes from one chronic granulocytic leukemia patient studied by Tough et al. The inset shows four acrosomal chromosomes, one of which has a visible deletion in its long arm (initially thought to be chromosome 21), termed the Philadelphia chromosome (Ph[1]).[19] (Reprinted from The Lancet, 1(7174) Tough IM, Court Brown WM, Baikie AG, et al. Cytogenetic studies in chronic myeloid leukaemia and acute leukaemia associated with mongolism, 411–417, Copyright © 1961, with permission from Elsevier.)

One of the explanations offered for this abnormality was that the missing segment had somehow simply been deleted, but it was also suggested that it may instead have been translocated to another chromosome.[18,19] However, given the small size of the missing genetic material, the size of the chromosome at the receiving end of this proposed translocation would not be noticeably altered, and the techniques of the times were not advanced enough to identify such small changes. As such, it would be another 12 years, in 1973, before techniques for chromosome staining were improved to the extent that a research group was able to notice an additional piece of DNA about the size of the missing fragment of Ph[1] attached to another chromosome.[20] This was the first translocation identified in humans—the significance and impact of which will be discussed later in this chapter—and in time, several more would be discovered.

Around the same time, a different translocation was consistently observed in Burkitt lymphoma cells, and Zech et al. suggested that this translocation occurred from the long arm of chromosome 8 to the long arm of chromosome 14.[21] A few years later, in 1982, Southern blotting was used independently by two groups to determine that the breakpoint on chromosome 14 occurred within the immunoglobulin variable heavy chain locus (V_H, one of the protein subunits of antibodies)[22,23] and the possibility was suggested by Croce's group that *the expression of malignancy in these cells results from activation of a gene located on the long arm of chromosome 8, by recombination with either a V_H region or a V_H promoter.*[22]

These groups also reported that a gene of unknown function, *c-myc*, which was known to be analogous to the avian transforming gene, *v-myc*, but present in normal, healthy human cells, was somehow involved in the oncogenic potential of this translocation. Like, *src*, *c-myc* expression was shown to be increased following viral infection, though neither the mechanism by which the expression was increased nor the role of the protein product were yet known. As such, these investigators again used Southern Blotting to elucidate the former and assigned the chromosomal location of the *c-myc* gene to chromosome 8 in normal cells, whereas it was located on chromosome 14 in their tumor samples; as such Dalla-Favera et al. concluded that the *c-myc* gene was located between the break-point and the end of the chromosome 8 in these cell lines.[24] Thus, the authors proposed that the *c-myc* oncogene was involved in Burkitt lymphoma, and specifically, that its translocation *may have caused a failure in mechanisms that control repression of this gene* (Fig. 16.2). Indeed, others had shown that *c-myc* expression was normally turned off during cellular differentiation, leading to the suggestion that *constitutive expression of c-myc resulting from a chromosomal translocation in Burkitt cells [...] may prevent normal cellular differentiation by maintaining the cells in an immature proliferative state.*[24] The next year, Croce's group used Northern blotting to assess the expression of *c-myc* mRNA and determined that although overall *c-myc* expression was not significantly

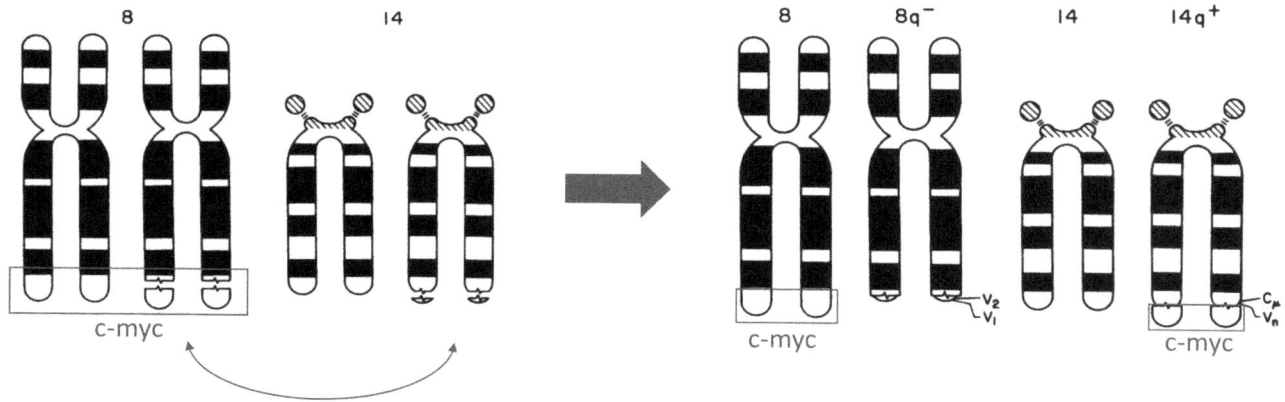

FIGURE 16.2 Translocation t(8:14) involving c-myc and immunoglobulin-related genes in Burkitt lymphoma. Shown are the breakpoints in wild type chromosomes 8 and 14 common in Burkitt lymphoma, as well as the *c-myc* gene location on chromosome 8 (red box). *c-myc* on chromosome 8 translocates to the end of chromosome 14 and the small piece of chromosome 14 translocates to the end of chromosome 8. Also shown is the arrangement of some of the genetic elements of the region surrounding the break point on chromosome 14 as a result of this translocation. The *c-myc* gene is translocated downstream of regulating regions of antibody genes (Cμ and V_n) which are constitutively active. Therefore, expression of one of the *c-myc* alleles is no longer controlled by its usual regulatory regions, which results in its constitutive expression.[22] (Adapted from Erikson J, Finan J, Nowell PC, Croce CM. Translocation of immunoglobulin VH genes in Burkitt lymphoma. Proc Natl Acad Sci U S A. 1982;79(18). Used with permission.)

increased in Burkitt Lymphoma, only the translocated allele was expressed in these cells while the wild-type allele was silent, which indicated that the two alleles *are under different control in Burkitt Lymphoma cells* and that *transcription from the translocated c-myc gene initiates by a new promoter.*[25] We now know *c-myc* to be a protein involved in cell proliferation; therefore, increased expression or expression at times when it should not be expressed leads to increased or deregulated cell proliferation, an important and necessary step in the genesis of cancer.

We saw in Chapter 14 that Bcl-2 is a protein involved in the regulation of cell apoptosis and that its overexpression resulted in resistance to a number of apoptosis stimuli. In 1984, Croce and his group determined that, like *c-myc* and the Philadelphia chromosome, *Bcl-2* was also involved in a translocation, but this time from chromosome 18 to chromosome 14, the latter being the same chromosome involved in the *c-myc* translocation we just discussed.[26,27] Also similar to c-myc, *Bcl-2* was found to be translocated just downstream of the V_H coding region, which explained why this translocation led to the overexpression of Bcl-2. However, since Bcl-2 is a pro-survival protein that inhibits pro-apoptosis proteins, the aberrant expression of this protein leads to increased survival of cells since the apoptotic pathway is partly blocked.

＊ ＊ ＊ ＊ ＊

In these last few examples, abnormal expression of a protein was the result of a specific translocation between two chromosomes, which led to the dysregulation of important genes. And up until the early 1980s, this was thought to be the *pivotal event in carcinogenesis*;[28] however, another mechanism for the initiation of cancer soon came to light. We saw in previous chapters that proteins fold into a specific three-dimensional shape, or conformation, which is dictated by the sequence of amino acids specified by a given gene, and that this conformation was instrumental in defining the manner in which a protein functions and interacts with other proteins. We also saw how changes in amino acids within a protein have profound effects on the function of proteins by disrupting their native conformation. For example, we saw in Chapter 10 how changing one amino acid in GFP resulted in the enhancement of its fluorescence, or even in a protein which fluoresced at a different wavelength, such a blue or red. Similarly, it was found that single amino acid substitutions in certain key amino acids could result in important changes in the function of proteins, leading to a strong survival advantage for cells and to tumor progression.

In the early 1980s, one of the about ten human oncogenes known to exist, H-ras, was discovered due to its homology to two other oncogenes of murine sarcoma viruses (called *v-has* and *v-bas*). H-ras was found to be the cause of some bladder carcinomas but the gene was determined to not have *undergone major genetic rearrangement* and no other gross differences, such as restriction enzyme analysis, were observed between the proto-oncogene (the wild type gene) and the oncogene.[29] Therefore, two groups independently set-out to elucidate the mechanism by which H-ras-mediated cellular transformation occurred. Tabin et al. reasoned that since no major genetic changes were found in this case, the genetic changes leading to cellular transformation must be relatively minor. In addition, the authors hypothesized that these changes could affect protein function either by altering the expression-regulating section of the gene, thereby changing the levels of expression, or by changing the *protein encoding portion of the gene*, which would result in the *synthesis of an altered protein.*[28] Since RNA and protein analysis determined that levels between normal and tumor cells were similar, the former was deemed unlikely.

As such, the two groups carried out cloning experiments with an aim to *localize genetically the regions of the oncogene that specify the altered migration rate of the protein and the change in gene function.*[28,29] To this end, Reddy et al.

excised DNA fragments from different locations within the H-ras oncogene (the mutated gene) and put them in place of the homologous region of the proto-oncogene (the WT gene) to determine which section of the gene was sufficient to impart transforming ability onto the cells (Figure 16.3). Results from both groups indicated that a region between

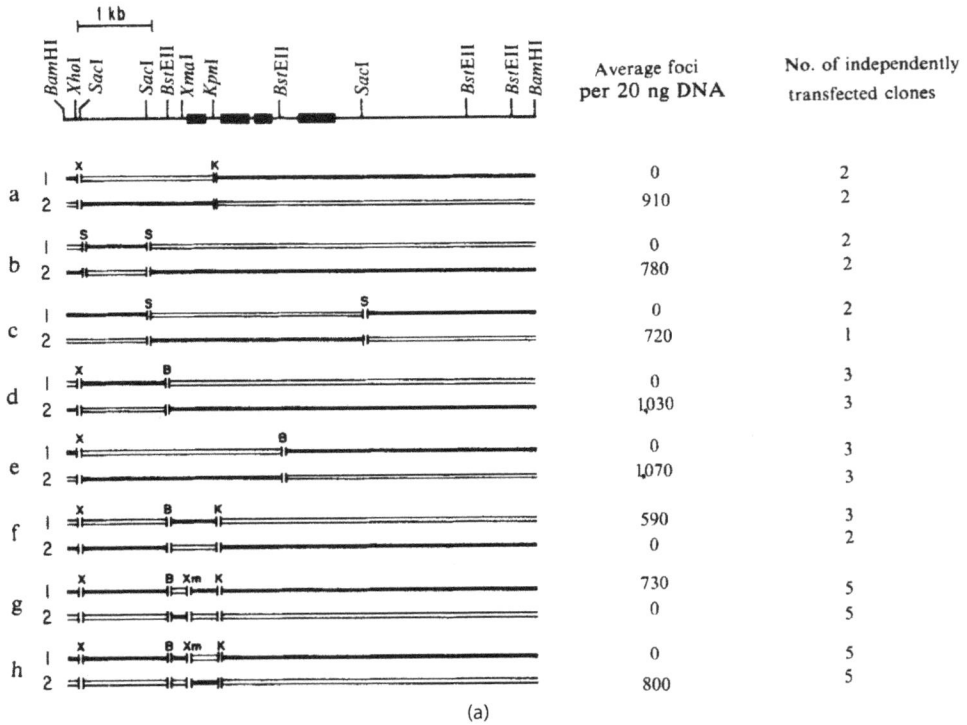

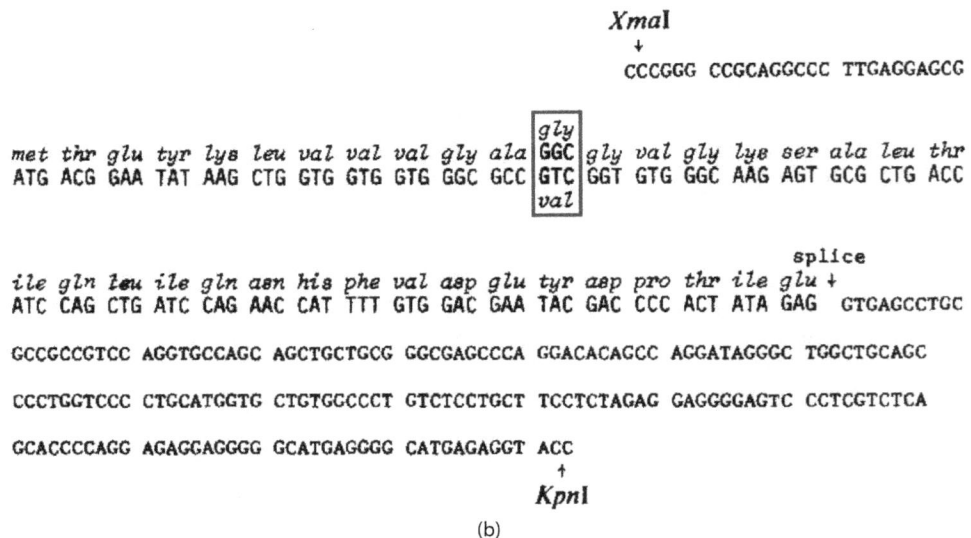

(a)

```
                                    XmaI
                                     ↓
                              CCCGGG CCGCAGGCCC TTGAGGAGCG
```

```
                                             gly
met thr glu tyr lys leu val val val gly ala  GGC  gly val gly lys ser ala leu thr
ATG ACG GAA TAT AAG CTG GTG GTG GTG GGC GCC  GTC  GGT GTG GGC AAG AGT GCG CTG ACC
                                             val
```

```
                                                              splice
ile gln leu ile gln asn his phe val asp glu tyr asp pro thr ile glu ↓
ATC CAG CTG ATC CAG AAC CAT TTT GTG GAC GAA TAC GAC CCC ACT ATA GAG  GTGAGCCTGC

GCCGCCGTCC AGGTGCCAGC AGCTGCTGCG GGCGAGCCCA GGACACAGCC AGGATAGGGC TGGCTGCAGC

CCCTGGTCCC CTGCATGGTG CTGTGGCCCT GTCTCCTGCT TCCTCTAGAG GAGGGGAGTC CCTCGTCTCA

GCACCCCAGG AGAGGAGGGG GCATGAGGGG CATGAGAGGT ACC
                                             ↑
                                           KpnI
```

(b)

FIGURE 16.3 (a) Scheme and results from cloning experiments which identified a 350-nucleotide sequence in H-ras that was responsible for its transforming abilities. The wild-type H-Ras proto-oncogene and mutant H-Ras oncogene were digested with the restriction enzymes shown in the map and the excised fragments were reciprocally ligated to replace the native segments as shown in the diagram. Solid boxes on the restriction map indicate protein-coding exons. The segments from which the various sequences were derived are indicated by white (wild type gene) or black (mutated gene) lines. Also shown is the transforming ability of each resultant DNA construct. Note how the transforming ability moves with the section between XmaI and KpnI. Indicating that this locus possessed the oncogenic properties in question. (b) A single nucleotide mutation in the glycine codon leads to a change to valine; italicized sequence is wild type.[28] (Reprinted by permission from Springer Nature Customer Service Centre GmbH: Springer Nature, Nature, Tabin CJ, Bradley SM, Bargmann CI, et al. Mechanism of activation of a human oncogene, Copyright © 1982.)

XmaI and KpnI cleavage site of the gene was responsible for the transforming abilities of H-ras. Comparison of this region between the WT and the mutant revealed the presence of a single-nucleotide mutation—guanidine to thymidine—which led to a change in amino acid from the small and compact glycine (the only amino acid without a side-chain) to the bulkier valine, and suggested that this alteration *represents abrupt changes in the local stereo-chemistry of* [the] *protein*. Reddy et al. also compared the amino acid sequence to homologous mutated proteins in other organisms and found that they too had a mutation in the same amino acid, though the new residue was different in each case; therefore, *this substitution appears to represent the critical agent of the conversion of the proto-oncogene into an active oncogene.*[28] Thus, the authors proposed that the ras homolog's *malignant properties are the result of the elimination of the glycine residue present in their normal counterpart, rather than due to a specific transition from one type of amino acid residue to another.*[29]

In addition, when the authors compared the predicted secondary structures of the normal and mutated protein, they found that the substitution of glycine at this position resulted in a significant decrease in the size of a hinge created by this region of the protein, due to the *prominent projection of the [...] domain [...] away from the central core of the molecule.*[29] Finally, Tabin et al. concluded that:

> Most amino acid sequence alterations are either neutral or deleterious to protein function. Few are able to actively potentiate the normal functions of a protein. We suggest that only a small number of sites of the *[Ras]* protein can be altered in a fashion leading to oncogenic activation. Most mutations will affect other residues whose alteration will be unproductive for oncogenic conversion. The target for oncogenic conversion may be exceedingly small [...]
>
> The present data suggest that the alteration of one nucleotide in one bladder cell leads to the creation of an activated oncogene. There are three possible point mutations at this position, and it is perhaps not coincidental that the G-T transversion observed here is precisely the mutation favoured by many suspected bladder carcinogens [...] The point mutation implicated as a central event in this oncogenic transformation represents the first demonstration of a lesion in cellular DNA whose occurrence is directly related to the carcinogenic process.[28]

The next year, Pincus et al. used a recently developed computer-assisted protein modeling system to predict the effects that this mutation might have on the conformation of H-ras. The amino acids surrounding glycine in the proto-oncogene were found to constitute a relatively flexible hydrophobic sequence that could assemble into 29 different conformations of an α-helix.[30] In contrast, the Gly to Val mutation resulted in a much more restricted structure composed of only nine different conformations. Interestingly, one of the 29 conformations in the wild-type protein was equivalent to the most stable conformation of the mutated protein, while none of the possible conformations of the mutated protein were equivalent to

the wild-type conformation. These results indicated that although a small fraction of the wild-protein assembled into a conformation that could lead to aberrant activity, the mutated protein never adopted the wild-type conformation. As such, Pincus et al. concluded that if *some function of the normal* [Ras] *(containing glycine-12) is dependent on this conformation, our results predict that the tumor* [Ras] *(containing valine-12) could not perform this function.*[30] It was subsequently discovered that Ras was a G-protein, and this point mutation, which was determined to occur in the GTPase domain of Ras, resulted in the insensitivity of Ras to inactivation by its GAPs. Therefore, replacement of glycine-12 for another amino acid leads to Ras being permanently "stuck" in the ON position (GTP-bound) and results in the hyperactivation of the Ras pathway.

* * * * *

We discussed in previous sections how in the 1960s, it was discovered that uncontrolled expression of certain proteins, called oncogenes—e.g., *c-myc* and *bcl-2*, proteins involved in cell growth and survival, respectively—could lead to tumors formation. In contrast, retinoblastoma was the first protein to be discovered whose function served as a tumor suppressor, that is, its expression suppresses cell growth and therefore, its loss results in the formation of tumors. Similarly, p53 was found to induce apoptosis upon detection of DNA damage and the loss of this protein was subsequently determined to be involved in the progression of many tumors. The discovery of the role of these proteins in the development of cancer usured in a new era of cancer research: cancer researchers now had specific mechanisms of cell functions to target. Thus, great effort has been expended in the last few decades in elucidating the precise components of each cellular pathway utilized by cells and to determine if and how these pathways are dysregulated in various cancers and how they could be used to eliminate cancers and treat patients.

In 2000, Hanahan and Weinberg published their seminal "Hallmarks of Cancer" (which was updated in 2011 to included more recent aspects of cancer), in which they reviewed what was known about the requirements for cancer formation and discussed the *rules that govern the transformation of normal human cells into malignant cancers.*[31,32] As such, the authors proposed six characteristics that cancers must acquire in order to thrive (in no particular order): *self-sufficiency in growth signals, insensitivity to anti-growth signals, evasion of apoptosis, limitless replicative potential, sustained angiogenesis,* and *tissue invasion and metastasis. Self-Sufficiency in Growth Signals* means that cells either no longer require growth signals to proliferate or that they have become hypersensitive to them. Growth signals, such as transforming growth factor α (TGFα) and platelet-derived growth factor (PDGF) bind to dedicated cell surface receptors and initiate the activation of a protein cascade—most often via phosphorylation—which results in the proliferation of the cell. This last step can be accomplished either by the activation of a protein that inhibits an inhibitor of proliferation (see next section), or by activating transcription factors—proteins which bind to a specific recognition sequence on DNA and results in the transcription of those genes—which, in this case, would activate genes involved in cell proliferation.

For example, we saw in the last section that a mutation in the GTPase, Ras, leads to its constitutive activation; and, as this protein is at a major intersection for incoming signals involved in proliferation, this mutation leads to aberrant regulation of cell growth (Figure 16.4, black box). As we saw in Chapter 13, Ras protein is normally in the GDP bound, "off" conformation; that is, it folds in a conformation that reduces its ability to interact with other proteins. The exchange of GDP for GTP, assisted by "guanine nucleotide exchange factors" (GEFs, such as SOS), leads to a change in conformation which allows Ras to bind to downstream signaling proteins and activate them. Termination of Ras activation can occur intrinsically since

Ras has built-in GTPase function (which hydrolyses GTP to GDP and P$_i$), but the rate at which this occurs is very slow; so GTPase activating proteins (GAPs, such as NF-1) are present to catalyze the reaction and better regulate the on/off state of Ras. As such, when the Ras gene is subject to a common point mutation which replaces gly-12 for another amino acid, it assumes a conformation which prevents the hydrolysis of GTP and therefore, it is "stuck" in the "ON" conformation. As we saw earlier, cells with this particular mutation divide continuously and have the potential to give rise to tumors. Similarly, mutations that result in the overexpression of genes that code for extracellular signaling molecules themselves, such as TGFα or PDGF, might also lead to cancerous growth since cells are continuously producing and receiving signals instructing them to grow. This is called autocrine signaling, *i.e.*, cells are activated by their own secreted signals—a phenomenon that occurs in normal cells but as one might expect, in a very regulated manner.

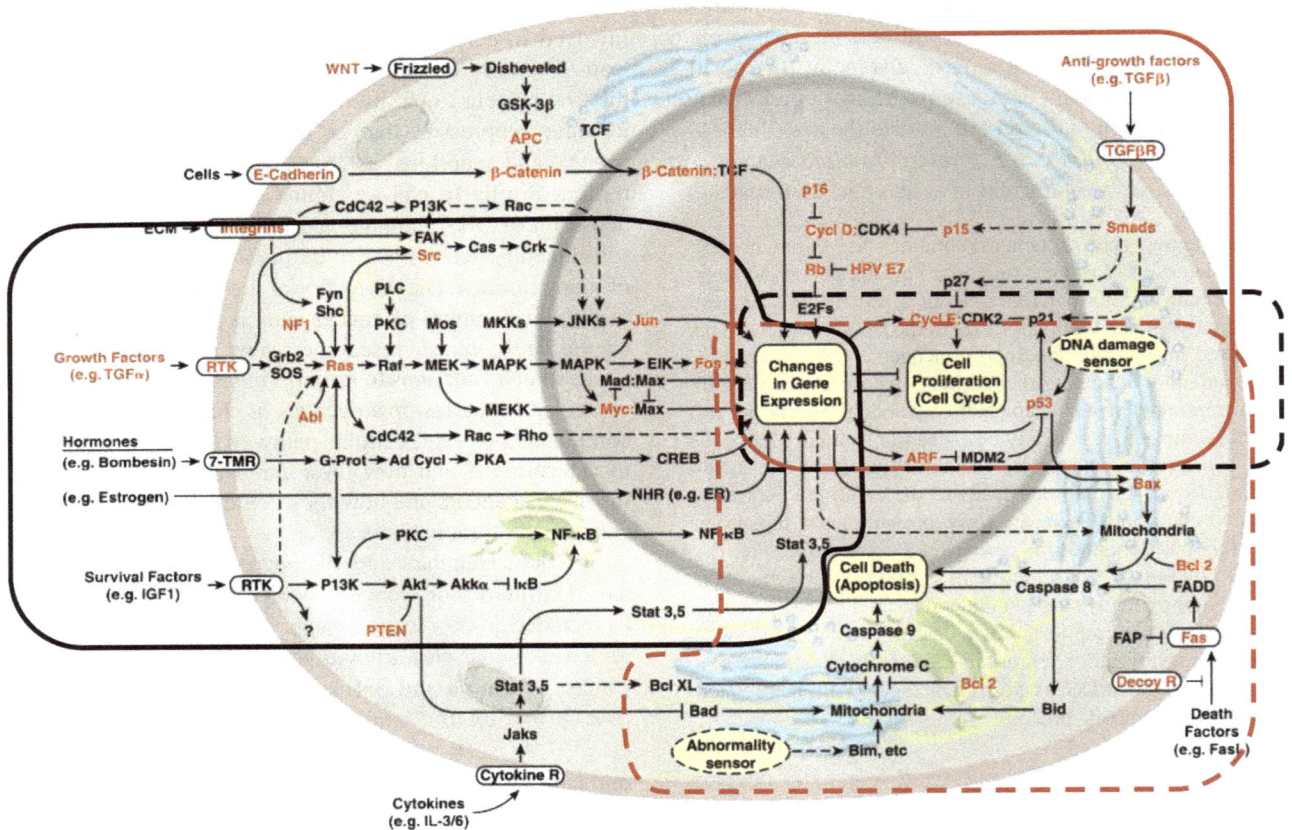

FIGURE 16.4 Diagram of important pathways contributing to the maintenance of cellular homeostasis. In red are shown proteins which are commonly mutated in cancers. **Black box: Cell proliferation pathway.** Note how Ras is at a major intersection of incoming signals leading in growth factor-mediated cell proliferation. Therefore, mutations in Ras are common in cancer and result in the hyperproliferation of cells. 7-TMR: Seven transmembrane receptor (i.e., G protein-coupled receptors); RTK: receptor tyrosine kinase. **Red box: anti-proliferation pathway.** TGFβ released by surrounding cells binds to its receptor and initiates a series of intracellular signals leading to the activation of Rb. This protein, in turn, inhibits transcription factors responsible for proliferation. **Red dashed box: The apoptosis pathway.** Mutations in proteins involved in apoptosis can lead to development of cancer by overprotecting the cell from cell death. **Black dashed box: Protein pathways involved in senescence.** p53 is a transcription factor that is activated upon DNA damage. It is responsible for the transcription of antiproliferative proteins, such as p21, until the damage repaired. Since DNA damage is common in cancers, inactivating mutations in p53 is likewise a major contributing factor to cancer growth.[31] (Adapted from Cell 100(1), Hanahan D, Weinberg RA. Hallmarks of cancer, 57–70, Copyright © 2000, with permission from Elsevier.)

Similar consequences would result if the intracellular domain of a receptor for growth signals, such as the TGFα receptor—which is usually in a conformation that prevents activation until a signaling protein binds to its extracellular domain—was modified with a genetic mutation which changed the protein such that it adopts a conformation similar to when it is bound by a ligand, resulting in the constitutive activation of the pathway (Figure 16.4, black box). Another alteration we saw earlier in this chapter leading overactivation of a pathway is that of the Philadelphia Chromosome, in which an exchange of genetic material occurs between one end of chromosome 22 and one end of chromosome 9. The significance of this translocation became obvious when it was discovered that Abl—the protein that is hyperactivated in this case—has a domain that normally forces it into a conformation that is auto-inhibitory. The translocation involved in Ph[1] is such that it separates this inhibitory domain from the rest of the Abl gene and renders the protein product of the gene constitutively active, thereby permanently activating Ras (and any of the other proteins regulated by Abl).

At the opposite end of TGFα and PDGF are cellular signals that inhibit cell growth. These are released, for example, when a cell population becomes too dense and signal that there are enough cells in the area and it is time to stop growing. As such, another characteristic that a cell needs to acquire to become cancerous is an *Insensitivity to Antigrowth Signals*. Examples of proteins involved in this pathway are transforming growth factor β (TGFβ) and retinoblastoma (Rb), the latter whose gene, as we already saw, is commonly mutated in cancers. Like its counterpart, TGFα, TGFβ serves as a mitogenic signal released by cells which binds to a dedicated cell surface receptor, triggering a chain of events leading to inhibition of proliferation. Rb is an intermediate in this pathway and it binds to transcription factors and prevents them from initiating gene transcription (Fig. 16.4, red box). As such, mutations that inactivate Rb result in transcription factors that are unhindered and leads to excessive proliferation, assuming, of course, that proliferation signals are present.

In Chapter 15, we discussed the mechanisms that regulate apoptosis and we saw how many proteins involved in this pathway were discovered due to the fact that their dysregulation often led to cancer. For example, *Bcl-2* is commonly upregulated as a result of a translocation in a number of leukemias; as such, the apoptosis pathway in these cells is largely unaffected by signals that trigger cell death and therefore, the life of these cells are significantly extended (Figure 16.4, red dashed box). However, the potential for oncogenesis would only be fully realized *if* these cells were also exposed to growth factors. This is why single mutations in proto-oncogenes are rarely enough to transform a cell into a cancer cell; additional mutations—in *Ras*, for example—would enable cells to become independent of growth signals and could lead to the development of tumor-forming cells. Furthermore, mutations in tumor-suppressing genes—usually requiring

one mutation on each allele since the protein product of one allele is often enough to adequately perform the duties of the protein—would also be required to enable cells to ignore "OFF" signals.

Fortunately for us, even mutations affecting all three of these aspects would likely not be enough to result in a thriving cancer since an additional requirement for cancer growth is that of *limitless replicative potential*. This is a DNA-based requirement in that the ends of each chromosome, called telomeres, get progressively shorter every time a cell divides; when the telomeres are too short, DNA replication can no longer take place and the cell stops dividing; this implies that each cell has a preset lifespan that is determined by the length of its telomere. This phenomenon was first discovered in 1961 by Leonard Hayflick and PS Moorehead, who used ten criteria to monitor cellular changes—such as chromosomal integrity and inability to grow in culture—in 25 different cell strains and found that most cells in culture *degenerate after about 50 subcultivations and one year in culture.*[33] Cells that stop growing due to this condition are said to be in "senescence", a nonproliferative but viable cell stage. It was later found that the tumor-suppressor genes *Rb* and *p53* were responsible for the entry of cells into senescence: the Rb protein helps monitor the conditions of a cell's environment and determines if they are apt for cell growth (Fig. 16.4, black dashed box). In contrast, p53 is involved in pathways that monitor intracellular conditions, such as DNA integrity and glucose levels. Inactivating mutations in either of these genes enables cells to bypass "senescence" and continue to grow for several more generations until they reach a new phase called "crisis", in which apoptosis is triggered. Of course, the presence of additional mutations in key proteins of the apoptosis pathway would result in a cancerous growth.

Telomeres are composed of multiple tandem repeats of hexanucleotide sequences (TTAGGG in vertebrates) that act to protect chromosomes from recombination events and degradation. Because of the mechanism of DNA replication by DNA polymerase, a portion of the ends of a chromosome are not replicated when a cell divides and therefore telomeres get progressively shorter with each cell division. As such, the shortening of telomeres *to a critical length results in loss of telomere protection, which leads to chromosomal instability and loss of cell viability.*[34] Some cells express an enzyme called telomerase which synthesizes hexanucleotide sequences of telomerase and helps maintain their integrity, but expression of this protein is normally reserved for specialized germ cells (i.e., ova and sperm). As expected, mutations in the regulation of expression of this protein enables cells to breach crisis, which results in their ability to grow beyond the normal lifespan of the cell.

The four characteristics we just saw are generally required for the initiation of all cancers: cells need to grow uninhibitedly, they need to ignore signals telling them to stop growing, they need to ignore signals telling them to die, and they need to continuously repair their telomeres to avoid going into crisis. And as far as blood cancers are concerned, that is, cancers derived from mobile cells that travel freely in the bloodstream and are not attached to any structural components, this might be the end of the story. However, most cells of an organism are immobilized since they form organs and surrounding structures, and consequently, abnormal growth results in the formation of solid tumors which have addition requirements in order to thrive. For example, if a brain cell—which is attached to surrounding cells and structural elements of the brain—starts to divide uncontrollably, the new cells will also be embedded in this space and thus a ball of cells will be produced. In this case, the cells at the periphery of this ball will have access to nutrients and oxygen from the surrounding blood vessels, but that is not so for the cells in the center. Under these conditions, a tumor would only be able to grow to a certain extent before the cells in the middle start to die from a lack of nutrients and oxygen (all cells need to be within 100 μm of a blood vessel to have access to these molecules and survive). Therefore, an additional characteristic that needs to be acquired by solid cancers is the ability to generate blood vessels that will "feed" cells at the center of the ball of cells; that is, they require *sustained angiogenesis*. This is accomplished by mutations in genes that code for proteins involved in stimulating the growth of blood vessels—called angiogenesis—such as "vascular endothelial growth factor" (VEGF) or the "fibroblast growth factor" (FGF). These soluble proteins bind to cell surface receptors on endothelial cells (a cell type which forms the walls of blood vessels) and trigger an "angiogenic switch", that is, the proliferation of these normally quiescent cells and the development and penetration of existing vessels into the center of the tumor, which will feed it as it gets larger.

The idea that the size of a tumor that grows in a particular area of the body would likely be limited—and even contained—by surrounding structures led to the proposition that cells need to be able to displace, or even eliminate, these structures if the tumors they form are to thrive in an organism. As such, an additional characteristic that solid tumors require is the ability for *tissue invasion and metastasis*. Cells which make up tissues are connected to one another by cell–cell contacts which are enabled via transmembrane proteins called cadherins. As such, common mutations observed in solid cancers result in the downregulation of E-cadherins and the upregulation of N-cadherins, the former resulting in the disassembly of surrounding structures that restrain cells and the latter increasing the ability of cells to migrate (through a mechanism which remains poorly understood). These changes, which only occur in a small fraction of cancer cells, enable these cells to dissociate from the mass and "walk around" until they encounter a blood vessel. They can then enter the vessel and travel in the bloodstream until they attach to its walls at a remote site, which is enabled by cell surface receptors called integrins—proteins involved in adhesion

between cells or with the surrounding structures—at which point the cell exits the blood vessel and enters the tissue. If and when this unlikely journey is successful, the cell is free to establish a neoplastic growth at this new location. This process is called metastasis and usually signals the end of effective treatment for cancer patients as there are too many areas to treat and the chances of removing all cancer cells that are distributed across the body is very unlikely (hence, the saying "it's not the tumor that kills, it is its metastasis").

* * * * *

We learned, in these last few chapters, about the complexity of the cellular machinery which is replete with redundant circuits regulating various cellular functions (Figure 16.4). The redundancy of the system is important in preventing cells from proliferating uncontrollably due to mutations in one or more key proteins. As such, disregulation of multiple proteins involved in different cellular processes are required for cancer to develop and thrive. These unlikely events increase with age simply because the genetic mutations which lead to them occur during cell division and the older one gets, the more divisions one's cells will have undergone. Some of us are born with mutations in important genes, some of which are proto-oncogenes while other are tumor suppressors. As we saw, the latter usually require one hit to both alleles to become oncogenic since the protein product of a single gene is usually enough to keep everything running smoothly. However, oncogenes often require only one mutation in a single allele to have detrimental effects since many of these mutations are activating mutations, that is, they lead to the constitutive activation or in a significant increase in the amount of a protein. Therefore, a single allele with aberrant function results in an increase in that protein's concentration and, consequently, the cell gains an increased ability to grow or survive.

The fact that cancers require several different malfunctioning proteins to thrive points to the difficulties encountered in treating cancers: therapies need to target multiple pathways to be effective. In addition, the drug needs to be able to reach all the cancer cells, including the ones that live in the center of the tumor. As such, the first step in the treatment of solid tumors is often to surgically remove the bulk of a tumor such that the drug treatment has an increased chance of being delivered to all remaining cells. In addition, the cells that make up any one cancer, even within one particular individual, are heterogeneous. That is, they are likely made up of cells that may have significant differences in their genetic mutations. Therefore, chemotherapy regimens usually include the administration of a combination of drugs in an attempt to increase the likelihood that all cancer cells will be sensitive to at least one of the drugs, thereby maximizing the effectiveness of the treatment. The advent of personalized medicine is a great boon to the treatment of cancer. At its performance apex, this would mean that an individual presenting with cancer would get his or her DNA sequenced to determine which of their genes are subject to translocations and mutations leading to aberrant protein functions. With this information in hand, treatment would include drugs that specifically target the affected protein pathways to stop the cancer in its track. However, basic research into the inner workings of cells is still required to continue to learn about the role of proteins in these pathways—and cells in general—before personalized medicine can be used to its full potential to inform about treatment.

* * * * *

REFERENCES

1. Hajdu SI. A note from history: Landmarks in history of cancer, part 1. *Cancer.* 2011;117(5):1097–1102. doi:10.1002/cncr.25553

2. Hajdu SI. A note from history: Landmarks in history of cancer, part 2. *Cancer.* 2011;117(12):2811–2820. doi:10.1002/cncr.25825

3. Hajdu SI. A note from history: Landmarks in history of cancer, part 3. *Cancer.* 2012;118(4):1155–1168. doi:10.1002/cncr.26320

4. Virchow R. *Cellular Pathology as Based upon Physiological and Pathological Histology.*; 1860. Accessed March 29, 2019. https://archive.org/details/dli.ministry.10993

5. Rous P. A transmissible avian neoplasm. (Sarcoma of the common fowl.). *Journal of Experimental Medicine.* 1910;12(5):696–705. doi:10.1084/jem.12.5.696

6. Rous P. Peyton Rous: Nobel Lecture: The Challenge to Man of the Neoplastic Cell. NobelPrize.org. Nobel Media AB © The Nobel Foundation 1966. https://www.nobelprize.org/prizes/medicine/1966/rous/lecture/

7. Rous P. A sarcoma of the fowl transmissible by an agent separable from the tumor cells. *Journal of Experimental Medicine.* 1911;13(4):397–411. doi:10.1084/jem.13.4.397

8. Rous P. Transmission of a malignant new growth by means of a cell-free filtrate. *Journal of the American Medical Association.* 1983;250(11):1445. doi:10.1001/jama.1983.03340110059037

9. Bernhard W. The detection and study of tumor viruses with the electron microscope. *Cancer Research.* 1960;20(5 Part 1):712–727. Accessed March 29, 2019. http://www.ncbi.nlm.nih.gov/pubmed/13799753

10. Huebner RJ, Todaro GJ. Oncogenes of RNA tumor viruses as determinants of cancer. *Proceedings of National Academy of Sciences U S A.* 1969;64(3):1087–1094. doi:10.1073/pnas.64.3.1087

11. Brugge JS, Erikson RL. Identification of a transformation-specific antigen induced by an avian sarcoma virus. *Nature.* 1977;269(5626):346–348. doi:10.1038/269346a0

12. Brugge JS, Collett MS, Siddiqui A, Marczynska B, Deinhardt F, Erikson RL. Detection of the viral sarcoma gene product in cells infected with various strains of avian sarcoma virus and of a related protein in uninfected chicken cells. *Journal of Virology.* 1979;29(3):1196–1203. doi:10.1128/jvi.29.3.1196-1203.1979

13. Collett MS, Erikson E, Purchio AF, Brugge JS, Erikson RL. A normal cell protein similar in structure and function to the avian sarcoma virus transforming gene product. *Proceedings of National Academy of Sciences U S A.* 1979;76(7):3159–3163. doi:10.1073/pnas.76.7.3159

14. Knudson AG Jr. Mutation and cancer: Statistical study of retinoblastoma. *Proceedings of National Academy of Sciences U S A.* 1971;68(4):820–823. doi:10.1073/pnas.68.4.820

15. Comings DE. A general theory of carcinogenesis. *Proceedings of National Academy of Sciences U S A.* 1973;70(12):3324–3328. doi:10.1073/pnas.70.12.3324

16. Nowell PC, Hungerford DA. Chromosome studies on normal and leukemic human leukocytes. *Journal of the National Cancer Institute.* 1960;25(1):85–109. doi:10.1093/jnci/25.1.85

17. Baikie AG, Court-Brown WM, Buckton KE, Harnden DG, Jacobs PA, Tough IM. A possible specific chromosome abnormality in human chronic myeloid leukæmia. *Nature.* 1960;188(4757):1165–1166. doi:10.1038/1881165a0

18. Nowell PC, Hungerford DA. Chromosome studies in human leukemia. II. Chronic granulocytic leukemia. *Journal of the National Cancer Institute.* 1961;27(5):1013–1035. doi:10.1093/jnci/27.5.1013

19. Tough IM, Brown C, Baikie WM AG, et al. Cytogenetic studies in chronic myeloid leukaemia and acute leukaemia associated with monogolism. *Lancet.* 1961;1(7174):411–417. doi:10.1016/S0140-6736(61)90001-0

20. Rowley JD. A new consistent chromosomal abnormality in chronic myelogenous leukaemia identified by quinacrine fluorescence and giemsa staining. *Nature.* 1973;243(5405):290–293. doi:10.1038/243290a0

21. Zech L, Haglund U, Nilsson K, Klein G. Characteristic chromosomal abnormalities in biopsies and lymphoid-cell lines from patients with Burkitt and non-Burkitt lymphomas. *International Journal of Cancer.* 1976;17(1):47–56. doi:10.1002/ijc.2910170108

22. Erikson J, Finan J, Nowell PC, Croce CM. Translocation of immunoglobulin VH genes in Burkitt lymphoma. *Proceedings of National Academy of Sciences U S A.* 1982;79(18):5611–5615. doi:10.1073/pnas.79.18.5611

23. Taub R, Kirsch I, Morton C, et al. Translocation of the c-myc gene into the immunoglobulin heavy chain locus in human Burkitt lymphoma and murine plasmacytoma cells. *Proceedings of National Academy of Sciences U S A.* 1982;79(24):7837–7841. Accessed March 29, 2019. doi:10.1073/pnas.79.24.7837

24. Dalla-Favera R, Bregni M, Erikson J, Patterson D, Gallo RC, Croce CM. Human c-myc onc gene is located on the region of chromosome 8 that is translocated in Burkitt lymphoma cells. *Proceedings of National Academy of Sciences U S A.* 1982;79(24):7824–7827. doi:10.1073/pnas.79.24.7824

25. ar-Rushdi A, Nishikura K, Erikson J, Watt R, Rovera G, Croce CM. Differential expression of the translocated and the untranslocated c-myc oncogene in Burkitt lymphoma. *Science (1979).* 1983;222(4622):390–393. doi:10.1126/science.6414084

26. Tsujimoto Y, Finger LR, Yunis J, Nowell PC, Croce CM. Cloning of the chromosome breakpoint of neoplastic B cells with the t(14;18) chromosome translocation. *Science (1979).* 1984;226(4678):1097–1099. doi:10.1126/science.6093263

27. Tsujimoto Y, Cossman J, Jaffe E, Croce CM. Involvement of the Bcl-2 gene in human follicular lymphoma. *Science (1979).* 1985;228(4706):1440–1443. doi:10.1126/science.3874430

28. Tabin CJ, Bradley SM, Bargmann CI, et al. Mechanism of activation of a human oncogene. *Nature.* 1982;300(5888):143–149. doi:10.1038/300143a0

29. Reddy EP, Reynolds RK, Santos E, Barbacid M. A point mutation is responsible for the acquisition of transforming properties by the T24 human bladder carcinoma oncogene. *Nature.* 1982;300(5888):149–152. doi:10.1038/300149a0

30. Pincus MR, van Renswoude J, Harford JB, Chang EH, Carty RP, Klausner RD. Prediction of the three-dimensional structure of the transforming region of the EJ/T24 human bladder oncogene product and its normal cellular homologue. *Proceedings of the National Academy of Sciences.* 1983;80(17):5253–5257. doi:10.1073/pnas.80.17. 5253

31. Hanahan D, Weinberg RA. The hallmarks of cancer. *Cell.* 2000;100(1):57–70. doi:10.1016/s0092-8674(00)81683-9

32. Hanahan D, Weinberg RA. Hallmarks of cancer: The next generation. *Cell.* 2011;144(5):646–674. doi:10.1016/ j.cell.2011.02.013

33. Hayflick L, Moorhead P. The serial cultivation of human diploid cell strains. *Experimental Cell Research.* 1961;25(3):585–621. doi:10.1016/0014-4827(61)90192-6

34. Blasco MA. Telomeres and human disease: Ageing, cancer and beyond. *Nature Reviews of Genetics.* 2005;6(8): 611–622. doi:10.1038/nrg1656

Index

Note: Locators in *italics* represent figures and **bold** indicate tables in the text.

Index

For Product Safety Concerns and Information please contact our EU
representative GPSR@taylorandfrancis.com
Taylor & Francis Verlag GmbH, Kaufingerstraße 24, 80331 München, Germany

www.ingramcontent.com/pod-product-compliance
Lightning Source LLC
Chambersburg PA
C3HW061404210326
41598CB00035B/6089